Reduced Thermal Processing for ULSI

NATO ASI Series

Advanced Science Institutes Series

A series presenting the results of activities sponsored by the NATO Science Committee, which aims at the dissemination of advanced scientific and technological knowledge, with a view to strengthening links between scientific communities.

The series is published by an international board of publishers in conjunction with the NATO Scientific Affairs Division

A	**Life Sciences**	Plenum Publishing Corporation
B	**Physics**	New York and London
C	**Mathematical and Physical Sciences**	Kluwer Academic Publishers Dordrecht, Boston, and London
D	**Behavioral and Social Sciences**	
E	**Applied Sciences**	
F	**Computer and Systems Sciences**	Springer-Verlag
G	**Ecological Sciences**	Berlin, Heidelberg, New York, London,
H	**Cell Biology**	Paris, and Tokyo

Series B: Physics

Reduced Thermal Processing for ULSI

Edited by

Roland A. Levy

AT&T Bell Laboratories
Murray Hill, New Jersey

Plenum Press
New York and London
Published in cooperation with NATO Scientific Affairs Division

Proceedings of a NATO Advanced Study Institute on
Reduced Thermal Processing for ULSI
held June 20–July 1, 1988,
in Boca Raton, Florida

Library of Congress Cataloging in Publication Data

NATO Advanced Study Institute on Reduced Thermal Processing for ULSI (1988:
Boca Raton, Fla.)
   Reduced thermal processing for ULSI / edited by Roland A. Levy.
     p.   cm.—(NATO ASI series. Series B, Physics; vol. 207)
   "Published in cooperation with NATO Scientific Affairs Division."
   "Proceedings of NATO Advanced Study Institute on Reduced Thermal
Processing for ULSI, held June 20–July 1, 1988, in Boca Raton, Florida"—T.p.
verso.
   Includes bibliographical references.

   ISBN-13: 978-1-4612-7857-3 e-ISBN-13: 978-1-4613-0541-5
   DOI: 10.1007/978-1-4613-0541-5

   1. Integrated circuits—Very large scale integration—Heat treatment—Con-
gresses. I. Levy, Roland A. (Roland Albert), 1944-   . II. North Atlantic Treaty
Organization. Scientific Affairs Division. III. Title. IV. Series: NATO ASI series.
Series B, Physics; v. 207.
TK7871.85.N377   1988   89-23011
621.39′5—dc20   CIP

© 1989 Plenum Press, New York

Softcover reprint of the harcover 1st edition 1989

A Division of Plenum Publishing Corporation
233 Spring Street, New York, N.Y. 10013

LECTURERS

G. AUVERT

CNET
BP 98
38243 Meylan
France

F. S. BECKER

Siemens
Otto Hahn Ring 6
D-8000 Munich 83
West Germany

J. O. BORLAND

Applied Materials, Inc.
3050 Bowers Ave.
Santa Clara, CA 95054

I. D. CALDER

Northern Telecom Electronics Ltd.
P. O. Box 3511
Station C
Ottawa K1Y4H7, Canada

R. DEKEERSMAECHER

IMEC
Kapeldreef 75
B3030 Leuven
Belgium

J. M. MARTINEZ-DUART

Universidad Autonoma Cantoblanco
Instituto de Ciencia de Materiales
Cantoblanco 28049
Madrid, Spain

G. FOTI

Universita di Catania
Dipartimento di Fisica
57 Corso Italia
I95129 Catania
Italy

T. O. HERNDON

MIT Lincoln Laboratory
P. O. Box 73
Lexington, MA 02173-0073

C. HILL

Plessey Research Casewell
Towcester
Northants NN128EQ
United Kingdom

D. C. JACOBSON

AT&T Bell Laboratories
600 Mountain Ave.
Murray Hill, NJ 07974

J. NULMAN

AG Associates
1325 Borregas Ave.
Sunnyvale, CA 94089

A. H. READER

Philips Research Laboratories
5600 JA Eindhoven
The Netherlands

J. G. WILKES

Mullard Ltd.
Millbrook Estate
Southhampton Hamphire SO97BH
United Kingdom

PREFACE

As feature dimensions of integrated circuits shrink, the associated geometrical constraints on junction depth impose severe restrictions on the thermal budget for processing such devices. Furthermore, due to the relatively low melting point of the first aluminum metallization level, such restrictions extend to the fabrication of multilevel structures that are now essential in increasing packing density of interconnect lines. The fabrication of ultra large scale integrated (ULSI) devices under thermal budget restrictions requires the reassessment of existing and the development of new microelectronic materials and processes.

This book addresses three broad but interrelated areas. The first area focuses on the subject of rapid thermal processing (RTP), a technology that allows minimization of processing time while relaxing the constraints on high temperature. Initially developed to limit dopant redistribution, current applications of RTP are shown here to encompass annealing, oxidation, nitridation, silicidation, glass reflow, and contact sintering.

In a second but complementary area, advances in equipment design and performance of rapid thermal processing equipment are presented in conjunction with associated issues of temperature measurement and control. Defect mechanisms are assessed together with the resulting properties of rapidly deposited and processed films. The concept of RTP integration for a full CMOS device process is also examined together with its impact on device characteristics.

The third interrelated area, devoted to novel microelectronic materials for low temperature processing, provides topics pertinent to the fabrication of multilevel structures as well as to the properties and applications of epitaxial, dielectric, and metal films produced by innovative chemistries and/or by energy-assisted deposition processes.

This book is the result of a NATO Advanced Study Institute (ASI) held in Boca Raton, Florida. The selected blend of topics is meant to shed light on the science as well as the "hardware" aspects of this emerging technology. A 22 hour video program providing an audiovisual documentation of the contents of this book has also been produced. Information about this program is available from the editor.

I am indebted to numerous people who have made this book possible. First, I wish to thank the authors for their outstanding contributions and patience during the lengthy editing process. I also with to thank D. C. Jacobson and N. Erdos for their valuable review comments. Thanks are due to E. D. Miller for the art work and G. Moore for the required text processing. I specially wish to thank Dr. L. V. da Cunha at NATO Headquarters without whose generous financial

support ant continued guidance this ASI and the events leading to this publication
would not have been realized. Last, but not least, thanks are due to my
management at AT&T Bell Laboratories for providing the opportunity of working
on this book.

Murray Hill, New Jersey   R. A. Levy

CONTENTS

Rapid Thermal Process Integration

I. Calder

Introduction to Direct Writing of Integrated Circuit

G. Auvert

Ion Beam Assisted Processes

G. Foti

Micrometallization Technologies

J. M. Martinez-Duart and J. M. Albella

Multilevel Interconnect Structures

T. O. Herndon

Interlevel Dielectrics for Reduced Thermal Processing

F. S. Becker

Low Temperature Silicon Epitaxy for Novel Device Structures

J. O. Borland

RAPID THERMAL PROCESSING WITH REACTIVE GASES

Jaim Nulman

Process Technology Department
AG Associates
1325 Borregas Ave *
Sunnyvale, CA 94089
U.S.A.

INTRODUCTION

Two major trends have always been consistent in silicon integrated circuit technologies: lateral and vertical scaling of the device area into the submicron regime and the increase in wafer size. The growth of high quality thin thermal silicon dielectrics on the order of 3 to 15 nm constitutes one of the major challenges associated with submicronmeter silicon ULSI. Mutually conflicting requirements the growth conditions arise from process integration. Growth times for thermal oxides in the 10 nm regime are very short at high temperatures. Consequently, oxide growth occurs under the initial phase growth kinetics, which is hard to control in furnaces and is currently not well understood. Silicon dioxide films with good electrical characteristics require high temperatures both during growth and post oxidation annealing[1,2]. Chlorine-based mobile ion gettering enhances the oxidation rate and requires activation temperatures on the order of 1100°C [3,4]. Two step oxide growth processes have been suggested as one partial solution to the above problems[5], but they seem to require excessively long high temperature steps for post-oxidation annealing and chlorine activation. On the other hand, shallow junctions on the order of 50 to 200 nm are needed for submicrometer ULSI. This prohibits the use of long high temperature steps after implantation[6].

Pure silicon dioxide films might not be suitable as gate dielectrics for submicron ULSI technologies due to their susceptibility to hot electron effects[7]. Furthermore, the need of dielectric materials with higher dielectric constants, particularly for megabit DRAM technologies[8], requires the exploration of alternative dielectric films, such as oxinitrides. Oxinitrides or nitreded oxides (SiO_xN_y) have shown good masking characteristics against impurity diffusion, excellent dielectric strength, larger dielectric constant, and

---

* Present Address: Applied Materials, Inc., 3050 Bowers Ave., Santa Clara, CA 95054

1

improved durability against both irradiation and carrier injection[9-15]. The nitridation process typically requires long furnace times or plasma enhanced processes, which are in conflict with the need for a low thermal budget. Furthermore, these techniques suffer from an inability to control the spatial nitrogen distribution in the oxides, and consequently have not exhausted the full potential of nitrided oxides.

The trend towards 150 and 200 mm wafers introduces handling and process control problems. Large wafers require bigger loading systems and large gas volumes; thus, furnace processing requires extended temperature stabilization times resulting in possible wafer warpage, an undesired problem for the high resolution lithography steps that follow in ULSI manufacturing. Minimized wafer handling is also a requirement for enhancing device yield, as mandated by larger IC's.

The use of Rapid Thermal Processing (RTP) for the processing of silicon dielectrics as encountered in several steps of ULSI fabrication is discussed here. RTP is described here to include Rapid Thermal Cleaning (RTC)[16], Rapid Thermal Oxidation (RTO)[17], Rapid Thermal Annealing (RTA)[18], Rapid Thermal Nitridation (RTN)[19], Rapid Thermal CVD (RT-CVD) of polysilicon[20] and silicon epitaxy[21], and in-situ multisequential processing combining, two or more of these RTP steps. In the following, the basics of RTP with reactive gases are described, followed by the kinetics of silicon wafer cleaning, oxide growth, nitridation of oxides and re-oxidation. The applications of RTP oxides and oxinitrides are then discussed for device isolation, gate, tunnel and trench capacitors. Further, dielectrics grown on poly-silicon are discussed, followed by a description of in-situ multisequential processing for the fabrication of bipolar devices and poly-silicon gate MOS devices.

Since the electrical characteristics of thin silicon dielectrics depend on the quality of the materials used, e.g., wafers, gases, chemicals, etc.; the magnitude of the electrical characteristics is given here as a figure of merit. However, since they are always compared to control furnace processes, they are indicative of the improved performance possible with materials processed by RTP.

RAPID THERMAL PROCESSING

In diffusion furnace processing, a batch of wafers is slowly loaded at a temperature between 700° and 800°C in a nitrogen ambient, Fig. 1a. The temperature is then ramped to the process temperature, in the 850° to 1100°C range, over a 20 minute period. A soak period follows to allow for the temperature stabilization of the large mass of wafers, quartz cassettes and automatic loading mechanism. Only then are the reactive gases, such as oxygen for SiO_2 growth, introduced. Therefore, in order to control a uniform dielectric growth not only across a wafer but from wafer to wafer, the temperature needs to be sufficiently low to allow for a reaction kinetic slower than the reactant gas diffusion time along the furnace tube and between wafers. This limits the temperature for growing 3 to 20 nm SiO_2 films to about 950°C. Furthermore, the large furnace tube volume requires long purge times if another type of process is sequentially required, such as a nitridation in an ammonia ambient. At the end of the growth time, an inert gas is introduced for a typical 10 minute soak before a ramp down to a lower temperature for wafers removal. The wafer temperature is inderectly monitored at different zones by a set of thermocouples located outside the process tube. A furnace may be categorized as a system in which process control is performed by switching gases at a constant temperature. Therefore, the processes are gas diffusion limited.

Figure 1b shows a cross section of a typical RTP system. This includes reflector plates, lamps, a low mass quartz tube and wafer support tray, a door that closes the loading port during processing, and a pyrometer for contactless direct wafer temperature monitoring. Gases are introduced into the quartz tube from one side, through diffusion baffles, and exhausted out at the other end. Due to the small chamber volume (typically less than one liter) ambient gases can effectively be purged in a matter of seconds. This, combined with the fact that the chamber walls (including the quartz) are kept cool, allows for the sequential combination of processes with different gases. In RTP, a single wafer is introduced into the nitrogen-filled quartz chamber at room temperature. After a typical 10 second purge time, pure process gases are introduced to replace the nitrogen ambient. The wafer temperature is then quickly ramped up at rates between 10° and 250°C/second to perform chemical reactions during typical time frames of between 1 and 120 seconds. The temperature is subsequently ramped down to stop the chemical reaction and an inert gas is introduced for either a final purge before removing the wafer or for a purge before introducing another reactive gas for next step. Temperature control is performed through a closed loop system. A computer analyzes the temperature monitored by the pyrometer, relative to the programmed values, and adjusts lamp

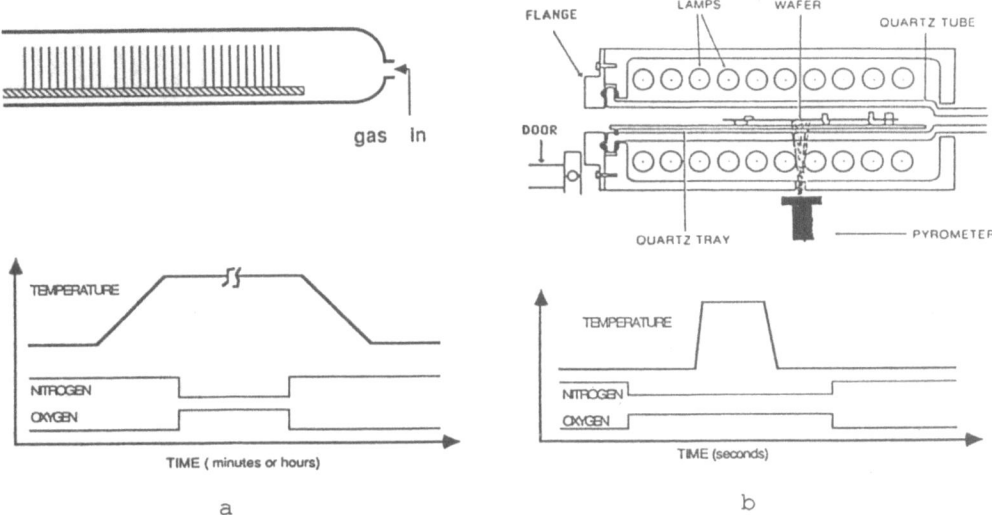

Fig. 1   Furnace (a) and RTP (b) processing systems.

intensity accordingly. Since the wafer heats up by absorbing the energy emitted from the lamps, and all other portions of the chamber are cool, fast temperature changes are possible. RTP may therefore be categorized as a system in which process control is performed by switching temperatures at a constant gas flow. Thus, the chemical reactions are not gas diffusion limited, and allow for 3 to 20 nm thick SiO_2 growth at temperatures above 1000˚C.

The quartz chamber in RTP systems allows the introduction of highly reactive and corrosive gases such as ammonia and hydrogen chloride. The latter nay be provided directly from a pure liquid source or inderectly, from the decomposition of 1,1,1-Trichloroethane (TCA)[22]:

$$C_2H_3Cl_3 + 2O_2 \ \text{------>} \ 3HCl + 2CO_2 \ .$$

Table 1 lists the gases used for wafer cleaning (RTC), oxidation (RTO), annealing (RTA), nitridation (RTN), and CVD (RT-CVD). It is important to note that these gases should be of the highest purity available. In the case of ammonia, it is necessary to ensure that the residual oxygen and moisture concentrations be below 500 ppb. Fig. 2 shows typical temperature/time profiles used for the growth of different silicon dielectrics by RTP: RTO/RTA, RTO/RTN, RTO/RTN/RTO, RTC/RTO/RTA, etc. For the discussion in this manuscript, RTP time is defined as the time "at the steady state temperature," that is, the temperature between ramp-up and ramp-down, (Fig. 2).

KINETICS OF SILICON DIELECTRICS BY RTP

Rapid Thermal Cleaning

The electrical properties of thin silicon dielectrics are strongly influenced by cleaning procedures. These procedures use solutions of ammonium hydroxide/hydrogen peroxide and hydrochloric acid/hydrogen peroxide, or sulfuric acid/hydrogen peroxide. These cleaning processes result in a thin chemical

Table 1. Gas ambients used for the RTP growth of silicon dielectrics and the CVD of poly-Si and epitaxy Si.

RTP CYCLE	RTC	RTO	RTA	RTN	RT-CVD
		O_2	N_2	NH_3	$SiH_4/B_2H_6/H_2$
AMBIENT	$Ar/10\%H_2/$ $0.2-1\%HCl$	$O_2/1-4\%HCl$	Ar		$SiH_4/AsH_3/H_2$
	H_2				$H_2/7\%SiH_2Cl_2/$ $2\%HCl$

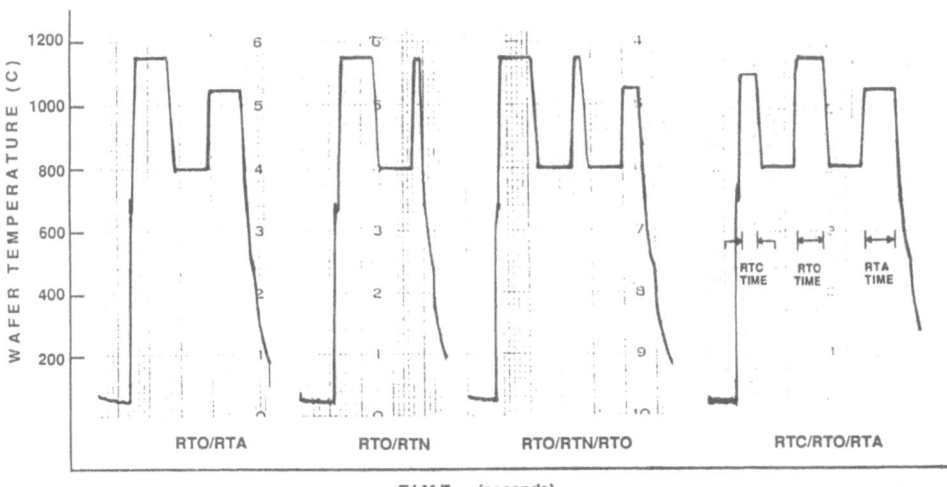

Fig. 2 Typical temperature/time cycles used for
the growth of silicon dielectrics.

oxide with a thickness between 1.0 and 1.5 nm. In some cases,
this native oxide is etched in buffered hydrofluoric acid,
followed by a de-ionized (DI) water rinse. In either case,
particles in the chemicals and DI water reduce device yields,
and bacteria in the water impact the dielectric electrical
endurance. Some processes call for a plasma etch before oxide
growth. In this case, surface damage might result in an
increase of the dielectric fixed charge density and interface
trapped charge density. These problems can be solved by the use
of *in-situ* cleaning prior to SiO_2 film growth.

Rapid thermal cleaning allows for the reduction of native
and chemical oxides, as well as chemical etching of small
amounts of silicon. It is typically performed at temperatures
between 1000° and 1150°C for 2 to 10 seconds. The gases used
are listed in Table 1 under the RTC column. Using RTC process,
an ultra clean silicon surface is prepared prior to film
growth. Therefore, wafers can be transported from a wet
cleaning station to the RTP equipment having a passivation
chemical oxide. Thus, avoiding the need for buffered oxide etch
and the effects of cleanroom ambient air on the wafer surface,
prior to oxidation.

The RTC process starts by the reduction of the native or
chemical oxide layer. Silicon dioxide films subjected to high
temperature heat treatments in oxygen free ambients (less than
100 ppb) are unstable if silicon is supplied to form volatile
silicon monoxide:

$$Si + SiO_2 \quad ---> \quad 2SiO\uparrow \ .$$

The thermodynamics for this decomposition reaction and its
dependence on the oxygen partial pressure in the process
ambient can be found elsewhere[23,24]. In RTP the use of $Ar/10\%H_2$

results in the reduction of chemical oxides at rates shown in Table 2. From this data, it is observed that RTC can be accomplished in a matter of 2 to 10 seconds at steady state temperature, while 10 to 40 % of the etching takes place during temperature ramp-up. Argon serves as a carrier gas, while H_2 reduces some silicon to gaseous SiH_x and also allows for the formation of SiO.

Higher rates than those listed in Table 2, are obtained when RTC is performed in $Ar/10\%H_2/0.2-1\%HCl$. However, due to the very thin chemical oxide (1 to 2 nm), it is very difficult to determine the etching rate because 90 % of the chemical oxide is etched during temperature ramp-up, when the final steady state temperature is 1050°C or higher. The mixture of $Ar/10\%H_2/0.2-1\%HCl$ allows a more effective reduction of silicon than pure forming gas or hydrogen ambients. This type of treatment has been widely used on silicon wafers prior to epitaxy because of its two-fold advantage of cleaning and surface polishing. This cleaning also eliminates metallic impurities, which adhere to the silicon surface, by the formation of the corresponding metallic hallides. The kinetics of silicon etch in H_2/HCl gas systems is represented by a series of adsorption-desorption reactions and subsequent chemical reaction of the absorbed species. The possible reactions are represented by:

$$nHCl* + Si \longrightarrow SiCl_n* + nH*,$$

$$1/2H_2(g) + * \longrightarrow H*,$$

$$H* + HCl* + Si(s) \longrightarrow SiH_xCl*_{4-x},$$

and

$$SiH_xCl*_{4-x} \longrightarrow SiH_xCl_{4-x}(g)$$

where * denotes a free site on the surface. These kinetics have

Table 2. RTC etch rate for chemical oxides in $Ar/10\%H_2$.

TEMPERATURE (°C)	RAMP-UP ETCH (nm)	ETCH RATE (nm/s)
1150	0.4	0.1
1100	0.3	0.04
1050	0.2	0.015
1000	0.05	0.005
950	----	0.002

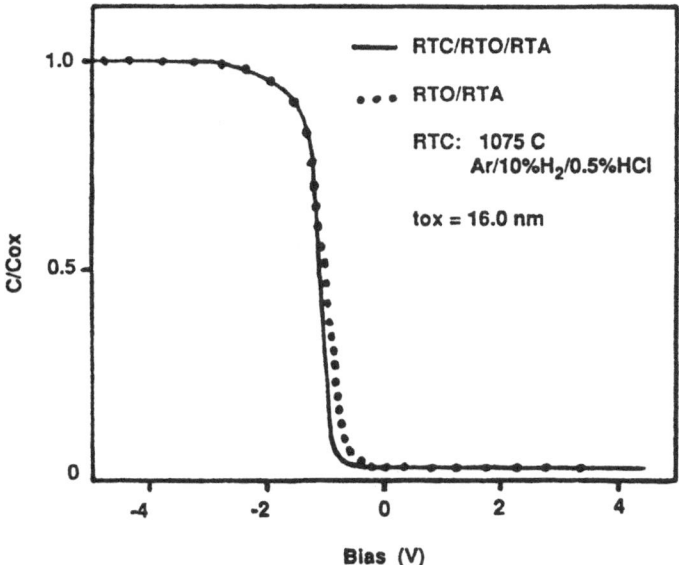

Fig. 3 Capacitance-Voltage characteristics for SiO₂
films grown by RTP with and without *in-situ*
cleaning.

been studied for conventional furnace processing[25-29], and are
the same for the RTP case because in both cases they are
thermally driven.

The effect of RTC on the electrical properties of SiO₂ films
was studied by comparing Capacitance-Voltage (C-V) curves for
films grown by RTO/RTA with and without RTC. Fig. 3 shows such
curves for 15.0 nm SiO₂ films grown at 1150°C in dry O₂
followed by post oxidation annealing at 1050°C for 40 seconds
in N₂. The RTC cycle was done at 1075°C in Ar/10%H₂/0.5%HCl.
These samples were cleaned in a plasma reactor prior to RTP.
The reduction of interface states is clearly indicated in the
C-V curves of Fig. 3 by the sharper transition from charge
accumulation to inversion for the wafer with RTC compared to
the one without[30]. Wafers with RTC cycle show a mid gap density
of states of less than 5×10^9 cm^{-2}eV^{-1}.

Rapid Thermal Oxidation

Silicon dioxide film thickness as function of RTO time and
temperature for one- and two-step dry O₂ oxidation cycles, are
shown in Figs. 4a and 4b, respectively. The two-step oxidation
cycle consists of a temperature soak of 30 seconds at 775°C
before a ramp-up of 200°C/second to the RTO temperature. In the
case of the one-step RTO, a direct ramp-up of 200°C/second from
room temperature to the RTO temperature is used. As shown in
Fig. 4, RTO provides for SiO₂ film growth in the 3 to 15 nm
range at temperatures above 1000°C. Since each wafer is
processed individually, its temperature is accurately
monitored, thus ensuring temperature repeatability in the order

of $\pm$ 3°C. In both cases, at zero RTO time, the SiO_2 film thickness increases with increased oxidation temperature. This is because in RTO process, temperature ramp-up is performed in the presence of the oxidizing ambient. Fig. 5 shows the corresponding Arrhenius plot of the total SiO_2 film grown during the ramp-up part of the RTO cycles. For the two-step process, the plot is not linear at temperatures below 1050°C. This is because of the 1.7 nm SiO_2 film that grows during the 30 seconds soak at 775°C. For temperatures above 1050°C, both the one- and two-step processes follow the linear relationship

$$tox = to \; e^{-Ea/kT},$$

where k is the Boltzmann constant and T the temperature. The values for to and Ea are 5.2×10^2 nm and 0.66 eV respectively for the one-step cycle, and 7.8×10^4 nm and 0.6 eV respectively for the two-step cycle. This data clearly indicates that for RTO kinetics studies, the temperature ramp-up conditions must be taken into account.

The SiO_2 film thickness curves in Fig. 4a show a delay of about 15 seconds in the oxidation kinetics for temperatures below 850°C. This delay is related to an incubation time for

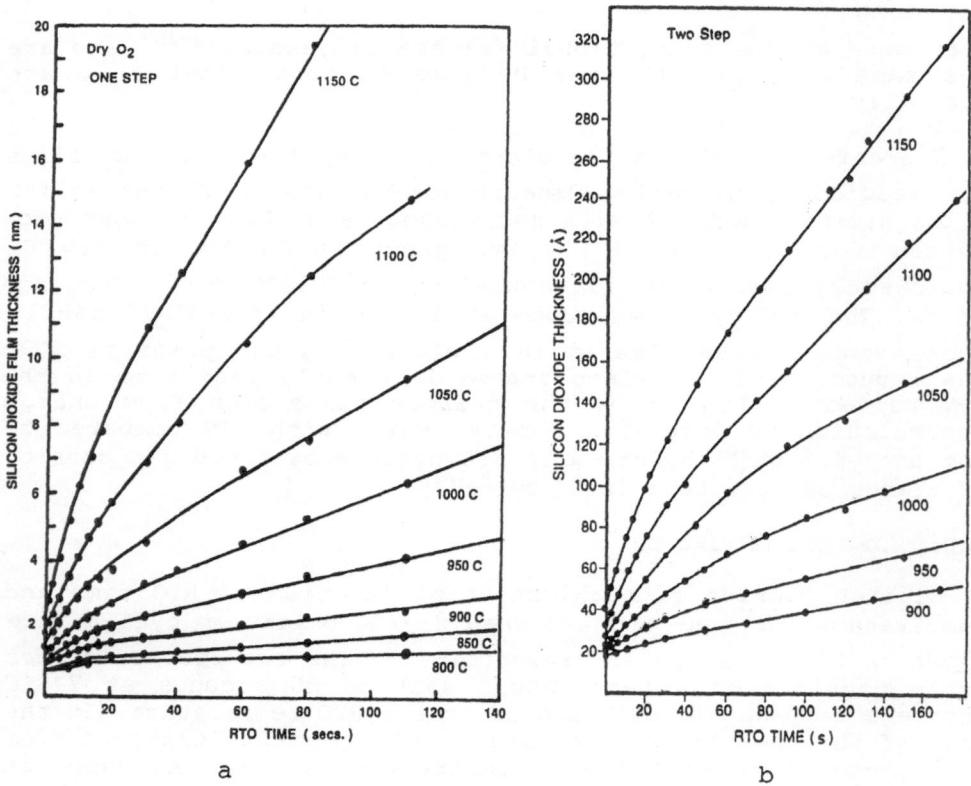

Fig. 4 Typical SiO_2 film thickness as function of RTO time and temperature for a one- (a) and two-step (b) cycle.

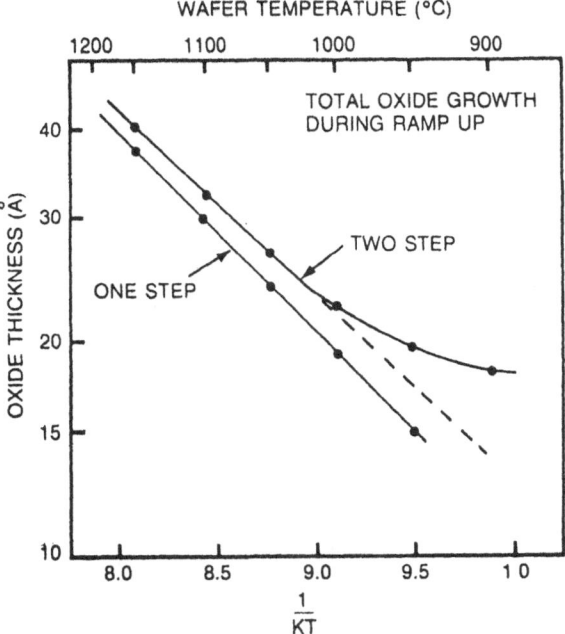

Fig. 5   Arrhenius plot of the total SiO$_2$ film
grown during temperature ramp-up for a
one- and two-step RTO cycle as a
function of temperature.

the oxidation to take place. It may be related to a need for
chemisorption of oxygen on the silicon surface as a precursor
to the chemical reaction to form oxide[31], and/or to the
diffusion of oxygen into silicon during initial exposure of
silicon to oxygen[32]. The higher the oxidation temperature, the
shorter the incubation time; therefore, it is not observed for
temperatures higher than 900°C. This effect is not observed in
the two-step oxidation process because it is masked by the soak
step at 775°C for 30 seconds.

Both the one- and two-step SiO$_2$ growth curves in Fig. 4 show
three distinctive oxidation kinetics after the initial delay: a
very fast initial linear growth regime, a transition regime and
a second, slower, linear growth regime. The first fast linear
oxidation regime could be due to the low resistance offered by
the ultra-thin SiO$_2$ layer to oxidizing species and to the
possible presence of an O$_2$ diffused zone in the silicon[32]. Other
possible explanations include the presence of micropores in the
oxide[33], a change in oxygen solubility in the oxide[34], the
accumulation of positive charge at the SiO$_2$/Si interface[35,36],
field-assisted oxygen diffusion in the oxide[37], stress in the
oxide[38,39], and atomic oxygen near the SiO$_2$/Si interface
enhancing the rate of oxidation[40,41]. Fig. 6a shows the SiO$_2$
film thickness for the initial linear growth regime for a
two-step RTO cycle. The oxidation rate (Rox) in this regime
follows the linear Arrhenius relationship

$$Rox = Ro \; e^{-Ea/kT},$$

where the oxidation rate constant, Ro, and the activation energy, Ea, are 4.6×10^4 nm/second and 1.44 eV respectively, Fig. 6b.

The third regime in the SiO_2 growth kinetics in Fig. 4, is the well-known Deal-Grove linear oxidation kinetics[37]. The second growth regime is the transition between the fast and slow linear oxidation regimes. This transition regime can be analyzed by looking at the thickness and time values obtained at the intersection point of two straight lines tangent to the two linear regimes[42], Table 3. The transition times seem to be independent of the oxidation temperature, while the transition thickness increases with increased RTO temperature. These results can be explained qualitatively in terms of the reaction zone silicon oxidation model[32]. The model predicts increased oxygen diffused zone widths, and hence, increased transition thickness with increasing temperature due to the associated increase in O_2 diffusivity in silicon and the interface and volumetric O_2/Si reaction rates. More light on the initial oxide growth mechanism can be obtained by analyzing the fixed oxide charge (Q_f), the mid gap interface density of states (D_{it}), and the oxide growth rate as function of the SiO_2 film thickness. Figs. 7a, 7b, and 7c show the corresponding values for RTO oxides grown at 1100°C in dry O_2[43]. Fixed charges and

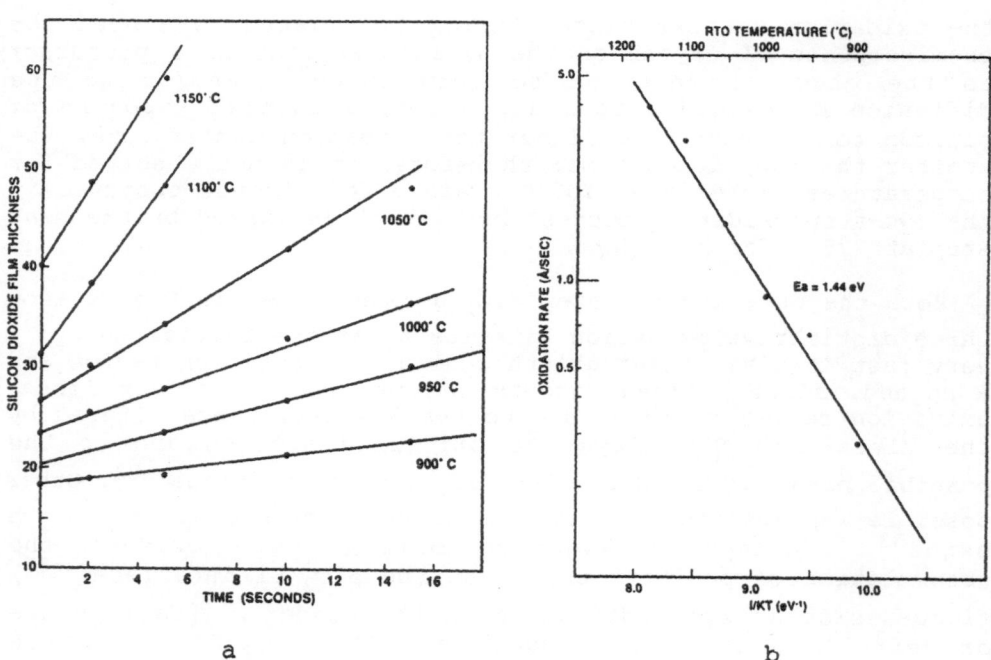

a

b

Fig. 6   (a) Initial SiO_2 film thickness as a function of RTO time and temperature for a two-step cycle. And (b) Arrhenius plot of the corresponding oxidation rate.

Table 3. Transition oxide thickness and time from the initial fast linear oxidation regime to the slower linear oxidation regime.

TEMPERATURE (°C)	TIME TO TRANSITION (s) DATA FROM		THICKNESS AT TRANSITION (nm) DATA FROM	
	Fig. 4b	Ref. 42	Fig. 4b	Ref. 42
1200	--	23	--	15
1150	30	27	12	11
1100	30	27	9	8.6
1050	33	30	7	7
1000	30	--	4.5	--
950	30	--	3.6	--

interface density states can be measured using surface photovoltage techniques which eliminate the need for processing that might affect the measured values[44]. The oxide growth rate is obtained by assuming linear growth between adjacent points on the oxide thickness vs. oxidation time curve and the local slope is calculated. All of the graphs in Fig. 7 have a break point at 9 nm in agreement with the data in Table 3, suggesting a close relationship between the oxide growth mechanism and the origin of fixed charges and interface charges.

The need for mobile ionic and metallic ion gettering demands the use of hydrogen chloride during the oxidation cycle. However, the use of HCl might result in silicon surface pitting. In order to avoid this problem two step furnace oxidations have been suggested, where during the first step a passivation layer is grown in the absence of HCl[15]. In the case of RTO, this passivation layer is grown during temperature ramp-up, since it is done in the presence of the oxidazing ambient. As shown in Fig. 5, about 2.0 nm thermally grown SiO_2 film is present when the wafer reaches 1050°C, thus eliminating the need for an extra step for the passivation layer growth. Fig. 8 shows the effect of adding 0 to 4% HCl to the O_2 ambient for RTO at 1150°C. Enhanced oxidation rate is observed, as in the case of furnace oxidation[3]. The addition of each 1% HCl results in an oxide film thickness equivalent to an oxide grown in dry O_2 at 12°C higher temperature. RTO in O_2 with 3% HCl has proven effective in gettering metallic contamination of the wafer surface, as well as to provide for an oxide film with better endurance than RTO in dry O_2 or low-temperature furnace O_2 with TCA oxides[45]. This is shown in Fig. 9, which lists the yield on the time to breakdown for 8 and 10 nm such SiO_2 films during stress at a current density of 100 mA/cm^2.

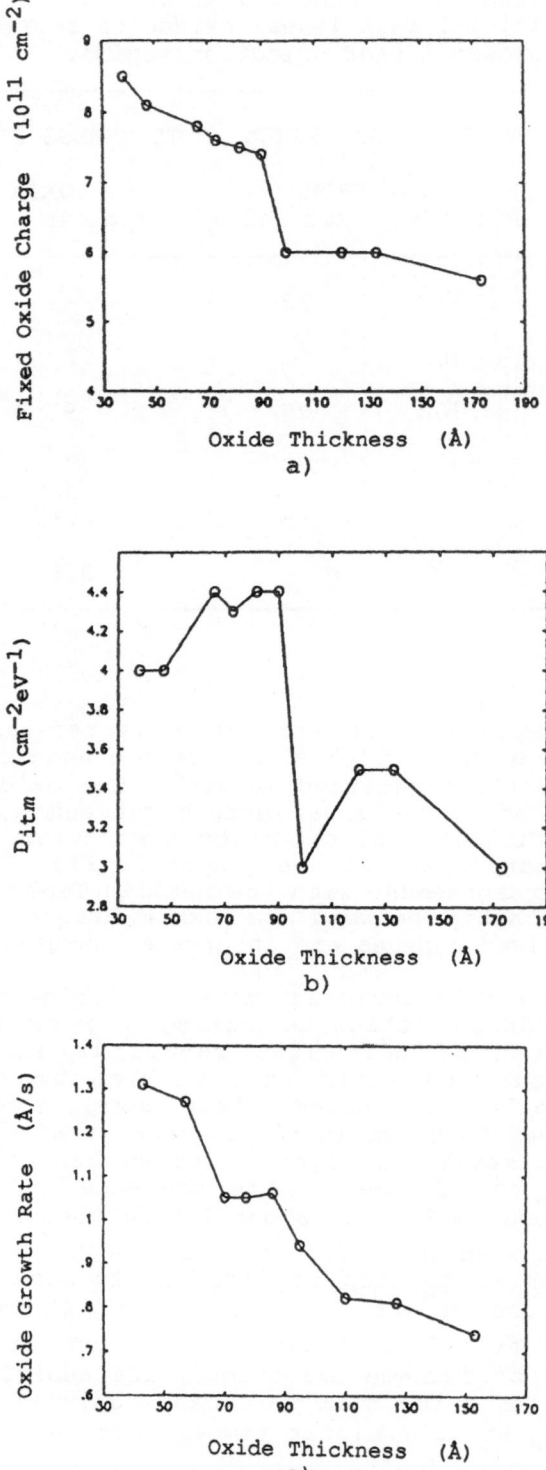

Fig. 7 (a) Fixed oxide charges (Q_f), (b) Interface density of states (D_{it}), and (c) oxidation rate as function of SiO_2 film thickness for RTO in dry O_2 at 1100°C [43].

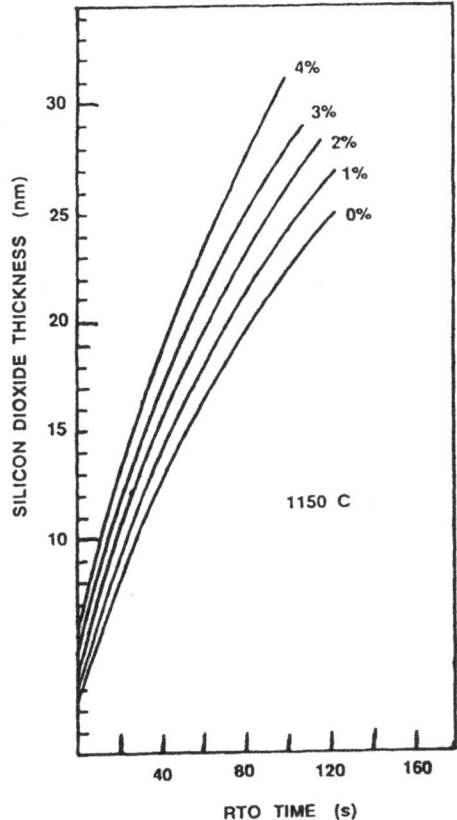

Fig. 8 Silicon Dioxide film thickness for RTO in O_2 + 0 to 4% HCl.

tox	Tanneal	RTO/HCl	RTO/O	FURNACE
80Å	None	100% 81.3%	68.8% 75.0%	81.3% 75.0%
	900°C	93.8% 100%	56.3% 68.8%	81.3% 56.3%
	1000°C	100%	87.5% 50.0%	75.0% 56.3%
	1050°C	93.8% 87.5%	75%	81.3%
100Å	None	100% 93.8%	100% 93.8%	68.8% 75.0%
	900°C	100% 93.8%	93.8% 81.3%	68.8% 56.3%
	1000°C	100% 93.8%	93.8% 81.3%	62.5% 62.5%
	1050°C	81.3% 100%	75.0% 93.8%	56.3% 68.8%

Fig. 9 Wafer yield as determined by time to breakdown under 100 mA/cm^2 constant current density stress[45].

Rapid Thermal Annealing

Post-oxidation annealing is a heat treatment required to reduce fixed charges (Q_f), interface charges (Q_{it})[46], and electron and hole trapping[18]. In a diffusion furnace, annealing is accomplished, in part, by the built-in inert gas purge after the oxidation, and, sometimes by an additional 900° to 1000°C temperature treatment. In the case of RTO, the oxidation is stopped by a fast temperature ramp-down, 10° to 150°C/second. Therefore, unreacted oxidizing species are frozen along the SiO_2 matrix resulting in localized tensile stress. Furthermore, since RTO temperatures are above the viscous flow temperature of SiO_2, there is, effectively, a layer of viscous SiO_2 on the silicon wafer at the end of the growth period. This is indicated by the presence of a six-membered ring structure typical of bulk SiO_2 close to the SiO_2/Si interface, as revealed by the Si_{2p} energy shifts of 4.5 eV compared to 4.4 eV for furnace-grown oxides measured by X-ray photo-electron spectroscopy (XPS). After RTA in N_2 or Ar, the local stress formed by the conjunction of the four-membered rings in the silicon crystal and the six-membered rings of the bulk SiO_2 is relaxed by forming a smooth distributed transition region and the XPS analysis shows a decrease in the average ring structure of the interface region. At the proper anneal time and temperature, RTO oxides show the Si_{2p} energy shift of 4.4 eV typical of furnace oxides[47].

The localized stress and the frozen oxidizing species are a source of charge trapping, both fixed charges and interface charges. Therefore, tests of the electrical properties of the SiO_2 films are the best indication of the effectiveness of the RTA process. These charges can be characterized by measuring the Flat-Band Voltage (V_{FB}) of MOS capacitors with RTO SiO_2 films and various RTA conditions[30],

$$Q_f = (\varepsilon \, \varepsilon ox/tox) \times \Delta V_{FB},$$

where ε is the permittivity in vacuum, εox is the SiO_2 dielectric constant, and tox is its thickness. Fig. 10 shows V_{FB} as function of the RTA time at 1050°C in N_2 ambient for 10 nm SiO_2 films grown at 1150°C in dry O_2. A very fast anneal process takes place during the initial 20 seconds, as indicated by the reduction of V_{FB} and the increase of the electric breakdown field (E_{BD}). The reduction of V_{FB} correlates with a reduction of the fixed oxide charge. Fixed charge density values of 1×10^{10} cm^{-2} are typical for RTO oxides after a 30 seconds RTA cycle. Post-oxidation RTA times longer than 40 seconds at 1050°C result in unnecessary processing time[48]. Charge trapping also results in a V_{FB} shift. Fig. 11 shows such a shift as a result of timed avalanche injection of electrons into a 50 nm RTO grown oxide, and a 50 nm RTO oxide with a 10 seconds anneal at 1000°C in argon ambient. It is clear that the RTA cycle is necessary in order to reduce the trapping centers in the grown film[49].

Long-term high-temperature post-oxidation furnace anneals in N_2 ambient shows a nitrogen peak at the SiO_2/Si interface. As a result, the density of fixed oxide charges and surface states increases[50]. Therefore, most furnace post-oxidation anneals are done at temperatures below 900°C or in Ar ambients. Fig. 12 shows SIMS nitrogen concentration profiles on 50 nm SiO_2 films after 5 minutes and 17 hour anneals in N_2 at 1000°C. It is clear that the short time RTA in N_2 does not introduce the nitrogen peaks observed after furnace annealing[18]. This is in agreement with the observed reduction in the density of fixed and interface charges described above.

In RTP, the use of Ar ambients needs special attention. Due to the leak-tight low volume chamber design, oxygen concentrations of 500 ppb or less are possible. High temperature anneals in such an "oxygen-free" environment results in degradation of the SiO_2 film due to the creation of localized $SiO^{23,51}$. This defect creation problem is eliminated by adding 1 to 2% O_2 to the annealing ambient.

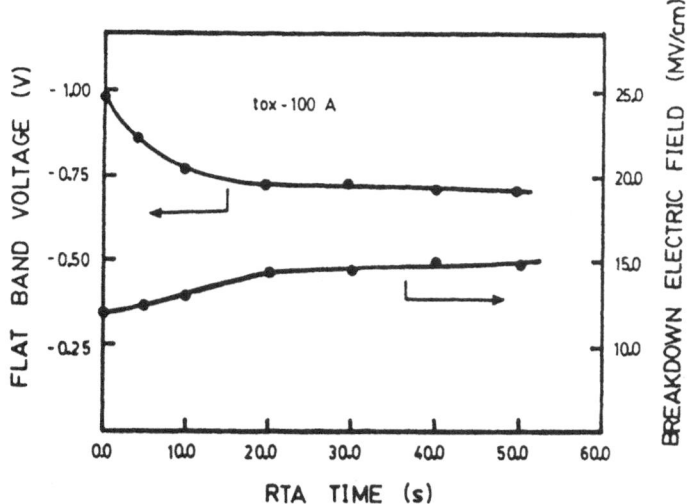

Fig.10   Effect of post-oxidation RTA time on the MOS flat-band voltage and electric breakdown field[48].

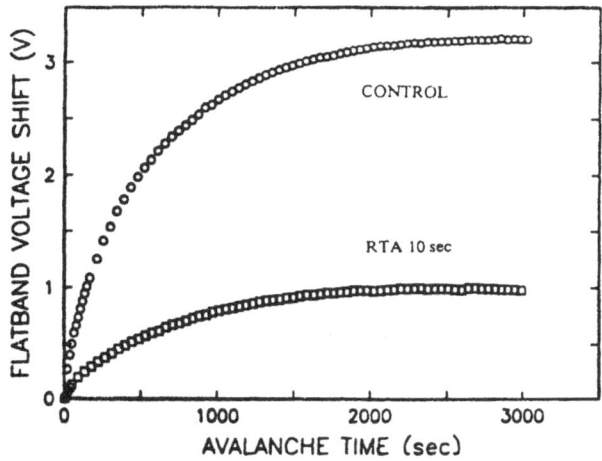

Fig.11   Flat-band voltage shift as a result of electron injection into 50 nm RTO oxide with and without RTA. Injection current density of 60 $\mu A/cm^2$ [49].

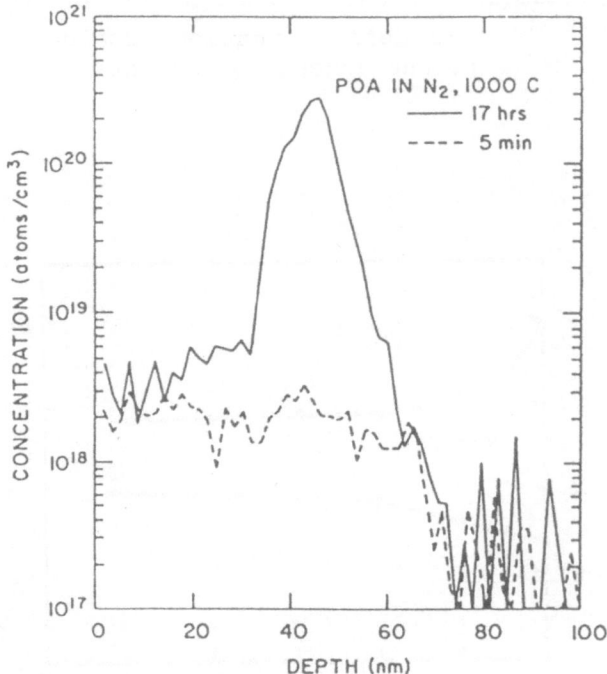

Fig. 12   Secondary ion mass spectroscopy (SIMS) of
^{14}N for 50 nm oxides after a long or a short
post-oxidation anneal in N_2 at 1000°C [18].

<u>Rapid Thermal Nitridation</u>

The short times, in the order of seconds, used in RTP, combined with fast temperature ramp-up and ramp-down, allows for accurate incorporation of nitrogen into the SiO_2 film[19,52]. The Rapid Thermal Nitridation (RTN) process uses ammonia as the source of nitrogen (Table 1). RTN may be performed on oxides thermally grown in a furnace or sequentially after an RTO cycle (Fig.2). The nitrogen incorporation into the SiO_2 film as function of the RTN time and temperature is investigated using Auger analysis and ellipsometry.

Fig. 13 shows the atomic concentration of nitrogen, oxygen and silicon as function of the Auger sputter etch time for nitrided oxides at 1150°C. For RTN times shorter than 5 seconds the nitrogen Auger signal reaches the noise level at a depth of approximately 3 nm and no nitrogen builds up at the SiO_2/Si interface (Fig. 13a). These short nitridations result in surface nitridation only or NO structure. Longer RTN times allow a gradual second nitrogen peak, through nitrogen diffusion, to build up at the SiO_2/Si interface, or NON structure. This effect continues until the interface region saturates with nitrogen (Fig. 13b, 60s). At this point, the nitrogen atomic concentration near the oxide silicon interface peaks to about 20%. For even longer RTN times, nitrogen incorporation starts at the bulk of the oxide and eventually a uniform nitrogen profile is observed, oxinitride (Fig 13b, 300s). The nitrogen atomic concentration reaches a peak value of approximately 22 % while the oxygen atomic concentration is monotonically reduced from 70% for pure SiO_2 to approximately 52% for a nitrogen-saturated oxinitride[19].

Because of the diffused nature of the nitrogen distribution in the oxide, it is only possible to calculate an effective index of refraction. The thickness of the nitrided oxide does not change with RTN time[53], and, since the nitrogen concentration increases monotonically as function of both the RTN time and temperature, the measurement of the effective index of refraction provides information about the nitridation process. Fig. 14a shows the effective index of refraction as function of the RTN temperature for 8 nm nitrided oxides at a constant RTN time of 120 seconds. Fig. 14b shows the corresponding values for a constant RTN temperature of 1150°C and varied times. An index of refraction of 1.46 is used for pure SiO_2 at the He-Ne laser frequency. Fig. 14a shows that the threshold nitridation temperature for reasonable RTN times is about 900°C. For temperatures in the 900° to 950°C range,

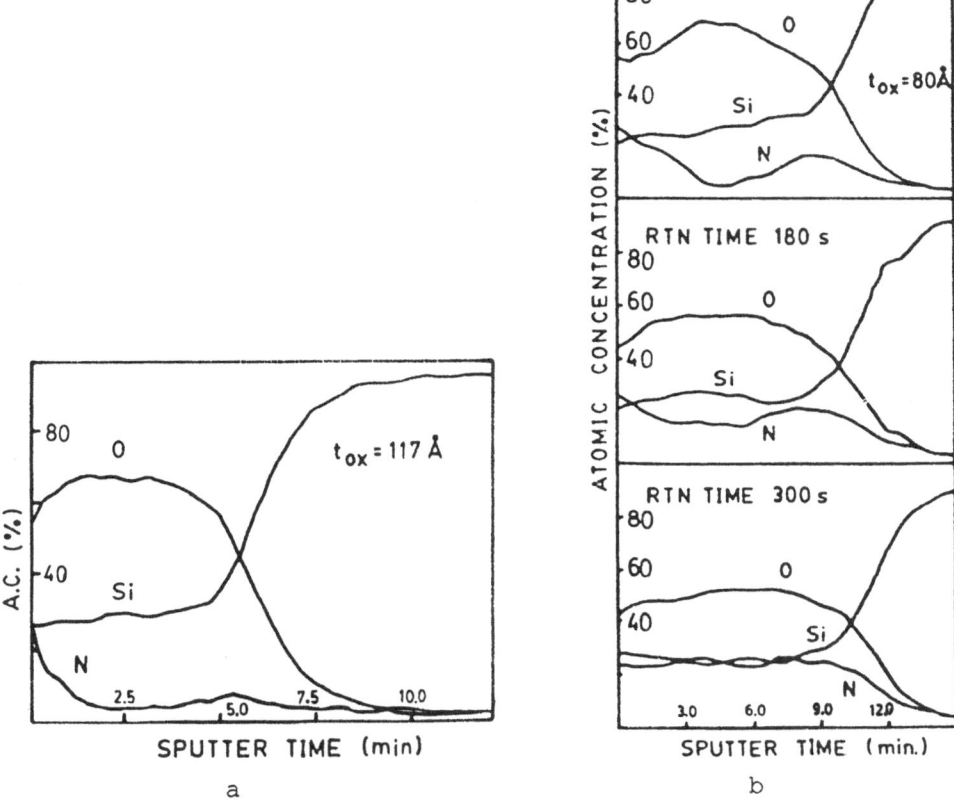

Fig. 13   Sputter Auger profiles for a 3 seconds RTN
           time (a), and longer RTN times (b), at an
           RTN temperature of 1150°C in pure NH_3 [19].

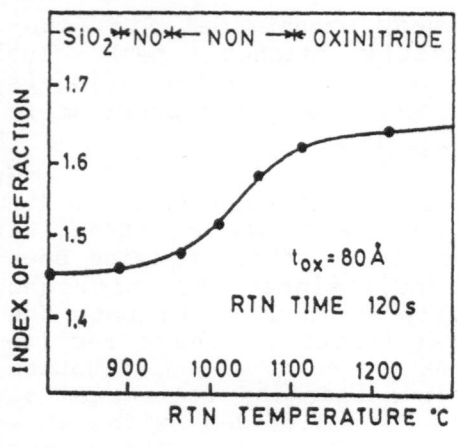 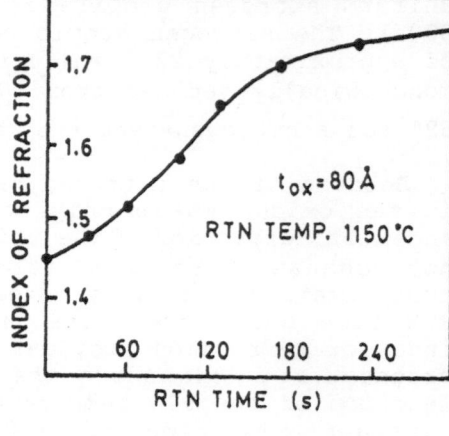

Fig. 14   Effective index of refraction at the He-Ne
           frequency for 8 nm nitreded SiO$_2$ films as
           function of (a) RTN temperature and (b)
           RTN time[19].

nitridation occurs only at the surface, NO. As temperature
increases, nitridation occurs at both the surface and the
oxide-silicon interface as shown by the rapid increase in the
index of refraction. For temperatures above 1100°C, the index
of refraction saturates at about 1.64, corresponding to a
saturated and uniform nitrogen distribution in the dielectric.
At a constant, high RTN temperature, the index of refraction
increases rapidly until saturation is reached at values of
1.72.

    For short RTN times, the nitridation process is transport-
limited even at high temperatures and does not show saturation.
The transport-limited reaction is also seen for thicker oxides.
At an RTN temperature of 1150°C and RTN time of 30 seconds, an
NON structure is observed for SiO$_2$ films up to 23 nm thick,
while thicker oxides result in NO structures. Since the RTN
process is carried out in ammonia ambient, the diffusion
species include atomic nitrogen as well as NH$_x$ and atomic
hydrogen. Therefore, hydrogen also incorporates into the
nitrided oxide layer. Fig. 15 shows hydrogen SIMS depth
profiles for a pure oxide and nitrided oxides at 1150°C for 5
and 300 seconds[53]. A dip near the surface in each profile is
an artifact, showing a Cs-deficient surface region*. Even with
this disturbance, the data in Fig. 15 shows that nitrided oxide
layers contain larger amounts of hydrogen than non-nitrided
thermal oxides. Furthermore, the hydrogen concentration in the
nitrided oxide increases monotonically as nitridation proceeds

---

*   A negatively charged hydrogen ion is generated only when a
        hydrogen atom is surrounded by Cs atoms. At the early stage of Cs$^+$
        sputtering, there are not sufficient Cs atoms implanted in the
        film, therefore the amount of emitted H$^-$ ions is small.

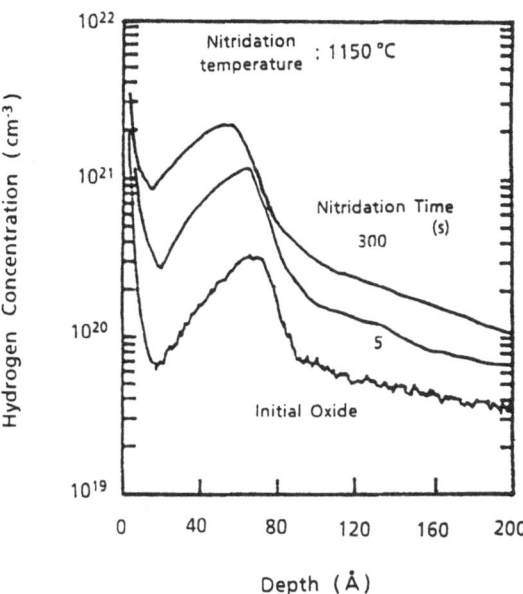

Fig. 15 Hydrogen SIMS depth profile for pure and
nitreded 8 nm SiO$_2$ films. Nitridation is
done in NH$_3$ at 1150°C for 5 and 300
seconds[53].

and some hydrogen also incorporates into the substrate. It is
believed that at the point in the process when the hydrogen
concentration reaches its maximum, the nitrogen concentration
can not increase any more due to the hydrogen saturation of
dangling silicon bonds. This is in agreement with the delay in
nitridation of thermal oxides observed when the oxides are
annealed in hydrogen prior to nitridation in ammonia[9].

The changes in the physical properties of nitrided oxides
are further clarified by the chemical shift of the Si$_{2p}$ lines
obtained from XPS analysis. The chemical shifts are larger than
for silicon nitride (2.5 eV) and smaller than for silicon
dioxide (4.6 eV). As nitridation time increases, the nitrogen
concentration increases (Fig. 13), and the Si$_{2p}$ peak moves
toward the silicon nitride peak position. The Si$_{2p}$ chemical
energy distribution for RTN at 1150°C for 300 seconds is very
narrow. This fact, combined with the uniform Auger sputter
profile, implies that this film is chemically homogeneous. In
contrast, an oxide nitrided for short times shows a broad Si$_{2p}$
chemical energy distribution, reflecting the non-uniform
nitrogen depth profile[53]. The RTN process, in fact, results in
SiO$_x$N$_y$ dielectric films with physical characteristics between
those of pure SiO$_2$ and those of pure Si$_3$N$_4$.

Rapid Thermal Re-Oxidation

Re-oxidation of Nitrided Oxides (RONO) can be accomplished by three sequential *in-situ* RTP steps: RTO/RTN/RTO, as shown by one of the temperature/time profiles in Fig. 2. The re-oxidation process results primarily in a change of the surface composition of the dielectric. Fig. 16 shows sputter Auger depth profiles for 15 nm dielectrics grown by RTO/RTN and RTO/RTN/RTO, with RTN at 1150°C for 15 seconds, and the re-oxidation at 1050 for 5 seconds in dry O_2. The short re-oxidation does not seem to affect the nitrogen at the SiO_2/Si interface, but, instead, it lowers the nitrogen concentration while increasing the oxygen concentration in the top 2.5 nm. Fig. 17 shows SIMS depth profiles for oxides nitrided at 1150°C for 60 seconds, and then re-oxidized for 60 seconds at 1050° and 1150°C. The hydrogen concentration decreases drastically. These decreases are monotonicall as the re-oxidation time or the re-oxidation temperature increases. Re-oxidation effectively reduces hydrogen-containing species in nitrides oxides[54]. If the nitridation step is done at temperatures below 1050°C, the re-oxidation step might affect both the surface of the dielectric and the dielectric-silicon interface (Fig. 18a). In any case, the surface effect is the reduction of the nitrogen concentration. The nitrogen concentration at the dielectric-silicon interface changes very little. Fig. 18b shows that the dielectric-silicon interface moves about 1 nm away from the initial position into the substrate by re-oxidation at 1150°C for 60 seconds[55].

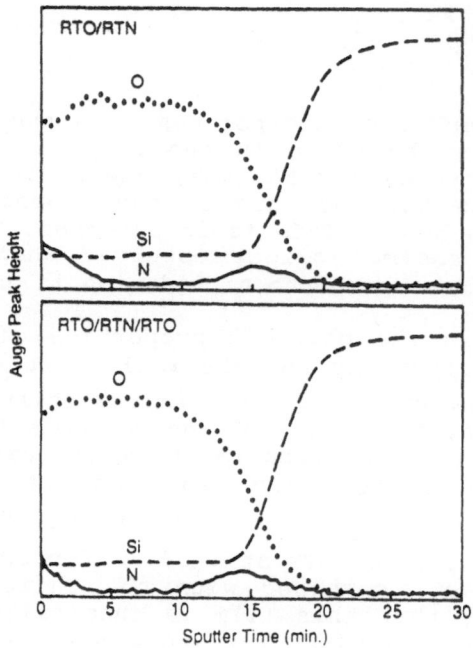

Fig. 16   Sputter Auger depth profiles for nitrided and re-oxidized nitreded 15 nm SiO_2 films.
RTN conditions are 1150°C for 15 seconds.
Re-oxidation is at 1050°C for 5 seconds.

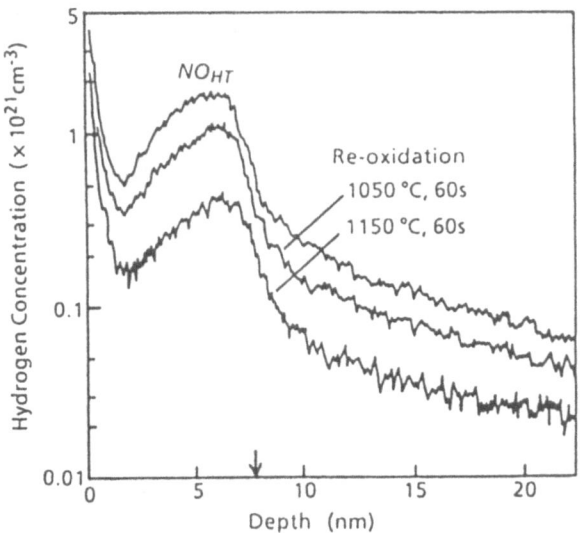

Fig. 17   Hydrogen SIMS depth profile for
nitrided and re-oxidized oxides.
The arrow indicates the
dielectric-silicon interface[54].

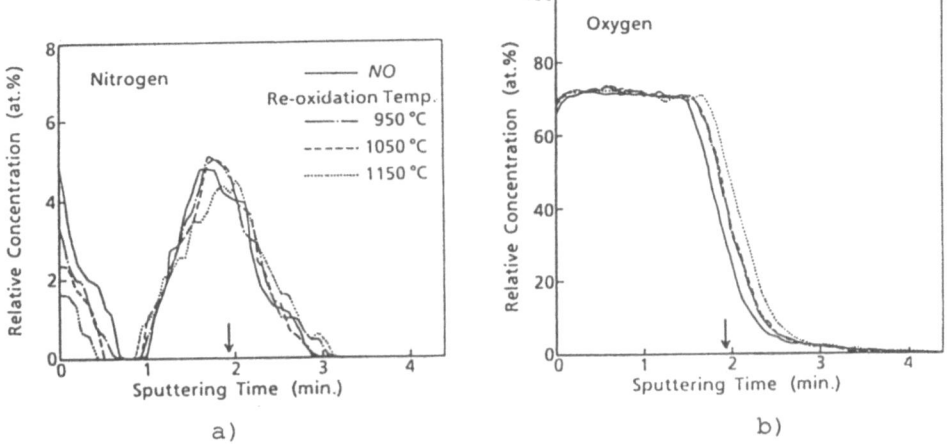

a)   b)

Fig. 18   (a) Nitrogen and (b) oxygen Auger depth profiles
for nitrided and re-oxidized oxide films.
Nitridation at 950°C for 60 seconds[55].

Device isolation

Isolation processes for ULSI technologies require that active device areas reproduce as closely as possible the reduced dimensions defined in device layouts. The bird's beak effect encountered in LOCOS-like isolation processes is a limiting factor in the active device area reduction. Typically, LOCOS processes use a layer of CVD Si_3N_4 on top of a pad dielectric as the active area masking material. The pad dielectric is a thermally grown layer used between the silicon and the CVD layers, usually SiO_2. The thinner this SiO_2 film is, the less lateral field oxide encroachment occurs. If this SiO_2 layer is replaced by Si_3N_4 (SILO process), the lateral encroachment is minimized. However, using CVD Si_3N_4 only as the mask pad dielectric may result in poor device yields.

The ability of RTP to grow dielectrics of different compositions as described in the previous section, allows for a variety of pad dielectrics. Table 4 lists five types of pad dielectrics, their thickness ranges and growth techniques, while Fig. 19 shows schematic cross-sections of completed masking structures used for the growth of 600 nm thick field oxide. The masks are patterned using reactive ion etching (RIE). For a pad dielectric of RTN Si_3N_4 (A), additional 20 nm of CVD SiO_2 for stress relief, and 100 nm CVD Si_3N_4 for masking, complete the LOCOS mask. In the case of pad dielectrics formed by RTO/RTN (B,C) or RTO (D), a 15 nm CVD Si_3N_4 needs to be added before the CVD SiO_2. This thin layer of Si_3N_4 film improves the masking properties of the pad dielectric. Finally, for a conventional furnace SiO_2 pad dielectric, a single layer of 100 to 150 nm Si_3N_4 completes the mask.

Table 4. Pad dielectrics used for masking the active
device area in LOCOS-like isolation processes.

CASE	PAD DIELECTRIC TYPE	THICKNESS (nm)	PROCESS
A	Si_3N_4	2.0 - 2.5	RTN
B	SiO_xN_y (saturated)	3.0 - 8.0	RTO/RTN
C	SiO_xN_y (surface/ interface)	3.0 - 8.0	RTO/RTN
D	SiO_2	3.0 - 8.0	RTO
E	SiO_2	25.0 - 35.0	FURNACE

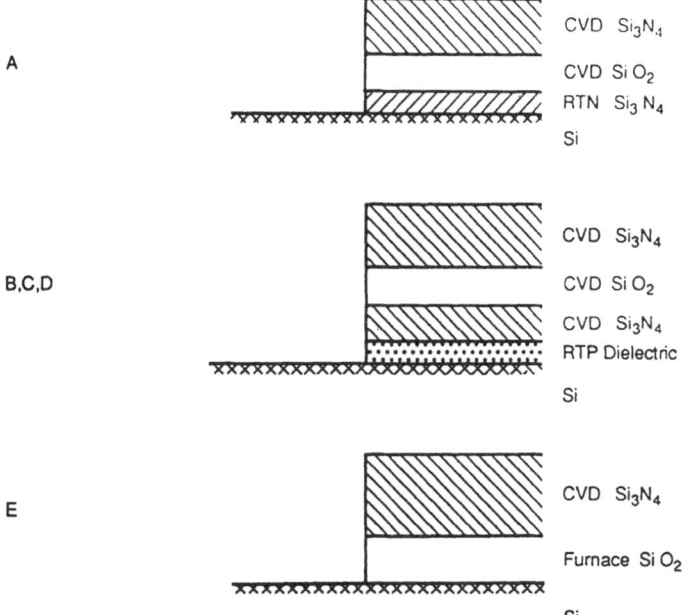

Fig. 19   Mask structures used for the growth of field oxide by LOCOS-like process. The pad dielectric type and growth technique is described in Table 4.

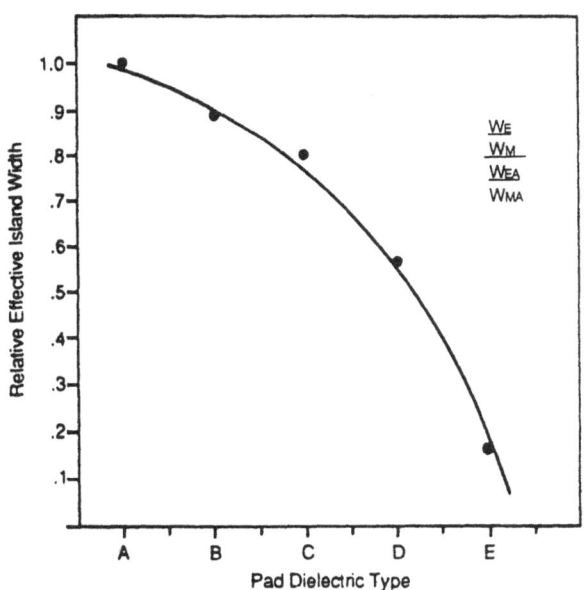

Fig. 20   Relative effective active silicon island width as a function of the mask pad dielectric type defined in Table 4.

23

The width of the active area can be obtained from the electrical characteristics of MOS transistors fabricated in this area after the field oxide is grown. A relative effective active area island width can be defined as the ratio of the electrically measured width (WE) to the layout or mask defined width (WM). Fig. 20 show this relative effective active island width as function of the pad dielectric type, normalized to the value for case A. The field oxide thickness is 600 nm. The larger the ratio, the smaller the bird's beak effect as shown by the cross sectional SEM pictures for cases B, D, and E in Fig. 21. This are obtained after deposition of a polysilicon cap and a selective etch of the field oxide. By changing from a

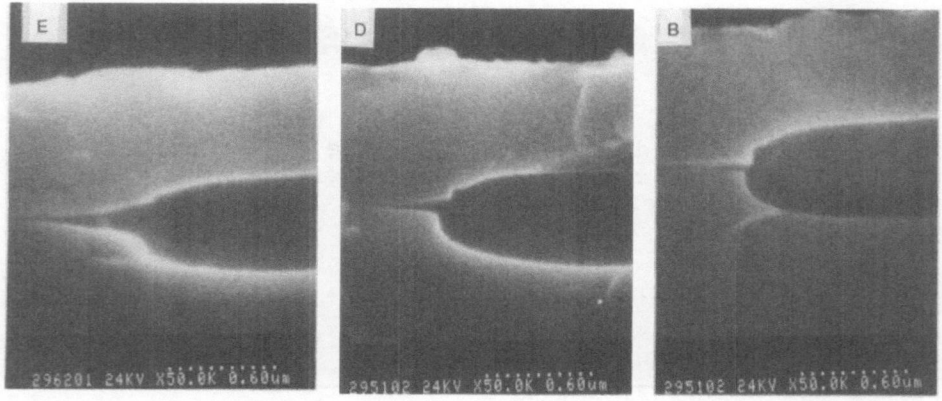

Fig. 21   Cross sectional SEM views of 600 nm field oxides showing the reduction of the bird's beak effect for three of the pad dielectrics defined in Table 4.

thick oxide (case E) to a thin oxide (case D) an improvement of more than 50% is obtained. The use of SiO_xN_y (cases B and C) and Si_3N_4 (case A), allows for active silicon islands that reproduce very close the layout value. This is a result of the enhanced diffusion barrier properties of nitrided oxides compared to pure oxides[12]. The heavier the nitridation, the better the diffusion barrier properties of the dielectric. The use of SiO_xN_y pad dielectrics is preferred over pure Si_3N_4 due to their ease of wet etching and reduced stresses.

Silicon Dioxide - RTO/RTA. Capacitance-Volatage (C-V), Current-Voltage (I-V), and Time To Breakdown (TTB) are characterization techniques that provide information about the quality of silicon dielectrics. In order to evaluate the quality of SiO_2 films grown by RTO, a controlled experiment is necessary. The following comparison between RTO and furnace grown 10 nm SiO_2 films uses p/p$^+$ type (100) epitaxial silicon wafers from a single cassette. The epi layer has a nominal resistivity of 1.5 ohm-cm. These wafers follow simultaneous processing steps, i.e. RCA wafer clean, buffered oxide etch, de-ionized water rinse, spin rinse/dry, gate electrode deposition, and sintering. Both Furnace and RTO films have the same post-oxidation RTA in nitrogen at 1050°C for 40 seconds. Thus, a direct comparison between these two SiO_2 film growth techniques is possible. Dry O_2 ambient at 1150°C is used for RTO. A two step process is used for furnace control oxides. In the first step 6 nm of oxide are grown at 850°C in an O_2/4%HCl ambient. The remaining 4 nm are grown during the second step at 1050°C in N_2/4%HCl/2%O_2 [48].

From C-V measurements[30], a midgap density of states (D_{itm}) below $1x10^{10}$ cm^{-2}eV^{-1} is obtained for RTO/RTA oxides compared to $5x10^{10}$ cm^{-2}eV^{-1} for the two-step furnace oxides. Since both furnace and RTO oxides had the same post-oxidation RTA, it can be concluded that the low D_{itm} values obtained for RTO films is a direct consequence of the higher oxidation temperature used in the RTO process[1]. Figs. 22a and 22b show typical current-voltage characteristics for the furnace and RTO oxides respectively. I-V measurements are made using a negative gate electrode bias, i.e. MOS capacitor accumulation mode, using a Hewlett-Packard 4145 semiconductor device analyzer. A fast voltage ramp is necessary in order to avoid extra voltage drop across the oxide resulting from charge trapping. The leakage current for gate voltages below -8 V, in Fig. 22, is mainly due to limitations of the test system. For the RTO oxide, results for two consecutive bias stress runs on an untested MOS capacitor up to 13 V show no degradation of the dielectric I-V characteristics. The third voltage bias run, in Fig. 22b, is taken up to 20 V, resulting in a catastrophic breakdown voltage of 16 V. An average catastrophic dielectric breakdown field of 15 MV/cm is typical of RTO oxides compared to 13.5 MV/cm for furnace-grown oxides. These dielectric brakdown values are calculated after taking into account the work function of the gate electrode.

The I-V characteristics for furnace control oxides (Fig. 22a) are qualitatively similar to the ones for RTO oxides (Fig. 22b), but significant voltage stress effects are seen in the Fowler-Nordheim tunneling current as represented by the shift and change in slop of the I-V curves for successive bias stress runs. These shifts are due to large charge trapping present in low-temperature furnace oxides compared to high-temperature RTO oxides. Non-catastrophic dielectric breakdown is defined as the gate electrode voltage at which a determined current density flows trough the dielectric. Fig. 23 shows breakdown yields at a current density of 3 mA/cm^2 for two 15 nm RTO oxides[56]. The dielectric breakdown field is in excess of 10 MV/cm. Using Poisson statistics on the capacitor yield data[57], a defect density estimate of 0.5 cm^{-2} is obtained. However, the breakdown yield is strongly dependent on the purity of the gases used for the process as shown by the breakdown statistics of 10 nm RTO oxides with RTA using purified and unpurified nitrogen in Fig. 24 [49].

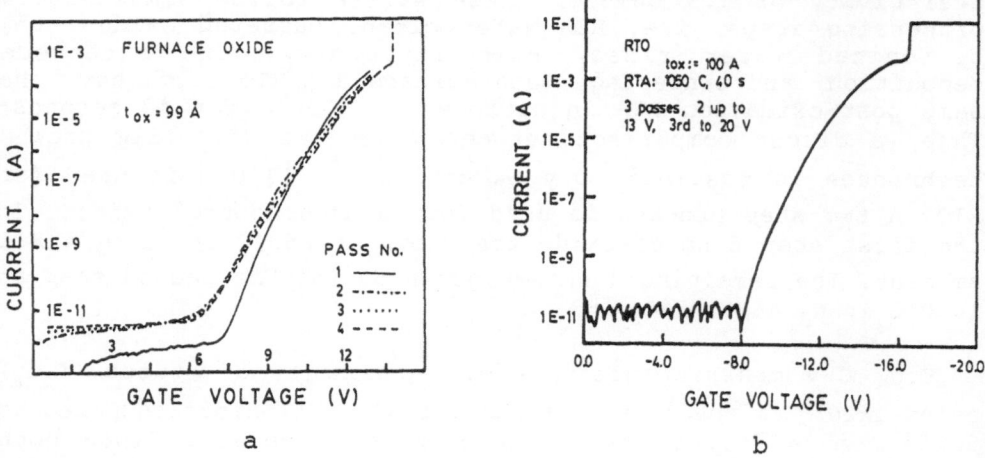

Fig. 22   Current-Voltage characteristics for 10 nm SiO$_2$ films grown by (a) two-step furnace process, and (b) RTO/RTA[48].

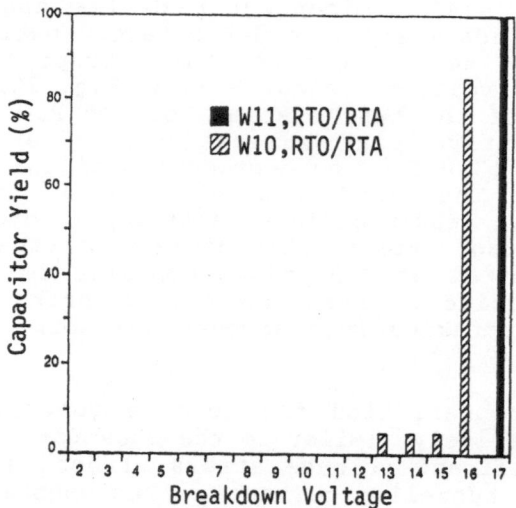

Fig. 23   Typical 15 nm SiO$_2$ film breakdown histogram yield at a current density of 3 mA/cm^2. The defect density is 0.5 cm^{-2}. RTO in O$_2$ + 4% HCl at 1150°C, RTA in ultra pure N$_2$ at 1050°C for 40 seconds[56].

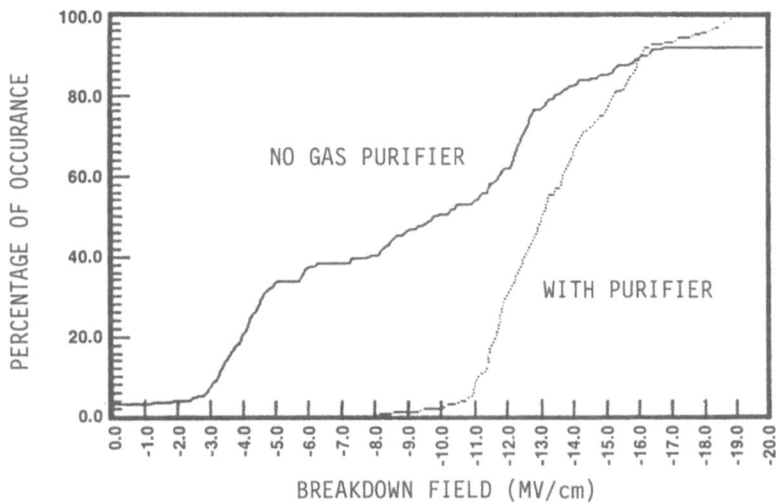

Fig. 24   Cumulative dielectric breakdown histogram
           for SiO_2 films annealed in nitrogen without
           purifier, left curve; and in nitrogen with
           purifier, right curve[49].

Dielectric stress at a constant current density results in Time-to-Breakdown (TTB) or Charge-to-Breakdown (Q_{BD}) information which emulates the conditions experienced in MOS devices that rely on charge tunneling, such as E^2PROM memories[58]. Fig. 25 shows the gate voltage needed to maintain a constant current density of 20 mA/cm^2. The gate voltage increases until breakdown. For RTO/RTA oxides, this increase is only 900 mV over a stressing period of 4000 seconds without dielectric breakdown, while furnace films show an increase of about 3 V over a stressing time of 1600 seconds, resulting in a Q_{BD} of 80 and 32 C/cm^2 for RTO/RTA and furnace-control oxides, respectively[48]. It has been demonstrated that the increase in the gate voltage during this type of stress is due to electron trapping in the SiO_2[58,59]. The smaller change on the gate holding voltage for RTO/RTA oxides combined with the lack of degradation on the Fowler-Norheim tunneling characteristics (Fig. 22b), indicates that electron trapping occurs at a much smaller rate than for the furnace-control oxides. Thus, for RTO/RTA films, the breakdown appears to be caused by positive charge generated by impact ionization in the SiO_2[60,61].

During the process integration of silicon gate dielectrics using RTP, the presence of implanted dopants must be taken into account, especially for CMOS and BICMOS technologies. For example, NMOS transistor threshold voltage (V_T) and punch-through voltage (V_P) adjustment implants merit special attention. For a 0.25 micron process with 10 nm gate SiO_2, typical boron implant energies for V_T and V_P are 35 KeV and 75 KeV through 120 nm sacrificial SiO_2 film, and doses of 5.4×10^{12} cm^{-2} and 1×10^{11} cm^{-2}, respectively[48]. After removal of the

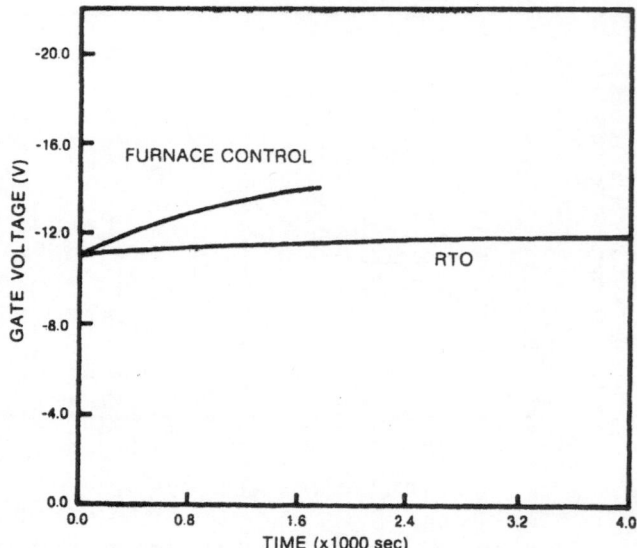

Fig. 25 Gate voltage necessary to maintain a
constant current density of 20 mA/cm^2
as a function of the stressing time for
RTO/RTA and furnace control 10 nm SiO$_2$
films. The gate area is 9×10^{-4} cm^2 [48].

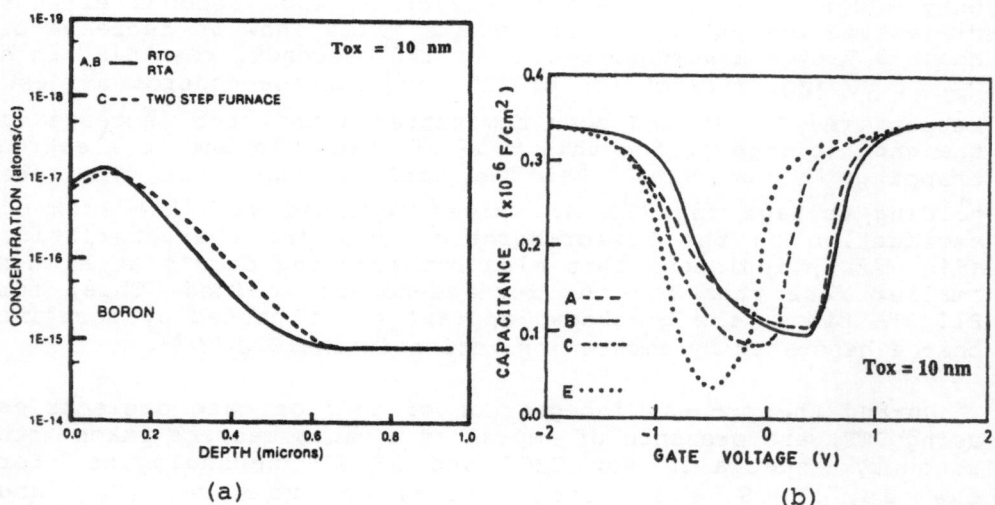

(a)   (b)

Fig. 26 (a) Boron SIMS depth profiles and (b)
corresponding C-V characteristics after
RTO (A), RTO/RTA (B), and two-step furnace
(C). Curve E is for a non-implanted RTO/RTA
reference device. RTO at 1150°C in dry O$_2$,
RTA at 1050°C for 40 seconds in N$_2$ [48].

sacrificial oxide, 10 nm SiO_2 films are grown by a simple RTO cycle (case A), RTO/RTA cycle (case B), and two-step furnace (case C). As a reference, 10 nm SiO_2 is also grown on a non-implanted wafer by RTO/RTA cycle (case E). Fig. 26a shows the boron concentration as function of the distance from the SiO_2/Si interface, for cases A, B, and C as obtained by SIMS analysis. No measurable difference is observed between the profiles for the RTO/RTA and RTO only cycles, while the two-step furnace oxidation process results in considerable diffusion of the implanted species. The diffusion of boron observed for the two step furnace oxidation is due in part to the relatively long process times and the enhanced diffusivity in oxidizing ambients. The lack of boron diffusion in the case of RTP-grown oxides is confirmed by the corresponding quasi-static C-V curves for test capacitors shown in Fig. 26b. The threshold voltage shifts 400 mV towards a negative value in the case of furnace-grown oxides, while, for RTP, V_T remains at its designed value of 450 mV. Comparison between doping profile values obtained from C-V and SIMS indicates a 100 % activation of the implanted species for the RTO/RTA cycle. Thus, device implant/annealing simulations could be simplified since no models for implant activation are needed.

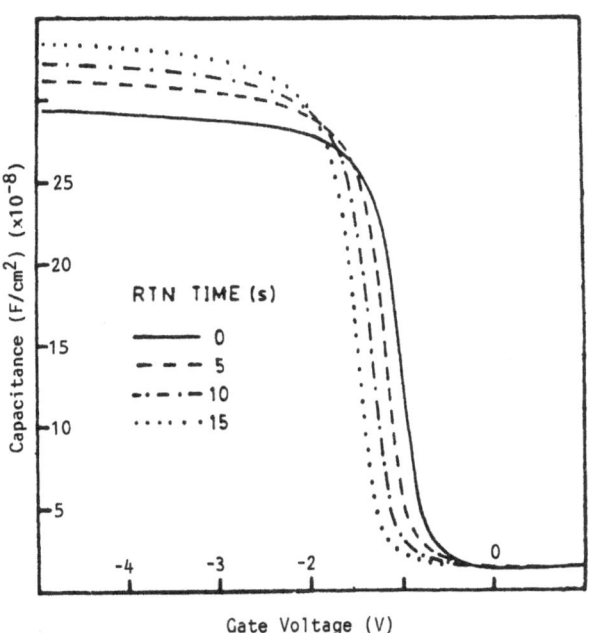

Fig. 27   High frequency C-V curves for 11.7 nm
           nitrided oxides at 1150°C for 0, 5,
           10, and 15 seconds. Every 5 seconds RTN
           induces a 140 mV negative C-V shift[19].

Oxinitrides - RTO/RTN and RTO/RTN/RTO. The effects of RTN on the electrical properties of nitrided SiO_2 films is best evaluated by looking at their C-V, dielectric breakdown, and time to breakdown characteristics as function of dielectric thickness, and RTN temperatures and times. The C-Vs for 11.7 nm SiO_2 films nitrided at 1150°C for 0, 5, 10, and 15 seconds are shown in Fig. 27[19]. Two distinct effects are observed: negative C-V shift and increase in the accumulation capacitance (Cox) with increase in RTN time.

The C-V curves, and therefore the flat-band voltage (V_{FB}), shifts monotonically towards negative values due to nitridation induced positive fixed charge buildup close to the dielectric-silicon interface. Every 5 seconds RTN at 1150°C results in about a -140 mV shift in V_{FB}, equivalent to an additional fixed charge density of 2×10^{11} cm^{-2}. However, this shift reaches a maximum as nitridation time increases, and the shift reverses to the point were the nitrided dielectric reaches the same V_{FB} as the original SiO_2 film, as shown in Fig. 28[62]. The C-V curves in Figs. 27 and 28 show little stretchout, indicating that the observed negative C-V shifts are due primarily to fixed charges. The turnaround in the C-V curve shifts can be further analyzed by calculating the density of fixed charges (N_f) and the interface state density at midgap (D_{itm}). Fig. 29a shows these parameters as function of the RTN time and temperature for an 8 nm thick dielectric. The N_f and D_{itm} of the initial oxides are -3×10^{10} cm^{-2} and 2×10^{10} $cm^{-2}eV^{-1}$, respectively*. For the nitrided oxides, both N_f and D_{itm} vary in a similar manner as nitridation proceeds at a given temperature: both increase at the early stage of nitridation, reach respective maxima at a certain nitridation time (t_{max}), and then decrease gradually showing turnarounds. At the final stage, N_f and D_{itm} have low values comparable with those of pure oxides, e.g. the N_f and D_{itm} after RTN of an 8 nm SiO_2 film at 1150°C for 300 seconds are -1×10^{10} cm^{-2} and 2.7×10^{10} $cm^{-2}eV^{-2}$, respectively[62].

The higher the RTN temperature, the faster the N_f and D_{itm} initial increasing rate is, the shorter the t_{max} is, and the faster the final N_f and D_{itm} decrease is. For nitridation at 950°C, both N_fs and D_{itm}s do not appear to reach respective maxima by RTN up to 300 seconds but certainly pass the maxima after 1200 seconds (Fig. 29a). By comparing the results for 8-nm-thick nitrided oxides with those for 5- and 12-nm-thick oxides (Fig. 29b), it is observed that the thinner the initial oxide thicknesses are, the faster the initial increase of both

---

* The negative sign of N_f represents negative polarity of effective fixed charges might be caused by the uncertainty in the gate work function which can translate into large offset errors in calculated N_f values, especially for ultra-thin dielectrics.

N_f and D_{itm} is, the shorter t_{max} is, and the faster both N_f and D_{itm} decrease at the final stage, in agreement with the kinetics of nitrogen incorporation by diffusion process.

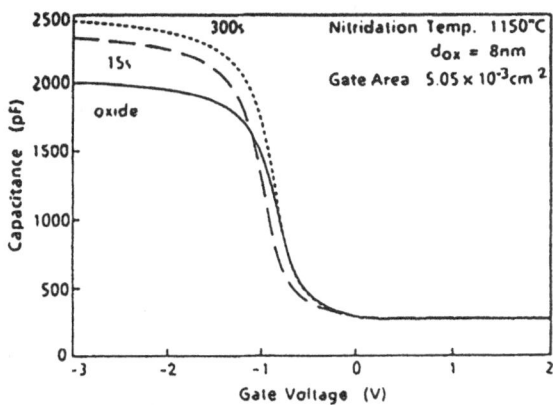

Fig. 28   High-frequency C-V curves of capacitors with oxides nitrided at 1150°C for 0, 15, and 300 seconds. The negative C-V shift disapears for long RTN times[62].

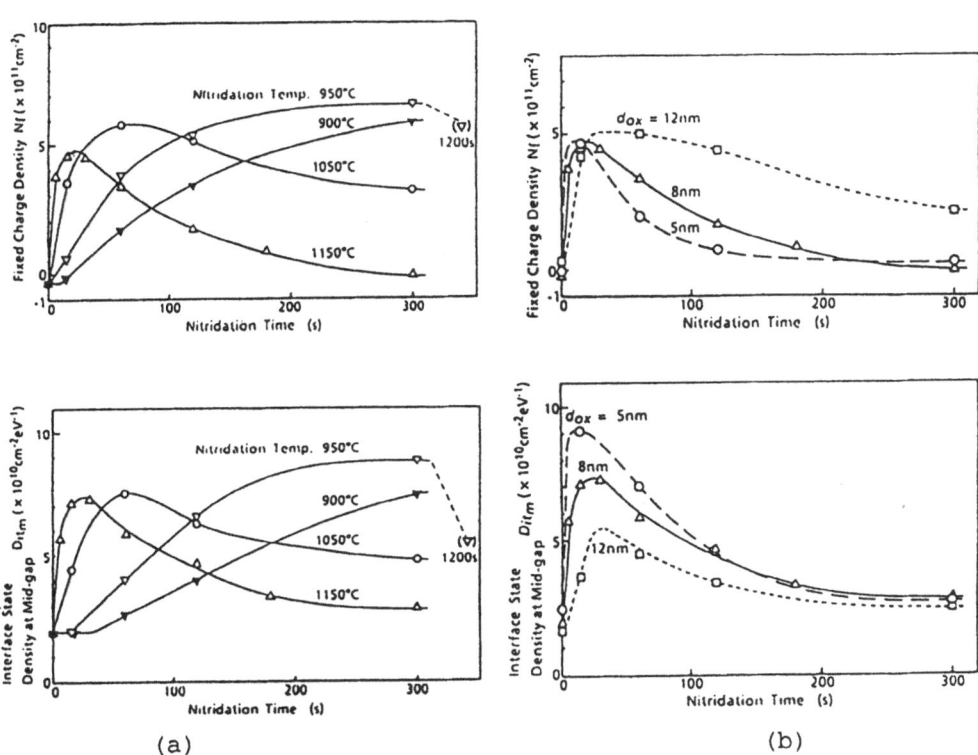

(a)   (b)

Fig. 29   Fixed charge density (N_f) and interface state density at midgap (D_{itm}) as function of the nitridation time of (a) 8-nm-thick oxides, for several nitridation temperatures; and (b) 5-, 8- and 12-nm-thick oxides nitrided at 1150°C [63].

The accumulation capacitance (Cox) increases with increasing RTN time (Fig. 27). Since the dielectric thickness (tox) does not increase with RTN time[53], the Cox increase is directly proportional to the increase in the dielectric constant εox:

$$Cox = \varepsilon \varepsilon ox/tox.$$

Fig. 30a shows the dielectric constant of 11.7 nm nitrided SiO_2 films at 1150°C as function of RTN time, while Fig. 30b shows the effect of RTN temperature on the dielectric constant of 8.0 nm SiO_2 films as function of RTN time. Due to the diffusion kinetics of the RTN process, the thinner the SiO_2 film is, the faster the rise in the dielectric constant value. The εox increases more rapidly with nitridation time as the nitridation temperature is raised. The dependence of εox on the RTN temperature and time is similar to that of the nitrogen concentrations near the interface shown in Fig. 13. Thus, RTN allows for fine tuning of the dielectric constant with values in the 4.0 to 5.2 range, i.e. εox increases up to 33% compared with the value for the initial oxide. Therefore, MIS transistors with nitrided oxides should have enhanced current drive capabilities because the transistor current at saturation (I_{Dsat}) is directly proportional to the dielectric constant[65]:

$$I_{Dsat} = ZCox(V_G-V_T)\upsilon_s,$$

where Z is the width of the device, V_G is the gate voltage, V_T is the threshold voltage, and υ_s is the carrier saturation velocity. In the linear regime, the current (I_D) is also proportional to the inversion layer carrier mobility (μ):

$$I_D = (Z/L)\mu Cox(V_G-V_T)V_D,$$

where L is the device gate length and V_D is the drain voltage. The inversion layer carrier mobility is known to degrade due to the combined effects of fixed charges and electron traps[66]. Since nitridation introduces fixed charges, μ is degraded by coulombic scattering. In addition, the presence of interfacial electron traps, which are believed to be located near the conduction band of silicon and charge neutral when empty[67], further reduce the effective electron mobility in an n-channel device as a result of reduced mobile channel charge due to trapping and coulombic scattering by the trapped inversion layer electrons. For a p-channel device, μ is not affected by the electron traps since the traps are empty and hence, neutral, when a p-channel device is biased in inversion. In the case of RTN oxides, the increase in εox can compensate for the decrease in μ. Furthermore, due to the nature of the fixed charge turnaround effect described above, certain RTN conditions result in very small fixed charges and therefore, little degradation of μ should be expected. Thus, larger current in the linear regime can be expected for devices with RTN oxides compared to pure oxides.

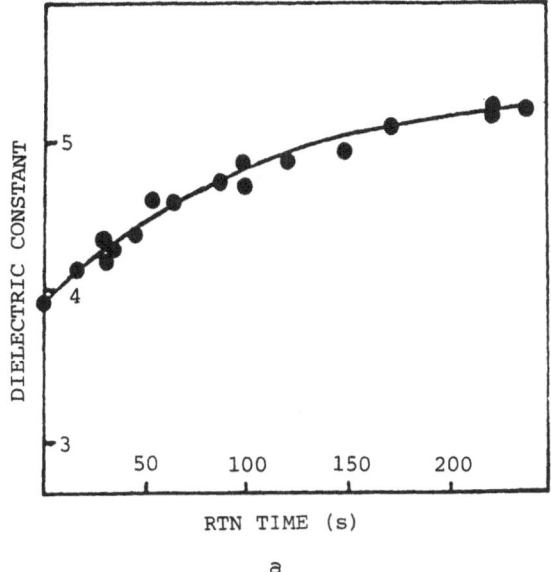

a

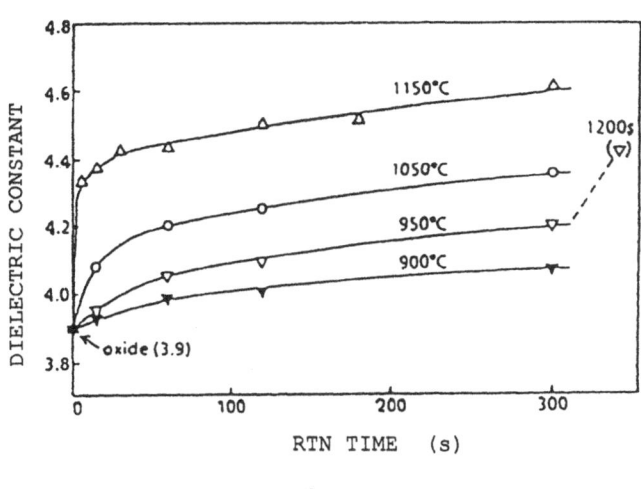

b

Fig. 30   Effective dielectric constant (ε_{ox}) for (a) 11.7 nm
RTN oxides as a function of RTN time at 1150°C [64],
and (b) 8.0 nm RTN oxides as function of RTN time
and temperature [63].

The current-voltage characteristics of RTN nitrided oxides
follow Fowler-Nordheim tunneling behavior for RTN times such
that the nitrogen concentration does not reach its maximum in
the bulk of the oxide. From the I-V studies, catastrophic
capacitor breakdown fields improve with increased RTN time[64]
similarly to furnace nitrided oxides[9]. Fig. 31a shows the
dielectric breakdown for capacitors formed on 15 nm pure and
nitrided SiO_2 films, at a current density of 3mA/cm^2. Over
1MV/cm improvement in the average dielectric breakdown field
for nitrided SiO_2 compared to pure SiO_2 films is observed,

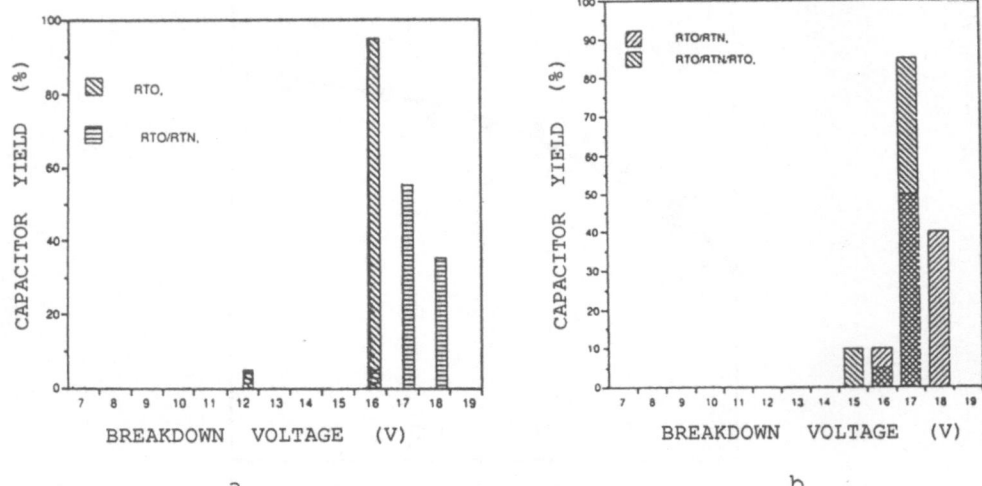

Fig. 31   Dielectric breakdown distribution for (a) pure
and nitrided 15 nm SiO_2 films, and (b) nitrided
and re-oxidized films, at a current density of
3 mA/cm^2. RTO at 1150°C in O2 + 4% HCl, RTN at
1150°C for 15 seconds, and re-oxidation in dry
O_2 at 1050°C for 5 seconds.

however this improvement is accompanied by a degradation in the
breakdown yield. These two effects are related to
ammonia-related traps induced by the nitridation process. This
traps are believed to reduce the role of localized breakdown
paths responsible for pure SiO_2 film breakdown, and thus
enhance the observed breakdown voltages[9]. A re-oxidation of RTN
nitrided oxides, shows a recovery on the breakdown dielectric
voltage yield with very little reduction on the average value,
as shown in Fig. 31b.

Charge trapping studies on RTO, RTO/RTN, and RTO/RTN/RTO
dielectrics give further information about the electrical
properties and nature of these dielectrics. Fig. 32 shows
quasi-static C-V curves for these type of dielectrics prior and
after electron charge injection from the substrate at a
constant current density of 10 mA/cm^2 [68]. These films had an
initial fixed charge density and midgap interface state density
in the low 10^{11} cm^{-2} range and the low 10^{10} cm^{-2}eV^{-1} range,
respectively. After electron charge injection, a large
distortion is observed in the C-V curve for a pure SiO_2 film
(Fig. 32a) showing its low resistance to interface-state
generation by high-field stress. For the nitrided SiO_2 film
(Fig. 32b), both a large positive shift and a large distortion
of the C-V curves is observed. The large positive shift
indicates a high electron trap density. The degradation
increases for larger charge injection for both pure and
nitrided SiO_2 films. By contrast, the re-oxidized nitrided SiO_2
film shows little distortion in the C-V curves independent of
the magnitude of the injected charges (Fig. 32c). This means

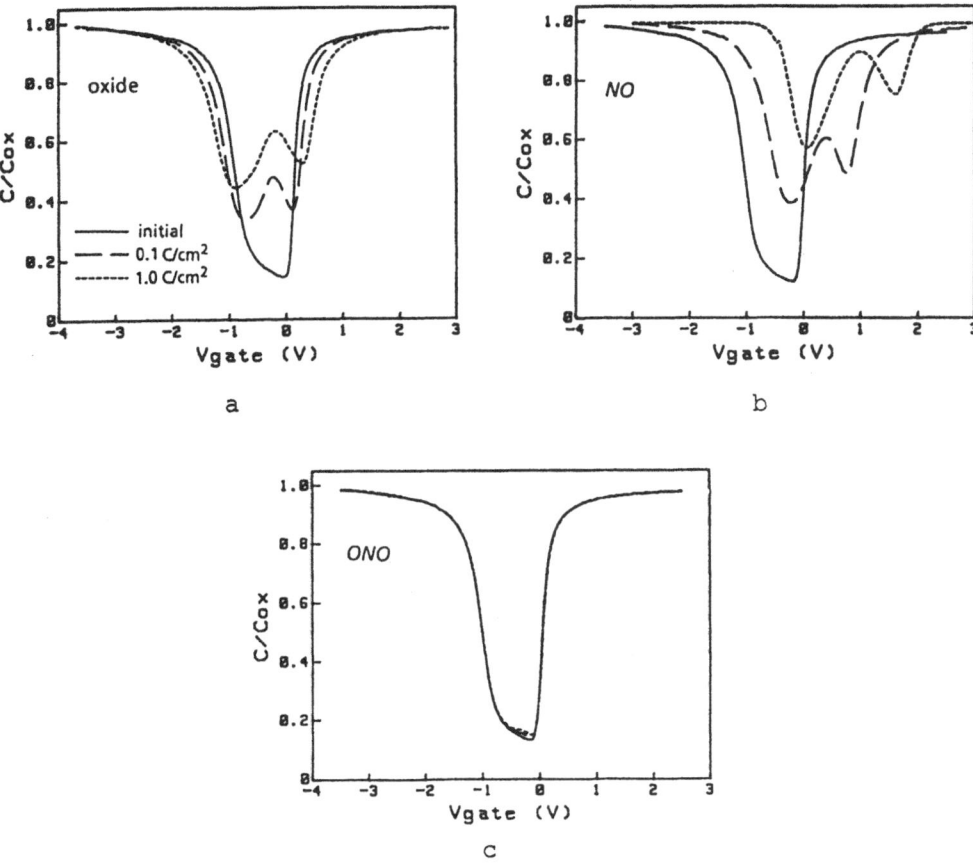

Fig. 32 Quasi-static C-V curves before and after electron injection for (a) an oxide, (b) nitrided oxide at 950°C for 60 seconds, and (c) re-oxidized film at 1150°C for 60 seconds. Dielectric thickness: 7.7 nm[68].

that re-oxidation results in a dielectric with high resistance to electron injection.

This high resistance to electron trapping observed in re-oxidized nitrided oxides also results in high charge-to-breakdown (Q_{BD}) values during time-to-breakdown stress at a constant current density of 10 mA/cm2, Fig. 33 [68]. The gate voltage for nitrided oxides increase more rapidly with stress time and the Q_{BD} is small compared to the pure oxide. Thus, the electron traps introduced during nitridation degrade the Q_{BD}. In contrast, the Q_{BD} of re-oxidized nitrided oxides is about 16 times higher than for pure oxide, 350 and 22 C/cm^2, respectively. The increasing rate of the gate voltage with stress time becomes monotonically smaller and the Q_{BD} becomes monotonically large with the progress of re-oxidation.

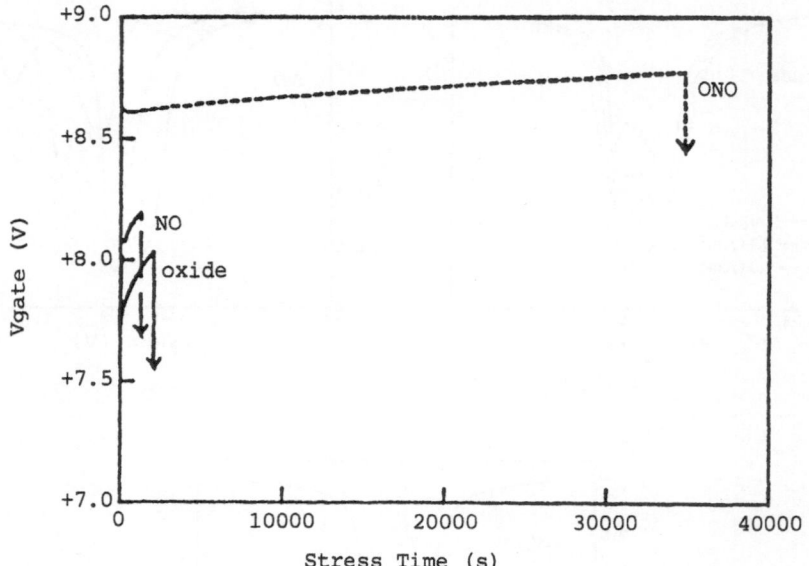

Fig. 33 Gate voltage necessary to maintain a constant current density of 10 mA/cm^2 versus stress time for 7.7 nm RTP dielectrics. Pure oxide in dry O$_2$ at 1100°C. Nitridation at 950°C for 60 seconds. Re-oxidation in dry O$_2$ at 1150°C for 60 seconds[68].

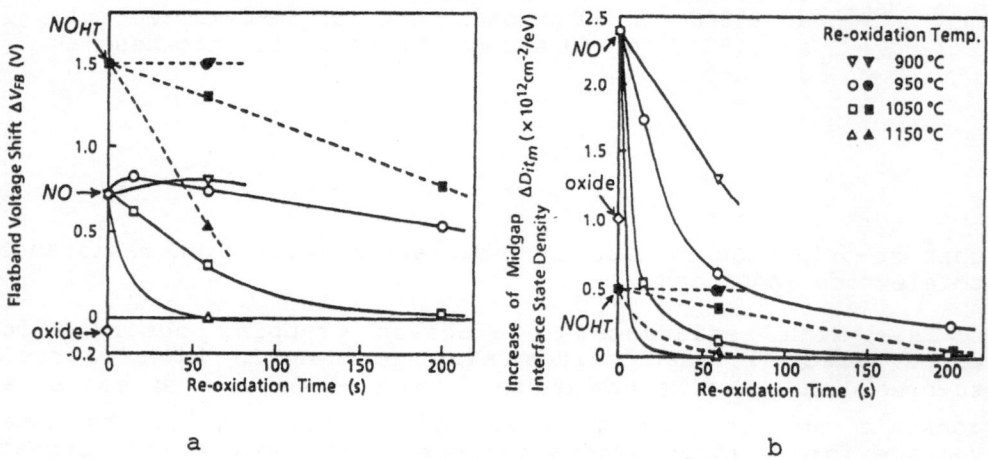

Fig. 34 (a) Flat-band voltage shift ΔV_{FB} and (b) increase of midgap density of states ΔD_{itm} induced by electron injection of 0.1 C/cm^2 as function of the re-oxidation time and temperature for nitrided oxides. The starting 7.7 nm nitrided oxides are NO (RTN at 950°C for 60 seconds) and NO$_{HT}$ (RTN at 1150°C for 60 seconds). The data for a pure oxide film is also shown[68].

The effect of re-oxidation time and temperature on the charge trapping of lightly (NO) and heavily (NO_{HT}) nitrided oxides on ΔD_{itm} and ΔV_{FB} is shown in Fig. 34 [68]. The data for pure oxides is also shown in Fig. 34. The ΔV_{FB} for the pure oxide is negative (-70 mV) and this is considered to be mainly due to the large generated donor-like interface states (ΔD_{itm} = 1.0x1012 cm-2eV-1). On the other hand, both ΔD_{itm} and ΔV_{FB} of NO and NO_{HT} are positive. This indicates that a large number of electron traps have been introduced in the nitrided oxides. The ΔV_{FB} of NO_{HT} is +1.5V which is about twice as large as that of NO films. On the other hand, the ΔD_{itm} of NO_{HT} is $4.8x10^{11}$ $cm^{-2}eV^{-1}$ and smaller than that for pure oxide.

Both ΔD_{itm} and ΔV_{FB} for NO and NO_{HT} films decrease monotonically as re-oxidation proceeds. Both decrease more rapidly as the re-oxidation temperature rises. The decrease is almost linear with re-oxidation time in the early re-oxidation stage and then the decrease slows down. Due to the diffusion barrier properties of nitrided oxides, these re-oxidation effects proceed faster for dielectrics nitrided at low temperatures (NO) compared to the ones nitrided at high temperatures (NO_{HT}). As re-oxidation can reduce both ΔD_{itm} and ΔV_{FB} at the same time, properly nitrided and re-oxidized oxides have improved charge trapping properties when compared to pure oxides. For NO films re-oxidized at 1150°C for 60 seconds, ΔD_{itm} and ΔV_{FB} take the values of $9x10^9$ $cm^{-2}eV^{-1}$ and +0.5 mV, respectively. These values are smaller by more than two orders of magnitude than those of a pure oxide[69].

These excellent properties of RTO/RTN/RTO dielectrics make them a perfect candidate for advance ULSI devices, especially devices with submicron dimensions where hot electron effects are one of the sources of device parameter degradation. However, because of the initial oxide thickness dependence on the nitridation and re-oxidation processes, each thickness requires nitridation and re-oxidation process optimization.

<u>Trench Capacitors</u>

Trench capacitor technology is required for four megabit and larger DRAM in order to increase packing density. Thin trench oxides of the order of 10 to 15 nm are needed to increase the storage capacitance relative to thick oxides. The controlled growth of these thin SiO_2 films requires low furnace temperature processing which results in non-uniform oxide film thickness on the trench walls due to the different oxidation rates observed for different silicon crystal planes[70]. Furthermore, trench capacitors have right-angled corners at the top and the bottom of the trench. Thus, stress concentration causes oxide thinning at the corners[71,72] which limits the dielectric breakdown voltage to values lower than for planar capacitors. In order to solve this thinning problem, oxidation needs to be done at temperatures above the viscous flow of SiO_2

relieving the stress[73]. Simultaneously, by high temperature oxidation, a uniform SiO_2 film thickness is possible, since the rate of oxidation become orientation independent[70]. Fig. 35 shows a cross sectional TEM view of one of the top corners of a trench after RTO at 1150°C to grow 11 nm thick SiO_2[74]. The same oxide thickness is observed on the two silicon planes, as well as a conformal and uniform covering of the sharp silicon corner.

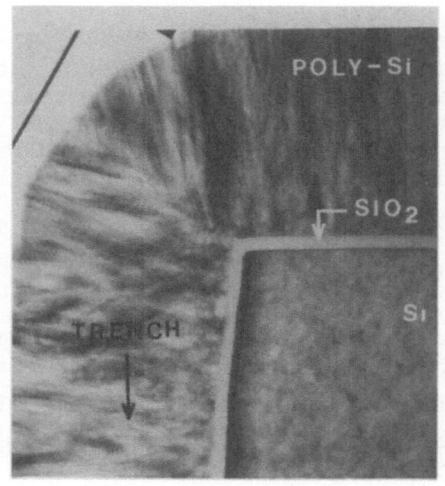

Fig. 35   Cross sectional TEM view of an 11 nm thick SiO_2 film grown on a silicon trench at 1150°C in dry O_2. A layer of CVD polysilicon has been deposited on top of the oxide. Uniform SiO_2 is observed on both silicon planes, and the step coverage is uniform[74].

Further information on the benefits of high temperature (greater than 1000°C) oxidation is obtained from electrical tests. The following paragraphs discuss the electrical characteristics of trench capacitors with 14 nm thick SiO_2 grown on trenches 4.0 μm deep and a 1x3 μm² top opening area[75]. All RTOs are in dry O_2 at 1000°, 1100° and 1150°C. Fig. 36 shows a cross-sectional SEM view of a single trench capacitor and a schematic cross-section of the test structure. By having multiple trenches connected, this test structure provides information on the yield per unit area. Fig. 37a shows the

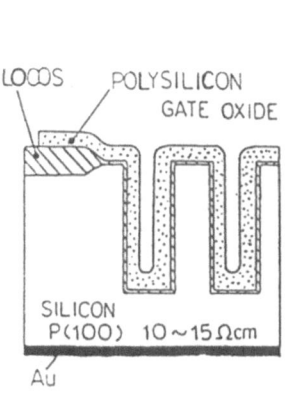 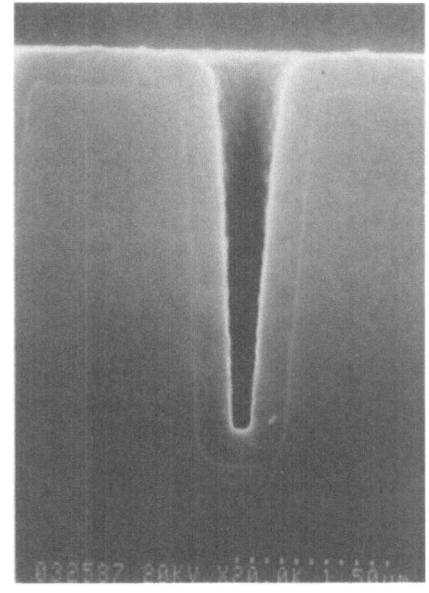

Fig. 36   Cross section SEM view of a trench
capacitor and schematic cross section
of par of the test structure. Trench
depth and top area are 4 μm and 1x3 μm^2
respectively. Trench oxide thickness is
14 nm and the polysilicon gate electrode
thickness is 300 nm[75].

trench capacitor breakdown dielectric yield for the three RTO
temperatures, determined at a tunnel current density of 0.1
mA/cm^2 on 272 samples. The average breakdown field is about 11
MV/cm and its yield increases with increased RTO temperature.
For comparison, Fig. 37b shows the corresponding dielectric
breakdown yield for planar capacitors with 10 nm thick oxides.
The improvement, as function of the increased RTO temperature,
is not as clear as in the case of trenches. Thus, the yield
increase with oxidation temperature, shown in trench
capacitors, is a direct indication of the improvement of the
oxide uniformity and stress relieve at the trench corners.

Calculation of the interface state density from
quasi-static C-V curves[30], shows that D_{it} decreases with
increased oxidation temperature and has a U-shape (Fig. 38). At
midgap, D_{it} decreases from 8×10^{10} cm^{-2}eV^{-1} for RTO at 1000°C to
about 2×10^{10} cm^{-2}eV^{-1} for RTO at 1150°C. In contrast, planar
capacitors do not show such a strong dependence of D_{it} to the
oxidation temperature[75].

The improvement in the quality of the oxide film as the
oxidation temperature rises is also shown by time to breakdown
studies. This test is useful in estimating the life time of the
dielectric. Fig. 39 shows the time to 50% cumulative failure as
a function of the stress current density. The lifetime of the

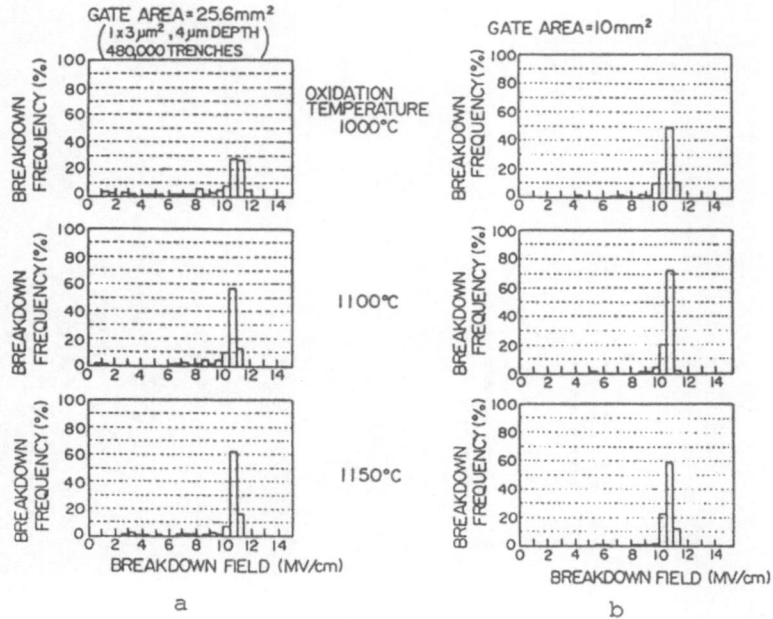

Fig. 37   Dielectric breakdown yield as function of the
           RTO temperature for (a) 14 nm SiO_2 trench
           capacitors, and (b) 10 nm SiO_2 planar
           capacitors[75].

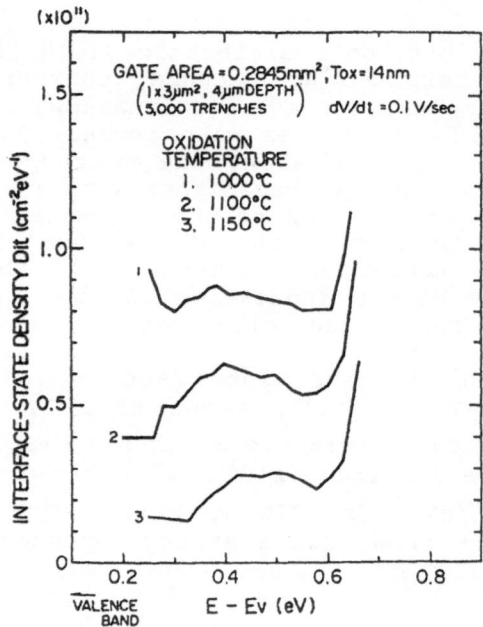

Fig. 38   Interface state density as a function of the
           energy in the silicon bandgap for trench
           capacitors with RTO oxides grown at 1000˚,
           1100˚ and 1150˚C [75].

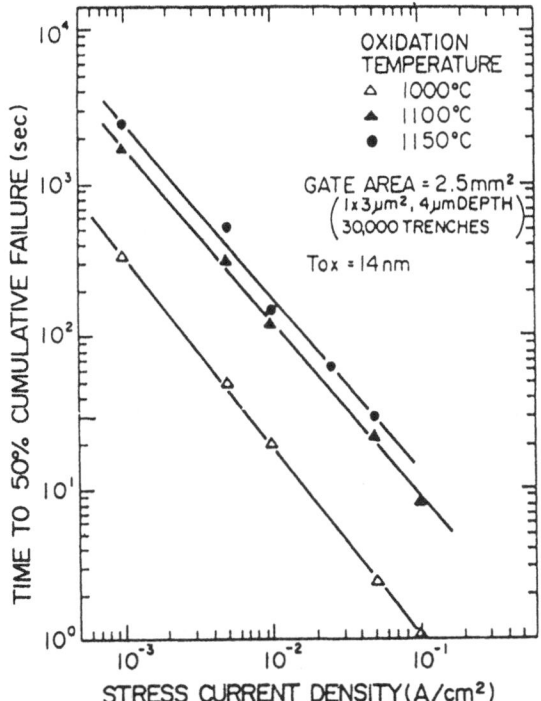

Fig. 39   Time to 50% cumulative failure as a
           function of stress current density for
           trench capacitors with RTO oxides grown
           at 1000°, 1100°, and 1150°C [75].

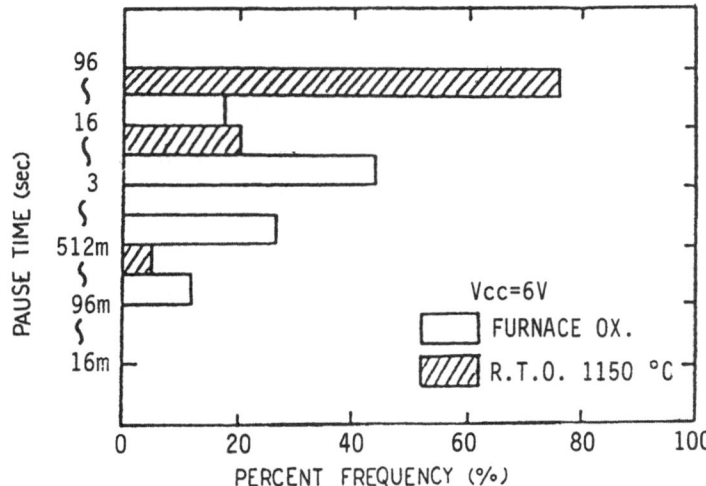

Fig. 40   Pause refresh time histogram of 1 megabit
           DRAM with trench capacitor cells with
           oxides grown by RTO at 1150°C and furnace
           at 1000°C [76].

silicon dioxide layer increases with oxidation temperature. A major improvement is observed when the RTO temperature is raised from 1000° to 1100°C. The electrical acceleration factor in the time to breakdown measurement is 1.5 decades/MV/cm for RTO trench oxides at 1150°C, more than 100 times longer than trench furnace oxidation at 1000°C. However, the life time for trench oxides is shorter than for planar oxides[75]. All these improved electrical characteristics for trench oxides grown at high temperature (greater than 1100°C) have a direct impact on the performance of charge retention. Fig. 40 shows a pause refresh time histogram for 1 megabit DRAM devices having 1150°C RTO trench oxides, compared to similar devices having 1000°C furnace trench oxides[76]. The pause refresh time is improved by more than an order of magnitude for the high temperature RTO. Therefore, simpler refresh circuits/cycles are possible when high temperature RTO oxides are used for the trench storage capacitors in megabit DRAM technologies.

POLYSILCON DIELECTRICS

Silicon dioxide films grown on polysilicon (polyoxides) are used as interpoly dielectric in EPROM and E^2PROM technologies and as storage capacitors in DRAM devices. The scaling down of polysilicon dielectrics has been limited because of the asperities at the polyoxide/polysilicon interface, which result in low breakdown dielectric fields and enhanced dielectric leakage[77,78]. The roughening at the interface is attributed to the irregular polysilicon grain growth which occurs during the furnace doping, typically phosphorus from $POCl_3$ at 900° to 1000°C as well as the varying oxidation rates of the differently oriented grains[79]. Even for *in-situ* doped or ion implanted polysilicon films, the asperities at the interface will be present due to the times involved in furnace oxidations.

These problems can be solved if RTP is used to grow polyoxides on polysilicon films if they have not been exposed to temperatures higher than the deposition temperature. Fig. 41 shows a cross sectional SEM view of a polysilicon layer with 10 nm SiO_2 film grown at 1150°C in dry O_2 and a layer of as deposited polysilicon on top of the polyoxide[80]. The polyoxide is conformal to the grains and no oxidation between the grains is observed. As in the case of silicon trench oxidations, a uniform oxide film is possible independent of the grain orientation because of the high temperature used in RTP oxidations. Polyoxides grown by RTO have show about 20% improvement in dielectric breakdown field[81]. The elimination of polysilicon surface roughness is also evident from the surface traces using a profilometer scan after different doping/oxidation conditions (Fig. 42). Furthermore since the temperature ramp-up is done in the presence of the oxidizing ambient, a thin passivation layer is grown during this stage of the process. This *in-situ* capping of the polysilicon surface reduces the sum of the grain boundary and surface energies so that silicon will not diffuse away from the intersection of a grain boundary and the surface, hence inhibiting the grain grooving phenomenon[82,83].

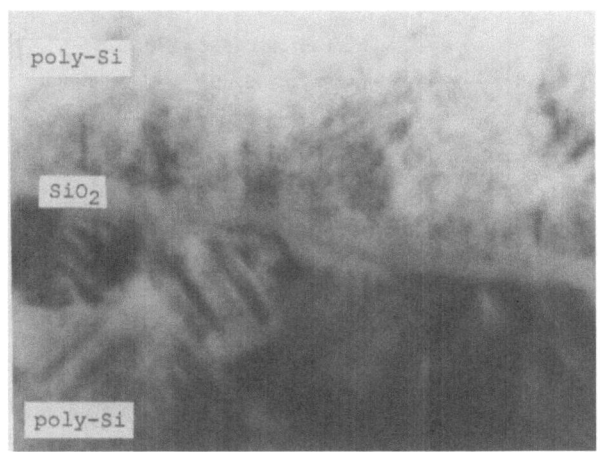

Fig. 41   Cross section TEM view of SiO$_2$ film grown
           by RTO at 1150°C in dry O$_2$ on as deposited
           polysilicon. Three films are observed from
           bottom to top: polysilicon after RTO cycle,
           grown SiO$_2$, and as deposited polysilicon[80].

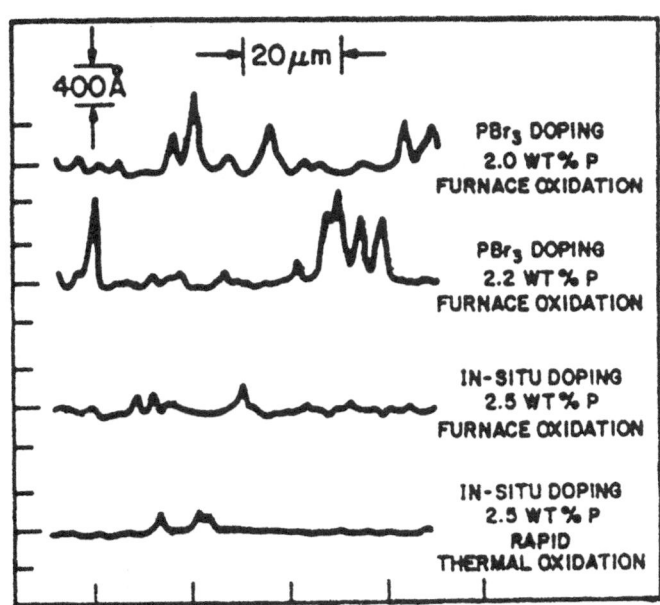

Fig. 42   Surface profilometer traces of polysilicon
           films oxidized under different conditions.
           Furnace oxidation at 950°C, and RTO at
           1050°C; both in dry O$_2$[81].

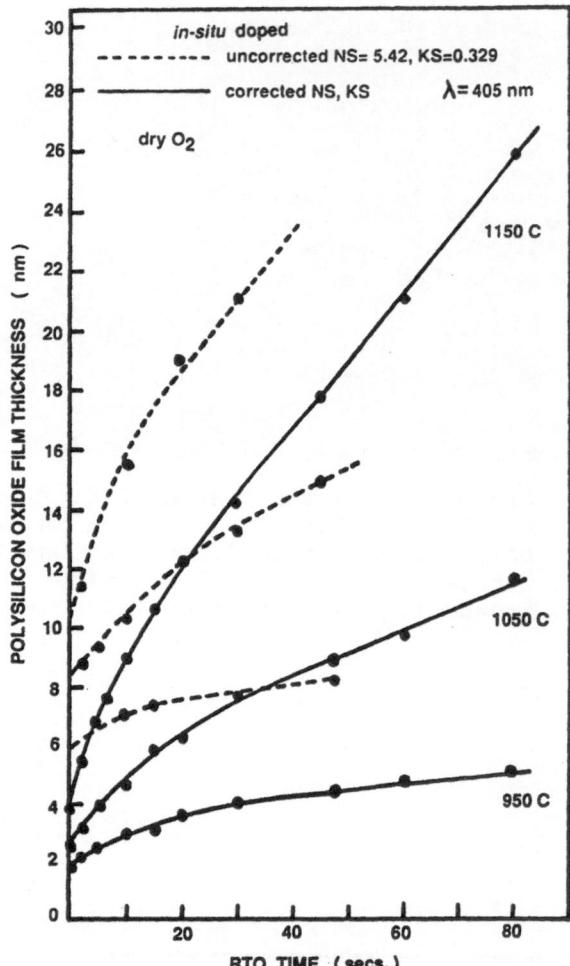

Fig. 43   Polyoxide film thickness as a function
of RTO time. Full and dashed lines
correspond to thicknesses calculated
with corrected and uncorrected polysilicon
optical constants respectively[80].

The kinetics of polyoxide film growth is similar to that of
single crystal silicon. Fig. 43 shows polyoxide film thickness
as a function of RTO time at three temperatures: 950°, 1050°,
and 1150°C. The polysilicon films were in-situ doped to yield a
sheet resistance of about 20 ohm/sqr. The polyoxide film
thickness is thicker than in single crystal silicon RTO (Fig.
4), in part, because of the high oxidation rates in phosphorous
doped polysilicon[84]. The polyoxide film thickness is determined
by ellipsometer measurements at 405 nm wavelength. At this
wavelength, polysilicon is opaque, and therefore, acts as the
substrate. However, special care has to be taken to measure the
substrate optical constants KS and NS, otherwise the calculated
oxide thickness is erroneous if the single crystal silicon
values for NS and KS are used, as shown by the dashed curves in
Fig. 43. Furthermore, because of the grain growth with
increased RTO time and/or temperature, NS and KS need to be

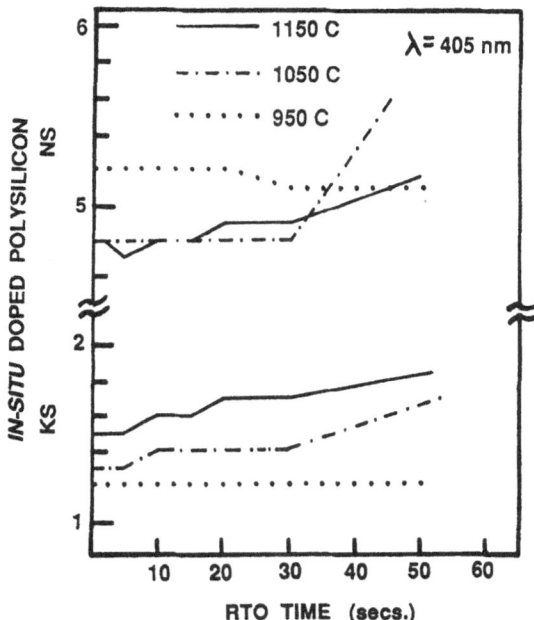

Fig. 44   Measured polysilicon optical constants
           as function of RTO time and temperature[80].

measured for each RTO condition. Fig. 44 shows the temperature
and time dependence of NS and KS for the polysilicon films used
for the polyoxide film thicknesses shown in Fig. 43. The higher
the RTO temperature, the stronger the dependence of NS and KS
on the RTO time. To accurately measure the polyoxide thickness,
first the ellipsometric values Δ and ψ are measured; then, the
polyoxide is etched away and NS and KS measured. Finally the
oxide thickness is calculated.

The oxidation of polysilicon is usually done after
patterning, therefore, corners and lateral walls need to be
oxidized as well as the top surface. Like in the case of single
crystal silicon trenches, the high temperature RTO process
allows for uniform oxide film to be grown on all the
polysilicon walls. In the case of EPROM (and E^2PROM) devices,
the voltages used during data write (and erase) are much larger
than the voltages encountered in DRAM silicon trench
capacitors, and therefore, are more for susceptible to
dielectric breakdown at the polysilicon corners. The enhanced
dielectric breakdown discussed for nitrided oxides on single
crystal silicon is also applicable to polysilicon. Fig. 45a
shows the dielectric breakdown for 15 nm SiO_2 and SiO_xN_y films
interpoly dielectrics as function of the test structures shown
in Fig. 45b[85].

The test structures shown in Fig. 45b allow for evaluation
of the polysilicon corner effects. The test structure labeled
'A' has no corner coverage by the second polysilicon, and
therefore, gives information about the intrinsic
characteristics of the interpoly dielectric. In the test
structure labeled 'B', the second polysilicon covers a single

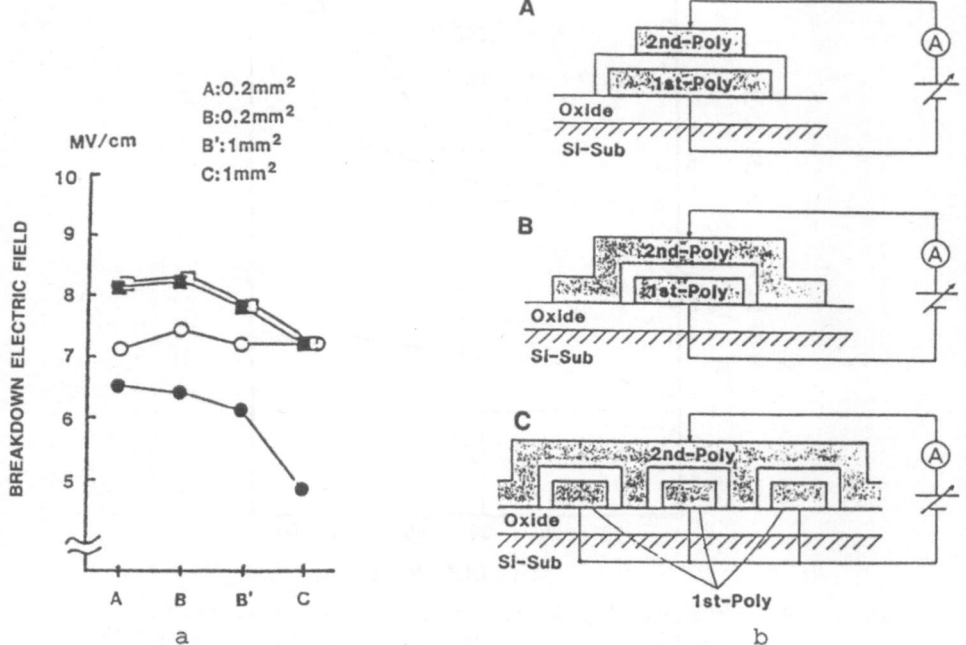

Fig. 45 (a) Dielectric breakdown field for 15 nm thick
SiO$_2$ (circles) and nitrided SiO$_2$ (squares)
interpoly dielectrics as function of the test
structure shown in (b). Solid data points
indicate positive polarity of the second
polysilicon. Gate area is indicated in (a)[85].

first polysilicon finger, while, in the structure labeled 'C',
covers multiple first polysilicon fingers giving information
about the corner effect and its yield. The first polysilicon
(lower capacitor electrode) with a thickness of 250 nm
deposited by LPCVD is implanted with 5×10^{14} cm^{-2} arsenic at an
energy of 50 KeV. After patterning, either SiO$_2$ or SiO$_x$N$_y$ is
grown by RTO/RTA or RTO/RTN/RTA. RTO at 1150°C in dry O$_2$, RTA
at 1100°C for 60 seconds in N$_2$, and RTN at 1150°C in NH$_3$ for 60
seconds. A second LPCVD polysilicon (upper electrode) with a
thickness of 300 nm completes the capacitor structure. Fig. 45a
shows that nitrided SiO$_2$ films have better breakdown
characteristics than that of pure SiO$_2$ independent of the test
structure and the polarity of the electrode bias. The
dielectric breakdown field for pure SiO$_2$ has a strong
dependence on the test structure when the upper electrode is
positively bias. This dependence is caused by the enhancement
of the electric field at the corner edges of the first
polysilicon. Nitrided SiO$_2$ films have very little dependence on
the test structure and the polarity of the capacitor electrode
bias. These results, combined with the enhancement in the
dielectric constant of nitrided oxides, should result in
improved device reliability and storage capacitance if applied
to single crystal trench capacitor technology.

The combination of rapid thermal process and chemical vapor deposition processes allows for a unique process technology in which the CVD process uses temperature, rather than gas flow, as a reaction switch: Limited Reaction Process (LRP)[86]. Thus, single chamber processes that combine epitaxial silicon with *in-situ* doping for *pn* junctions; or, epitaxial silicon, thin oxide growth, and polysilicon deposition for MOS devices, are possible. This is the newest of the RTP techniques, and its potential is currently being investigated. Its implementation requires some RTP chamber modifications to optimize the gas flow dynamics and temperature uniformity.

In the case of MOS capacitors, C-V measurements show midgap interface states and fixed charge densities of about 5×10^9 $cm^{-2}eV^{-1}$ and 1×10^{10} cm^{-2} respectively[20]. The *in-situ* deposition of polysilicon results in a passivation layer for the thin gate oxide, thus inhibiting the diffusion of impurities from the ambient air during transfer of wafers between separate processing reactors. The *in-situ* fabrication of epi-oxide-polysilicon structures for *n*- and *p*-channel MOS devices allow for process optimization with reduced CMOS latchup susceptibility. Preliminary evaluations show *n*- and *p*-channel devices with subtreshold slopes of approximately 90 mV/decade, and source/drain-to-substrate breakdown voltages in the 12 to 20 V range. This breakdown voltage range corresponds to the values expected for epitaxial layer doping of about 5×10^{16} cm^{-3} [87]. One of the key advantages of this technology is the minimization of the substrate/epitaxial layer outodoping. Fig. 46 shows a SIMS depth profile of the boron doping at the substrate/epitaxial layer interface after completion of MOS transistor fabrication. The boron concentration at the interface changes nearly two orders of magnitude in only 0.3 μm. Hence the epitaxial layer thickness can be below 2 μm without adversely affecting the transistors characteristics[87].

The *in-situ* fabrication of epi-oxide-polysilicon MOS transistors implies that any implants for threshold voltage adjustment have to be done through the gate polysilicon. Such a process results in gate oxide performance degradation[48]. Therefore, as an alternative to such implants, a tight control of dopant profile in the epitaxial layer is required during the layer growth.

Defects and generation-recombination centers can lower carrier lifetimes. Thus, the measurement of the minority carrier lifetime gives information about the quality of epitaxial layers. Good films require accurate alignment of the first epitaxial layers that nucleate during the temperature ramp-up. This is important because these layers serve as templetes for subsequent growth, and poor initial layers would result in a low quality film. The simplest minority carrier device is a *pn* junction diode. A commonly used measure of semiconductor junction quality is the diode ideality factor commonly referred to as "n". The ideality factor describes how fast the diode current increases in forward bias compared to the ideal slope of 59 mV/decade at room temperature, $n = 1.00$. Diodes fabricated in material with very low lifetimes, the

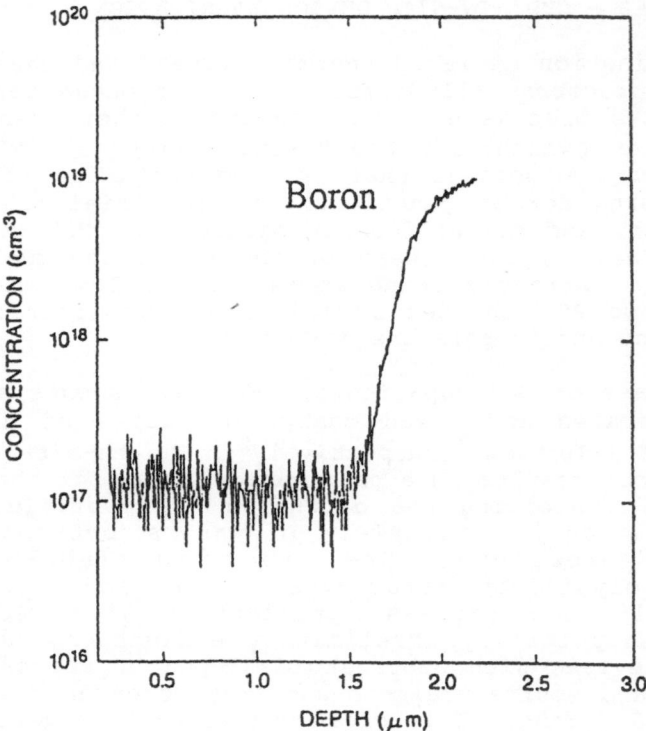

Fig. 46   SIMS depth profile of the boron doping in
the epitaxial layer after completion of
MOSFET transistor processing[87].

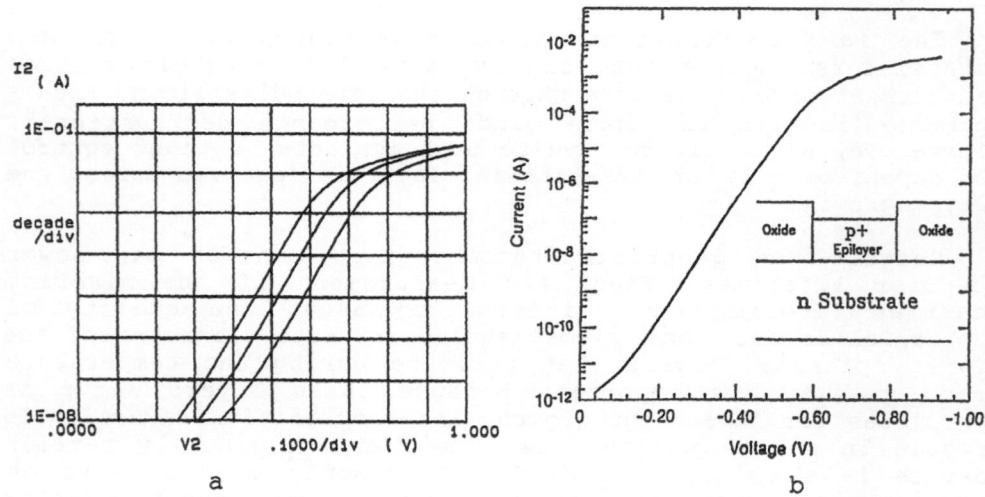

Fig. 47   Semilogarithmic plot of the forward bias
characteristics for (a) substrate/epitaxial
layer *pn* diodes. The diodes area are 4.9×10^{-2},
5.1×10^{-3}, and 4.8×10^{-4} cm^2. And (b) p^+n epitaxial
layer diode with an area of 3.4×10^{-4} cm^2 [21].

forward current is dominated by recombination at defects in the depletion region, leading to slope of 2x 59 or 118 mV/decade or $n = 2$ [88].

Fig. 47a shows the forward-bias characteristics for *pn* diodes with a *p* epitaxial layer doping of about 5×10^{17} cm^{-3} having different areas, the *n* substrate resistivity is about 0.02 ohm·cm [21]. Therefore, the depletion region associated with these junctions include the substrate/epitaxial layer interface. The diode current scales as the area and shows an ideality factor of 1.05, indicating a minimum of defects in the depletion region and, hence, negligible defects at the substrate interface. Thus, the temperature transient during the beginning of the epitaxial layer growth cycle has no adverse effects on the electrical properties of the substrate/epitaxial layer interface. The quality of the epitaxial layer itself is evaluated in a similar method, but this time the diode is formed by a junction on the grown layer only. Fig. 47b shows the forward-bias current for a p^+n diode formed by ion implantation of 1×10^{15} cm^{-2} dose of BF$_2$ ions at 50 KeV into an n-type epitaxial layer[21]. Because of the heavy dose implant, the junction depletion region is primarily in the lightly doped, unimplanted epitaxial material under the heavily doped surface layer. The diode shows an ideality factor of $n = 1.05$, indicating a minimum of defects in the bulk of the epitaxial layer. Refinements of the epitaxial process should result in ideality factors closer to one, and hence, higher quality epitaxial layers.

SUMMARY

Rapid Thermal Processing with reactive gases has been shown to be an effective technique for growing high quality thin dielectrics for ULSI technologies. RTP offers a unique control for fine tuning of the nitrogen composition in nitrided silicon dioxide films resulting in optimum dielectric electrical characteristics: low charge trapping, low interface density of states, low fixed charges, high charge to breakdown, and low leakage current. Nitrided oxides, applied as a dielectric pad in oxidation mask for LOCOS-like isolation technologies, results in reduced bird's beak effect. The high-temperature oxidation allows for excellent step coverage in DRAM silicon trench capacitors, enhancing device reliability. RTP dielectrics grown on as-deposited polysilicon films result in elimination of the asperities commonly observed in polysilicon technology, hence enhancing the breakdown characteristics of these dielectrics. The combination of RTP and CVD technologies, and the multisequential processing capabilities of RTP, such as RTC/RTO/RTN/RTO, opens a new way of manufacturing MOS devices with minimum wafer handling and enhanced yield.

ACKNOWLEDGMENTS

The author wishes to thank the material contributed by Dr. E. Bussmann, Siemens; Dr. T. Hori and Dr. M. Inoue, Matsushita Electric; Dr. N. Ajika, Mitsubishi; Dr. A. Manocha, AT&T; Dr. Z. Weinberg, IBM; Prof. J.C. Sturm, Princeton University; Prof.

M. Buhrman, Cornell University; and Dr. M. Moslehi, Stanford university. The author is also grateful for the cooperative research and discussions with Dr. V. Muraly and Mr. P. Freiberger, Intel; Prof. J.P. Krusius, Cornell University; Mr. D. Flowers and Dr. M. Burnham, Motorola; and Mr. C. Hanrahan, IDT.

REFERENCES

1. R.R. Razouk, B.E. Deal, J. Electrochem. Soc., 122, No. 8, p. 1573 (1979).
2. T. Hattori, Solid State Technol., p. 83, July 1982.
3. R.S. Ronen, P.H. Robinson, J. Electrochem. Soc., 119, No. 6 p. 747 (1972).
4. R.J. Kriegler, Y.C. Chen, D.R. Colton, ibid., p. 388.
5. C. Hashimoto, S. Muramoto, N. Shiono, O. Nakajima, ibid, 127, No. 1, p. 129 (1980).
6. I.D. Calder, This Book.
7. C. Hu, S.C. Tam, F.C. Hsu, P.K. Ko, T.Y. Chan, K.W. Terrill, IEEE Trans. Elec. Dev., ED-32, No. 2, p. 375 (1985).
8. W.P. Noble, W.W. Walker, IEEE Circuits and Devices Magszine, p. 45, January 1985.
9. T. Ito, H. Arakawa, T. Nozaki, H. Ishikawa, J. Electrochem. Soc., 127, p. 2248 (1980).
10. T. Ito, T. Nakamura, H. Ishikawa, IEEE Trans. Elec. Dev., ED-29, p. 498 (1982).
11. F.L Terry, Jr., R.J. Aucoin, M.L. Naiman, S.D. Senturia, IEEE Elec. Dev. Lett., EDL-4, p. 191 (1983).
12. S.S. Wong, C.G. Sodini, T.W. Ekstedt, H.R. Grinolds, K.H. Jackson, S.H. Kwan, J. Electrochem. Soc., 130, p. 1139 (1983).
13. S.K. Lai, J. Lee, V.K. Dham, IEEE IEDM Tech. Digest, p. 190 (1983).
14. S.S. Wong, W.G. Oldham, IEEE Elec. Dev. Lett., EDL-5, p. 175 (1984).
15. F.H.P.M. Habrakev, A.E.T. Kuiper, Y. Tamminga, J.B. Theeten, J. Appl. Phys., 53, p. 6996 (1982).
16. J. Nulman, The Electrochem. Soc. Proc. of Symp. on ULSI Science and Technol., 87-11, p. 141 (1987).
17. J. Nulman, J.P. Krusius, A. Gat, IEEE Elec. Dev. Lett., EDL-6, p.205 (1985).
18. Z.A. Weinberg, D.R. Young, J.A. Calise, S.A. Cohen, J.C. DeLuca, V.R. Deline, Appl. Phys. Lett., 45, No. 11, p. 1294 (1984).
19. J. Nulman, J.P. Krusius, L. Rathbun, IEEE IEDM Tech. Digest, p. 169 (1984).
20. J.C. Sturm, C.M. Gronet, J.F. Gibbons, IEEE Elec. Dev. Lett., EDL-7, pp. 282 (1986).
21. J.C. Sturm, C.M. Gronet, J.F. Gibbons, J. Appl. Phys., 59, No. 12, p. 4180 (1986).
22. J.C. Shcumacher Co., Newsletter No. 34, Rev. 1, J.C. Shcumacher Co., Oceanside, CA 92054-0233, U.S.A.
23. C. Gelain, A. Cassuto, P. LeGoff, Oxid. Met., 3, p. 139 (1971).
24. E.A. Gulbransen, S.A. Janson, ibid, 4, p. 181 (1972).
25. G.A. Lang, T. Stavish, RCA Rev., 24, p. 488 (1963).
26. J. Bloem, L.J. Giling, in "Current Topics in Material Science," 1, E. Kaldis, editor, Chap. 4, North Holland Publishing Co., Amsterdam (1978).

27. T.J.M. Kuijer, L.J. giling, J. Bloem, J. Cryst. Growth, 22, p. 29 (1974).
28. P. van der Putte, L.J. Giling, J. Bloem, ibid, 41, p. 133 (1977).
29. W.H. Shepherd, J. Electrochem. Soc., 112, p. 988 (1965).
30. E.H. Nicollian, J.R. Brews, in "MOS Physics and Technology," Chap.8, John Wiley & Sons, NY (1982).
31. S.M. Hu, J. Appl. Phys., 55, p. 4095 (1984).
32. V. Murali, S.P. Murarka, ibid., 60, p. 2106 (1986).
33. V.K. Samalam, Appl. Phys. Lett., 47, p. 736 (1985).
34. S.T. Lee, D. Nichols, ibid., p. 1001.
35. S.A. Schafer, S.A. Lyon, ibid., p. 154.
36. M. Hamasaki, Solid State Electron., 25, p. 479 (1982).
37. B.E. Deal, A.S. Grove, J. Appl. Phys., 36, p. 3770 (1965).
38. A. Fargeix, G. Ghibaudo, G. Kamarinos, ibid., 54, p. 2878 (1983).
39. R. Doremus, A. Szewczyk, The Electrochem. Soc. Proc. of Symp. on Silicon Nitride and Silicon Dioxide Thin Insulating Films, 87-10, p.350 (1987).
40. J. Blanc, Appl. Phys. Lett., 33, p. 424 (1978).
41. H.Z. Massoud, J.D. Plummer, E.A. Irene, J. Electrochem. Soc., 132, p. 2694 (1985).
42. V. Murali, S.P. Murarka, The Electrochem. Soc. Proc. of Symp. on ULSI Science and Technol., 87-11, p. 133 (1987).
43. V. Murali, A.T. Wu, D.B. Fraser, E. Kamieniecki, J. Nulman, The Electrochem. Soc., Extended Abstracts, 88-2, p. 385 (1988).
44. Optical Diagnostics Systems, 46 Manning Rd., Billerica, MA 01821.
45. M.E. Burnham, R.N. Legge, J. Nulman, P.L. Fejes, J.F. Brown, Mat. Res. Soc. Symp. Proc., 92, p. 115 (1987).
46. B. Deal, J. Electrochem. Soc., 114, No. 3, p. 266 (1967).
47. S.A. Nelson, H.D. Hallen, R.A. Buhrman, J. Appl. Phys. 1988.
48. J. Nulman, J. Scarpulla, T. Mele, J.P. Krusius, IEEE IEDM Tech. Digest, p.376 (1985).
49. Z.A. Weinberg, T.N. Nguyen, S.A. Cohen, R. Kalish, Mat. Res. Soc. Symp. Proc., 52, p. 327 (1986).
50. M. Revitz, S.I. Raider, R.A. Gdula, J. Vac. Sci. Technol., 16, p. 345 (1979).
51. N.E. McGruer, R.A. Oikari, IEEE Trans. Elec. Dev., ED-33, p. 929 (1986).
52. M.M. Moslehi, S.C. Shatas, K.C. Saraswat, The Electrochem. Soc. Proc. of Symp. on Silicon Materails Science and Technology, 86-4, p. 379 (1986).
53. Y. Naito, T. Hori, H. Iwasaki, H. Esaki, J. Vac. Sci. Technol., 5, p. 663 (1987).
54. T. Hori, H. Iwasaki, Y. Yoshioka, M. Sato, Appl. Phys. Lett., 52, No. 9, p. 736 (1988).
55. T. Hori, H. Iwasaki, IEEE IEDM Tech. Digest, p.570 (1987).
56. D.L. Flowers, J. Nulman, J.P. Krusius, Mat. Res. Soc. Symp. Proc., 92, p. 127 (1987).
57. D.L. Flowers, J. Electrochem. Soc., 134, p. 472 (1987).
58. E. Harari, J. Appl. Phys., 49, No. 4, p. 2478 (1978).
59. M.S. Liang, C. Hu, IEEE IEDM Tech. Digest, p. 396 (1981).
60. I. Chen, S. Holland, C. Hu, IEEE Trans. Elec. Dev., ED-32, p. 413 (1985).
61. Y. Nissan-Cohen, J. Shappir, D.Frohman-Bentchkowsky, J. Appl. Phys., 54, No. 10, p. 5793 (1983).

62. T. Hori, H. Iwasaki, Y. Naito, H. Esaki, IEEE Elec. Dev.Lett., EDL-7, p. 669 (1986).

63. T. Hori, H. Iwasaki, Y. Naito, H. Esaki, IEEE Trans. Elec. Dev., ED-34, p. 2238 (1987).

64. J. Nulman, J.P. Krusius, Appl. Phys. Lett., 47, No. 2, p. 148 (1985).

65. S.M. Sze, in "Physics of Semiconductor Devices," Chap. 8, 2nd edit.John Wiley & Sons, NY (1981).

66. S.C. Sun, J.D. Plummer, IEEE Trans. Elec. Dev., ED-27, p.1497 (1980).

67. F.L Terry, P.W. Wyatt, M.L. Naiman, B.P. Mathur, C.T. Kirk, S.D. Senturia, J. Appl. Phys., p. 2036 (1985).

68. T. Hori, H. Iwasaki, IEEE IEDM Tech. Digest, p. 570, (1987).

69. T. Hori, H. Iwasaki, IEEE Elec. Dev. Lett., EDL-9, p. 168 (1988).

70. E.H. Nicollian, J.R. Brews, in "MOS Physics and Technology," Chap. 13, John Wiley & Sons, NY (1982).

71. K. Imai, K. Yamabe, Y. Tsunashima, K. Iwai, T. Hashio, H. Tango, IEEE IEDM Tech, Digest, p. 702 (1985).

72. R.B. Marcus, T.T. Sheng, J. Electrochem. Soc., 129, p. 1278 (1982).

73. E.A Irine, E. Tierney, J. Angillelo, ibid., p. 2594.

74. H. Wendt, E. Bussmann, SEMI SEMICON/EUROPA Tech. Proc., p. 120 (1987).

75. Y. Miyai, K. Yoneda, H. Oishi, H. Uchida, M. Inoue, J. Electrochem. Soc., 135, p. 150 (1988).

76. K. Yoneda, T. Taniguchi, H. Uchida, H. Okada, H. Oishi, Y. Miyai, M. Inoue, Technical Digest VLSI Symp. on Technology in Karuizawa, p. 95 (1987).

77. R.M. Anderson, D.R. Kerr, J. Appl. Phys., 48, p. 434 (1977).

78. P.A. Heimann, S.P. Murarka, T.T. Sheng, ibid., 53, p. 6250 (1982).

79. T.I. Kamins, E.L. MacKenna, Metal Trans., 2, p. 2291 (1971).

80. J. Nulman, Mat. Res. Soc. Symp. Proc., 74, p. 641 (1987).

81. A. Maury, S.C. Kim, A. Manocha, K.H. Oh, D. Kostelnick, S. Shive, IEEE IEDM Tech. Digest, p. 676 (1986).

82. R.A. Swalin in "Thermodinamics of Solids," 2nd Edit., p. 248, John Wiley & Sons, NY.

83. C.Y. Chang, H.L. Huang, J. Appl. Phys., 54, p.2287 (1983).

84. H. Sunami, J. Electrochem. Soc., 125, p. 892 (1978).

85. N. Ajika, M. Shimizu, K. Tsukamoto, M. Hirayama, T. Matsukawa, Japan Soc. of Appl. Phys. 19th Conf. on Solid Satate Dev. and Mat., Extended Abst., p. 211 (1987).

86. J.F. Gibbons, C.M. Gronet, K.E. Williams, Appl. Phys. Lett., 47, p. 721 (1985).

87. J.C. Sturm, C.M. Gronet, C.A. King, S.D. Wilson, J.F. Gibbons, IEEE Elec. Dev. Lett., EDL-7, p. 577 (1986).

88. A.S. Grove, in "Physics and Technology of Semiconductor Devices," p. 186, John Wiley & Sons, NY 1967.

SILICIDATION BY RAPID THERMAL PROCESSING

L. Van den hove and R.F. De Keersmaecker

Interuniversity Microelectronics Center (IMEC)
Kapeldreef 75
3030 Leuven (Belgium)

1. INTRODUCTION

The development of microelectronics silicon technology in the last decade or so, has been characterized by an impressive trend towards higher complexity and enhanced circuit performance. This was made possible by the use of larger chip areas, but also by a continuous drive towards miniaturization, based upon the classical laws of scaling as introduced by Dennard et al. [1]. In these laws, all device dimensions are reduced by a factor λ, and other parameters are adjusted accordingly.

One of the shortcomings of device scaling lies in the interconnections responsible for the propagation of a signal within a circuit. It can be shown that classically, although all dimensions are scaled by λ, the RC time constant of an interconnection line will remain unaffected, while the device speed is increased by at least a factor of λ. Therefore signal delay on a chip becomes increasingly dominated by the interconnections, and will not be improved by further device scaling. In fact, the picture is even more gloomy, if one considers that, while the device dimensions are scaled by λ, the average interconnect length increases by a factor α, when the chip area increases by α^2. Therefore, the line resistance will increase by $\alpha.\lambda^2$ and the line capacitance will increase by a factor of α. As an end result, the RC time constant of the line will increase by $\alpha^2.\lambda^2$. It is clear that this is one of the reasons why interconnections are a major obstacle on the road towards further miniaturization.

The technology for interconnections should be divided into two categories: long-range and short-range interconnections. The increasing wiring complexity of long-range interconnects leads to multi-level metallization schemes, based upon planarized insulators, contact and via filling, a.s.o.; these issues are addressed in other contributions to this volume. We will limit our discussion here to short-range interconnections at the device level and between devices.

Since the early 1970's, polycrystalline silicon has been used as the gate material in MOS transistors, largely because of its excellent thermal and chemical stability, allowing device fabrication with a high packing density. The high sheet resistance of poly-Si, typically 20 Ω/sq, however, is often a limiting factor. Other parasitic resistances, such as the contact resistance and the series resistance of shallow source and drain diffusions are contributing to the degradation of device characteristics when dimensions are scaled down to the micron or submicron range. In order to solve the problem of high poly-Si gate resistivity, the 'polycide' structure was proposed, which combined the low resistivity of a silicide with the reliable and well-characterized stack consisting of poly-Si on top of the gate oxide. Refractory silicides, such as WSi_2, $MoSi_2$, $TaSi_2$ and $TiSi_2$ have received most attention for this application.

Recently, the *self-aligned silicidation technology* (also known as SALICIDE) was introduced. In this technology, silicide is formed on source and drain regions of the MOS transistors, thereby solving the problem of high parasitic series resistance and also of high contact resistance between the metallization and the device. Simultaneously, silicide is formed on the gate region, which solves the problem of interconnect resistance, just as in the case of the polycide technology. This technology was first proposed by Shibata et al. [2] in 1981 using PtSi, and later by Ting et al. [3] and others [4-6, ...] using $TiSi_2$.

In the salicide technology a silicide is formed on source, drain and gate regions simultaneously, by the reaction of a deposited metal layer with the Si regions. To achieve this, a conventional MOS process flow is followed until gate patterning is completed. Oxide spacers are then formed along the poly-Si sidewalls. A thin metal layer is deposited and reacted with the exposed Si regions. This is followed by a selective metal etch from those regions (usually SiO_2) where the metal did not react with the substrate to form a silicide. The oxide sidewall spacers are required to avoid shorting between the silicide on source-drain and gate regions. In this way the silicide is formed self-aligned to the Si regions. A schematic cross-section of an MOS transistor at various stages of the salicide process is given in Fig. 1.

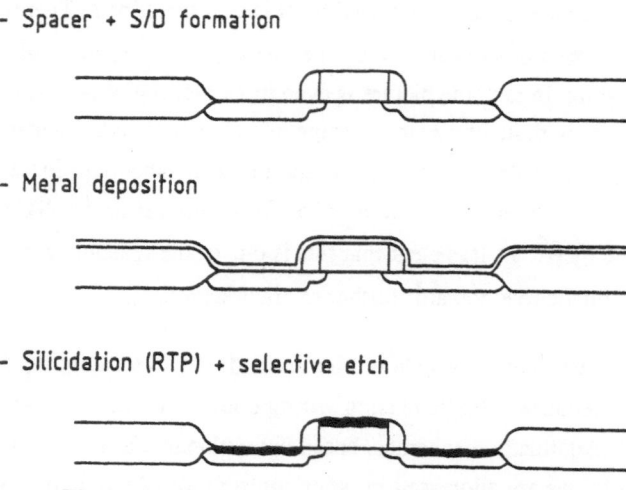

Fig. 1   *Process steps in the SALICIDE technology.*

The purpose of forming a silicide on the entire source and drain regions is to achieve a reduction of the parasitic series resistance. This is achieved (i) by a reduction of the junction sheet resistance and (ii) by a reduction of the contact resistance. The latter results from a low contact resistivity between silicide and n^+- and p^+-Si and from an increase of the contact area (silicidation of the entire source and drain). The current flow in a silicided MOS transistor is schematically illustrated in Fig. 2.

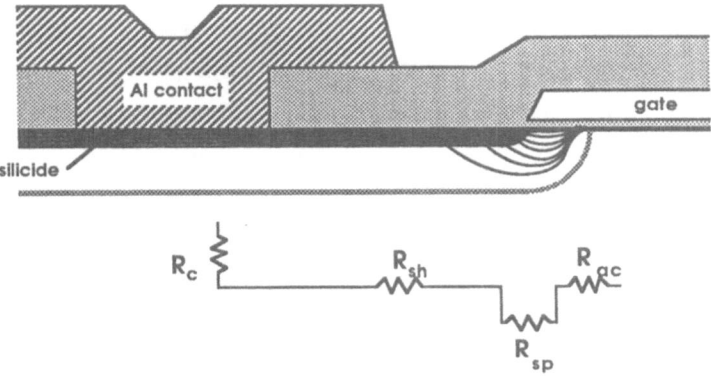

Fig. 2   *Schematic representation of the current flow and of the parasitic series resistances in the source/drain region of a MOS transistor with silicided source/drain.*

Since the polycide technology has already found widespread application in several production lines, whereas in the salicide technology the effort is largely spent on process development and optimization at this moment, we will concentrate in this paper on the latter technique. Moreover, as will be indicated below, rapid thermal processing is an attractive technique to carry out the silicide reactions in case of silicide formation by metal-Si reaction. In case of polycide technologies, the silicide is usually formed in a conventional furnace.

Some important criteria when selecting a metal for salicide applications are:

- low resistivity of its silicide;
- ease of silicide formation;
- absence of reaction with SiO_2 at the silicidation temperature;
- absence of lateral silicide formation;
- existence of an etchant, which removes the metal selectively with respect to the silicide;
- minimal silicon consumption during silicide formation;
- smoothness of the interface between silicide and silicon;
- thermal stability of the silicide;
- MOS compatibility (e.g. chemical stability);
- limited interaction with the dopants;
- limited stress.

Table I lists some metals which could be used in a salicide process [7].

Table I

Candidate metals for salicide process [7] and related data.

metal	silicide	reaction temperature [°C]	selective etchant	resistivity [μΩ.cm]
Mo	MoSi$_2$	IBI	NH$_4$OH:H$_2$O$_2$	120 - 130
Nb	NbSi$_2$	IBI	NH$_4$OH:H$_2$O$_2$	60
Ti	TiSi$_2$	600	NH$_4$OH:H$_2$O$_2$	13 - 16
Co	CoSi$_2$	550	HCl:H$_2$O$_2$	15 - 20
Pt	PtSi	500	HCl:HNO$_3$	28 - 35
Pd	Pd$_2$Si	300	KI:I$_2$	30 - 35
Ni	NiSi	400	HNO$_3$	12 - 15

IBI: ion beam induced silicidation

Since the introduction of the technology, both refractory metals and near-noble metals have been used for application in the salicide technology. In this paper, we will concentrate on an attractive representative from both categories, TiSi$_2$ and CoSi$_2$, respectively.

First, the various reactions occurring in the salicide process will be discussed (section 2). Then the implementation of these silicides into a process will be described (section 3) and the device-related aspects will be covered in section 4.

2. REACTIONS IN THE SALICIDE PROCESS

As an alternative for conventional furnace processing, rapid thermal processing (RTP) has been introduced [8], mainly to lower the thermal budget needed for activation of implanted dopant profiles. Although little information is available on the formation of silicides using RTP, some studies have indicated that RTP systems are very suited for carrying out silicide reaction steps. Okamoto et al. [9] have used RTP treatments in Ar ambient to form TiSi$_2$. The reaction of Ti with SiO$_2$ and the possibility of lateral silicidation leave only a limited temperature window to perform the silicidation reaction. Morgan et al. [10] and others [11-13] have demonstrated that the use of a N$_2$ ambient can broaden this process window.

In the following paragraphs the various reactions occurring in the salicide process will be described. Figure 3 shows schematically the areas which will be considered. First the reaction between the metal (Ti and Co) and Si is studied. For the case of Ti the influence of impurities originating from deposition and reaction ambient is investigated in great detail. In a salicide process the metal is in contact with both Si and SiO_2 during the silicide formation step. Therefore, the behavior of the metals on SiO_2 is described next. Finally the self-alignment of the process is examined in relation to the main moving species during silicide formation. It will be shown that the ability to form the silicides in a self-aligned way with respect to the Si regions at SiO_2 edges is closely related with the reactions of the metal with the Si and SiO_2 and is largely influenced by the ambient. The successful application of these silicides in a salicide process will largely depend on the precise understanding of the phenomena occurring at these three locations.

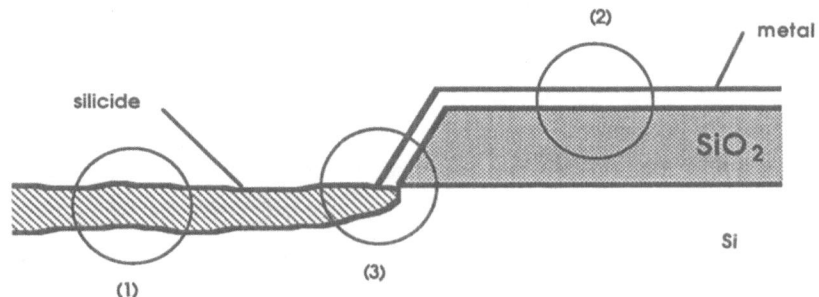

Fig. 3 *Schematic representation of the various reactions which will be considered in this section: (1) metal-Si reaction, (2) metal-SiO_2 reaction and (3) self-alignment of the process.*

2.1. Rapid thermal processing for silicide formation

As Ti is so reactive with oxygen often vacuum systems or modified tube furnaces had to be used in which wafers can be loaded without oxidizing them [14, 15]. Also for the case of Co oxidation has been observed. Our attempt to form $CoSi_2$ in a conventional furnace lead to Co_3O_4 formation as measured by X-ray diffraction (XRD). In an RTP system, wafers can be loaded cold. Because of the small processing chamber the atmosphere can be controlled more easily. The main advantage is that such RTP systems are commercially available and do not need any modification. Figure 4 gives a schematic representation of such an RTP system. After loading the wafer, the chamber is purged for some seconds with nitrogen, argon or a mixture of these with hydrogen. The wafer is then heated by a light source, which in the system used in this work consists of two banks of W halogen lamps. In this way oxide-free silicides can easily be obtained for both $TiSi_2$ and $CoSi_2$. An additional advantage of the use of an RTP system is that the silicidation reaction is performed with a minimal thermal budget, leading to minimal dopant diffusion.

2.2. Sample preparation and mesurement techniques

100 mm n- or p-type Si wafers were used, with a <100> orientation and a resistivity of 1-100 Ω.cm. In the case where the metal-SiO_2 interaction was studied, 100 nm of SiO_2 was grown

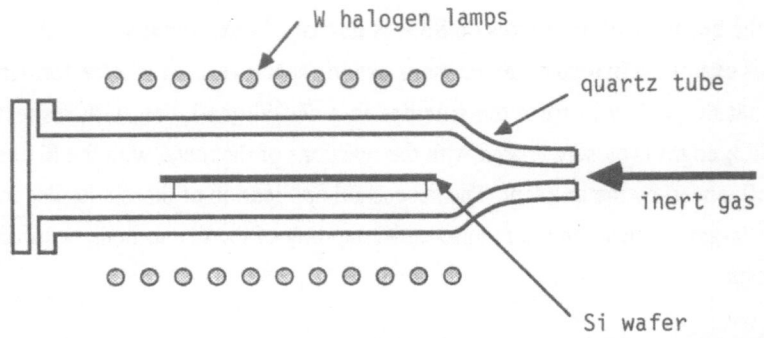

Fig. 4 *Cross-sectional view of a W-halogen lamp heated rapid thermal processing system.*

in dry O_2 at 950° C. The SiO_2 was removed from the back-side of the wafer, in order not to affect the temperature during RTP. After a cleaning step and a dilute HF-dip the wafers were immediately loaded in the sputtering system. Metal films were sputtered using DC-magnetron sputtering at a rate of 20 nm/min. The base pressure was always below $2x10^{-7}$ mbar. The wafers were subsequently heat-treated using an RTP system in either N_2 or Ar ambient.

Deposited and reacted films were evaluated using various analytical methods. Rutherford Backscattering Spectroscopy (RBS) measurements were performed using He particles with an energy around 2 MeV under glancing angle detection. Auger Electron Spectroscopy (AES) combined with sputter depth profiling was used for the detection of impurity elements. The overlap between the N and Ti signal makes AES less attractive when N is present in only small quantities. Moreover, the techniques in which profiling by ion bombardment is applied, can also be affected by the difficulty of preferential sputtering of N or O [16]. Because of these difficulties and because of the lack of sensitivity of RBS for O and N, Elastic Recoil Detection (ERD) was used. It has an excellent sensitivity for low mass elements like O, N and H. For these measurements Si^{5+} ions are accelerated to an energy of 30 MeV using a van de Graaff tandem accelerator [17]. Furthermore, Scanning and Transmission Electron Microscopy (SEM and TEM) were used to determine the layer microstructure and morphology. The sheet resistance of the layers was measured by means of a conventional four point probe (4PP). When the metal-SiO_2 interaction was studied, a first estimate of the magnitude of SiO_2-loss was obtained from ellipsometry measurements.

ERD spectra also give an indication of the purity of the as-deposited metal layers. The total amount of O in a 150-nm Ti-layer is $4x10^{16}$ at/cm^2, corresponding to 6 nm equivalent TiO_2 thickness. Since the larger amount of O is present at the Ti-surface, it is assumed that it is incorporated upon unloading the sample from the vacuum chamber. This has also been observed by other investigators [e.g. 18]. Furthermore, H appears to be present to an amount of $2.3x10^{16}$ at/cm^2 and is distributed through the whole film (about 2 at%). Only very little nitrogen is observed in the as-deposited layer ($1.9x10^{15}$ at/cm^2).

2.3. Metal-Si interaction

2.3.1. TiSi₂ formation using RTP in Ar ambient

RBS and ERD spectra of 150 nm Ti on Si, annealed for 10 s in Ar at 500, 600, 700 and 800°C are shown in Figs. 5 and 6, respectively. From the RBS spectra it is deduced that silicidation starts around 600°C. At 800°C roughly the entire layer is transformed into $TiSi_2$, as is deduced from the ratio of the Si and Ti peak heights in the RBS spectrum.

Information about the depth distributions of N and O and the total amount of H in the layers is obtained from the ERD spectra (Fig. 6). The areal densities, obtained by integrating the respective signals in the ERD spectra with respect to depth, are represented in Fig.7.

H is initially present over the entire Ti film in a concentration of about 0.5 at%. At larger RTP temperatures the amount of H decreases. Apparently, a small amount of N is incorporated during the RTP period, despite the fact that Ar is being used. The N is probably originating from the Ar gas, with N_2 as the most important contaminant. The N in the film is mainly located in a thin surface layer (about 10 nm), and its content depends slightly on the RTP temperature, as will be explained later.

The total amount of oxygen in the sample at various temperatures remains the same as in the as-deposited sample, which indicates that no measurable amount of O is taken up from the RTP ambient. The O depth distribution in the layer, however, changes drastically, as will be shown below.

The AES spectrum of the sample reacted at 700°C (Fig. 8a) shows some Si at the surface. In the RBS spectrum of the 800°C sample, a dip in the Si signal can be observed around channel 370, while the Ti-signal has a plateau. In the ERD spectrum, a shift of the N and O features to lower energies, corresponding to larger depths, can be seen. This is confirmed by the AES measurement shown in Fig. 8b. It also clearly shows a dip in the Si signal and an increase in the O and N content at the same location. The N signal is deduced from the comparison of the Ti_1 and the Ti_2 signal. This AES spectrum shows that a thin $TiSi_2$ layer is present on top of this O and N rich layer. Assuming that all Si in the film is bonded to $TiSi_2$, it can be calculated from the RBS spectrum that only ±93% of the Ti is consumed in the silicidation reaction. The remaining ±7% has most probably reacted with N and O in this subsurface layer, yielding a dip in the Si signal and a more or less constant Ti concentration.

The aforementioned observations may be explained by a rapid vertical transport of Si at temperatures around 700°C. This transport is supposed to proceed along Ti grain boundaries. It is accompanied by a preferential formation of silicide along the vertical grain boundaries. At 700°C the surface is only partially covered with silicide (on top of the former vertical Ti grain boundaries). A complete coverage of the surface must therefore be caused by a (horizontal) surface diffusion. This rapid horizontal supply of Si allows the Ti at the surface to convert to

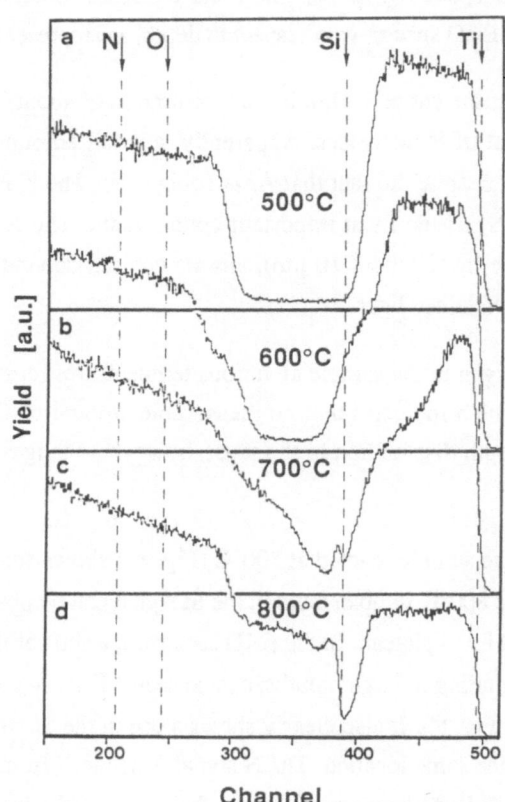

Fig. 5 *RBS spectra of 150 nm thick Ti films on Si after RTP treatment in Ar for 10 s at the indicated temperatures. The arrows indicate the surface positions of the elements.*

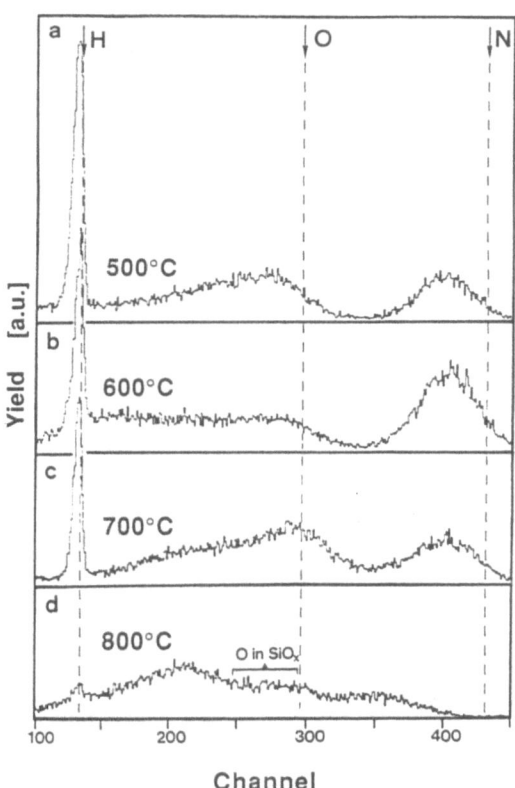

Fig. 6 *ERD spectra of 150 nm thick Ti films on Si after RTP treatment in Ar for 10 s at the indicated temperatures. The arrows indicate the surface positions of the elements.*

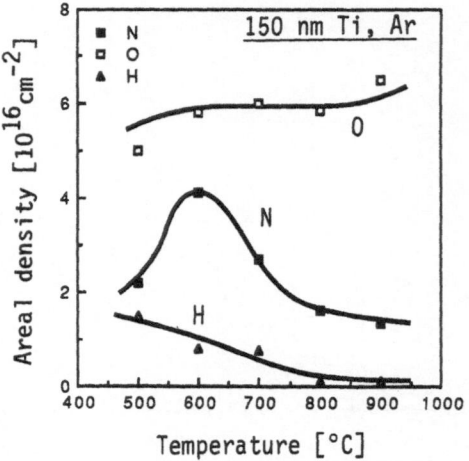

Fig. 7 *Areal densities of O, N and H in the 150 nm films after anneal in Ar, as obtained by integrating the respective signals in the ERD spectra over depth.*

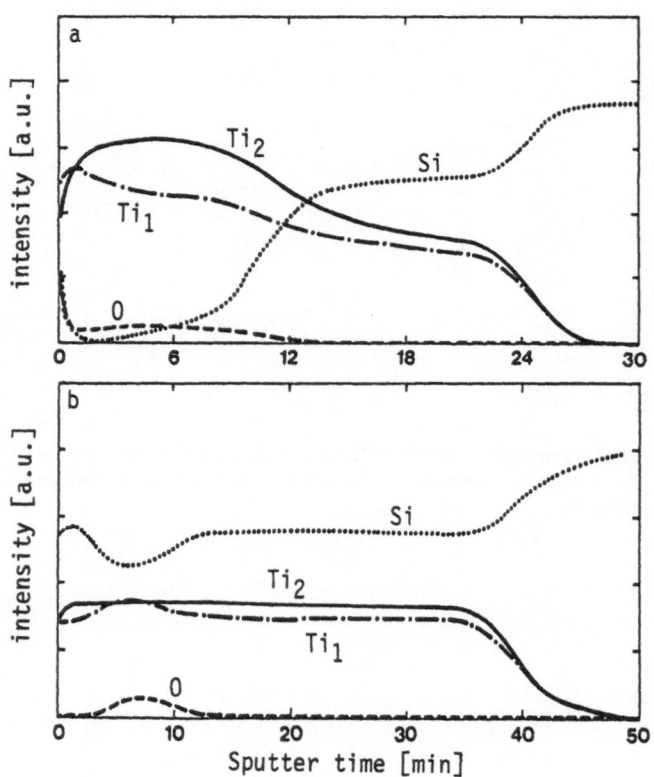

Fig. 8 *AES measurements of 150 nm Ti films on Si after a 10 s reaction step in Ar ambient at 700°C (a) and 800°C (b). Ti_1 represents the peak-to-peak intensity of the 385 eV transition, representative for Ti + N; Ti_2 represents the intensity of the 418 eV transition of Ti only.*

$TiSi_2$. The O and N, initially dissolved in the Ti near the surface, will then be snow-plowed ahead of the reaction front deeper into the unreacted Ti layer. It will be squeezed in between the two approaching $TiSi_2$ layers and react with Ti as soon as the solid solubility limit is exceeded (corresponding to the ±7% Ti which is not bonded to Si). This process is schematically illustrated in Fig. 9. At 800°C the surface is completely covered with $TiSi_2$, resulting in a complete inhibition of the N uptake.

2.3.2. $TiSi_2$ formation using RTP in N_2 ambient

A cross-sectional TEM micrograph of a silicide layer formed at 700°C for 30 s (starting from 65 nm Ti) is shown in Fig. 10. It clearly shows the presence of two layers. To examine the composition of this top layer ERD and AES measurements were performed. The results from the ERD measurements are represented in Figs. 11 and 12. They clearly reveal that N diffuses into the film in much larger quantities than in the case of Ar ambient. Around 750°C a saturation of the N uptake is observed.

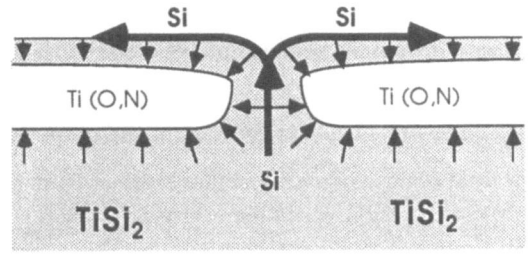

Fig. 9   Schematic representation of the rapid silicidation along the grain boundary and surface noticed at 700°C using RTP in Ar ambient (a). The evolution of the layered structure at various temperatures is represented in (b).

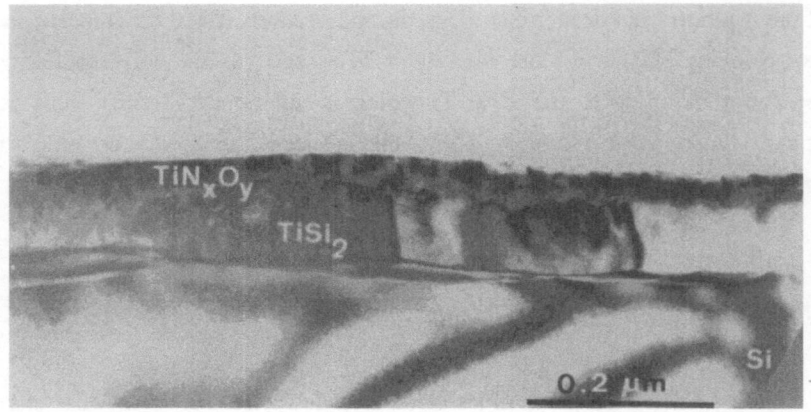

Fig. 10   *Cross-sectional TEM micrograph of the layer resulting from the reaction of 65 nm Ti with Si during a 30 s RTP treatment at 700 °C in N_2 ambient.*

As in the case of Ar RTP, the total amount of O remains more or less constant within the entire temperature range, but its distribution changes with RTP temperature. Upon increasing the RTP temperature, O, initially present at the surface, redistributes over the entire layer. Around 700°C it is snow-plowed ahead of the silicidation front as in the case of the Ar annealing. It is also pushed ahead of the nitridation front. As a result, O piles up at a depth corresponding to 20 - 25 nm below the surface. Apparently, at temperatures above 750°C a stable bilayer structure results from the competition between silicidation and nitridation: a $TiSi_2$ layer at the bottom and at the top a TiN-like layer with a thickness of 20 - 25 nm; in between both layers, oxygen is piled up in a thin layer with unknown stoichiometry.

A simulated RBS spectrum of this tri-layer structure together with the measured spectrum for a sample treated at 1100°C are given in Fig. 13. In the simulation the quantities and depth distributions of O and N as derived from the ERD spectra are used. Excellent agreement is obtained between the simulated and the measured spectrum.

2.3.3. $CoSi_2$ formation using RTP

Figure 14 shows the sheet resistance of 80 nm Co layers deposited on Si after annealing at various temperatures for 30 s in N_2 ambient. The initial sheet resistance of 2.5 Ω/sq corresponds with an as-deposited Co resistivity of $\pm$ 20 $\mu\Omega$.cm. From 350°C on the sheet resistance starts to increase, reaching its maximum value around 500°C. At higher temperatures it decreases again. Around 700°C, the minimum value of 0.7 Ω/sq is reached, remaining constant up to 1100°C (for this silicide thickness). From this value and the thickness as determined by RBS, a resistivity of 16 $\mu\Omega$.cm can be calculated, which is very close to the value determined for $TiSi_2$ (14 $\mu\Omega$.cm).

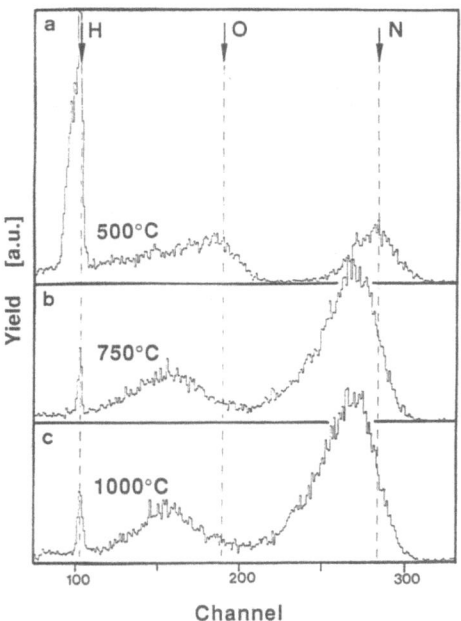

Fig. 11   *ERD spectra of 150 nm thick Ti films on Si after RTP treatment in N_2 for 10 s at the indicated temperatures. The arrows indicate the surface positions of the elements.*

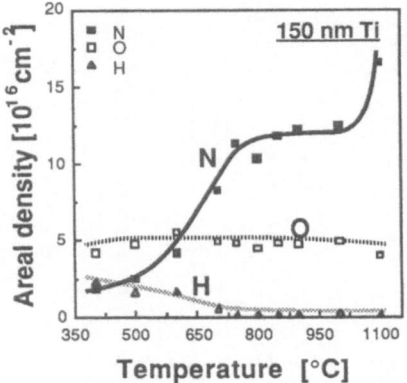

Fig. 12   *Areal densities of O, N and H for 150 nm thick Ti films after a 10 s anneal in N_2, as obtained by integrating the respective signals in the ERD spectra over depth.*

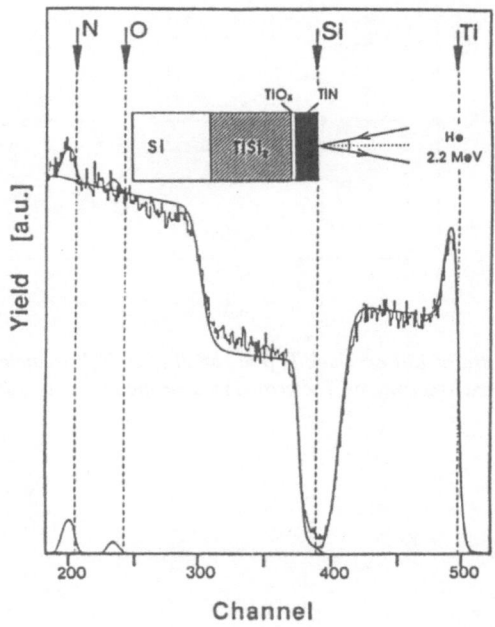

Fig. 13   *Measured and simulated RBS spectrum of the Ti sample on Si after RTP at 1100°C for 10 s in N_2 ambient. The O and N concentrations as determined from the ERD measurements were used for the simulation.*

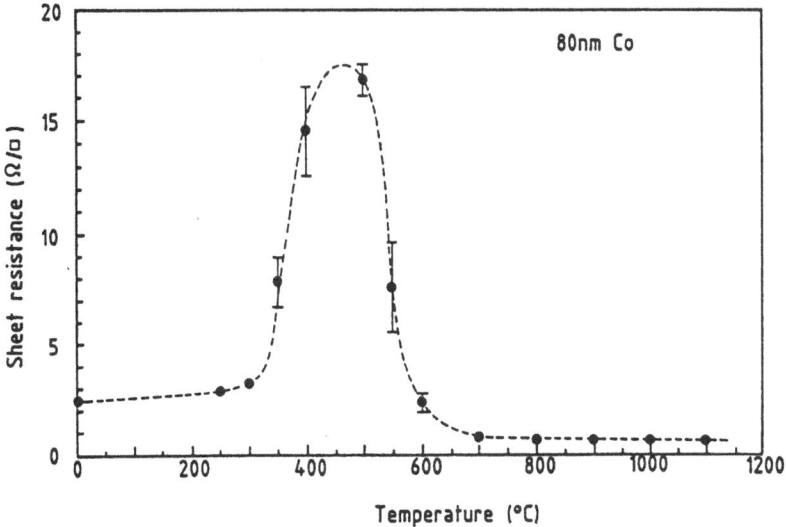

Fig. 14   *Sheet resistance as a function of reaction temperature during RTP (30 s) in N_2 for 80 nm Co on Si.*

This sheet resistance behavior can be explained by considering the phase sequence, as obtained from X-ray diffraction spectra. At 400°C Co_2Si and CoSi are detected, although the Co_2Si phase is difficult to assign. At this temperature some peaks are also noticed which can be assigned to CoO. At 500°C the entire layer is converted to CoSi. Some traces of $CoSi_2$ can already be found. At 700°C only $CoSi_2$ is detected. The observed phase sequence

$$Co \rightarrow Co_2Si \rightarrow CoSi \rightarrow CoSi_2$$

agrees with the measured temperature behavior of the R_{sh} and the sequence using RTP is identical to that observed using conventional furnace processing [19,20].

A cross-sectional TEM micrograph of a sample with 38 nm Co reacted for 30 s at 700°C (Fig. 15) indicates that extremely smooth silicide layers with relatively large grains (0.1-0.3 μm) can be obtained by RTP. The interface roughness amounts to only 20 nm for a 110 nm thick $CoSi_2$ layer. For the case of silicidation by means of RTP, temperatures sufficiently above the nucleation temperature are used, so that more nuclei are formed. This may be the reason for the observed smooth interfaces in this case. From the silicide thickness and the measured sheet resistance a resistivity of 16-17 μΩ.cm is calculated.

2.4. Metal - SiO^2 interaction

2.4.1. Behavior of Ti on SiO_2

The possible interaction between Ti and SiO_2 is dramatically illustrated by the cross-sectional TEM micrograph of Fig. 16 for a 150 nm Ti layer after RTP in N_2 for 10 s at 1100°C. Only 10-20 nm of the original 100 nm SiO_2 is left after the RTP cycle. This indicates that the

Fig. 15   Cross-sectional TEM micrograph of a silicide layer obtained from 38 nm Co
after an RTP cycle at 700 °C for 30 s.

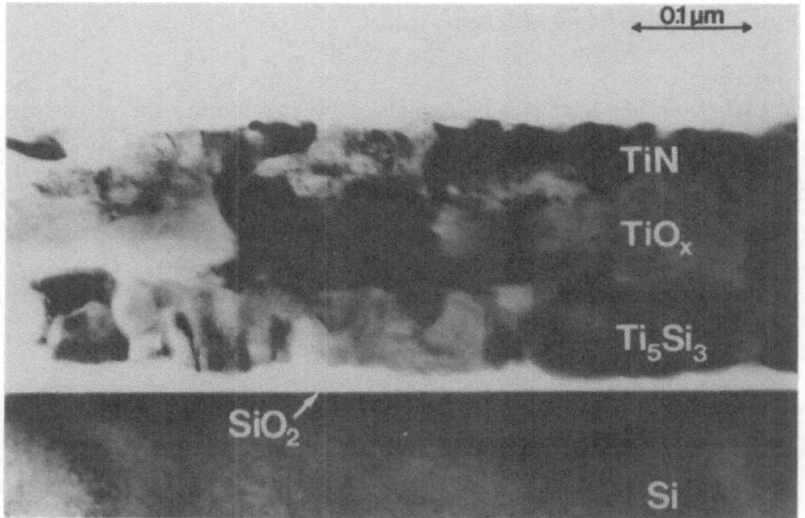

Fig. 16   Cross-sectional TEM micrograph of the layered structure resulting from the
reaction of 150 nm Ti with 100 nm SiO_2 at 1100 °C for 10 s in N_2 ambient.
Note the nearly complete consumption of the SiO_2 layer.

structure obtained after a high-temperature treatment in N_2 ambient consists of three layers: a
Ti_5Si_3 layer at the bottom, a TiN layer at the top and a TiO_x layer in between (phases were
determined using RBS and XRD). RTP in Ar ambient results in a two-layer structure: Ti_5Si_3
and TiO_x.

The reduction of SiO_2 under RTP conditions, although already detectable from 500°C (10 s),
is substantial at temperatures of 700°C and beyond. At these temperatures this reduction leads
to the formation of a Ti_5Si_3 layer at the previous SiO_2/Ti interface. The reduction reaction
proceeds as follows:

$$5 \text{ Ti} + 3 \text{ SiO}_2 \rightarrow \text{Ti}_5\text{Si}_3 + 6 \text{ O}.$$

The O liberated during this reaction is dissolved in the upper Ti layer since Ti has a very large solid solubility for O (34 at%, [21]). Figure 17 displays the amount of O in the Ti layer as a function of the RTP temperature. This is obtained by integrating the O signal from ERD spectra over the width of the Ti layer.

The considerable reaction between Ti and SiO_2 at temperatures beyond 700°C is the basic motivation for establishing a two-step reaction process in the Ti-salicide technology.

2.4.2. Behavior of Co on SiO_2

Phase detection is performed using XRD. As deposited the Co layers on SiO_2 are finely grained Co with the hcp crystal structure (α-Co). Above 400°C the high temperature fcc Co phase (β-Co) is detected. At high temperature only Co is detected (no silicide or silicate).

This suggests that no serious interaction took place between Co and SiO_2 at 900°C, although RBS spectra indicate only a partial coverage of the surface by Co, so that SiO_2 is detected at the surface.

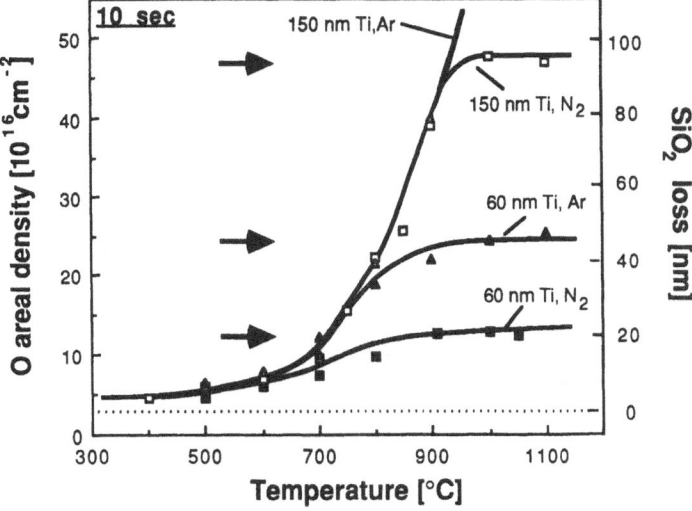

Fig. 17   Areal densities of O for the 150 nm and the 60 nm Ti films on SiO_2 after a 10 s anneal in Ar or N_2 ambient, as obtained by integrating the O signals in the ERD spectra with respect to depth. The amount of SiO_2 lost during the reduction reaction (right scale of the plot) can be calculated from the areal density of O, assuming all but the O initially present in the Ti, originating from this reaction.

Figure 18 shows a series of SEM micrographs of Co layers with an initial thickness of 40 nm on SiO$_2$ after RTP for 30s at 700°, 900° and 1100°C. These pictures clearly reveal that the layers tend to agglomerate with increasing temperature. Thin layers ball up into discrete islands, while for thicker layers only a roughening is observed. It is known that this balling up is a result of the tendency of the system to lower its energy (a reduction of the surface energy is obtained by a reduction of the surface and interface area). This behavior is typical for metals which have a low affinity for SiO$_2$ [22].

When the SiO$_2$ layers are evaluated after the Co has been etched off, a remarkable morphological change of the SiO$_2$ surface can be noticed at the high temperatures (see SEM-pictures of Fig.19).

Within the measurement limits of RBS, the amounts of O and Si in the SiO$_2$ layers remain constant with anneal temperature up to 1000°C. This indicates no loss of SiO$_2$.

Figure 20 shows a cross-sectional TEM micrograph of the sample with 40 nm Co after RTP at 900°C (the Co is not etched). This micrograph clearly demonstrates the agglomeration of the Co layer in discrete globules. More surprisingly, these globules have penetrated into the SiO$_2$. This penetration is most pronounced near the edges of the globule.

The above observations might be explained by a reduction of SiO$_2$ to volatile SiO by residual H$_2$ present in the RTP ambient. It is known that a Co surface can act as a catalytic area on which molecular H$_2$ is dissociated into atomic H [23]. This model, however, cannot account

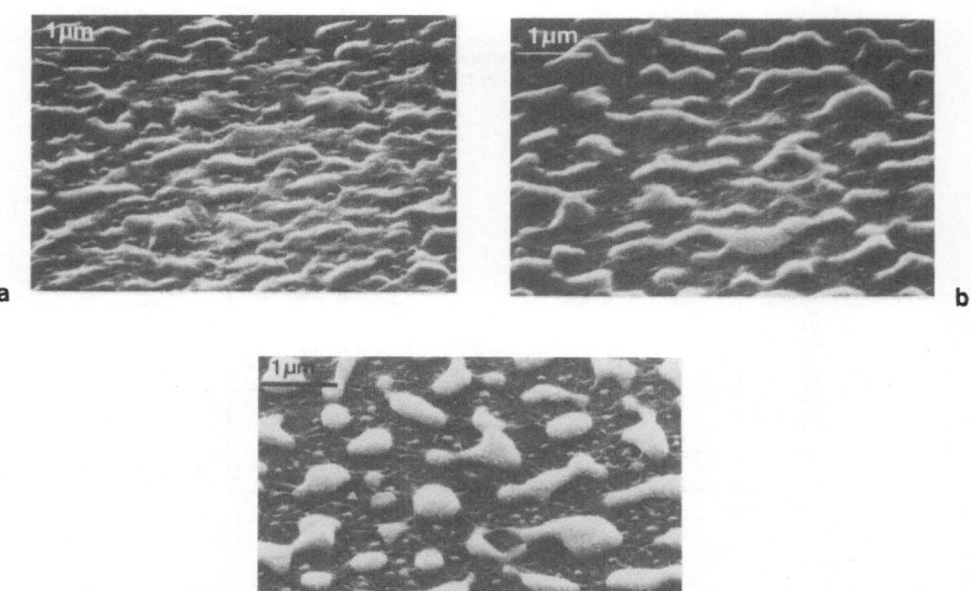

Fig. 18 *Scanning electron micrographs of the surface of Co layers after heat treatment at (a) 700, (b) 900 and (c) 1100°C on SiO$_2$. The initial Co thickness was 40 nm.*

for the observed morphological change of the SiO$_2$ layers at high temperatures. An evaporation of such large amounts of SiO should be detectable using RBS in the channeling mode. Since no large amounts of SiO$_2$ have disappeared and deep groves are detected, a certain amount of SiO$_2$ had to be displaced from under the Co globule to the area next to it. In oxidation studies [24-25] the intrinsic stress built up during oxidation has been observed to

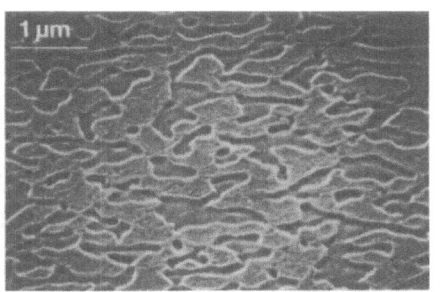

a
b

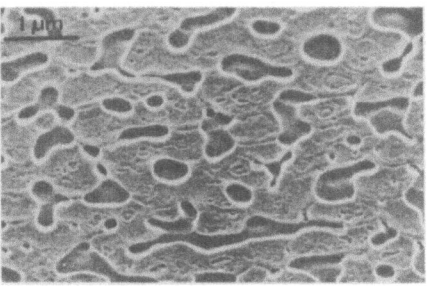

c

Fig. 19   Scanning electron micrographs of the surface of the SiO$_2$ layer after etching off the Co. 40 nm Co layers were heat treated on SiO$_2$ at (a) 700, (b) 900 and (c) 1100 °C for 30 s.

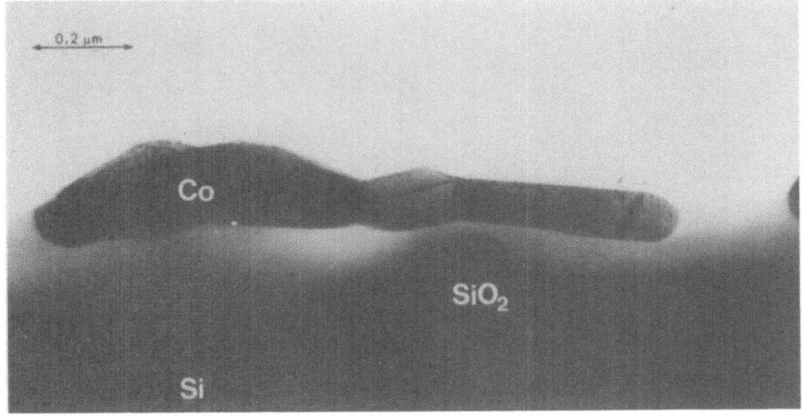

Fig. 20   Cross-sectional TEM micrograph of a Co globule on SiO$_2$ originating from the agglomeration of a 40 nm thick Co layer deposited on SiO$_2$ and heat treated at 900 °C for 30 s in N$_2$ ambient. Note the penetration of the globule in the SiO$_2$.

decrease with oxidation temperature above 900°C. This was attributed to visco-elastic flow of the oxide, the driving force being the amount of internal stress. The balling up of Co into globules is a consequence of the presence of large surface tensions. Therefore interface tensions at the Co/SiO_2 interface might be the driving force for the morphological change of the Co/SiO_2 interface. These forces apply locally and can therefore have a strong impact.

2.5. Self-aligned silicidation

2.5.1. General considerations

In a salicide process, a silicide has to be formed in a self-aligned way on silicon regions. The key parameter influencing the lateral silicidation at oxide edges is the identity of the moving species during silicide formation. In general when a reaction starts in a metal-silicon diffusion couple (M-Si) a metal-silicide (MSi_x) is formed, thereby creating two new interfaces: a silicon/silicide interface and a silicide/metal interface. If the silicide grows at the silicide/metal interface, Si has to diffuse through the layer. If the silicide grows at the silicon/silicide interface, a transport of metal is required.

When the silicide is formed in a laterally confined region by *Si diffusion*, Si may diffuse through the already formed silicide near the oxide edge and react with the metal on top of the oxide, so that the silicide layer actually grows over the oxide edge (Fig. 21). This will be referred to as *"lateral overgrowth"*. Voids or spikes may be created due to the enhanced Si consumption near the oxide edge. Both phenomena, lateral overgrowth and spiking, can cause a drastic reduction of device yield: the former by shorts across a side-wall spacer, the latter by a penetration of shallow junctions under the silicided regions, causing leakage problems.

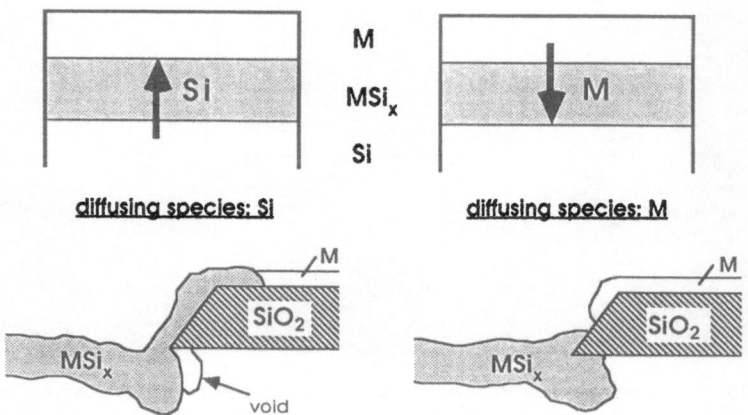

Fig. 21   Schematic representation of the influence of the diffusing species during silicide formation on the lateral silicidation.

When the silicide is formed by *metal diffusion,* no lateral overgrowth is expected. Some silicide can, however, be formed under the SiO_2 edge (Fig. 21). This again can cause shallow junction leakage problems. This phenomenon will be referred to as *"encroachment".*

Chu et al. [26,27] have demonstrated, using implanted Xe as a marker, that Si is the main moving species during $TiSi_2$ formation. For the Co-Si system, the situation is far more complex. As indicated three silicide phases are formed in sequence: Co_2Si, CoSi and $CoSi_2$. van Gurp et al. [28] have shown that Co is the moving species for Co_2Si formation and Si is diffusing during CoSi formation. d'Heurle and Petersson [29] and Lien et al. [30] found Co to be the main moving species during $CoSi_2$ formation. These phase transitions and the various moving species make the Co-Si system extremely complicated with respect to lateral silicidation.

2.5.2. Self-alignment of the TiSi₂ process

Silicidation in Ar ambient

In the study of $TiSi_2$ formation using RTP in Ar ambient a rapid formation of silicide was noted to occur along the Ti grain boundaries at temperatures around 700°C (FIG.9). This preferential grain boundary silicidation is accompanied by a rapid vertical diffusion of Si. In combination with a horizontal surface migration it leads to the formation of a silicide surface layer, before the silicidation front which has started from the Si/Ti interface, has reached the surface. In the region near an oxide step this rapid transport of Si may continue horizontally along grain boundaries of Ti on SiO_2. The Si supplied by this rapid diffusion may then convert all the Ti grains on the SiO_2 near the contact edge into $TiSi_2$ almost simultaneously, starting from each grain edge inwards. This is illustrated by the schematic in Fig. 22. This is further evidenced by the morphology of the lateral silicidation. If the lateral silicidation were to proceed by a silicide formation based on bulk diffusion a smooth silicide/Ti interface would be expected. The irregular shape of the silicidation front suggests a growth process based on rapid diffusion paths.

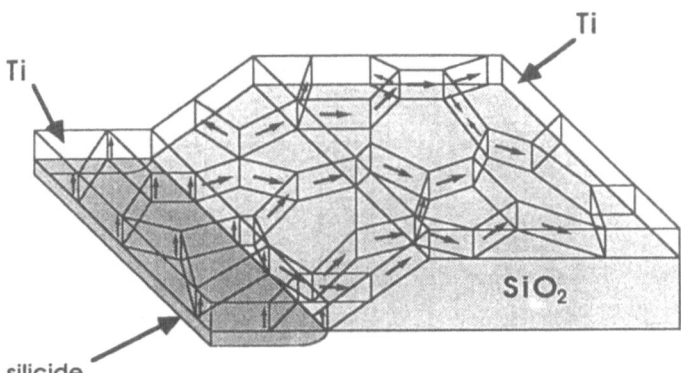

Fig. 22   *Model explaining the rapid lateral silicidation on top of SiO_2, due to the presence of the Ti grain boundaries.*

We have demonstrated that the rapid vertical diffusion of Si along former Ti grains can be avoided by performing the silicidation in a N_2 ambient. In this case a double layer results from a competition between a silicidation reaction starting from the interface and a nitridation reaction from the surface. The uptake of N apparently inhibits the preferential silicidation along the grain boundaries, possibly by blocking the vertical rapid diffusion paths. This is clearly illustrated by the TEM cross-section of Fig. 23 which indicates the perfect self-alignment of the silicidation. In the contact hole the double layer $TiSi_2$/TiN is seen. The silicide grain near the SiO_2 edge has not grown over the oxide at all.

The ERD spectra of Fig. 24 obtained from samples consisting of 150 nm Ti on SiO_2 annealed for 10 s at 700°C in Ar (a) and in N_2 (b) ambient show evidence for the above model. The N profile in the Ti layer after anneal in N_2 ambient (Fig. 24b) consists of two parts which typically result from two different diffusion mechanisms: a large N concentration near the surface (corresponding to lattice diffusion) and a small, but nearly constant concentration in the bulk of the Ti film (corresponding to grain boundary diffusion). Apparently, the indiffusion of N along the grain boundaries blocks the rapid silicidation along the grain boundaries, which is responsible for the lateral silicidation in case of Ar ambient. For the latter case nearly no N is observed in the layer (Fig. 24a).

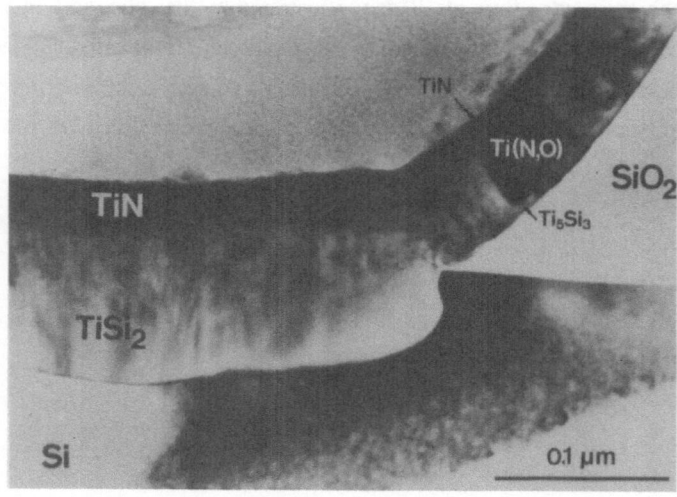

Fig. 23   *Cross-sectional TEM of $TiSi_2$ formed near an SiO_2 edge. The silicidation was performed at 700°C for 30 s in N_2 ambient. Note the presence of a TiN/$TiSi_2$ bilayer on top of Si and of a trilayer structure TiN/Ti(O,N)/Ti_5Si_3 layer on top of SiO_2.*

2.5.3. Self-alignment of the CoSi$_2$ process

A cross-sectional TEM micrograph of a CoSi$_2$ layer selectively formed near an SiO$_2$ edge is given in Fig. 25. The silicidation reaction was performed using RTP at 700°C for 30 s in N$_2$ ambient. A perfect self-alignment can be noticed. Unlike for the case of TiSi$_2$, no difference

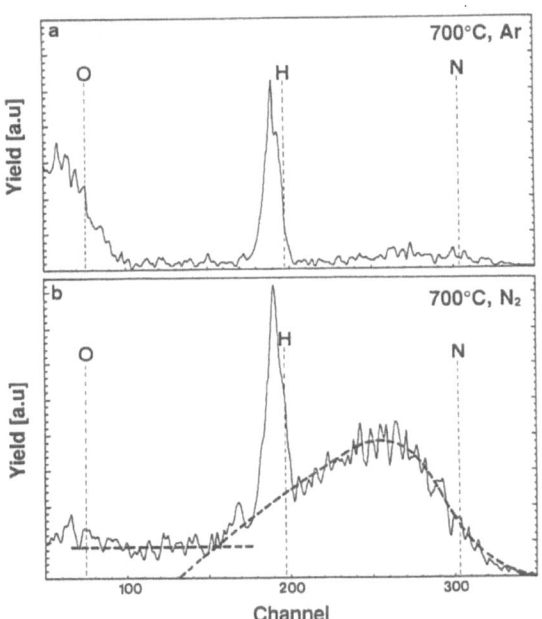

Fig. 24   *ERD spectra of samples consisting of 150 nm Ti on 100 nm SiO$_2$ after anneal at 700°C for 10 s in Ar (a) and N$_2$ (b) ambients.*

has been observed when using N$_2$ or Ar ambients. The absence of any lateral silicidation in the Co-Si system is often loosely ascribed to the fact that Co is the moving species during silicide formation. This, however, should be considered as an oversimplification. Moreover, Co as the moving species causes concern for silicide encroachment under the Si-SiO$_2$ step. Surprisingly, this was never observed in our studies.

A study has been performed [31], which indicates that under certain experimental conditions lateral silicide formation can indeed occur as is illustrated for example in Fig. 26. In this case the silicidation was performed at 750°C, starting from a layer with a composition corresponding to Co$_2$Si deposited by cosputtering. This picture clearly reveals that large amounts of Si have diffused over the oxide edge to react with Co$_2$Si leading to an enhanced Si consumption at the contact hole edge and resulting in void formation. EDAX microspot

Fig. 25 *Transmission electron micrograph of CoSi₂ formed selectively on Si with respect to SiO₂. 38 nm of Co was used to form the silicide at 700°C in N₂ ambient.*

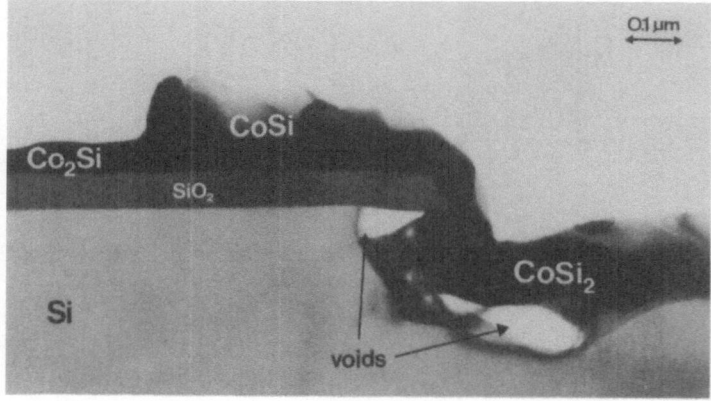

Fig. 26 *Cross-sectional TEM micrographs of CoSi₂ formed in a contact hole etched in SiO₂, starting from a cosputtered "CoSi₀.₅" layer (reaction at 750°C for 30 s in N₂).*

analysis was performed in order to determine the composition of the different phases on this sample. Three different compositions were obtained: Si/Co = 2.1 on Si, 0.7 on SiO₂ and 1.1 in the region of lateral silicidation. The somewhat higher Si/Co ratio obtained for the layer on SiO₂ (expected to be 0.5 as determined from RBS) can be explained by the fact that part of the

Si signal is originating from Si in SiO$_2$ due the straggling of electrons out of the thin Co$_2$Si layer. It is therefore safe to conclude that the three different silicide phases are present as labeled on the TEM picture.

In the study of ref. 31 the following model was presented to describe the absence of lateral silicidation for the case of the normal salicide process (see fig. 27). Initially, the Co layer reacts to form Co$_2$Si and CoSi. A reduced reaction near the SiO$_2$ edge prevents the CoSi from growing laterally over the edge at this stage of the process. At higher temperatures (600°-700°C) Co starts to ball up near the SiO$_2$ edge. The further reaction between these Co globules and the CoSi$_2$ layer is prevented by a diffusion barrier, which is thought to originate from an accumulation of (a) impurities in the Co layer, (b) O from the ambient and (c) O from the native oxide originally present between Co and the Si substrate. At higher temperatures or by the addition of H$_2$ to the ambient, this barrier fails to prevent the reaction and lateral overgrowth is observed.

3. TECHNOLOGICAL IMPLEMENTATION

Up to now, the various reactions in the salicide process were discussed from a fundamental point of view and independent of possible applications. We will now focus on the implementation of silicides in self-aligned silicidation technologies for MOS applications. The various issues are indicated on a schematical cross-section of an MOS transistor in Fig. 28:

1. *Influence of the native oxide on the Si on the silicidation reactions*
2. *Influence of dopants on the silicidation reactions*
3. *Thermal stability of the silicides*
4. *Chemical stability of the silicides (against wet and dry etching)*
5. *Interaction with Al-based metals*
6. *Consumption of Si during the silicidation reactions*
7. *Generation of stress during silicidation and formation of stress-induced defects*
8. *Techniques to form local interconnects.*

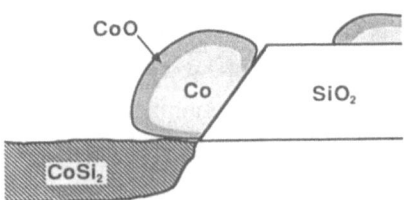

Fig. 27   *Schematic illustration of the model describing the absence of lateral silicidation for the case of the standard salicide process.*

3.1. Influence of the Si native oxide

Since Ti reduces SiO_2 at the silicidation temperature, the presence of a native oxide only results in an incubation period before the silicidation starts as was demonstrated by van Houtum et al. [32]. After this SiO_2 reduction the silicidation rate was not influenced by the initial (native) oxide thickness. However, different oxide thicknesses lead to different $TiSi_2/TiN$ ratios and thus to different final sheet resistances.

Since Co is not able to reduce SiO_2 under normal processing conditions, the presence of a native oxide on the silicon surface, before Co deposition, might strongly affect the silicidation reaction. A very smooth interface is, however, obtained when the wafers are given a diluted HF dip immediately before Co deposition. Thicker oxides ($\pm$ 2 nm), as obtained after a cleaning step without HF dip, can completely impede the silicidation reaction.

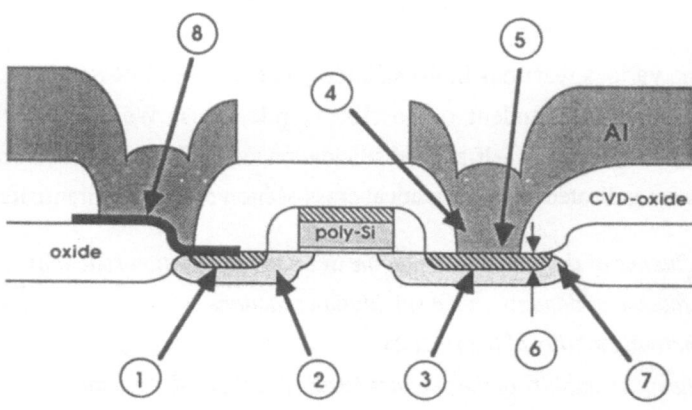

Fig. 28   *Schematic cross-section of an MOS transistor indicating the various issues which will be treated in this section.*

The observation that thin native oxides do not hinder the silicide formation although Co is not able to reduce SiO_2, may be explained by observing the initial phase of the reaction. A cross-sectional TEM micrograph of a Co/Si structure heat treated at 350°C for 30 s is shown in Fig. 29. The top layer consists of unreacted Co and some CoO. The bottom layer was identified as Co_2Si. Both layers are separated by a thin amorphous layer, which could not be identified using AES nor by any other technique. It is, however, likely that this layer originates from the native oxide, initially present between Si and Co. The presence of this layer strongly suggests that Co has diffused through it to react with the Si. The remnants of this oxide can thus be considered a marker, agreeing with Co being the main moving species during Co_2Si formation [33].

3.2. Influence of dopants on the silicidation

The interaction between dopants and silicide formation has two aspects. *Firstly*, the diffusion of dopants may be largely influenced by the simultaneous occurrence of a silicidation reaction. *Secondly*, the presence of dopants in the Si may also affect the silicidation reaction.

A retardation of $TiSi_2$ formation has been observed on heavily As doped regions, the exact mechanism still being somewhat unclear [34]. It results in a considerable increase of the final sheet resistance, due to the $TiN/TiSi_2$ competition. Although this was not observed for $CoSi_2$, the CoSi-to-$CoSi_2$ transition temperature was found to increase with As implant dose. This was tentatively attributed to the influence of the presence of As at the CoSi/Si interface on the nucleation of the $CoSi_2$ phase. While it results in a more pronounced interface roughness for high implantation doses, no roughness increase was observed for a practical dose of $2x10^{15}$ /cm^2.

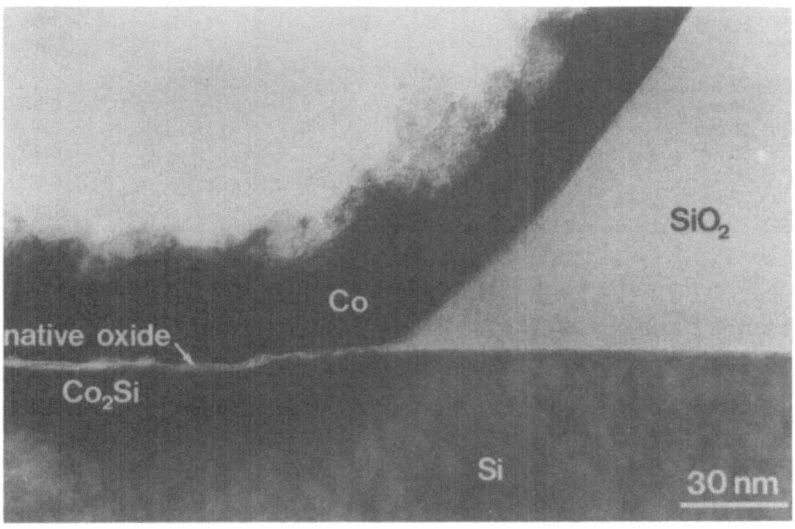

Fig. 29   *Cross-sectional TEM of the edge of a contact hole after silicidation at 350°C for 30 s. Note the thin amorphous layer (native oxide) between the Co_2Si layer and the unreacted Co.*

3.3. Thermal stability

One of the motives to choose for silicides in the search for interconnection materials for microelectronic device fabrication was their good thermal stability. A good thermal stability allows to flow doped glass layers deposited on top of the silicide. It is also required, when one wants to form shallow junctions by diffusion of dopants from doped silicide layers.

$TiSi_2$, belonging to the category of the refractory metal silicides, is generally considered to have a better thermal stability than $CoSi_2$, which belongs to the near noble metal silicides.

Although the respective melting points of $TiSi_2$ and $CoSi_2$ are 1540°C and 1326°C, stability problems have been observed for both silicides at temperatures above 900°C [35,36].

$TiSi_2$

$TiSi_2$ layers were subjected to various heat treatments and the variation in sheet resistance is shown in Table II. Figure 30 compares the top surface of the silicide and the silicide/Si

Table II

Sheet resistance of $TiSi_2$ layers after heat treatments from 800°C to 1200°C for 30 s or 30 min in N_2 or Ar ambient. Three silicide thicknesses were used (starting from 40, 60 and 100 nm Ti, yielding a silicide layer of ±50, 90 and 150 nm, respectively). Values are in Ω/sq.

	40 nm Ti Ar	40 nm Ti N_2	60 nm Ti capped (N_2)	100 nm Ti N_2
as formed	3.1	3.1	1.6	0.88
RTP				
900°C, 30 s	2.9	3.2	-	-
1000°C, 30 s	2.7	3.8	1.6	0.88
1100°C, 30 s	30	26	1.6	1.04
1200°C, 30 s	-	-	2.4	-
Conventional furnace				
800°C, 30 min	2.8	2.9	-	0.87
850°C, 30 min	-	3.3	1.7	
900°C, 30 min	5.5	7.1	1.7	0.90
950°C, 30 min	-	63	2.0	-
1000°C, 30 min	-	-	3.5	-

interface of a sample without additional heat treatment with samples which received an 1100°C RTP step for 30 s in N_2 or Ar or a 900°C anneal for 30 min in N_2. The morphology of the layers after the RTP treatment in N_2 or Ar is completely different. After the heat treatment in Ar an agglomeration of the silicide in relatively large islands can be observed (Fig. 30 b_1).

After removal of the silicide, wide ridges are left between the areas where the agglomerates were residing (Fig. 30 b_2). For the case of a N_2 anneal at the same temperature, this phenomenon is not observed (Fig. 30 c). The interface, however, roughens locally, which is due to the fact that at this high temperature $TiSi_2$ reacts with N_2 to form TiN. This leads to a very irregular structure. The formation of the TiN layer and the irregular structure with large amounts of Si, precipitated in the layer, are responsible for the observed increase of the sheet resistance. Although for the 900°C, 30 min sample (Fig. 30 d) the formation of ridges can be observed, similar as for the RTP case, the top surface appears smooth in this case. This is expected to be related to the formation of a thin oxide on the silicide surface (most probably during the loading of the wafer in the furnace tube).

When a $TiSi_2$ layer is capped with a deposited oxide before the heat treatment, the agglomeration of the silicide and the formation of Si ridges in between the $TiSi_2$ agglomerates is also observed after high temperature treatments as in the case for the uncapped samples annealed in Ar ambient. This is clearly illustrated by the cross-sectional TEM micrograph of Fig. 31. It indicates that these ridges consist of Si realigned epitaxially to the substrate.

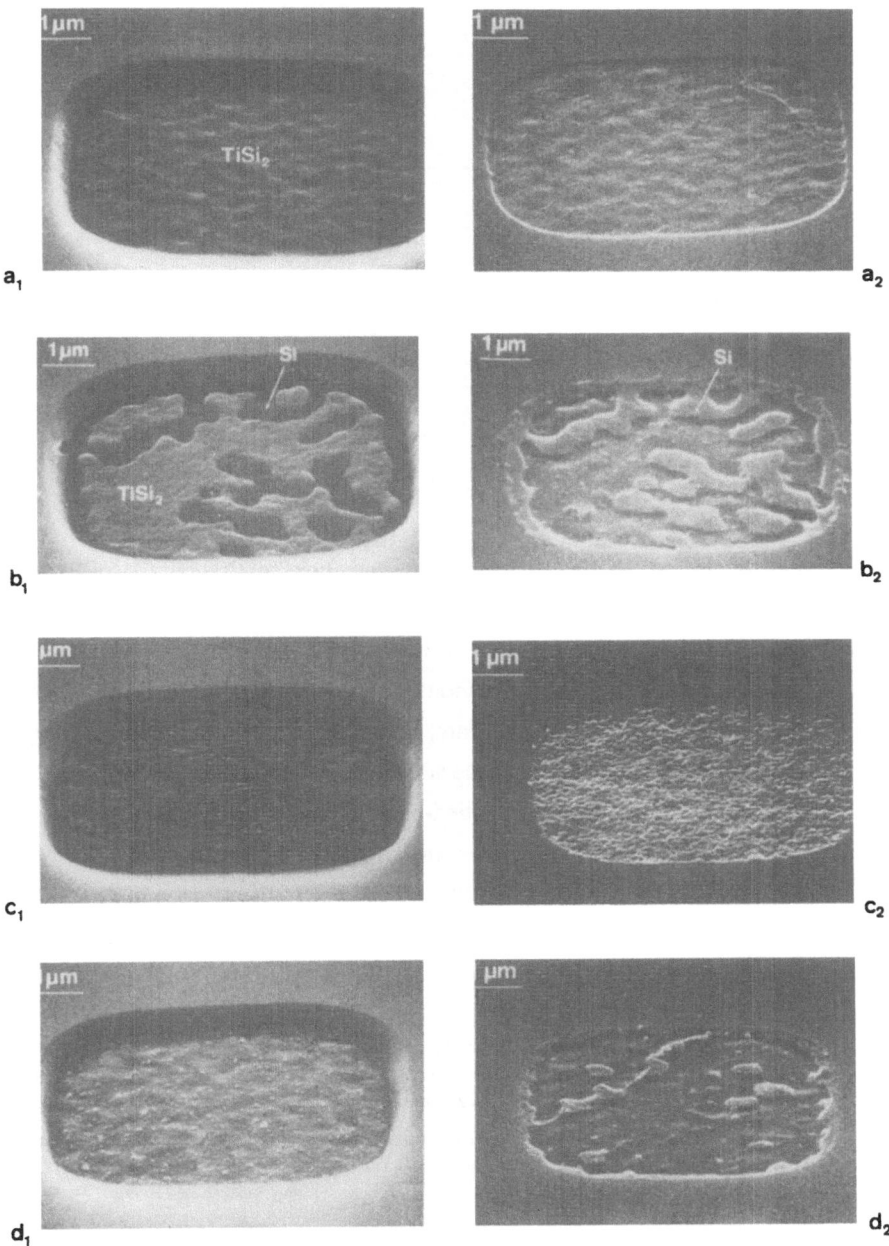

Fig. 30 Scanning electron micrographs of $TiSi_2$ formed in a contact hole etched in SiO_2 after various heat treatments (± 50 nm $TiSi_2$): (a_1) as formed; (b_1) after 1100 °C, 30 s in Ar; (c_1) after 1100 °C, 30 s in N_2; (d_1) after 900 °C, 30 min in N_2; (a_2), (b_2), (c_2) and (d_2) are SEMs of the $TiSi_2/Si$ interface of the same samples after removing the silicide and the oxide using an HF-solution.

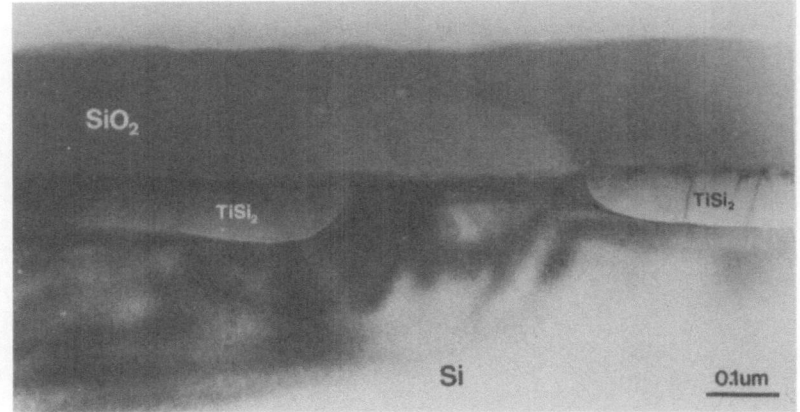

Fig. 31   *Cross-sectional TEM micrograph showing the agglomeration of 50 nm thick TiSi$_2$ layer after heat treatment at 1100 °C for 30 s in N$_2$. The silicide was covered with a 200 nm thick deposited oxide layer before the heat treatment. Note the formation of ridges between the silicide grains consisting of Si realigned epitaxially to the substrate.*

CoSi$_2$

Similar experiments are performed for CoSi$_2$. The sheet resistance of CoSi$_2$ layers after various heat treatments is given in Table III. Although no drastic sheet resistance increase is observed, unless for high temperature anneals in Ar ambient, the SEM micrographs of Fig. 32 reveal severe morphological changes. Previously no interaction has been observed between the N$_2$ ambient and Co or CoSi$_2$. It appears, however, that at high temperature (1000°C - 1100°C) the N$_2$ from the ambient retards the silicide degradation (compare Fig. 32c with 32b). In Ar ambient the thin CoSi$_2$ layer has balled-up completely into discrete islands. After heat treatment in N$_2$, the CoSi$_2$ agglomerates are still more or less connected. For the high temperature anneals using conventional furnace processing (for example at 950°C) the top surface remains smooth as was the case for TiSi$_2$.

Table III

Sheet resistance of CoSi$_2$ layers after heat treatments from 800 °C to 1100 °C for 30 s or 30 min in N$_2$ or Ar ambient. Two silicide thicknesses were used (starting from 20 nm and 40 nm Co, yielding a silicide layer of ±60 and 120 nm, respectively). Values are in Ω/sq.

	20 nm Co Ar	20 nm Co N$_2$	40 nm Co N$_2$
as formed	2.6	2.6	1.29
RTP			
900°C, 30 s		2.5	1.18
1000°C, 30 s	2.4	2.4	1.18
1100°C, 30 s	13	2.9	1.19
Conventional furnace			
800°C, 30 min		2.4	1.27
850°C, 30 min		2.4	1.23
900°C, 30 min	5.5	2.6	1.19
950°C, 30 min		2.7	1.26

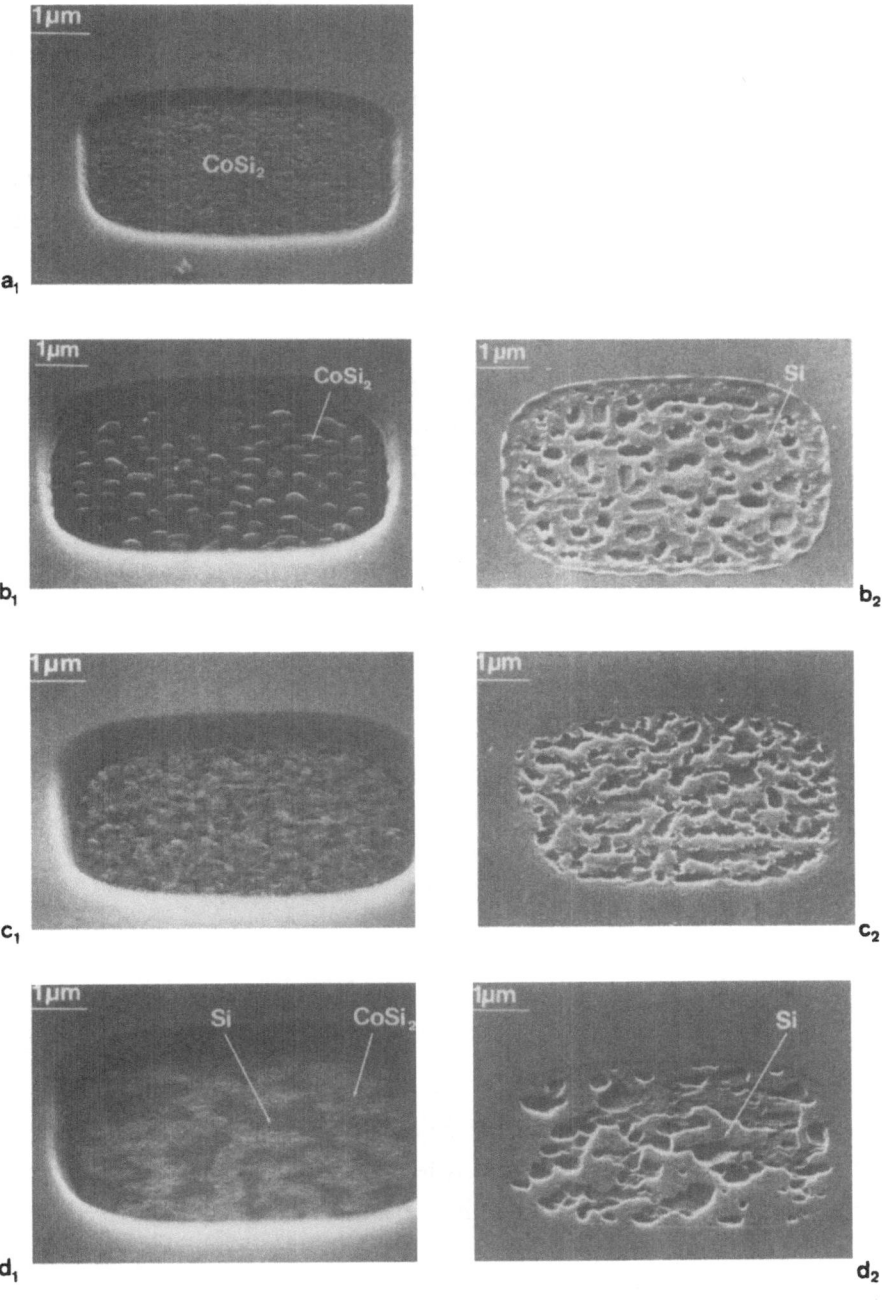

Scanning electron micrographs of $CoSi_2$ formed in a contact hole etched in SiO_2 after various heat treatments (± 60 nm $CoSi_2$): (a_1) as formed; (b_1) after 1100 °C, 30 s in Ar; (c_1) after 1100 °C, 30 s in N_2; (d_1) after 950 °C, 30 min in N_2; (b_2), (c_2) and (d_2) are SEMs of the $CoSi_2/Si$ interface of the same samples after removing the silicide and the oxide using an HF-solution.

Discussion

Thermodynamic and kinetic factors can be used to explain the silicide degradation. The driving force for silicide degradation is the reduction of surface and interface energy. The degradation is nucleated at the intersections of the grain boundaries with the top surface and the bottom interface, where the interface tensions will tend to equilibrate as is depicted in Fig. 33a ($2.\gamma_s \cos \beta/2 = \gamma_{gb}$). This phenomenon is known as "thermal grooving" [37-41]. It can be demonstrated that eventually the grooves will grow through the entire silicide layer and the grains will disconnect. These silicide islands will further agglomerate and the surface tensions

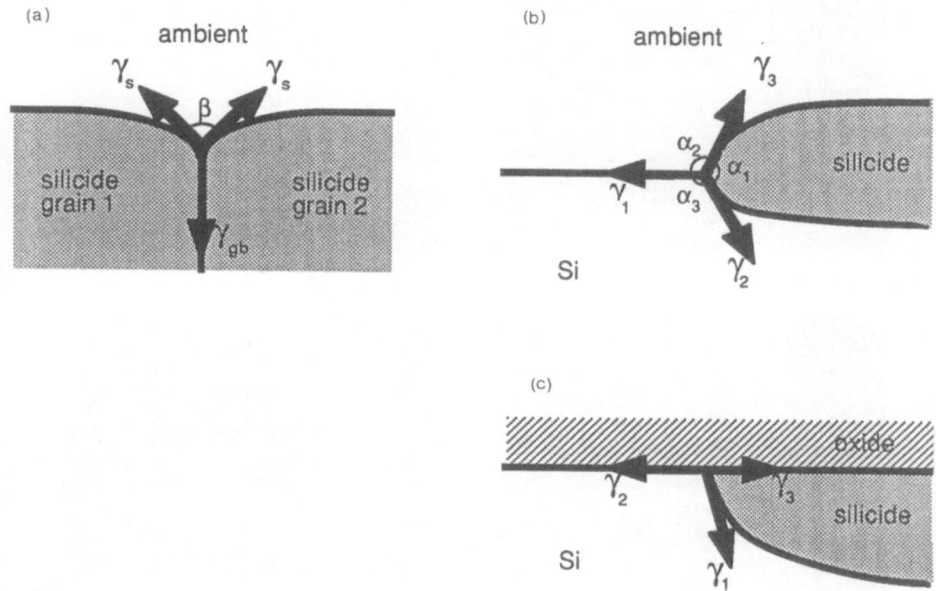

Fig. 33   *Equilibration of the surface tensions at the grain boundary or interface intersections: (a) thermal grooving, (b) and (c) agglomeration.*

will equilibrate at the edges as shown in Fig. 33b and c. The agglomeration is accompanied by the formation of Si ridges or mesas, which are believed to grow through a mechanism of solid phase epitaxy (SPE) with the silicide or its interface with Si as fast transport medium [42,43]. Apparently, when an oxide cap is present on the silicide, the surface migration is reduced. Hence, the surface remains flat. A similar degradation phenomenon does, however, occur at the bottom of the layer. Also, the use of N_2 ambient may influence the surface migration and consequently the surface morphology. Because the degradation is nucleated at the grain boundaries, the size of the final agglomerates is related to the grain size of the initial silicide film, which explains the smaller size of the agglomerates for $CoSi_2$ compared to $TiSi_2$. $TiSi_2$ was found to have a slightly better thermal stability than $CoSi_2$, which agrees with the SPE experiments carried out in other studies [44].

3.4. Chemical stability

In the fabrication sequence of semiconductor devices several chemicals are used for cleaning and etching purposes. $CoSi_2$ and $TiSi_2$, as well as most other commonly used silicides, are fairly resistant to the standard chemicals used for cleaning, such as the $H_2SO_4 : H_2O_2$ mixture and the RCA-cleaning solutions. $TiSi_2$, however, has the large disadvantage of being soluble in HF-containing solutions. In a buffered HF solution (10% HF), an etch rate of ≈ 200 nm/min and ≈ 5 nm/min can be measured for $TiSi_2$ and $CoSi_2$, respectively. This excludes the use of all HF based solutions, when they are likely to come in contact with $TiSi_2$.

A high resistance to dry etch processes is desired, particularly to fluorine based plasmas commonly used for oxide etching. In a salicide process, holes have to be etched in a deposited oxide layer on top of the silicide, allowing it to make contact to the Al interconnection pattern. Based on the comparison of sheet resistance measurements before and after dry etching with various overetch times, a selectivity (defined as the ratio of oxide etch rate to silicide etch rate) in the range of 50:1 to 200:1 was easily achieved for $CoSi_2$. For the case of $TiSi_2$ selectivities around 10:1 are typically measured. The difference can probably be attributed to a different volatility of the fluorides of Ti and Co.

The higher selectivity of (oxide) dry etch processes to $CoSi_2$ and the lower etch rates in HF solutions are considerable advantages of $CoSi_2$ for its use in a salicide process.

3.5. Interaction with Al

Since silicides have to come in contact with Al-based metallization schemes, the stability of the Si/silicide/aluminum contact should be investigated. This structure is commonly subjected to a $400°$-$500°C$ heat cycle at the end of the Si processing sequence. Several studies are available concerning the silicide/Al interaction [44-46, ...].

All silicides interact with Al in a similar way: the silicides decompose and the metal reacts with Al to form intermetallic compounds. In the case of $CoSi_2$, the entire layer is consumed after 30 min at $450°C$ and a large amount of intermetallics are observed in XRD, of which most can be attributed to Co_2Al_9 and some also to poly-Si. The latter results from a precipitation of Si in large islands which accompanies the compound formation as was also observed by van Gurp et al. [46] (see Fig. 34).

For $TiSi_2$ a similar compound formation has been observed, but it only starts around $550°C$ [45]. At lower temperatures, the $Al/TiSi_2/Si$ contact structure exhibits another failure mechanism. The silicide grain boundaries are permeable for an upwards diffusion of Si to the Al layer, satisfying the Si solid solubility in Al at the considered temperature and a downwards diffusion of Al filling the spikes under the silicide layer. Similar spiking of Al under $TiSi_2$ has been observed by Ting and Wittmer [45]. When pure Al is used, spiking along silicide grain boundaries is already observed at 400 or $450°C$. No contact spiking was observed for Al(1% Si) layers, deposited on $TiSi_2$ and heat treated at $450°C$. A downwards diffusion of Al to the $TiSi_2/Si$ interface has also been observed (even for Al/1%Si). The presence of Al at this interface has been shown to affect the contact resistance [47].

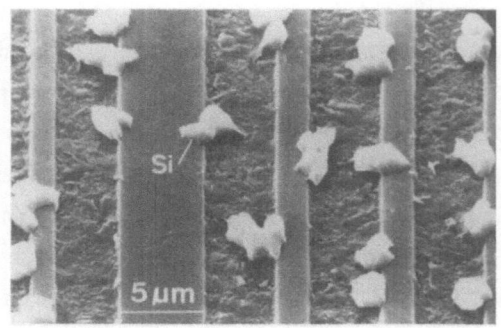

Fig. 34 *SEM micrograph of a sample on which Al (without Si) was sintered at 450°C for 30 min on top of CoSi$_2$ formed in lines. The Al layer was removed before SEM inspection. Large Si precipitates are clearly revealed.*

Therefore, to avoid the possible interaction between the silicide and the Al metallization, it is a good practice to use a diffusion barrier (TiN, TiW, W, ...).

3.6. Silicon consumption

In order to achieve a sheet resistance as low as possible, the silicide thickness has to be maximized. With increasing silicide thickness diode integrity problems may start to occur, because of local junction penetration due to the Si consumption during the silicidation.

Co and Ti layers of various thicknesses were deposited on wafers with patterned oxides and with large uniform Si areas. The silicide was formed selectively on the Si regions. After measuring the sheet resistance on the large Si regions, the Si consumption was measured using two techniques. On some samples RBS measurements were performed, which allow determining the areal density of Si atoms used in the silicide. Through the density of Si (5×10^{22} at/cm^3) one can calculate the corresponding Si thickness. From the samples with the silicide formed in windows etched in oxide, the silicide and oxide were removed by an HF containing solution. The Si consumption was then obtained through measurement of the silicon step using a surface profilometer. In Fig. 35 all the data are summarized for CoSi$_2$ and TiSi$_2$ and compared as a function of the sheet resistance of the silicide. Whereas the Si consumption is often quoted as a major disadvantage for CoSi$_2$, the measurements clearly indicate that the Si consumption *for equal silicide sheet resistance* is only marginally larger for Co. Other factors are therefore equally or even more important when comparing both silicides.

3.7. Stress and defect generation

3.7.1. Stress generation during CoSi$_2$ and TiSi$_2$ formation

It is known that during silicide formation high stress levels can be generated [48-50]. The stress occurring during silicide formation may have very important technological implications,

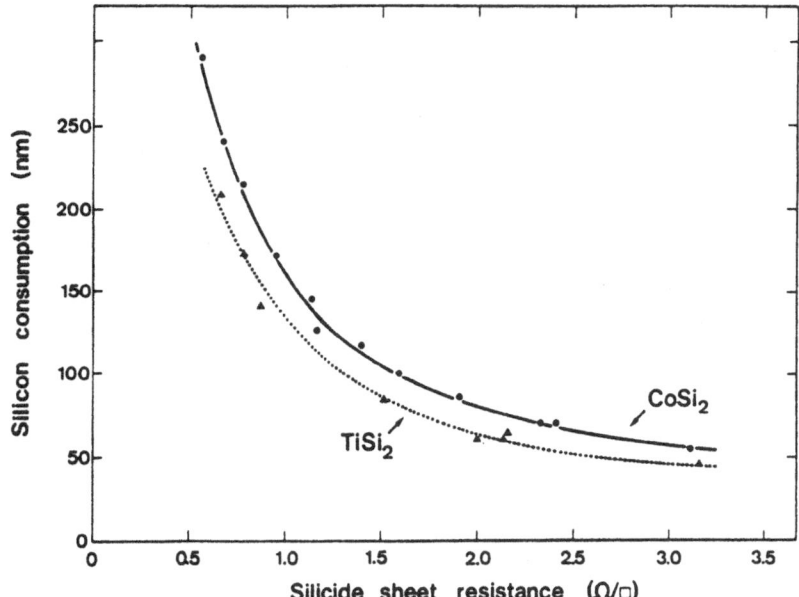

Fig. 35   *Silicon consumption during silicide reaction versus silicide sheet resistance for CoSi₂ and TiSi₂.*

firstly, because the magnitude of the stress is very large and secondly, because the silicide layers are grown in very close proximity to the active areas of the devices. The generation of stress in $CoSi_2$ and $TiSi_2$ is studied by measurements at room temperature and during silicide formation (*in situ* measurements) [51].

Measurement technique

Stress measurements were performed using an X-ray topography (Lang) camera. The wafer curvature was determined from the angular adjustment required to obtain Bragg reflection at two positions of the sample for transmission diffraction of the (110) Si planes for (001) oriented Si. In order to perform stress measurements at elevated temperatures, a small oven was mounted on the X-ray system. The maximum temperature in this oven was limited to 600°C.

Results

Wafer curvature measurements were performed *in situ* during the reaction of Co and Ti layers with the Si substrate. Figure 36 shows the results for the reaction of 65 nm Co with Si. Since the metal layer transforms gradually into a silicide layer with increasing thickness during the experiment, it is impossible to calculate the stress from the measured curvature (unknown film thickness). Therefore the y-axis is labeled with the product of stress and film thickness which is proportional to the inverse of the wafer curvature and which is a measure for the interface force exerted on the Si substrate.

Figure 36 shows that the Co layer, which was deposited by DC magnetron sputtering starts off with a small tensile stress, which reduces with increasing temperature. Between 400°C and 550°C some intrinsic (growth) stress is measured. At 600°C almost no stress is detected. During the cooling tensile stress is built up linearly, the latter clearly being a result of the difference in thermal expansion of film and substrate.

Similar results are obtained for TiSi$_2$. In this case compressive stress is measured starting from 550°C, corresponding with the onset of TiSi$_2$ formation. During cooling tensile thermal stress is built up.

Since the temperature range in the high temperature stress measurement system was limited to 600°C, information about the stress in the silicide layers after higher temperature (RTP) treatments is acquired through room temperature measurements. Co layers, 65 nm thick, were subjected to RTP cycles ranging from 300°C to 1100°C for 30 s. The results are represented in Fig. 37. In the temperature range below 700°C the product of stress and film thickness

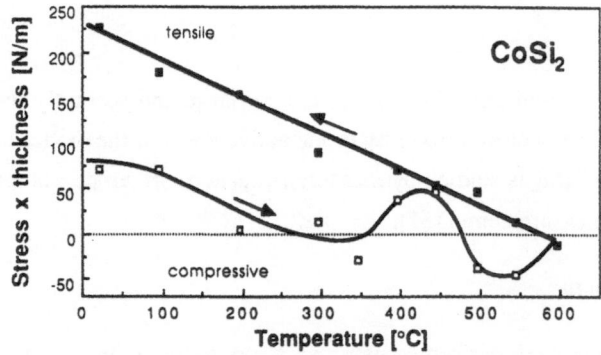

Fig. 36 *In situ stress measurement of a 65 nm Co layer on Si. The wafer curvature was measured after a 15 min heating period at each temperature. The sample was held at 600°C for 2 hr before the cooling was started.*

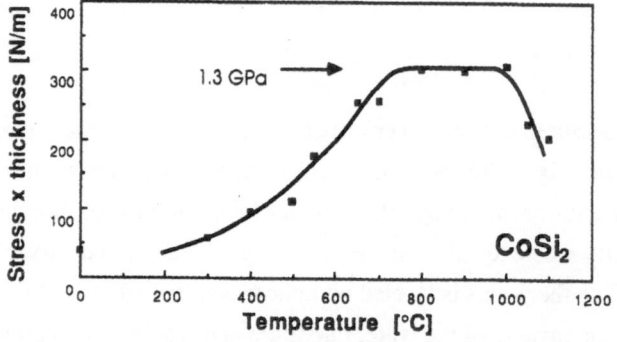

Fig. 37 *Room temperature stress in films resulting from the reaction of 65 nm Co layers with Si during RTP cycles for 30 s in the temperature range from 300°C to 1100°C.*

increases gradually for the case of Co (corresponding with the increasing silicide thickness). At 700°C the layer is completely transformed into $CoSi_2$. For reactions between 750°C and 1000°C the stress level remains remarkably constant. This level corresponds to a stress of *1.3 GPa*. This flat stress level confirms that at these high temperatures all stress is relieved and the room temperature stress consists of thermal stress built up during cooling. Even during the extremely fast cooling from 1000°C, the stress only starts to build up below 700°C. At high temperatures (1050°C and 1100°C) the measured wafer curvatures are lower. This is due to the limited thermal stability of the silicide resulting in local thinning of the film (see section 3.3).

Similar results are obtained for the case of $TiSi_2$. A more or less flat stress level, corresponding to *2.3 GPa*, is obtained between 800°C and 1000°C. For $TiSi_2$, this plateau starts at a higher temperature, compared to $CoSi_2$, which agrees with the higher silicide formation temperature for $TiSi_2$.

To understand the build-up of thermal stress and to have an idea of the slope of the stress curve, the stress of silicide layers which are formed by RTP was measured at different temperatures using the high temperature stress measurement system. As can be seen from Fig. 38, the stress follows a linear behavior during heating, typical for thermal stress, as pointed out earlier. It is believed that during cooling after the silicide formation using RTP, the silicide layer has followed the stress curve obtained in this experiment during heating. During the slow cooling in this experiment no stress was built up in $CoSi_2$ down to 500°C. Starting from this temperature, thermal stress is built up during further cooling with the same slope as observed during the heating cycle of this experiment. Again, similar measurements are performed for $TiSi_2$ (also shown in Fig. 38). In this case no stress relaxation was observed during cooling below 600°C. The steeper slope for the case of $TiSi_2$ reflects a higher thermal expansion coefficient or a higher biaxial elastic modulus.

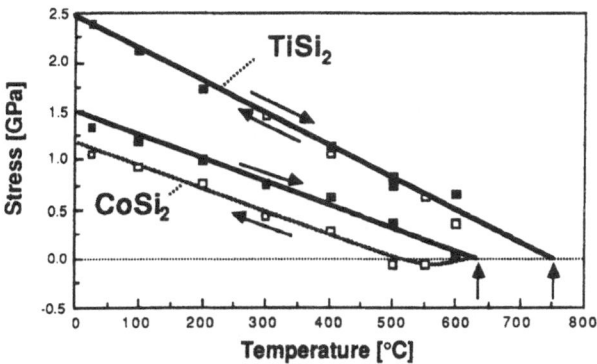

Fig. 38 *High temperature stress measurement of a $CoSi_2$ layer and a $TiSi_2$ layer formed by RTP at 850°C for 30 s.*

By extrapolating the curves of Fig. 38 to zero stress level one can obtain the temperature (stress relaxation temperature) to which the silicide is able to relieve its stress during the fast cooling after silicide formation (by RTP). The curves clearly indicate that the $TiSi_2$ structure freezes in at higher temperatures than the $CoSi_2$ structure (750°C for $TiSi_2$ and 630°C for $CoSi_2$). Also the influence of the cooling rate on the stress relaxation temperature is clearly demonstrated. For slow cooling the stress is relieved down to 500°C for $CoSi_2$, whereas for RTP stress is built up starting from 600°C. This is one of the few areas where RTP shows a disadvantage as a technique to perform the silicidation reactions.

Discussion

The origin of the observed intrinsic stress is not fully understood. It is assumed to result from volume reductions or expansions during the reactions [49]. The experimental results, however, clearly indicate that at sufficiently high temperature the intrinsic stress as well as the thermal stress is relieved. The experiments of Fig. 38 illustrate the influence of the cooling rate on this stress relaxation. The room temperature stress of properly formed $CoSi_2$ and $TiSi_2$, therefore, consists of thermal stress built up below the relaxation temperature only. The larger stress for $TiSi_2$ is mainly due to its higher thermal expansion coefficient and/or biaxial elastic modulus *and* due to the higher stress relaxation temperature.

3.7.2. Silicide-edge induced dislocation generation

The interface stresses of thin films are concentrated at the film edges. At these locations the yield strength of Si can be exceeded, resulting in the generation of extended lattice defects. A detailed theoretical study of dislocation generation was performed recently by Vanhellemont et al. [52] and applied for silicon nitride films. The stress levels observed during silicide formation are of the same order of magnitude as those induced by nitride films. Although silicide-edge induced defect generation received only limited attention in the literature, it will be demonstrated here that it is a potential problem for self-aligned silicidation technologies.

Experimental

Several experiments were performed to investigate defect formation along silicide edges. The wafers used in this study are (001) oriented Czochralski Si wafers. $TiSi_2$ and $CoSi_2$ were grown using rapid thermal processing (RTP) up to various thicknesses in the temperature range of 600°C to 900°C in windows etched in SiO_2 (the SiO_2 was grown at 975°C in steam ambient). Optical and scanning electron microscopy, combined with preferential etching using Secco etchant, and High Voltage Transmission Electron Microscopy (HVEM) were used to examine the defect formation. For plan view HVEM investigation the silicide and oxide layers were removed by wet etching using HF based solutions.

Results

Figure 39 shows plan-view and cross-section HVEM micrographs of defects formed along $TiSi_2$ lines. The silicide film edges are oriented along a <110> direction. The edges of the active

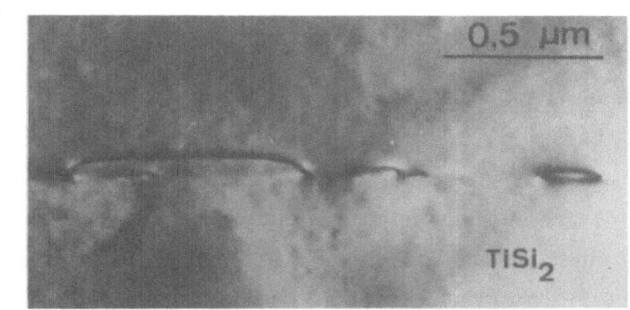

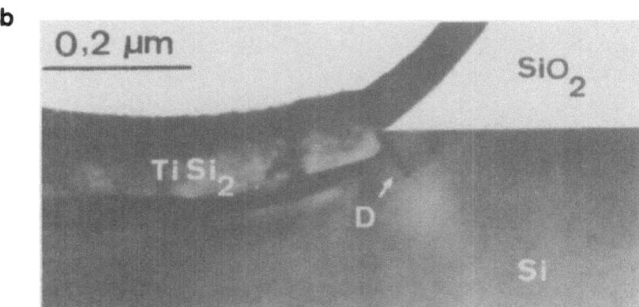

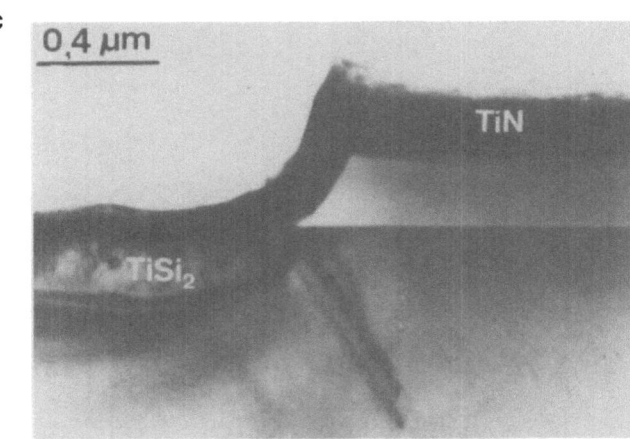

Fig. 39 *Plan-view (a) and cross-section (b) HVEM micrographs of dislocation half-loops nucleated along 100 nm TiSi$_2$ lines (silicidation at 700°C during 30 s). Thicker TiSi$_2$ layers (200 nm) result in a higher density of larger defects (c). (The silicide and oxide layers were removed using wet etching.)*

areas in most designs are aligned along this direction. Figure 39a reveals the presence of dislocation half-loops along a 100-nm-thick TiSi$_2$ line formed at 700°C during 30 s in N$_2$ ambient. The dislocations (D) are present in a {111} plane close to the <110> oriented edge (Fig. 39b). The dislocation segments parallel to the <110> oriented edge are the well-known 60°-dislocations as also observed along Si$_3$N$_4$ lines after local oxidation [52]. Increasing the initial Ti thickness and/or the reaction temperature, more and larger dislocation half-loops are formed (Fig. 39c). The defects were found to be statistically distributed over all silicide edges as determined by optical and scanning electron microscopy combined with Secco etching.

For CoSi$_2$ similar experiments with varying silicide thickness and silicidation temperature are performed. For silicide thicknesses up to 300 nm no defects are observed. A very low density of small defects is observed in the sample with 500 nm of silicide as is illustrated by the HVEM micrograph of Fig. 40a for lines with a similar spacing as in Fig. 39. For very narrow lines and spacings, a higher density of defects is observed (Fig. 40b), due to an overall increase of the stress field. A similar influence of the spacing between silicide lines on defect formation was also observed for TiSi$_2$. It should be emphasized here that the latter CoSi$_2$ layers are unrealistically thick for application in a salicide technology.

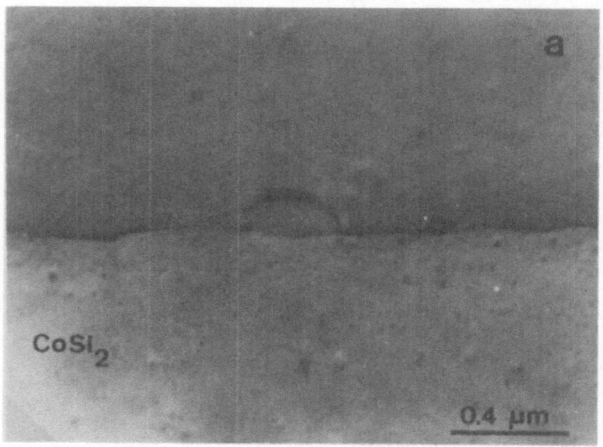

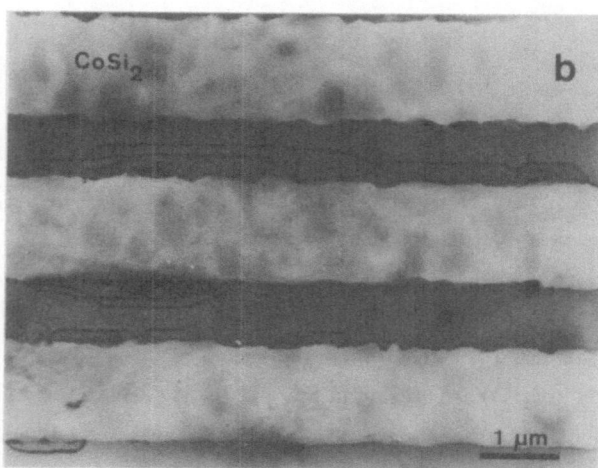

Fig. 40 Plan view HVEM micrographs of dislocation half-loops along CoSi$_2$ lines formed at 800°C during 30 s starting from 160 nm Co (yielding a silicide thickness of ±0.5 μm); in (a) a region with wide lines and spacings is shown (as in Fig. 39 and in (b) a region with fine lines and spacings is shown. (The silicide and oxide layers were removed using wet etching.)

HVEM two-beam diffraction-contrast analysis allows to determine the sign of the stress field responsible for the formation of the observed defects [53]. For all HVEM observations made in this study the stress field was found to be tensile; it is therefore believed that the defects are generated during cooling after silicide formation.

It is remarkable that no defects are observed along $CoSi_2$ edges for thicknesses up to 300 nm, although these thicknesses are several times larger than the $TiSi_2$ thickness of 100 nm from which defects start to form. From the above experiments the room temperature stress was determined to be 1.3 and 2.3 GPa for $CoSi_2$ and $TiSi_2$, respectively. At room temperature the interface force (silicide stress x thickness) is therefore considerably larger for 300 nm $CoSi_2$ (390 N/m) than for 100 nm $TiSi_2$ (230 N/m). As plastic deformation of silicon by dislocation generation and multiplication occurs at temperatures above 500°C, one should compare the interface forces at these higher temperatures. For the samples considered in this study, the interface force F is represented in Fig. 41 as a function of the temperature, as it occurs during cooling. These curves are determined from the *in situ* stress measurements presented above. The curves corresponding with the samples in which defects were observed are plotted as dotted lines. An interesting point to note is the difference between the temperature from which thermal stress builds up for TiS_2 and $CoSi_2$ (750°C compared to 630°C). Defect nucleation is closely related to the critical glide force F_c [52] required for dislocation generation. Below 600°C F_c increases drastically with decreasing temperature. Although some information is available in the literature on the yield force of Si [54] and on the critical interface force for defect nucleation at nitride edges [55], only qualitative information can be transferred to our study, because of different experimental conditions. Therefore, superimposed on Fig. 41 is a qualitative curve of the yield force, required for defect nucleation, versus temperature. This curve divides the plot in two areas. If the force along a silicide edge exceeds this curve during cooling after a heat treatment, defects may be formed. Because of the lower stress relaxation temperature for $CoSi_2$, much higher stress levels are allowed at room temperature without plastic deformation. The observations of defect generation can thus be correlated with the *in situ* stress measurements for $TiSi_2$ and $CoSi_2$. It must be remarked, however, that the critical film thickness resulting in defect nucleation depends strongly on the width and the spacing of the silicide lines as is illustrated in Fig. 40.

The higher probability of dislocation generation in the case of $TiSi_2$ may have a severe impact on the use of relatively thick $TiSi_2$ layers ($\geq$ 100 nm) for salicide applications.

3.8. Local interconnection technologies

Recently, local interconnection technologies in combination with the salicide process have been proposed in the literature [56,57]. These processes basically realize an extension of the source or drain region over SiO_2 (also called "strap"). This technology allows the designer to

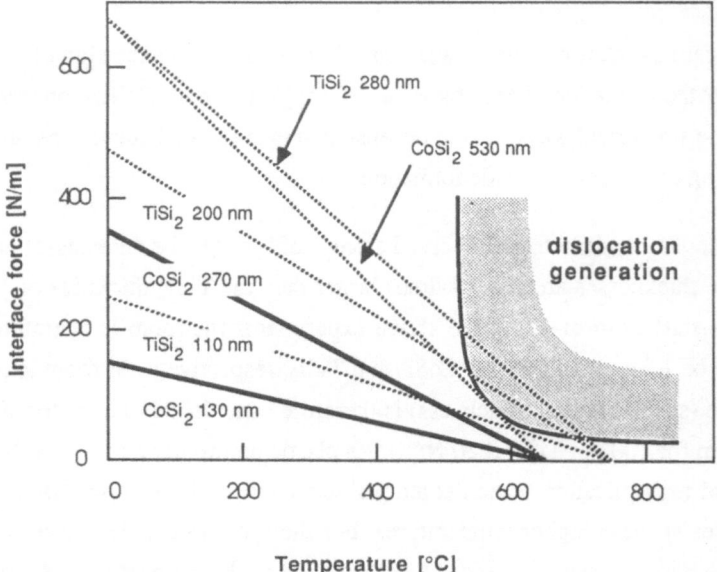

Fig. 41 *Force at the silicide edge (silicide stress x thickness) versus temperature as occurs during cooling after silicide formation for the various samples used in this study. The dotted lines correspond with the samples in which defects are observed. A qualitative curve of the yield force required for dislocation generation is also represented.*

contact source and drain regions partly or completely on top of SiO_2. In this case a higher packing density is obtained due to a drastic reduction of the source and drain area.

Several schemes have been presented to implement local interconnections (Fig. 42). In a first scheme (Fig. 42b$_1$ and c$_1$) Wong et al. [56] have sputtered an amorphous Si (α-Si) layer on top of the Ti layer used for the salicide process. Prior to silicide formation the α-Si is patterned. $TiSi_2$ is then formed by a reaction between the Ti layer and the Si from both the substrate or poly-Si and the sputtered layer on these locations where the latter layer was not removed. The α-Si layer thickness is adjusted in such a way that $TiSi_2$ is obtained on top of SiO_2 from the reaction of Ti with α-Si.

A second local interconnect technology was presented by Tang et al. [57]. It was demonstrated that Ti layers on SiO_2, when heat-treated in an N_2 ambient, are partly converted to TiN. A sheet resistance of 10 to 20 Ω/sq can be achieved after a 900°C reaction step for the metal thicknesses commonly used in a salicide process (50-100 nm). Although this sheet resistance is not as low as what can be obtained using a silicide layer, it is acceptable for the application as local interconnect. To achieve this, the salicide process is modified in the following way (Fig. 42b$_2$ and c$_2$): the selective etch of TiN between the two heat treatments of the salicide process is preceded by a photolithographic step which allows one to remove the TiN using a combination of a dry and wet etching step, except on those areas where local interconnects are desired.

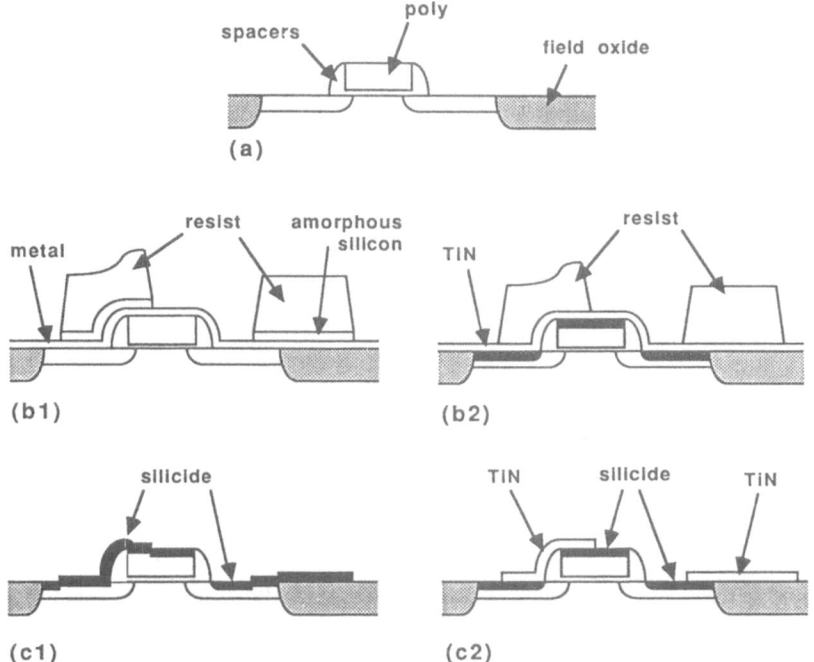

Fig. 42 *Local interconnection technologies based on the deposition of an α-Si layer (1) and on the patterning of TiN local interconnects (2).*

Due to the agglomeration of Co on SiO_2 a technique similar to the use of TiN on SiO_2 which was proposed in combination with the $TiSi_2$ process, is excluded for $CoSi_2$. The technique based on the deposition of an α-Si layer on top of Co was examined [58]. The presence of some small voids, detected at the $CoSi_2/SiO_2$ interface, may hinder the applicability of this technology. Local interconnection techniques can, however, also be based on the use of a deposited conductor like TiW or TiN. After carrying out the salicide process (with either $TiSi_2$ or $CoSi_2$) this conductor can be deposited and patterned using conventional techniques. Promising results were obtained using TiW(N) on $CoSi_2$ [58].

4. DEVICE IMPLEMENTATION OF TiSi₂ AND CoSi₂

4.1. Introduction

In the previous two sections we have discussed the materials aspects related to the implementation of $TiSi_2$ and $CoSi_2$ in a self-aligned silicide technology. The following section will focus on the influence of these silicides on the electrical performance of the devices.

The influence of silicidation on shallow junction integrity will be discussed. This is one of the critical issues in a salicide process, since during the silicidation reaction part of the Si from the shallow source and drain junctions is consumed. In order to avoid the penetration of the shallow junction by the silicide consuming some silicon, a technology was proposed in which shallow junctions are formed by diffusion of dopants from the silicide. In this work CoSi$_2$ and TiSi$_2$ will be compared as a source of both As and B diffusion.

Finally, some results will be shown concerning incorporation of silicides in MOS transistors.

4.2. Formation of shallow silicides junctions

4.2.1. Silicidation of pre-formed junctions

Dopant profiles

In order to study the effect of *CoSi$_2$* formation on dopant profiles, about 55 nm of CoSi$_2$ was grown on Si substrates implanted with As and B to various depths. The implantations were annealed prior to metal deposition. Figure 43a compares SIMS measurements of As with and without silicidation. The profiles are shifted in such a way that the depth scale represents the distance from the original Si surface (only dopants in the Si are shown). In this case 5x10^{15} As/cm^2 was implanted with an energy of 150 keV through an oxide of 25 nm and annealed at 950°C for 30 min. Analogous measurements were carried out for B dopant profiles (Fig. 43b). Boron was implanted to a dose of 2x10^{15} /cm^2 at 20 keV through 25 nm SiO$_2$ and annealed at 1100°C for 10 s (RTP) in N$_2$. The silicide was formed at 700°C during 30 s in both cases. According to SIMS, no appreciable impurity movement is observed in the Si under the silicide. Dopants residing in the Si which is consumed during silicide formation, either accumulate at the silicide-metal interface or leave the sample by evaporation.

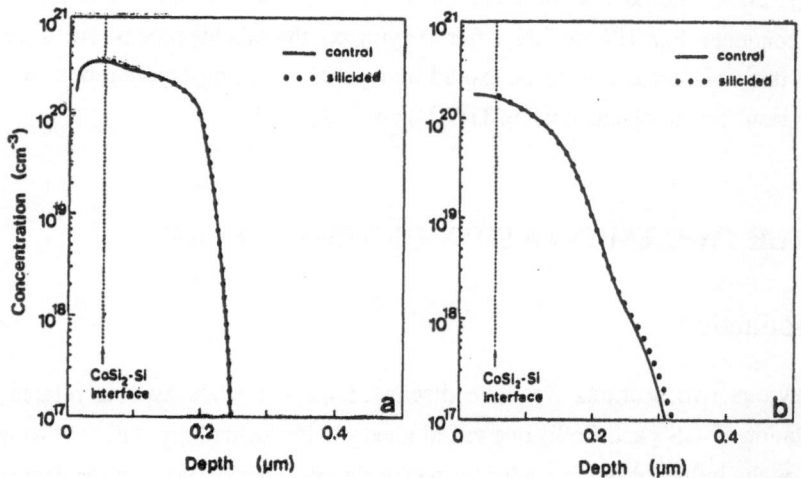

Fig. 43   SIMS measurement of the As (a) and B (b) profile in the Si for a silicided and a non-silicided sample. The profiles are shifted in such a way that the depth scale represents the distance from the original Si surface. 20 nm of Co was used to form the silicide at 700 °C.

Similar results were found for *TiSi₂*. It was found that the As and B profiles in the Si remain essentially unchanged during silicidation reactions up to 900°C [59].

Diode integrity

Since the silicon from the source and drain regions is partially converted to silicide, the silicide-Si interface can come within close vicinity of the junction space charge layer, especially since the silicide/silicon interface is not atomically flat. Therefore junction depth and silicide thickness have to be carefully optimized to obtain the lowest sheet resistance, without degrading junction performance. The junction depths quoted below are measured with respect to the original Si surface (before silicidation), which is about the same as the gate oxide level in a MOSFET.

For the experiment of Table IV, all diodes have an equal junction depth (0.25 μm), but the Co thickness varies from 20 to 55 nm, leading to a silicon consumption of 55 to 170 nm. It can be seen that the yield is drastically reduced when the Co thickness is increased to 55 nm (which is an extreme case).

Table IV

Mean and standard deviation of leakage current density measured at 5 V reverse bias (100 devices measured) for arsenic diodes with varying silicide thicknesses. Junction depth before silicidation was 0.25 μm (diode area: 4.4x10⁻³ cm²)

As junction depth: 0.25 μm							
Co	SI cons.	SILICIDED DEVICES			CONTROL DEVICES		
thickness [μm]	[nm]	mean [nA/cm²]	st.dev. [nA/cm²]	%good devices	mean [nA/cm²]	st.dev. [nA/cm²]	%good devices
20	55	1.6	0.3	99	1.3	0.1	98
30	85	1.5	0.1	98	1.2	0.3	100
38	105	3.4	1.2	89	2.9	0.8	96
55	170	2.7	1.8	52	1.2	0.4	100

I-V curves were recorded at various temperatures for a diode of this wafer with a high leakage current. Figure 44 shows Arrhenius plots of the reverse current. Whereas a diode with a low leakage current exhibits an activation energy of ≈ 1.1 eV for the entire temperature range (on a *log* I_r/T^3 plot) characteristic for generation current, highly leaking diodes typically show a deviation from the straight line for lower temperatures [60]. On a *log* I_r/T^2 plot the lower part of the curve corresponds with an activation energy of 0.44 eV, which agrees fairly well with the barrier height of CoSi₂ to p-type Si. This strongly suggests that the failure mechanism results from a parallel circuit of the n⁺-p junction and small CoSi₂/p-Si Schottky barriers, due to local penetrations of the junction by the silicide. This is schematically illustrated in Fig. 45.

97

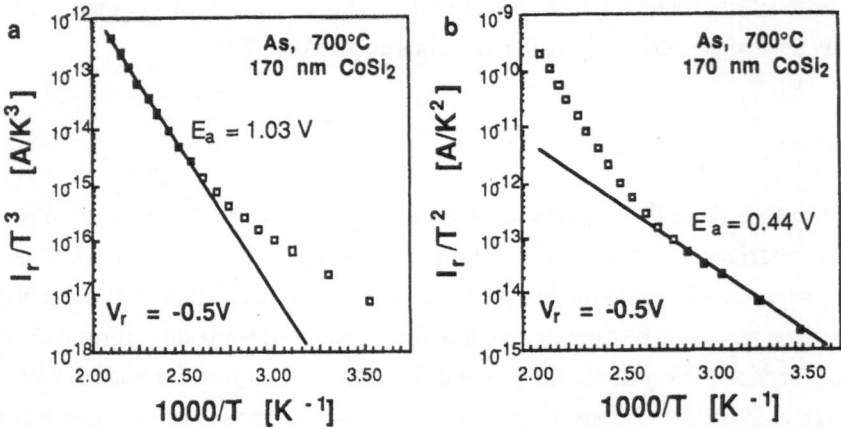

Fig. 44   *Arrhenius plots of the leakage current at 0.5 V reverse bias of a diode with a thick silicide (170 nm of CoSi₂ on a 0.25 μm deep As diode). (a) log I_r/T^3 and (b) log I_r/T^2 versus 1000/T plots.*

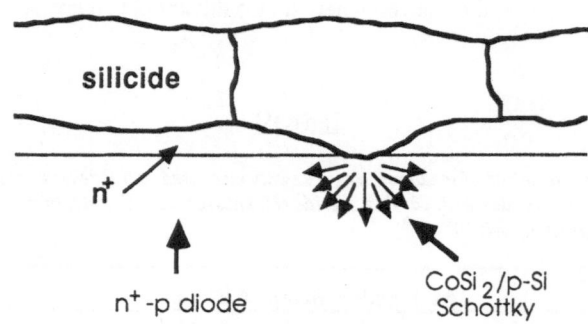

Fig. 45   *Schematic representation of the leakage mechanism.*

From a more detailed investigation it can be concluded that highly yielding diodes with leakage current densities as low as 1 nA/cm² can easily obtained for both CoSi₂ and TiSi₂, provided a thin buffer layer is allowed between the silicide/Si-interface and the metallurgical junction.

Contact resistance

The contact resistance of TiSi₂ and CoSi₂ was measured. For that purpose, cross-bridge Kelvin resistor structures [61-63] were fabricated. To measure the interfacial contact resistance between the silicide and the highly doped n⁺ and p⁺ regions the self-aligned silicide process was applied after contact hole opening as is shown in Fig. 46. In this way the silicide/Si and the metal/silicide contact resisitances are measured in series. When the entire active area (source/drain area) is silicided, only the silicide/metal contact resistance is measured. In this way contact resistivities in the range of 10^{-8} Ω.cm² are obtained for well cleaned silicide/metal contacts. The contribution of the silicide-metal contact to the total resistance can generally be neglected.

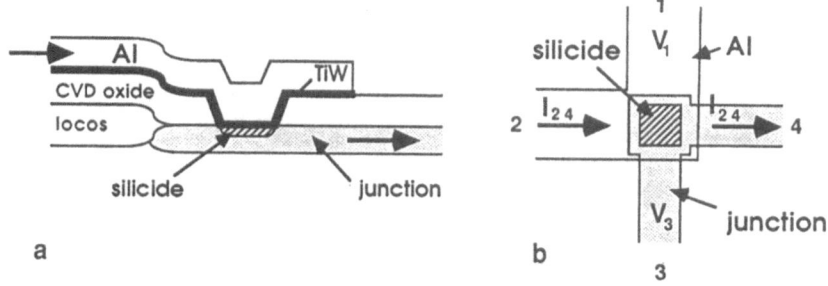

Fig. 46   *Four terminal Kelvin structure to measure the interfacial contact resistance: (a) cross-section and (b) top view.*

Figure 47 compares the contact resistivities for $CoSi_2$, $TiSi_2$ and Al/Si. $5x10^{15}$ As/cm² at 100keV and $3x10^{15}$ B/cm² at 50 keV were used to form the highly doped regions. The As implants were annealed at 975°C for 20 min and at 950°C for 30 min. The B implants received only the second annealing step. The surface concentrations associated with these implantations are about $1.5x10^{20}$ cm⁻³ and $6x10^{19}$ cm⁻³ for As and B, respectively. After contact hole opening, $CoSi_2$ was formed in an RTP step at 700°C on a first split of the batch and $TiSi_2$ was formed on a second split using the two-reaction step process at 730°C and 850°C for 30 s. The metal thicknesses were chosen in such a way as to result in a similar Si consumption (± 60 nm) for $TiSi_2$ and $CoSi_2$. After a sputter etch to remove any possible oxide on top of the silicide, 100 nm of TiW was deposited as a barrier material to avoid the interaction between the silicide and the Al/Si layer, which was subsequently deposited and patterned.

Ideally ρ_c should be independent of the contact window size and the curves should be straight horizontal lines. However, it is widely recognized now that the bending observed in Fig. 47 is due to lateral current crowding effects, which are influenced by (a) the ratio of the sheet resistance of the diffusion under the contact and next to the contact, (b) the contact window size, (c) the value of ρ_c and (d) the distance between the contact edge and the diffusion region edge [64,65].

$CoSi_2$ and $TiSi_2$ show very low specific contact resistance values. For $CoSi_2$ a value of $1.5x10^{-7}$ Ω.cm² is obtained for both n⁺ and p⁺ Si measured on 1 µm² contact holes (this contact window size is representative for the considered technologies). For $TiSi_2$ a slightly higher value is measured on p⁺ Si ($3x10^{-7}$ Ω.cm²).

The influence of the doping concentration on the contact resistivity is shown in Fig. 48. As was implanted at 130 keV and B at 20 keV with a dose ranging from $5x10^{14}$ cm⁻² to $1x10^{16}$ cm⁻². For all measurements the contact resistivity strongly decreases with increasing doping level. It is known that for the doping range where current across the contact is dominated by tunneling, the contact resistivity changes exponentially with the inverse of the square of the interface doping concentration [66].

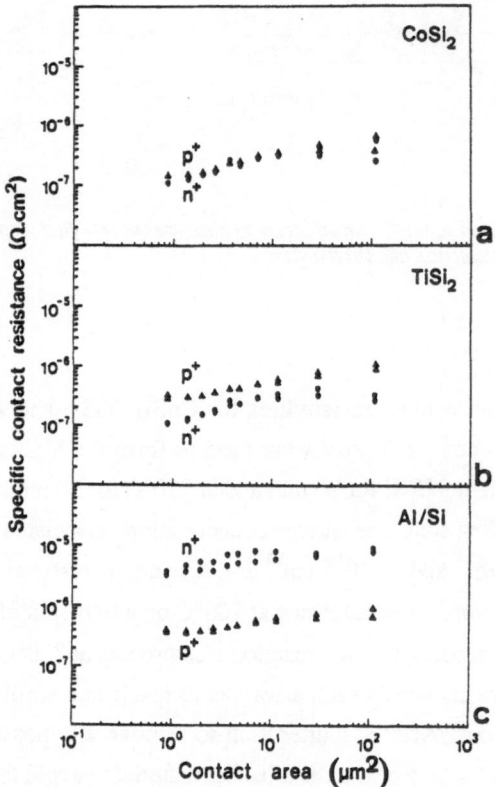

Fig. 47   *Contact resistivity versus contact area from (a) CoSi₂, (b) TiSi₂ and (c) Al/Si to n⁺- and p⁺-Si. See text for doping concentrations.*

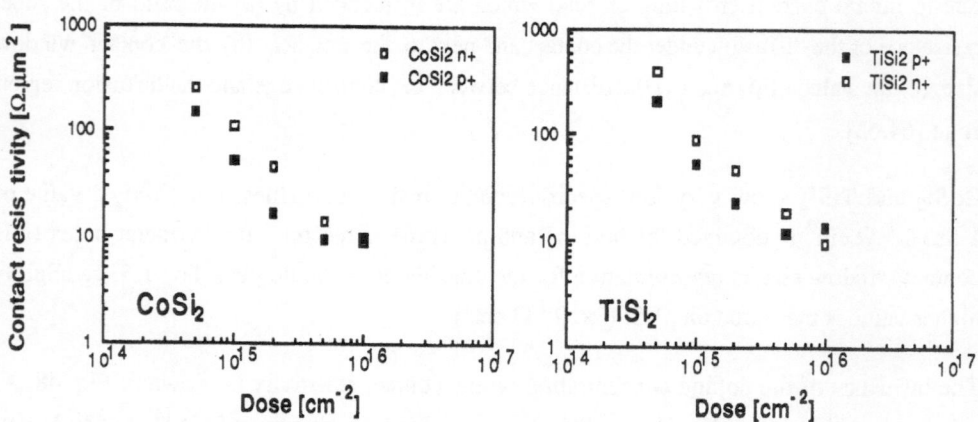

Fig. 48   *Contact resistivity of CoSi₂ and TiSi₂ to n⁺- and p⁺-Si versus implantation dose. In (a) the results are shown for CoSi₂ and (b) for TiSi₂.*

To investigate the influence of subsequent thermal treatments on the contact resistivity, As and B junctions were formed and $TiSi_2$ and $CoSi_2$ were grown in the contact holes in a similar way as described above. Subsequently, the wafers were subjected to thermal cycles at 800°C and 900°C for 30 min or at 900°C, 1000°C and 1100°C for 30 s in a N_2 ambient. Then, the structures are metallized as in above experiments. The contact resistivities are summarized in Table V. A dramatic increase of the contact resistivity can be observed with increasing temperature, which is most pronounced for the B doped samples.

This increase most likely results from a depletion of dopants in the Si region right under the silicide/Si interface. The width of this depletion is determined by the diffusion length in the Si for the considered thermal cycle. Some diffusion lengths as taken from [59] for the thermal cycles used in the above experiment are listed in Table VI. Because of the higher diffusion coefficient of B in Si, the depletion width is the largest for this dopant, corresponding with the highest observed contact resistance increase. The driving force for this diffusion may result (i) from the concentration gradient caused by the rapid diffusion of dopants in the silicide as observed by many investigators [59, 67, 68] and the subsequent evaporation and/or segregation of dopants at the surface or (ii) from a dopant sink caused by compound formation between the dopant and the metal from the silicide (TiB_2, see below).

Table V

Increase of the silicide/metal contact resistivity during subsequent thermal processing measured on $3x3$ μm^2 contact holes (results are in $\Omega.cm^2$)

treatment	CoSi$_2$		TiSi$_2$	
	As	B	As	B
reference	$2.1x10^{-7}$	$1.1x10^{-7}$	$1.8x10^{-7}$	$2.2x10^{-7}$
800°C, 30 min	$3.1x10^{-7}$	$7.6x10^{-7}$	$1.9x10^{-7}$	$7.5x10^{-7}$
900°C, 30 min	$2.2x10^{-7}$	$3.3x10^{-6}$	$9.3x10^{-7}$	$5.2x10^{-6}$
900°C, 30 s	-	$7.4x10^{-7}$	$1.3x10^{-7}$	$1.0x10^{-6}$
1000°C, 30 s	$6.7x10^{-7}$	$4.6x10^{-6}$	$6.1x10^{-7}$	$1.1x10^{-5}$
1100°C, 30 s	$3.3x10^{-6}$	$9.5x10^{-6}$	$3.2x10^{-6}$	$1.6x10^{-5}$

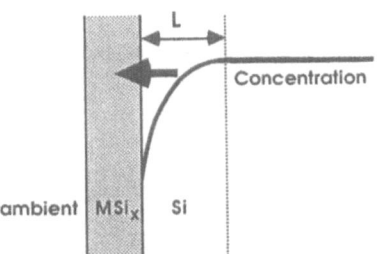

Fig. 49. *Schematic representation of the local depletion of dopants under the silicide as occurs during thermal treatments.*

Table VI

Diffusion constant D and width of depleted region (L
≈ 2.λ) for As and B for a heat treatment at 1100°C
for 30 s.

		Arsenic	Boron
D	[cm^2/s]	10^{-14}	2×10^{-13}
L= 2.$\sqrt{D.t}$	[nm]	11	50

4.2.2 Formation of shallow junctions by dopant diffusion from the silicide

An attractive technology which copes with most of the aforementioned problems connected with silicidation of preformed junctions, was presented recently [59, 68, 70]. As is schematically illustrated in Fig. 50b, it consists of first forming the silicide (using the self-aligned technique described in this work), followed by the implantation of the desired dopants into the silicide. Finally, the structure is subjected to a thermal treatment during which shallow junctions are formed under the silicide by the outdiffusion of dopants into the silicon.

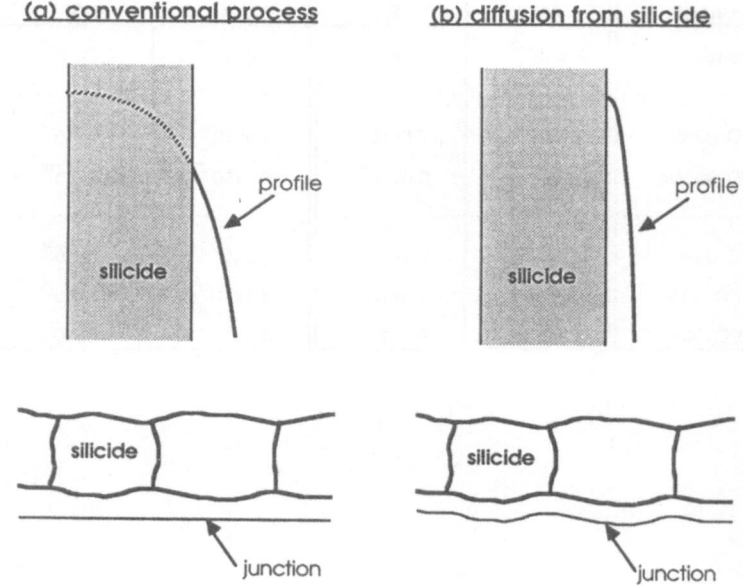

Fig. 50   *Comparison between (a) the conventional process of formation of shallow silicided junctions and (b) the process based on diffusion of dopants from the silicide. The dashed curve in (a) represents the dopant profile in the silicon before silicidation.*

This scheme of shallow junction formation exhibits some interesting advantages:

- it is expected that the junction forms conformal to the silicide/Si interface (Fig. 50b), which reduces the possibility of junction shorting by silicide inhomogeneities or at the diode edges as observed for the conventional process;
- high dopant concentrations are expected at the silicide/Si interface leading to low contact resistances (Fig. 50b), whereas in the conventional process the consumption of a large part of the very shallow junction leads to lower interface concentrations (Fig. 50a);

- by implanting the dopants in the silicide, the ion damage is confined to the silicide and no damage is created in the Si substrate; in this way the thermal cycle can be minimized;

- moreover, it was suggested recently [60], that due to the extremely rapid diffusion in the silicide, dopants are homogenized in the silicide nearly instantaneously, eliminating the implantation shadowing effects, which become more important in ULSI type of structures;

- doping after silicide formation eliminates the need to form a silicide on highly doped As regions, which was found to result in rougher silicide/Si interfaces for $CoSi_2$ and in thinner silicides for $TiSi_2$ (using implantation doses above 2×10^{15} cm^{-2});

- finally, the thermal cycle applied for the diffusion process may also be used to simultaneously reflow doped glass layers on top of the silicide.

The major issues to investigate are: (i) the *diffusion of dopants* in and from the silicide; (ii) the *thermal stability* of the silicide (section 3.3) and (iii) the possibility of the creation of electrically active *generation-recombination centers* during the high temperature treatment.

Dopant profiles

In order to investigate $CoSi_2$ and $TiSi_2$ as sources for B and As diffusion, some experiments were set up to measure dopant profiles after various diffusion cycles. 120 nm of $CoSi_2$ was grown using RTP at 700°C and 100 nm or 160 nm of $TiSi_2$ was formed using the two-step process at 730°C and at 850°C on unpatterned Si-wafers. The silicide layers were implanted with B and As up to doses of 5×10^{15} or 1×10^{16} cm^{-2} with energies of 20 and 50 keV, respectively. Capping of the silicides with a 200-nm thick deposited oxide layer was used to prevent dopant loss due to evaporation.

Figures 51 a and b give the profiles obtained by diffusion of B and As from $CoSi_2$ at 950°C (30 min) and 1050°C (10 s). Both diffusion cycles result in similar junction depths: ± 0.2 µm for B and ± 0.15 µm for As. Shallower junctions were obtained (≤ 100 nm) at lower temperatures.

The scanning electron micrographs of Fig. 52 clearly reveal that the diffusion front follows the CoSi$_2$/Si interface at a nearly constant distance. Even for the extreme case of Fig. 52c (1100°C/60s), where the CoSi$_2$ has agglomerated as described earlier, the As junction follows the CoSi$_2$/Si front conformally.

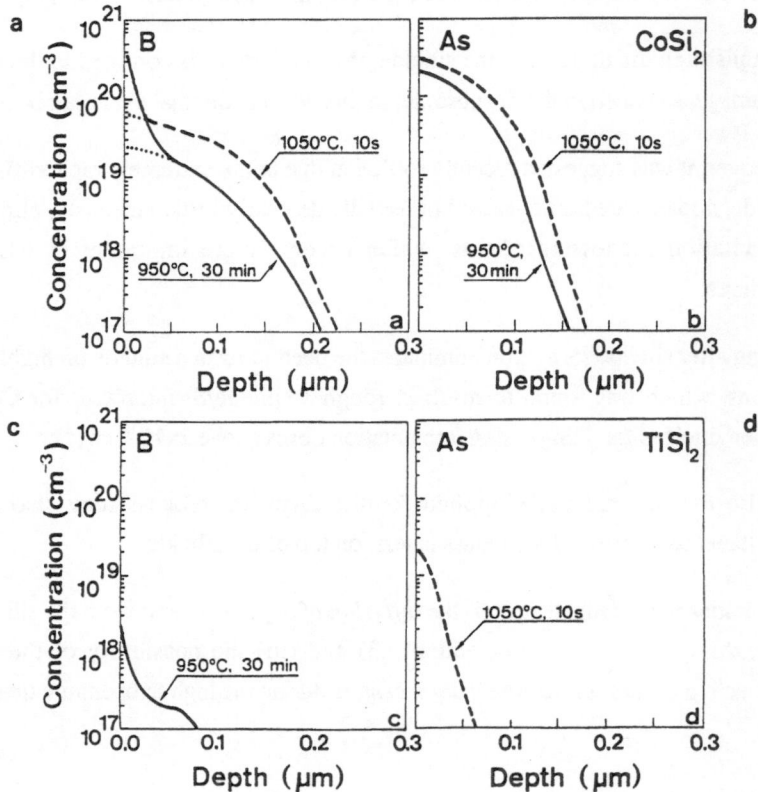

Fig. 51 *SIMS measurements of dopant profiles in the Si substrate resulting from the diffusion of B (a) and As (b) from CoSi$_2$ and B (c) and As (d) from TiSi$_2$. The silicide was removed before the SIMS measurements using an HF-based etch solution.*

Similar experiments were performed for TiSi$_2$. Figure 50 c and d show SIMS measurements of the dopant profiles in the Si resulting from the diffusion of B at 950°C for 30 min and of As at 1050°C for 10 s from TiSi$_2$. Especially for the case of B, only a very small quantity of dopants is detected in the Si.

The SIMS measurements of Fig. 51c were performed after removing the silicide using HF, followed by an etching in an NH$_4$OH/H$_2$O$_2$/H$_2$O-solution. The second etching was added since a careful SEM evaluation of the sample surface after HF-etch only, revealed the presence

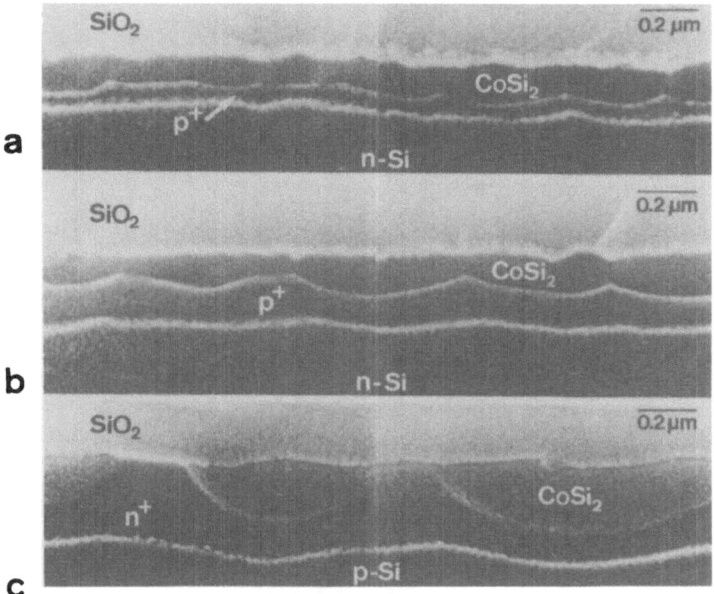

Fig. 52   *Cross-sectional scanning electron micrographs showing the diffusion of dopants from the silicide. The CoSi$_2$ and highly doped regions (> 10^{18} cm^{-3}) are delineated using an etching technique. (a) B diffusion at 900°C/30 min, (b) B diffusion at 1050°C/30 s and (c) As diffusion at 1100°C/60 s.*

of small bright particles at the former silicide grain boundaries (cfr. Fig. 53). A cross-sectional TEM micrograph of such a particle is given in Fig. 54. These particles were removed using the second etching. SIMS measurements of the samples after the HF-etch only, revealed large amounts of B and Ti [72]. From the comparison of the SIMS profiles of the samples with and without additional etching it can be concluded that the particles at the grain boundaries consist of a compound containing Ti and B. The particles were identified as TiB$_2$ using XRD. The formation of TiB$_2$ was also used by Gas et al. [67] to explain the apparent immobility of B implanted into TiSi$_2$, but no prove of the existence of TiB$_2$ was given in their work.

The profile of Fig. 51d for As diffusion at 1100°C also indicates a limited indiffusion only. In this case no As containing precipitates could be separated. However, using higher implantation doses (1x10^{17} cm^{-2}), TiAs has been detected using XRD [71]. This compound was also reported by Beyers et al. [73], when investigating the formation of TiSi$_2$ on heavily As doped Si.

Diode integrity

Perfectly yielding diodes with low leakage current (≤ 1nA/cm^2) can be obtained using this technique in case of CoSi$_2$ for diffusion cycles ranging from 800°C up to 1100°C. As an example an Arrhenius curve of a diode formed by diffusion of B from CoSi$_2$ is given in Fig. 55. For diffusion from TiSi$_2$, high leakage currents are observed for heat treatments at lower temperatures (800°C) because of the limited diffusion.

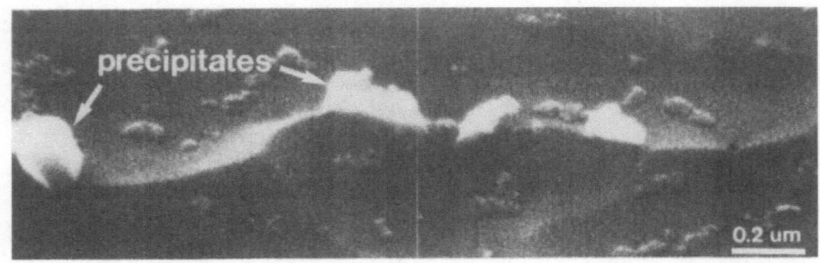

Fig. 53 *SEM micrograph of the sample of Fig. 51c. Bright particles are observed along the former grain boundaries of the TiSi₂.*

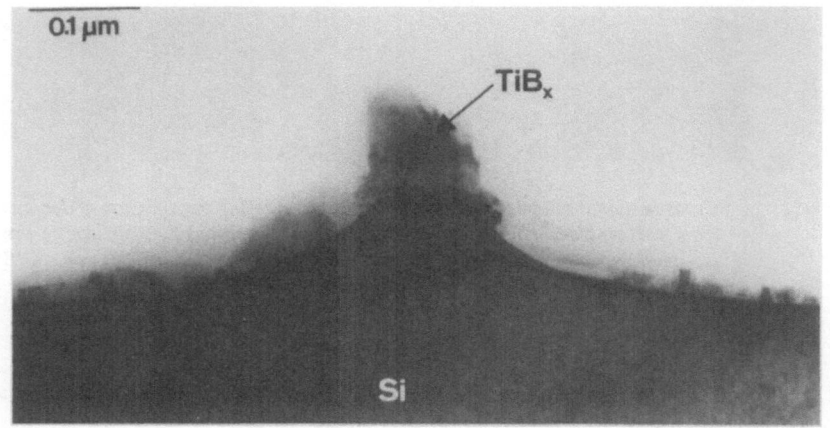

Fig. 54 *Cross-sectional TEM micrograph of the sample of Fig. 53.*

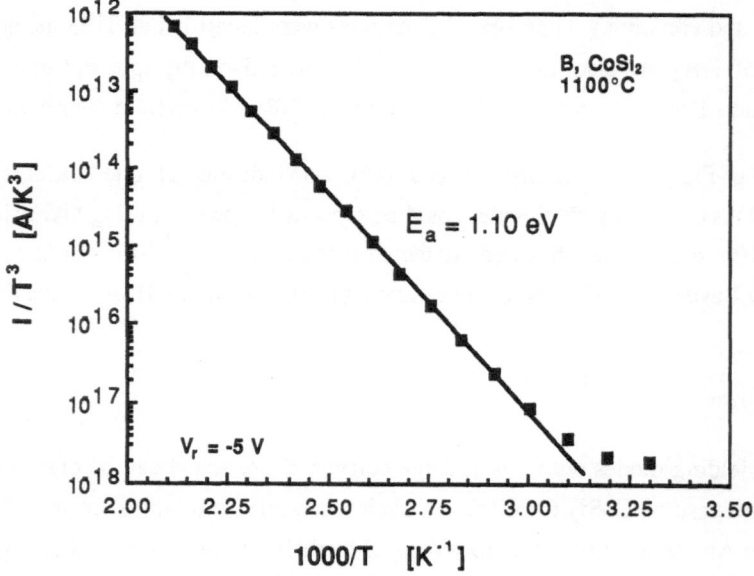

Fig. 55 *Arrhenius plots (log Iᵣ/T³) of the leakage current at 5 V reverse bias of diodes formed by diffusion of B from CoSi₂ at 1100°C for 10 s.*

Contact resistance

Some contact resistivity values measured on 3x3 μm^2 contacts are listed in Table VII. As a reference the contact resistivity from $CoSi_2$ and $TiSi_2$ to junctions implanted to a dose of $5x10^{14}$ cm^{-2}, corresponding to the initial situation without diffusion cycle, are added to the table. The results illustrate very clearly the difference between $CoSi_2$ and $TiSi_2$ for application as dopant diffusion source. For $CoSi_2$, resistivities around $5x10^{-7}$ $\Omega.cm^2$ can be obtained for As and B junctions. With increasing temperature, the contact resistivity decreases, due to the increasing solid solubility.

Table VII

Influence of dopant diffusion from the silicide on the contact resistivity of $CoSi_2/Si$ and $TiSi_2/Si$ interfaces measured on 3x3 μm^2 contacts. An implant of $5x10^{14}$ cm^{-2} was used in the entire active area to define the conducting paths. Implantations with doses up to $1x10^{16}$ cm^{-2} were performed in the silicide before the diffusion treatment. The results are in $\Omega.cm^2$.

	$CoSi_2$	$TiSi_2$
Arsenic $5x10^{14}$ As/cm^2 only:	$3x10^{-6}$	$5x10^{-6}$
900°C, 30 min	$9x10^{-7}$	$9x10^{-6}$
1100°C, 10 s	$2x10^{-7}$	$9x10^{-6}$
1100°C, 120 s	$4x10^{-7}$	-
Boron $5x10^{14}$ B/cm^2 only:	$2x10^{-6}$	$2x10^{-6}$
900°C, 30 min	$8x10^{-6}$	$5x10^{-5}$
1100°C, 10 s	$5x10^{-7}$	$6x10^{-5}$
1100°C, 120 s	$3x10^{-7}$	-

Diffusion from $TiSi_2$ leads to unacceptably high contact resistivity values for all considered cases. For B, the contact resistivity increases over more than an order of magnitude with respect to the starting situation. This suggests that the boride particles in the $TiSi_2$ act as a strong B sink, which results in a dopant depletion under the silicide/Si interface, due to a reverse diffusion of B. This model also agrees with the contact resistivity increase observed for preformed junctions after a high temperature treatment (see Fig. 49).

Deep levels by transition metals

It is known that transition metals exhibit large diffusivities and high solid solubilities in Si at the temperatures used for dopant diffusion or for glass reflow. The formation of generation/recombination centers in the depletion layer, due to Co or Ti diffusion, is therefore not unlikely. DLTS measurements performed on $CoSi_2$ samples (since the problem is expected to be more severe for the case of Co) have indicated that electrically active deep levels are created up to concentrations of 10^{11} cm^{-3} after an RTP cycle at 1100°C, which is four orders of magnitude lower than the solid solubility of Co in Si. The leakage current of diodes is therefore still dominated by diffusion current after heat treatments at 1100°C. For application in bipolar technologies the presence of these concentrations of deep levels may, however, have some implications on the lifetime and thus on the current gain of the bipolar transistors.

4.3. Influence of silicidation on MOS trasistor characteristics

N-channel transistors with and without silicide were compared thoroughly. The results presented here are for TiSi$_2$, but similar results were obtained for CoSi$_2$. Figure 56 shows typical subthreshold characteristics recorded at $V_{ds} = 5$ V of a silicided and a non-silicided transistor with effective channel length of 1.15 μm.

Figure 56 reveals no change in the subthreshold characteristic due to the silicidation process, which indicates that neither threshold voltage, nor subthreshold slope are affected (0.47 V and 90.5 mV/decade, respectively, for transistors with $L_{eff} = 1.15$ μm).

Typical I_{ds}-V_{ds} characteristics are compared in Fig. 57. They clearly indicate that the current driving capability is improved by the silicidation. This can be explained by the reduction of the parasitic series resistance of the source and drain regions.

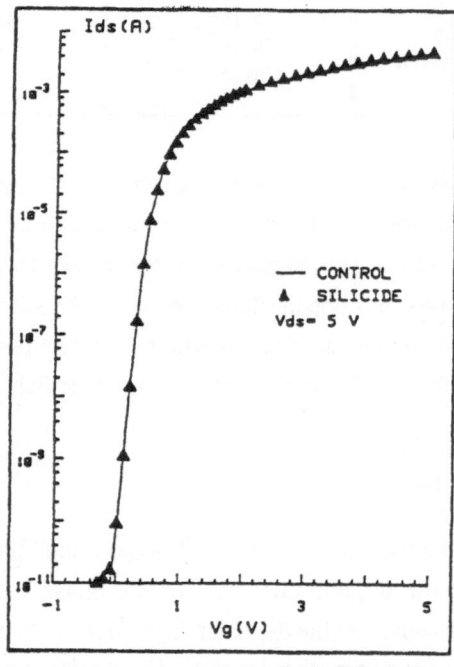

Fig. 56 *Typical subthreshold characteristic of a silicided (TiSi₂) and a non-silicided n-channel transistor at $V_{ds} = 5V$ (W = 20 μm, $L_{eff} = 1.15$ μm).*

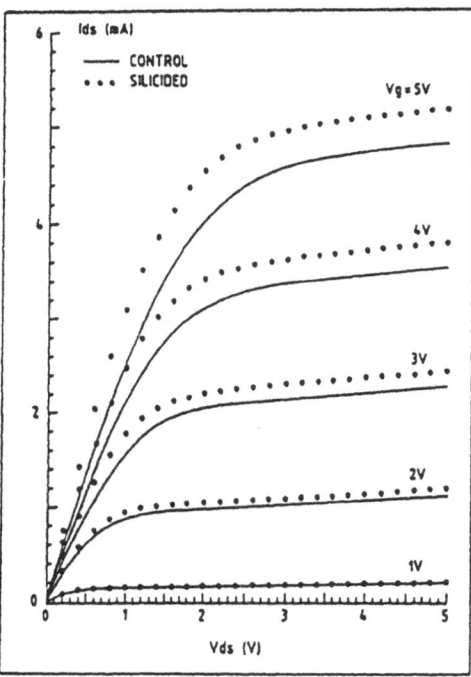

Fig. 57   *Typical I_{ds}-V_{ds} characteristic of a silicided (TiSi$_2$) and a non-silicided n-channel transistor for $V_{gs} = 1V$ - $5V$ (W = 20 μm, L_{eff} = 1.15 μm).*

5. CONCLUSIONS

One of the main technological obstacles limiting the further miniaturization of integrated circuits is the formation of interconnections. In this work the focus was on short-range interconnection technologies based on the use of silicides. The self-aligned silicidation (salicide) approach using rapid thermal processing (RTP) was described. TiSi$_2$ has been the most widely used material for the salicide process, but CoSi$_2$ exhibits some attractive properties. Therefore, both materials have been compared.

The basic reactions occurring in the salicide process were first described in detail. The reaction of the metal (Ti or Co) with Si and SiO$_2$ was investigated. Much attention was also paid to the self-alignment of the formation of these silicides along oxide edges.

It was shown that when performing the silicidation in N$_2$ ambient, the reaction between Ti and Si results in the formation of a double layer consisting of a TiSi$_2$ bottom layer and a TiN top layer. The rapid diffusion of N in Ti on SiO$_2$ 'stuffs' the grain boundaries, retarding the previously observed grain boundary diffusion of Si, which suppresses the lateral formation of silicide over oxide edges.

Although Co is the moving species during CoSi$_2$ formation, it was demonstrated that under certain processing conditions lateral silicidation can occur. In the classical silicide process based upon Co, however, any lateral silicidation is absent and a model for this behavior was presented.

Having gathered sufficient information on the basic reactions, the behavior of $CoSi_2$ and $TiSi_2$ in self-aligned silicidation technologies was assessed and the most important technological issues were discussed.

Various items related with the technological implementation of $TiSi_2$ and $CoSi_2$ in a self-aligned silicide process were discussed, and a synoptic comparison is given in Table VIII. The ratings are quantified by one or more "+" or "-" signs, depending on the importance of the issue. "x" indicates that the silicide presents no distinct advantage or disadvantage with respect to this issue.

Table VIII

Comparison of the technological implementation of $CoSi_2$ and $TiSi_2$ in a self-aligned technology.

	$CoSi_2$	$TiSi_2$
influence of native oxide	+	-
influence of heavy As doping	-	-
process window	x	x
thermal stability	- -	-
chemical stability	+++	- - -
interaction with Al	-	-
deposition of CVD W on silicide	++	- -
Si consumption	-	+
Stress	+++	- - -
Local interconnection	-	+

In the final section the incorporation of $CoSi_2$ and $TiSi_2$ into microelectronic devices and the influence on device performance were discussed.

It was shown that excellent diode integrity can be maintained using both silicidation technologies.It was demonstrated that the limitations of the technique of siliciding preformed junctions can be avoided when the junctions are formed by diffusion of dopants from the silicide.

The influence of the incorporation in processing technology of either $TiSi_2$ or $CoSi_2$ on MOS transistor performance was evaluated. An enhanced current driving capability was demonstrated, whereas no secondary effects degrading the electrical parameters were observed.

This work intended to review the present status and discuss some of the major issues concerning a silicide interconnect technology based upon the self-aligned metal-silicon interaction stimulated by rapid thermal processing. The basic materials aspects, process implementation and device aspects of this technology were covered. The novel salicide process using $CoSi_2$ was compared with the more established $TiSi_2$-process.

Although $TiSi_2$ has found a much more widespread application than $CoSi_2$ in self-aligned silicidation technologies, it was shown that the latter silicide exhibits some highly attractive features. The reason why $TiSi_2$ has received a more widespread application, is partly historical. $TiSi_2$ was one of the silicides used in the polycide approach. A large amount of information was, therefore, available when the salicide process was introduced. Based on the results of this work it can be concluded that $TiSi_2$ is a possible material for the salicide process in CMOS technologies with minimum geometries around 1 μm. Indeed, the disadvantages of $TiSi_2$ can be bypassed by carefully tailoring the processing conditions. It is, however, expected that, when the geometries shrink to 0.5 μm, the processing windows will reduce to a level where $CoSi_2$ will be a much more attractive material for a self-aligned silicidation technology. Especially, the technique of forming shallow silicided junctions will be a crucial factor determining the choice of the silicide. For this generation of technologies it may be expected that the silicidation of the source and drain regions may be decoupled form the silicidation of the gate regions. More stringent sheet resistance requirements on the gate may demand either thicker silicides (polycide) or refractory metals which is an important area for further research.

6. ACKNOWLEDGMENTS

This overview is the result of research to which many coworkers and colleagues have contributed by providing experimental assistance, pilot-line operations, analytical support and stimulating discussions.

We would like to express our special gratitude to D. Avau, H. Bender, W. Chen, B. Coenegrachts, G. Declerck, R. De Koninck, G. Groeseneken, C. Jehoul, P. Lippens, L. Liwu, K. Maex, M. Reybrouck, W. Vandervorst, D. Vanhaeren and J. Vanhellemont at IMEC; to M. Geyselaers, C. Langereis, A. Mauwen, T. Nellisen, T. Theunissen, W. van der Wijgert and R. Wolters at the Philips Research Laboratories, Eindhoven ; to W. Claassen and A. Goemans at Philips Elcoma, Nijmegen ; to V.Probst and H. Schaber at Siemens, Munich ; to F. Habraken, G. Krooshof and W. van der Weg at the State University, Utrecht ; and to F. d'Heurle at IBM Research, Yorktown Heights, New York. May those who recognise their contribution and remain anonymous here be assured of our gratitude at heart.

We thank M. Mestdag for her help in typing and organizing parts of the manuscript.

One of the authors (L. Van den hove) is indebted to the Belgian National Research Foundation for his successive fellowships as research assistant and senior research assistant.

7. REFERENCES

1. R.H. Dennard, F.H. Gaensslen, H.N. Yu, V.L. Rideout, E, Bassous and A.R. Leblanc, "Design of ion implanted MOSFET's with very small physical dimensions", IEEE J. Solid State Circuits, vol. SC-9, p. 256, 1974.

2. T. Shibata, K. Hieda, M. Sato, M, Konaka, R.L.M. Dang and H. Iizuka, "An optimally designed process for submicron MOSFET's", IEDM Techn. Digest, p. 647, 1981.

3. C.Y. Ting, S.S.Iyer, C.M. Osburn, G.J. Hu and A.M. Schweighart, "The use of TiSi$_2$ via self-aligned silicide technology", Proc. of the Electrochem. Soc. Meeting, vol. 82-2, p. 224, 1982.

4. C.K. Lau, Y.C. See, D.B. Scott, J.M. Bridges, S.M. Perna and R.D. Davies, "Titanium disilicide self-aligned source-drain + gate technology", IEDM Techn. Digest, p. 714, 1982.

5. M.E. Alperin, T.C. Holloway, R.A. Haken, C.D. Gosmeyer, R.V. Karnaugh and W.D. Parmantie, "Development of the self-aligned silicide process for VLSI applications", IEEE Trans. Electron Devices, ED-32, p. 141, 1985.

6. L. Van den hove, "Advanced interconnection and contact schemes based on TiSi$_2$ and CoSi$_2$: Relevant materials issues and technological implementation", Ph. D. Thesis, K.U. Leuven, Belgium, June 1988.

7. C.Y. Ting, "Silicides for contacts and interconnects", IEDM Techn. Digest, p. 110, 1984.

8. T.O. Sedgwick, "Rapid Thermal Processing: How well is it doing and where is it going?", Materials Research Society Proceedings, Pittsburgh, Vol. 92, p. 3, 1987.

9. T. Okamoto, K. Tsukamoto, M. Shimizu and T. Matsukawa, "Titanium silicidation by halogen lamp annealing", J. Appl. Phys. Vol. 57 (12), p. 5251, 1985.

10. A.E. Morgan, E.K. Broadbent and A.H. Reader, "Formation of Titanium Nitride/Silicide Bilayers by Rapid Thermal Anneal in Nitrogen", in "Rapid Thermal Processing", Eds. Sedgwick, Seidel and Tsau, Mat. Res. Soc. Proc., Pittsburgh, Vol.52, p. 279, 1985.

11. L. Van den hove, N. Saks, K. Maex, R. De Keersmaecker and G. Declerck, "A self-aligned titanium silicide technology using Rapid Thermal Processing", presented at the European Solid State Research Conference, Aachen, Germany, Sept. 1985.

12. T. Brat, C.M. Osburn, T. Finstad, J. Liu and B. Ellington, "Self-Aligned Ti Silicide Formed by Rapid Thermal Annealing", J. Electrochem. Soc., Vol. 133 (7), p. 1451, 1986.

13. U.N. Mitra, P.W. Davies, R.K. Shukla and J.S. Multani, "Material characterization of selectively formed Titanium Nitride and Silicide thin films", in "Semiconductor Silicon", Eds. H.R. Huff, T. Abe and B. Kolbesen, The Electrochem. Soc., Pennington, N.J., p.316, 1986.

14. F. Runovc, H. Norstrom, R. Buchta, P. Wiklund and S. Petersson, "Titanium disilicide in MOS technology", Physica Scripta, Vol. 26, p. 108, 1982.

15. M.S. Wang and J.B. Anthony, "Oxygen contamination control for a TiSi$_2$ atmospheric thermal anneal process", in *Multilevel Metallization, Interconnection and Contact technologies*, Eds. L.B. Rothman and T. Herndon, The Electrochem. Soc., Pennington, N.J., 1987, p. 114.

16. B.J. Burrow, A.E. Morgan and R.C. Ellwanger, "A correlation of Auger Electron Spectroscopy, X-ray Photoelectron Spectroscopy and Rutherford Backscattering Spectrometry measurements on sputter deposited titanium nitride thin films", J. Vac. Sci. Technol.A, Vol. 4, p. 2463, 1986.

17. G. Krooshof, F. Habraken, W. van der Weg, L. Van den hove, K. Maex and R. De Keersmaecker, "Study of the rapid thermal nitridation and silicidation of Ti using Elastic Recoil Detection, Part I and II", J. Appl. Phys. , Vol. 63 (10) , p 5104-5110, 1988.

18. G.G. Bentini, R. Nipoti, A. Armigliato, M. Berti, A.V. Drigo and C. Cohen, "Growth and structure of titanium silicide phases formed by thin Ti films on Si crystals", J. Appl. Phys. Vol. 57 (2), p. 270, 1985.

19. G.J. Van Gurp and C. Langereis, "Cobalt silicide layers on Si. I. Structure and growth," J. Appl. Phys., Vol. 46 (10), p. 4301, 1975.

20. S.S. Lau, J.W. Mayer and K.N. Tu, "Interactions in the Co/Si thin-films system. I. Kinetics", J. Appl. Phys. Vol. 49 (7), p. 4005, 1978.

21. M. Hansen, in Constitution of Binary Alloys, McGraw Hill, New York, 1958, p. 1069.

22. R. Pretorius, J.M. Harris and M.-A. Nicolet, "Reaction of thin metal films with SiO$_2$ substrates", Solid State Electronics, vol. 21, p. 667, 1978.

23. S. Nishiyama, private communication.

24. E. Kobeda and E.A. Irene, "Intrinsic SiO_2 film stress measurements on thermally oxidized Si", J. Vac. Sci. Technol. B, Vol. 5 (1), p. 15, 1987.

25. E.A. Irene, E. Tierney and J. Angilello, "A viscous flow model to explain the appearance of high density thermal SiO_2 at low oxidation temperatures", J. Electrochem. Soc., Vol. 129 (11), p. 2594, 1982.

26. W.K. Chu, J.W. Mayer, H. Muller, M.-A. Nicolet and K.N. Tu, "Identification of the dominant diffusing species in silicide formation", Appl. Phys. Lett. Vol. 25 (8), p. 454, 1974.

27. W.K. Chu, S.S. Lau, J.W. Mayer, H. Muller and K.N. Tu, "Implanted noble gas atoms as diffusion markers in silicide formation", Thin Solid Films, Vol. 25, p. 393, 1975.

28. G.J. Van Gurp, W.F. van der Weg and D.Sigurd, "Interactions in the Co/Si thin-film system. II. Diffusion-marker experiments", J. Appl. Phys, Vol. 49 (7), p. 4011, 1978.

29. F.M. d'Heurle and C.S. Petersson, "Formation of thin films of $CoSi_2$: nucleation and diffusion mechanisms", Thin Solid Films, Vol. 128, p. 283, 1985.

30. C.-D. Lien, M. Bartur and M.-A. Nicolet, "Marker experiments for the moving species in silicides during solid phase epitaxy of evaporated Si", Mat. Res. Soc. Symp. Proc., Vol. 25, p. 51, 1984.

31. L. Van den hove, R. Wolters, R. Geyselaers, R. De Keersmaecker and G. Declerck, "Key issues to the self-aligned formation of $CoSi_2$ in a salicide process", Mat. Res. Soc. Symp. Proc. Vol. 100, p. 99, 1988.

32. H.J.W. van Houtum and I.J.M.M. Raaijmakers, "First phase nucleation and growth of titanium disilicide with an emphasis on the influence of oxygen", in *Thin Films Interfaces and Phenomena*, Eds. R.J. Nemanich, P.S. Ho and S.S. Lau, Mat. Res. Soc., Vol. 54, p. 37, Pittsburg, 1986.

33. G.J. Van Gurp, W.F.van der Weg and D.Sigurd, "Interactions in the Co/Si thin-film system. II. Diffusion-marker experiments", J. Appl. Phys, Vol. 49 (7), p. 4011, 1978.

34. R. Beyers, D. Coulman and P. Merchant, "Titanium disilicide formation on heavily doped silicon substrates", J. Appl. Phys. Vol. 61 (11), p. 5110, 1987.

35. C.Y. Ting, F.M. d'Heurle, S.S. Lyer and P.M. Fryer, "High temperature process limitation on $TiSi_2$", J. Electrochem. Soc., Vol. 133 (12), p. 2621, 1986.

36. S. Vaidya, S.P. Murarka and T.T. Sheng, "Formation and thermal stability of $CoSi_2$ on polycrystalline Si", J. Appl. Phys., Vol. 58 (2), p. 971, 1985.

37. L.E. Murr, in *Interfacial phenomena in metals and alloys*, Adisson-Wesley Publishing Company, Massachusetts, p.240, 1975.

38. W.W. Mullins, "Theory of thermal grooving", J. Appl. Phys., Vol. 28 (3), p. 333, 1957.

39. W.W. Mullins, "Grain boundary grooving by volume diffusion", Trans. of the metallurgical society of AIME, Vol. 218, p. 354, 1960.

40. W.M. Robertson, "Grain boundary grooving and scratch decay on copper in liquid lead", Trans. of the metallurgical society of AIME, Vol. 233, p. 1232, 1965.

41. N.A. Gjostein, "Measurement of the surface self-diffusion coefficient of copper by the thermal grooving technique", Transactions of the metallurgical society of AIME, Vol. 221, p. 1039, 1961.

42. S.S. Lau and W.F. van der Weg, "Solid phase epitaxy", in *Thin Films Interdiffusion and Reaction*, Eds. J.M. Poate, K.N. Tu and J.W. Mayer, The Electrochem. Soc., Princeton, NJ, p. 433, 1978.

43. S.S. Lau, Z.L. Liau and M.-A. Nicolet, "Solid phase epitaxy in silicide-forming systems", Thin Solid Films, Vol. 47, p. 313, 1977.

44. C.Y. Ting and M. Wittmer,"The use of titanium-based contact barrier layers in silicon technology", Thin solid films, Vol. 96, p. 327-345, 1982.

45. C.Y. Ting and M. Wittmer, "Investigation of the $Al/TiSi_2/Si$ contact system", J. Appl. Phys., Vol. 54 (2), p. 937, 1983.

46. G.J. van Gurp, J.L.C. Daams, A. Van Oostrom, L.J.M. Augustus and Y. Tamminga, "Aluminum-silicide reactions. I. Diffusion, compound formation and microstructure", J. Appl. Phys., Vol. 50 (11), p. 6915, 1979.

47. J. Hui, S. Wong and J. Moll, "Specific contact resistivity of $TiSi_2$ to p^+ and n^+ junctions", IEEE Electron Device Letters, EDL-6 (9), p. 479, 1985.

48. J.T. Pan and I. Blech, "Stress measurement of refractory metal silicides during sintering", J. Appl. Phys., Vol. 55 (8), p. 2874, 1984.

49. F.M. d'Heurle, "Material properties of silicides for VLSI technology", in *Solid State Devices 1985*, Eds. P. Balk and O.G. Folberth, Elsevier, Amsterdam, 1986, p. 213.

50. V.L. Teal, S.P. Murarka, "Stresses in TaSi$_x$ films sputter deposited on polycrystalline silicon", J. Appl. Phys., Vol. 61 (11), p. 5038, 1987.

51. L. Van den hove, J. Vanhellemont, R. Wolters, W. Claassen, R. De Keersmaecker and G. Declerck, "Correlation between stress and film-edge induced defect generation along TiSi2 and CoSi2 lines formed by metal-Si reaction", in *Proceedings of the First International Symposium on Advanced Materials for ULSI*, Eds. M. Scott, Y. Akasaka and R. Reif, The Electrochem. Soc., Pennington, NJ, Vol. 88-19, p. 165, 1988.

52. J. Vanhellemont, S. Amelinckx and C. Claeys, "Film-edge-induced dislocation generation in silicon substrates", J. Appl. Phys., Vol. 61 (6), p. 2170, 1987.

53. J. Vanhellemont and L. Van den hove, presented at the 9th European Congress on Electron Microscopy, York, England, 4-9 September 1988.

54. Y. Kondo, "Plastic deformation and preheat treatment effects in Cz and FZ silicon crystals", in *Semiconductor Silicon 1981*, Eds. H.R. Huff and R.J. Kriegler, The Electrochem. Soc., Pennington, p. 220, 1981.

55. S. Isomae, Y. Tamaki, A. Yajima, M. Nanba and M. Maki, "Dislocation generation at Si$_3$N$_4$ film edges on silicon substrates and viscoelastic behavior of SiO$_2$ films", J. Electrochem. Soc., p. 1014, 1979.

56. S.S. Wong, D.C. Chen, P. Merchant, T.R. Cass, J. Amano and K.Y. Chiu, "HPSAC - A silicided amorphous-silicon contact and interconnect technology for VLSI", IEEE Trans. Electron Devices, ED-34 (3), p. 587, 1987.

57. T.E. Tang, C.-C. Wei, R.A. Haken, T.C. Holloway, L.R. Hite and T.G. Blake, "Titanium nitride local interconnect technology for VLSI", IEEE Trans. Electron Devices, ED-34(3), p. 682, 1987.

58. R. Wolters and L. Van den hove, "The feasibility of CoSi$_2$, TiW and TiW(N) as local interconnection in a self-aligned CoSi$_2$ technology", presented at the 1988 VLSI Multilevel Interconnection Conference, June 1988, Santa Clara, California.

59. K. Maex, "Interactions of implanted dopants and the metal/Si system: Fundamental aspects and applications of ion beam mixing and dopant redistribution", Ph.D. Thesis, K.U.Leuven, Sept. 1987.

60. R. Liu, D.S. Williams, W.T. Lynch, "Mechanisms for process-induced leakage in shallow silicided junctions", International Electron Devices Meeting, Techn. Dig., p. 58, 1986.

61. S.J. Proctor and L.W. Linholm, "A direct measurement of interfacial contact resistance", IEEE Electron Device Letters, Vol. EDL-3 (10), p. 294, 1982.

62. S. J. Proctor, L.W. Linholm and J.A. Mazer, "Direct measurements of interfacial contact resistance and interfacial contact layer uniformity", IEEE Trans. Electron devices, Vol. ED-30 (11), p. 1535, 1983.

63. J.A. Mazer, L.W. Linholm and A.N. Saxena, "An improved test structure and Kelvin-measurement method for the determination of integrated circuit front contact resistance", J. Electrochem. Soc., p. 440, 1985.

64. A. Scorzoni, M. Finetti, K. Grahn, I. Suni and P. Cappelletti, "Current crowding and misalignment effects as source of error in contact resistivity measurements: Part I. Computer simulation of conventional CER and CKR structures", IEEE Trans. Electron Devices, ED-34 (3), p. 525, 1987.

65. R.L. Gillenwater, M.L. Hafich and G.Y. Robinson, "The effect of lateral current spreading on the specific contact resistivity in 2D-resistor Kelvin Devices", IEEE Trans.Electron Devices, ED-34 (3), p.537, 1987.

66. S.M. Sze, in *Physics of Semiconductor Devices*, 2nd edition, J. Wiley & Sons, New York, p. 305, 1981.

67. P. Gas, V. Deline, F.M.d'Heurle, A. Michel and G. Scilla, "Boron, phosphorus and arsenic diffusion in TiSi$_2$", J. Appl. Phys., Vol. 60 (5), p. 1634, 1986.

68. A.H. Van Ommen, H.J.W. Van Houtum and A.M.L. Theunissen, "Diffusion of ion implanted As in TiSi$_2$", J. Appl. Phys., Vol. 60 (2), p. 627, 1986.

69. F.C. Shone,K.C. Saraswat and J.D. Plummer, "Formation of 0.1μm N$^+$/P and P$^+$/N junctions by doped silicide technology", International Electron Devices Meeting, Techn. Dig., p. 407, 1985.

70. R. Liu, F.A. Baiocchi, L.A. Heimbrook, J. Kovalchick, D.L. Malm, D.S. Williams and W.T. Lynch, "Formation of shallow p$^+$/n and n$^+$/p junctions with CoSi$_2$", in *ULSI Science and Technology 1987*, Eds. S. Broydo and C.M. Osburn, The Electrochem. Soc., Pennington, N. J., p. 446, 1987.

71. V. Probst, P. Lippens, L. Van den hove, K. Maex, H. Schaber and R. De Keersmaecker, "Shallow junction formation using CoSi$_2$ as a diffusion source", Proceedings of the European Solid State Device Research Conference, Bologna, Italy, Sept. 1987, p. 437.

72. V. Probst, P. Lippens, L. Van den hove, H. Schaber and R. De Keersmaecker, "Limitations of TiSi$_2$ as a source for dopant diffusion", Appl. Phys. Lett., Vol. 52 (21), p. 1803, 1988.

73. R. Beyers, D. Coulman and P. Merchant, "Titanium disilicide formation on heavily doped silicon substrates", J. Appl. Phys. Vol. 61 (11), p. 5110, 1987.

MICROSTRUCTURAL DEFECTS IN RAPID THERMALLY PROCESSED IC MATERIALS

Alec H. Reader

Philips Research Labs., P.O. Box 80.000
5600 JA Eindhoven, The Netherlands

1. INTRODUCTION

The most important trend in the current development of integrated circuits is the reduction in device dimensions which has lead to a number of specific requirements. The first of these requirements is that the "active" insulators in MOS devices must be ultra thin, typically below 10 nm thick. Using conventional furnace oxidation it is difficult to produce such thin oxides because the oxidation times are too short to be reproducibly controlled. As short annealing times can be carefully controlled in a rapid thermal processing (RTP) system, it can be understood that such systems can be usefully employed to carry-out reproducible short-period oxidations at temperatures typically used for furnace oxidation. Another requirement is that, as lateral device dimensions are reduced, there must be a concomitant shrinkage of the source/drain junction depths in the devices - forming so-called shallow junctions. Also the dopant profiles in the junctions must be precisely controlled. This requires that excessive temperatures and annealing times are reduced, so that less dopant diffusion occurs. However, after implantation of the dopant, a high temperature annealing step is required for dopant activation and reconstruction of the substrate's crystal-lattice. It must be pointed out here that it is the high temperature that is important; extended annealing times are not required for activation and lattice reconstruction. Extended annealing times are frequently required though for the removal of crystal-lattice defects, formed during lattice reconstruction. In many cases, these lattice defects must be removed for optimal electrical performance of the junction. Extended annealing times can, however, cause excessive dopant diffusion, which conflicts with the desire to form shallow junctions as pointed out above.

Figure 1 illustrates (1.1) how optimum annealing temperatures and times can be chosen to produce dislocation-free junctions in which limited dopant diffusion has occurred. In the figure, annealing temperatures and times required to produce dislocation-free junctions are indicated by the region containing the vertical "hatching". The region of the figure containing horizontal "hatching" indicates annealing temperatures and times required for the production of dopant profiles in which a limited amount of dopant diffusion has occurred. The area in the figure denoted as the "optimum annealing region" corresponds to high temperature anneals and short annealing times. The high

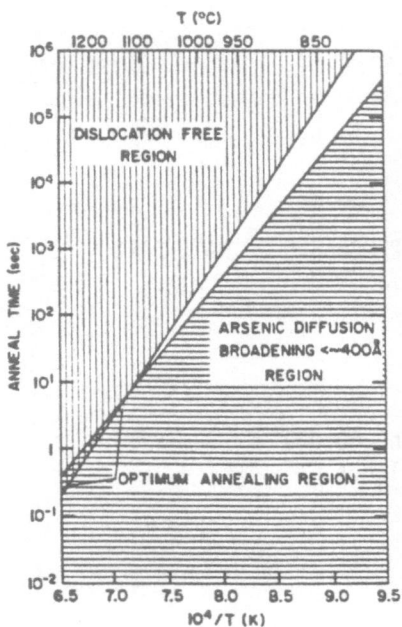

Fig. 1 Optimum annealing region to remove TEM defects and limit As diffusion for 5E15/cm² 100 keV As implant in Si

temperature being required to reconstruct a defect-free crystal lattice and concomitantly activate the dopant. Short annealing times are required so that limited dopant diffusion occurs .

In the case of modern high-speed bipolar devices with small lateral dimensions, shallow emitters and thin bases (1.2,1.3) are required. The simultaneous formation of shallow emitters and thin bases can only be achieved by careful control of diffusions anneals. Again such diffusions can be conveniently carried out in a rapid thermal processing system.

Another requirement of modern integrated circuits is that the resistivity of the interconnect conductors, which connect the various devices in the circuits together, be reduced. The switching time of an integrated circuit is limited by the product of the resistance (R) and capacitance (C) of the circuit - the so-called "RC" time. By reducing the value of R (by reducing the resistivity of the conductor), a circuit can be designed which has a fast switching time. Until recently, doped polycrystalline silicon (polysilicon) layers were used as interconnect material. Highly doped polySi has a resistivity of approximately ·1000 $\mu\Omega$cm. In modern circuits, metal silicides are used as a replacement for (or in conjunction with) polysilicon layers as the resistivity of silicides can be as low as 15 $\mu\Omega$cm. Here again, rapid thermal processing frequently plays an important role in the production of silicides.

In addition to the above requirements, rapid thermal annealing is frequently used in the planarization of deposited oxides (1.4,1.5,1.6) over devices, the improvement of Al contact resistance between Al lines and the substrate (1.7,1.8), and grain growth in thin polySi films (1.9) to name only a few examples. Such subjects will be discussed in other articles in these proceedings.

In this article, the microstructural defects created and annihilated in silicon-based IC materials during rapid thermal processing will be discussed. These materials, and the defects they contain, will be treated in successive sections. Each section will begin with an introductory literature review, followed by the results of various studies on materials subjected to rapid thermal processing treatments. Emphasis will be placed on the interpretation of transmission electron microscope (TEM) observations. In some cases these observations will be compared with those obtained from materials subjected to conventional furnace processing.

2. MICROSTRUCTURAL DEFECTS IN IMPLANTED AND SUBSEQUENTLY RAPID THERMALLY ANNEALED SILICON WAFERS

In this section lattice defects created during ion implantation and then annihilated during annealing will be discussed . Emphasis will be on TEM observations and specific problems encountered in forming shallow boron junctions. As mentioned in the introduction, after implantation, an activation anneal is required to make the dopant atoms electrically active and remove crystal-lattice damage (lattice defects). Frequently the anneal must be carried out at a high temperature (>1000°C) to remove the lattice defects. This is the case for silicon implanted with a high dose of boron as would be required for shallow p⁺ junctions. At such temperatures, the anneal must be short (~few seconds) to limit dopant diffusion. Such short, high temperature anneals are conveniently performed in a rapid thermal processing system.

In the following, typical implantation energies used in the production of junctions in integrated circuits will be considered - that is energies up to about 150 keV. Implantation into silicon (001) wafers nominally at room temperature will be examined. Additionally, the discussion will be sub-divided into low, medium and high dose implantations. The stated doses used in this division of the text should not be considered to be exact. They are given as an <u>indication</u> of the doses at which various microstructural defect are observed. Obviously, the energy of the ions influences the dose at which a particular defect occurs - in general, a lower dose of highly energetic ions is required to produce the same effect as that of low energy ions. In the text the defects found, by TEM observation, after implantation and after annealing will be outlined in turn, firstly for the case of the implantation of ions lighter than silicon. The consequences of these observations for the production of shallow p⁺ junctions will be discussed subsequently.

2.1 Light ions (e.g. B⁺).   a) low and medium doses (<5 x 10¹⁵cm⁻²).

<u>After implantation</u>. The implantation of light ions into silicon at low and medium doses causes the formation of small (~5nm) amorphous zones embedded in clouds of point defects and small (2-3 nm) dislocation loops, collectively known as damage clusters, in the implanted region of the silicon - see figure 2a. The density and mean diameter of these damage clusters increases with increasing dose.

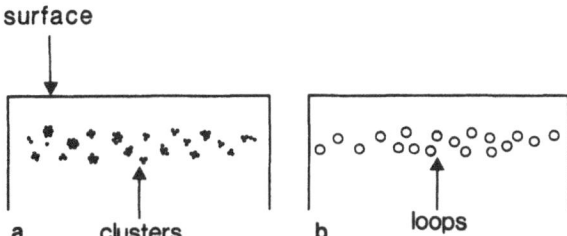

surface

a   clusters   b   loops

Fig. 2   Schematic cross-sectional diagram showing the damage distribution resulting from the implantation of light ions at low or medium doses into (100) or (111) Si: (a) after implantation and (b) after anneal (2.1).

<u>After anneal.</u> After annealing wafers containing damage clusters, scattered dislocation loops are observed in the materal as depicted in figure 2b. Here, during annealing, the clouds of point defects agglomerate (cluster) forming (additional) dislocation loops. There is evidence (2.5,2.6) that a lower density of dislocation loops is formed during rapid thermal annealing than during furnace annealing. It appears that high heating rates favour the dissolution of a certain amount of implantation damage while low heating rates favour the clustering of point defects.

b) high doses (>5 x 10^{15}cm^{-2})

   After implant. The implantation of light ions at high doses into silicon causes a high density of damage clusters to be formed in the implanted region of the silicon, as depicted in figure 3a. The density of the damage clusters· is dependent on the implantation dose; the higher the dose, the larger the defect cluster density.

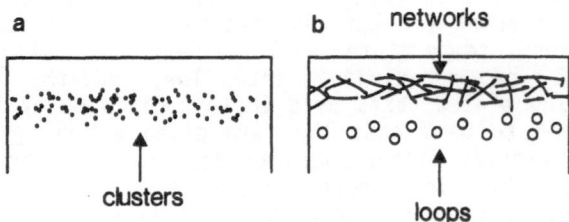

Fig. 3 Schematic cross-sectional diagram showing the damage distribution
     resulting from the implantation of light ions at high doses into
     (100) or (111) Si: (a) after implantation and (b) after anneal
     (2.1).

   After anneal. Again annealing causes the damage clusters formed during implantation to agglomerate to form dislocation loops. In this case, as a high density of damage clusters was created during implantation, a high density (>10^{11}cm^{-2}) of dislocation loops forms during annealing. If the anneal is carried out at temperatures greater than about 900°C, the dislocation loops interact to form dislocation networks, as depicted in figure 3b. Rapid thermal annealing, at temperatures of 1100°C and greater, reduces (2.8) the dislocation density in such networks.

   The problems related to the formation of shallow high-dose boron junctions can now be understood. As described above, silicon wafers implanted with a high boron dose normally contain, after annealing, a high density of dislocations in the neighbourhood of the p$^+$/n junction. The presence of such defects in this junction region can lead to high reverse-bias leakage currents during the operation of devices constructed with such junctions. Obviously, if the existence of these dislocations leads to the electrical malfunctioning of a device, it is a great advantage if these dislocations can be removed or avoided. As pointed out earlier, dislocations can be removed by high temperature annealing. However, in the cases of boron, it is difficult to remove lattice damage while maintaining a minimum amount of diffusion, as is suggested by figure 4.

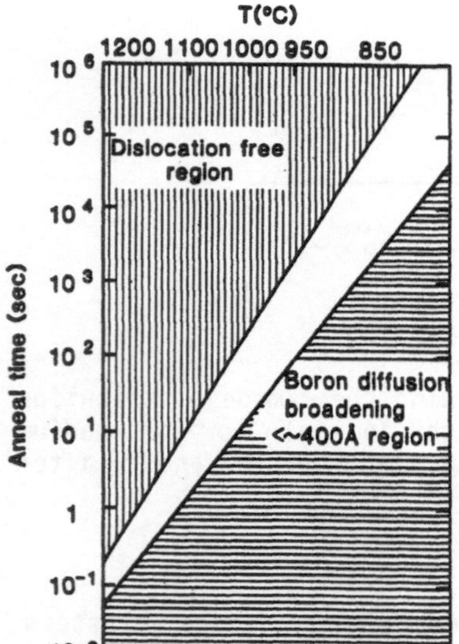

   In figure 4, the dislocation-free region is indicated,(as before- cf. figure 1), by vertical "hatching". The region in which boron diffusion broadening of less than 40 nm occurs is

Fig. 4 No optimum annealing
region to remove TEM defects
and limit B diffusion (for
high dose implantation in Si)
-compare with fig. 1.

indicated by horizonal "hatching". An overlap between the dislocation-free region and the boron diffusion region does not occur. This suggests that forming boron junctions, free from lattice defects and having a limited amount of diffusion, will be difficult to achieve.

2.2 Heavy Ions (e.g. As⁺, Sb⁺)   a) Low doses only (<5x10¹³cm⁻²)

After implant. The implantation of heavy ions at low doses into silicon results in the creation of damage clusters in the implanted region. In this case, small (about 5 nm) isolated amorphous clusters and point defects are formed as is depicted in figure 5a.

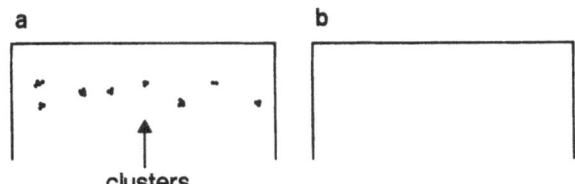

Fig. 5   Schematic   cross-sectional   diagram   showing   the   damage distribution in (100) or (111) Si due to low dose, heavy ion implantation:(a) after implant and (b) after anneal (2.1).

After anneal. After annealing, perfect defect-free substrate-silicon is normally formed by the epitaxial crystallization of the amorphous material in the damage clusters.   Sometimes, however, isolated (interstitial) dislocation loops are found - see figure 5b.

b) medium doses (5 x 10¹³ - 5 x 10¹⁴cm⁻²)

After implant. After implanting heavy ions into silicon at medium doses, a buried amorphous layer is observed to have formed. During implantation, the heavy ions severely damage the crystalline structure of the wafer by transferring large amounts of momentum to the Si atoms and thus displacing them from their lattice sites. This leads to the formation of the amorphous material. The position of the buried amorphous layer corresponds to the peak of the implantation profile. Increasing the dose, increases the amount of damaged material i.e. the buried amorphous layer becomes thicker. It should be pointed out that this amorphous layer is truly buried between two crystalline regions as depicted in figure 6a.

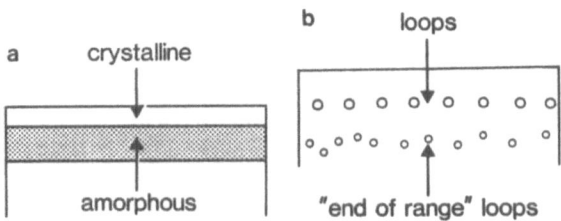

Fig. 6   Schematic   cross-sectional   diagram   showing   the   damage distribution in (100) or (111) Si due to medium dose, heavy ion implantation: (a) after implant and (b) after anneal (2.1).

After anneal. After annealing two distinct layers of interstitial dislocation loops are observed. The position of one layer of loops corresponds to approximately the middle of the previous amorphous layer. During the annealing of the buried amorphous layer, epitaxial regrowth begins at the two amorphous/crystalline interfaces. After some time during the anneal, these two epitaxially regrowing interfaces impinge on each other. As extra atoms are present at this impingement plane (extra atoms have been "implanted" in the material), interstitial loops form. This is the origin of the upper plane of loops depicted in the figure.

The second, deeper layer of dislocation loops (known as "end of range loops") are located at a depth approximately corresponding to the lower amorphous/crystalline interface that existed before annealing. These loops, shown in figure 6b, are formed by inelastic scattering events between the implanted ions and the host atoms in the silicon lattice. When the implanted ions have nearly lost all of their energy i.e. they are nearly at the end of their "range" (or "travel") in the Si, large transfers of energy occur during scattering events causing the host silicon atoms to be displaced out of their lattice sites. During annealing, some of these so-formed interstitial point defects and the additional implanted atoms condense together to form, the so-called, "end of range" interstitial dislocation loops. Figure 7 is a TEM micrograph obtained from an annealed arsenic-implanted silicon wafer. The two distinct layers of interstitial loops are clearly visible.

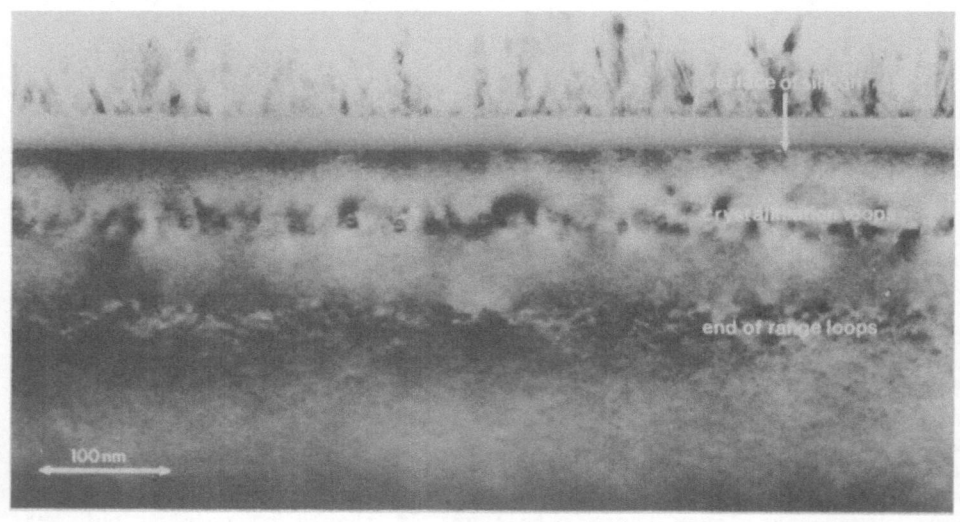

Fig.7 Cross-sectional TEM micrograph displaying two distinct layers of loops in silicon (5 x 10^{14} As$^+$cm^{-2}, 100 keV, 800°C, 10s).

c) High doses (>10^{15}cm^{-2})

After implant. In the early stages of the implantation of heavy ions at high doses, a buried amorphous layer is formed in the silicon substrate (as previously discussed). During prolonged implantation this buried amorphous layer thickens with increasing dose, until, at a certain dose, a continuous amorphous layer is formed, as depicted in figure 8. Thus when a certain threshold dose (about 5 x 10^{14} atoms cm^{-2} for 100 keV As ions) is surpassed, a continuous amorphous layer forms from deep in the substrate to its surface.

After anneal. During the annealing of the continuous amorphous layer, epitaxial crystallization of the amorphous silicon occurs on the underlying crystalline material in a layer by layer process. Complete

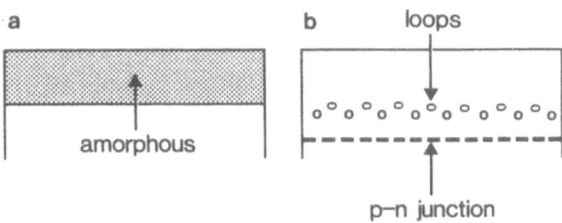

Fig. 8   Schematic cross-sectional diagram showing the damage distribution in (100) or (111) Si due to high dose, heavy ion implantation: (a) after implant and (b) after anneal (2.1).

crystallization by crystal growth into the amorphous material occurs very rapidly at temperatures greater than 600°C. TEM observations have shown that the regrown material has a good crystalline quality.

At a depth corresponding to just below the previous amorphous/ crystalline interface, a layer of dislocation loops is frequently observed. These are end-of-range loops, the formation mechanism of which was described in the previous section. It has been shown, by rapid thermal processing at temperatures greater 1100°C, that the density of these dislocation loops can be reduced to below $10^7 cm^{-2}$. This was pointed out in the case of arsenic in the introductory section - see figure 1. If heavy ions such as Sn and Sb are implanted, the end-of-range loops formed during annealing are found to be confined to a thinner (depth) band than in the case of arsenic implanted with the same ion range. This observation can be explained by considering the fact, that the heavier the ions, the smaller the "projected straggle" (2.22) at the end of the ion's range. The displacement of atoms that occurs during this "projected straggle" is therefore confined to a thinner band. During annealing, interstitial loops form by condensation (as described earlier) only in this thin band and thus the extent of the end-of-range loops is confined.

2.3 Summary

1. Light ions (Z<Si)

During nominal room temperature implantation, complete amorphization does not normally occur. After annealing silicon implanted with a high dose of ions, a large amount of lattice damage is normally present.

2. Heavy ions (Z > Si)

During implantation, amorphization occurs. During high dose implantation, an amorphous layer is formed which is continuous to the surface of the wafer. After the subsequent annealing of the implanted material, it is possible to obtain a low density of lattice defects in the substrate by choosing suitable annealing conditions.

2.4 Significance of complete amorphization

a) Absence of amorphous layer

With the absence of an amorphous layer after implantation, no solid phase epitaxial crystal regrowth can occur during subsequent annealing. Therefore, local agglomeration of point defects occurs leading to the observation of dislocation loops. These lattice defects in the wafer can act as preferential sites for the segregation of the dopant which reduces the amount of dopant available for substitutional sites in the perfect crystalline lattice. Less of the implanted dopant therefore becomes electrically active.

b) Presence of amorphous layer

When an amorphous layer is formed during implantation, solid phase epitaxy can occur during annealing. Solid phase epitaxy is a layer by layer process. Silicon and dopant atoms position (or locate) themselves in substitutional sites; there being no discrimination between the two sorts of atoms. The result is that, after annealing, approximately 95 % of the dopant atoms are electrically active as they sit in substitutional sites in a perfect crystalline lattice. Also, by the careful control of the annealing conditions, perfect epitaxial crystal regrowth can occur without the formation of lattice defects which sometimes form (2.23) during the crystallization process.

A solution to the problem of forming defect-free shallow boron junctions can now be given and understood. The solution involves first amorphizing the silicon wafer by, for example, employing a germanium or tin implant to form a thin amorphous film from the surface into the silicon wafer. By employing such heavy ions, a continuous amorphous layer can be formed as explained earlier. The second stage involves implanting the boron into this continuous amorphous layer, ensuring that the boron implantation profile is contained within the amorphous film.

Sb^+ 100 keV, $1.2 \cdot 10^{14} cm^{-2}$ + (equiv.) B^+ 5 keV; 900 °C, 30 min, N_2

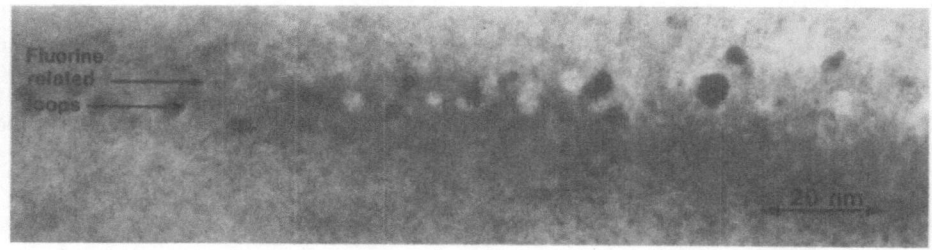

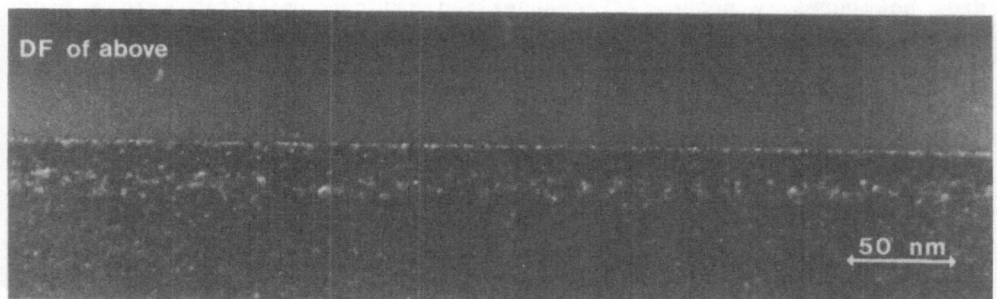

Fig. 9   Cross-sectional TEM micrograph showing defect structure in a shallow boron junction region in an annealed (100) silicon substrate. Shallow junction formed by pre-amorphizing silicon (by Sb^+) followed by BF_3 implantation. The dislocation loops displayed here were the only lattice defects detected in the specimen. The position of these loops coincided with the position of a fluorine peak in a SIMS spectrum obtained from the sample.

All the dopant is now situated in the amorphous layer. During subsequent annealing, solid phase epitaxial crystal regrowth begins. During the epitaxial regrowth process the dopant will be incorporated into substitutional sites in the silicon lattice. Thus after annealing most of the dopant atoms are located in electrically active sites. Also, as pointed out earlier, by choosing an annealing temperature around 600 ^{o}C and a heavy amorphization ion (such as Sn), the number of lattice defects created during the epitaxial crystallization process and the extent of end-of-range loops formed due to "projected straggle" effects will be small. Obviously the existence of a minimal number of lattice defects will reduce the possibility of the degradation of the electrical properties of shallow junctions. There are two additional advantages of the above method. First as implantation occurs into an amorphized material, boron ion channelling can not occur which is required for the formation of shallow junctions. Secondly, by employing a low temperature anneal (typically around 600°C) for the epitaxial regrowth process, dopant diffusion is minimized which also leads to shallower junctions.

The micrograph in figure 9 was obtained from a silicon substrate which was first amorphized by a shallow, heavy ion implantation. Boron in the form of BF_3^+ was then implanted into the amorphized substrate. After an anneal to simulate heat treatments during processing, a relatively shallow, low defect density boron junction was formed. No extended lattice defects were detected (by TEM) in the p$^+$/n junction region (approximately 200 nm from the surface of the substrate) of the annealed sample. Small loops were however detected at a depth of 18 nm. Secondary ion mass spectrometry (SIMS) indicated these loops were "decorated" with fluorine. It is unlikely that these loops will have a detrimental affect on the electrical properties of the junction (in a MOS device) as they are not in the p$^+$/n transition region.

3. SHALLOW EMITTERS (FOR BIPOLAR TRANSISTORS) FROM DOPED POLY-SILICON CONTACTS

In advanced bipolar circuits, shallow emitters and narrow bases are required (3.1,3.2). Such shallow emitters are normally formed by depositing a doped polysilicon layer directly onto the exposed silicon in a structured substrate. Before the polysilicon layer is deposited, the native oxide on the surface of the silicon regions is removed by wet (chemical) etching. If the layer is not in situ doped, then the polysilicon is doped by implantation. An anneal, typically at 950°C for 30 minutes, is then used to drive-in the dopant into the silicon substrate, thus forming the shallow emitter. In this way the polysilicon serves as a diffusion source and a self-aligned emitter contact. The polysilicon emitter contact resistance must be low for good transistor performance. Also, the out-diffusion of arsenic from the polysilicon into the silicon must be homogeneous from the entire area of the polySi/Si interface but minimized in depth so as to form a shallow emitter. Boron diffusion must also be minimized so that the base thickness remains homogeneously thin for good transistor characteristics.

Frequently, two problems are encountered in the production of these shallow emitters. The first of these problems is that the polySi/substrate contact resistance has a high value and the second is that the diffusion of the dopant from the interface is not reproducible "batch to batch". Both problems stem from the existence of a thin oxide at the polySi/substrate interface. One would imagine that, in principle, no oxide should be present at the interface as the purpose of the wet etch carried out just prior to polySi layer deposition is to remove any (native) oxide on the silicon regions. However, during the transport of the wafers (in "boats") into the polySi deposition system (typically at

625°C), an oxide grows on the exposed silicon. Polysilicon deposition then occurs. A thin oxide is therefore left "sandwiched" between the polysilicon layer and the silicon substrate. This "microstructural defect" acts as a diffusion and an electrical barrier between the doped polysilicon and the substrate. As the quality and thickness of this oxide varies between process "batches", arsenic diffusion during the drive-in anneal is not reproducible "batch to batch".

It has been found (3.1) that high temperature (> 1000 °C) annealing can break-up the native oxide between polysilicon and silicon, ultimately forming small islands of oxide along the interface. This effect is commonly known as "oxide ball-up". By employing such a high temperature anneal, the contact and diffusion barrier (in the form of the thin oxide) can be removed. Thus good electrical contact and better control of the diffusion, during the drive-in anneal, can be achieved, making the process reproducible. However, there is an inherent problem here as a furnace anneal at high temperature would lead to a wide base (caused by excessive boron diffusion at high temperatures) and would not result in a shallow emitter (for similar reasoning). Obviously, a short rapid thermal anneal at high temperature could solve the problem. During such a RTA step the thin oxide would break-up (ball-up), enabling epitaxial growth of the silicon in the layer onto the substrate by re-alignment of the crystals in the layer . Such a situation would provide a good electrical contact between the layer and the substrate. There is also the advantage of the inherently short diffusion anneal encountered during an RTA step. Shallow emitters and narrow bases would thus be formed. Additionally, epitaxial attachment of a polySi layer on a Si substrate has been shown (3.9) to promote the formation of shallow emitters.

Figure 10 displays micrographs obtained from three similar silicon wafers covered with a polySi layer which were subjected to different treatments. Figure 10a shows the interfacial native oxide formed during the transport of the wafer into the polysilicon deposition system. The native oxide can be seen sandwiched between the silicon substrate and the polysilicon layer. This wafer was not subjected to any further annealing treatment. Micrograph (b) was obtained from a wafer which had encountered the same treatment as the wafer in micrograph (a) plus a RTA step at 1050 °C for 10 seconds in nitrogen. It can be clearly seen in this micrograph that the oxide is no longer continuous. In fact, epitaxial growth of silicon in the layer onto the silicon substrate has occurred as can be seen in the micrograph (between arrows). It should be noted that the remaining oxide/substrate interface is still relatively flat. Micrograph (c) was obtained from a wafer subjected to the same treatment as that wafer in micrograph (a) plus a step at 1100 °C for 10 seconds. Epitaxial realignment of the silicon in the layer on the substrate, as described above has clearly occurred at many points. The original interface between the layer and the substrate is no longer flat but very rough. Oxide no longer appears to be present. Additionally in this latter example, arsenic was implanted into the polysilicon layer. It has been shown (and confirmed in this experiment) that epitaxial re-alignment rates (3.9) of silicon onto substrate silicon are dependent on arsenic doping levels.

The occurrence of epitaxial alignment in the polysilicon layer on the substrate during annealing is important for the formation of shallow emitters. Results of experiments at our laboratories (3.2) have suggested that suppression of epitaxial alignment (i.e. the preservation of the polycrystalline structure in the layer) maintains a high arsenic diffusion rate from the layer into the substrate and hence produces deeper junctions. It is the presence (3.5, 3.6) of the grain boundaries

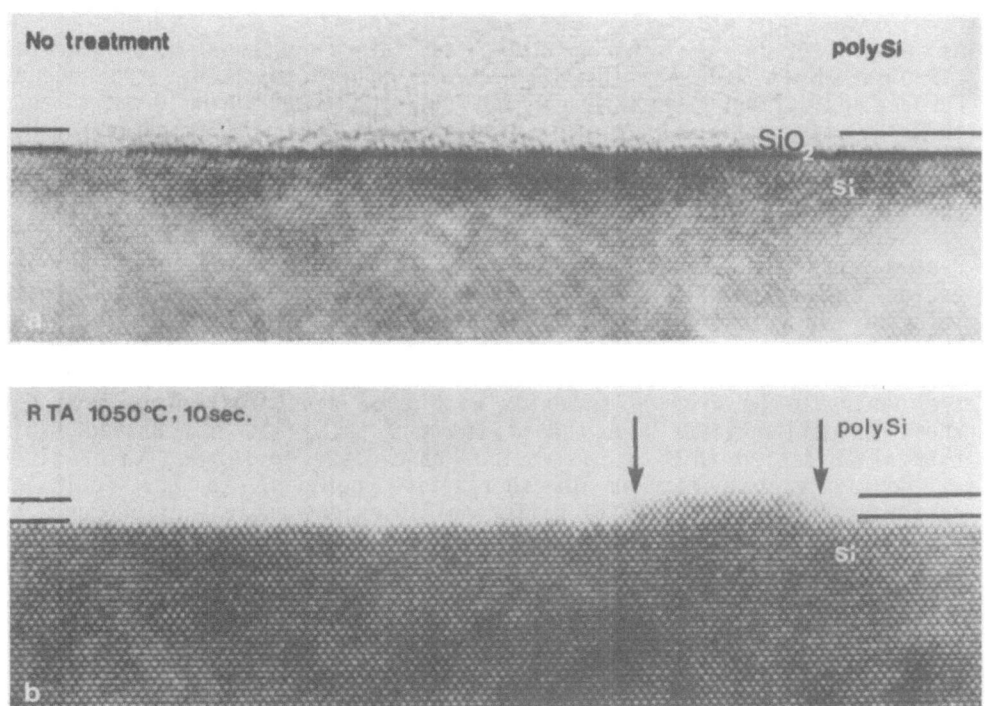

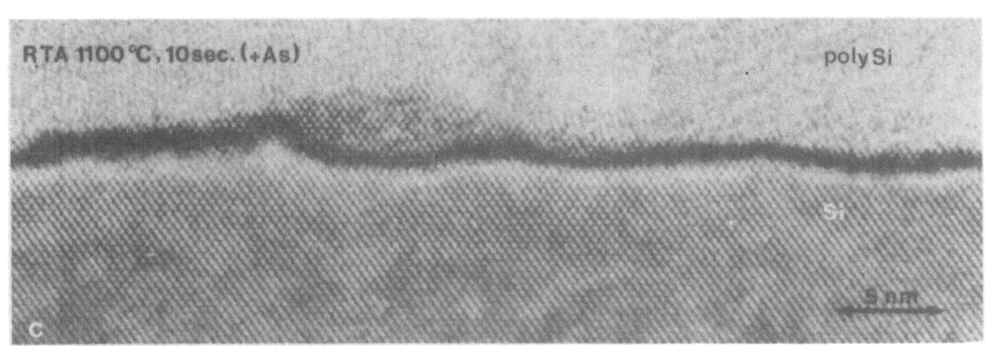

Fig. 10   Shallow emitter structure. Cross-sectional TEM
          micrographs of: a) thin oxide sandwiched between polySi
          and Si substrate; b) the same sample as (a) subjected to
          a 1050°C RTA for 10 secs. in N_2; c) similar to (a) plus
          2×10^{16}As cm^{-2} (60 KeV) into polySi then 1100°C RTA 10
          secs. in N_2.

in the polysilicon layer that maintains the high diffusion rates into the
monosilicon. Thus for the formation of shallow emitters, epitaxial
alignment is desirable as the diffusion rate through the realigned
material is low due to the lack of grain boundaries. Some authors (3.7.)
have suggested that epitaxial (re)alignment may be undesirable. Yet, in a
later publication (3.8) by the same authors, their results suggested that
a rapid thermal anneal at 1100 °C for 5 seconds is indeed well suited to
the formation of shallow emitters.

4. RAPID THERMAL OXIDATION (RTO)

As described in the introduction, as the dimensions of integrated circuits shrink, ultra thin (5 - 10 nm) active insulators are needed. Furnace oxidation, normally carried out at temperatures greater than 1000 ^{o}C for active insulator production, occurs very rapidly. For example, during an oxidation at 1100°C, a 50nm thick oxide (4.1) will grow during the first 15 minutes required to transport the wafers into the hot zone of the oxidation furnace. Reduced oxidation rates, obtained at lower furnace oxidation temperatures, aids the production of reproducible thin oxides as the resulting longer oxidation times are easier to control. However, oxides grown at lower temperatures have inferior electrical properties to those produced at higher temperatures. These low temperature oxides contain more electrical trapping sites. It can therefore be understood that normal ($\geq$1000°C) oxidation temperatures are desirable to produce an oxide with good electrical properties but short oxidation times are needed to grow thin (<10 nm) oxide. Rapid thermal oxidation (RTO) is an obvious candidate for forming thin oxides at desirable temperatures in short time periods. In the following paragraphs, the properties of differing RTO oxides and oxides produced by conventional furnace oxidation will be compared. Firstly, however, a brief description of some of the characteristics of RTO oxides will be given.

a) RTO oxides

The growth rates of oxides in a RTO system have been found to be constant, with an activation energy of 1.4 eV (4.5), after an initial period of rapid growth. Oxide thickness uniformities of $\pm$ 1.5% have been claimed (4.2) and, for oxidation times of less than 60 seconds, dislocation slip (in the substrate) has not been observed (4.2, 4.8). As with furnace oxidation, the presence of a small percentage (about 3%) of HCl during RTO (4.1) helps to remove unwanted metals ions from the system and increases the oxidation rate at a specific temperature.

Comparison between RTO and conventional oxides

Some authors (4.2)claim that sharper silicon dioxide/silicon interfaces are produced with RTO than with low temperature furnace oxidation, and that RTO oxides have better electrical properties than low temperature furnace oxides. No significant difference in interface roughness between the two sorts of oxide was also claimed. Other workers (4.12) who have also examined oxides formed by RTO, disagree with these findings. A part of a TEM study which was carried out on the latter oxides will be given in the following sections.

Figure 11 compares two RTO oxides with a low temperature furnace oxide.Micrographs (a) and (c) respectively were obtained from samples subjected to an RTO at 1150°C for 92 seconds in oxygen (plus 4% HCl) and a conventional furnace anneal at 900°C in 30% oxygen plus 3% HCl (in nitrogen) for about 20 minutes. It can be seen, by comparing the two micrographs, that the oxide formed by RTO (only) appears to have a rougher SiO$_2$/Si interface than that of the furnace oxide. We have also found that such RTO oxides have less sharp interfaces than furnace oxides. Additionally, the electrical properties of RTO oxides were, in general, found to be inferior to those of furnace oxides. The RTO oxide sufferred from large variations in flat-band voltage over the wafer and extrinsic break-downs during testing. The intrinsic break-down voltage of the RTO oxide displayed in figure 11b was determined to be 9.5 MV/cm.

b) Post oxidation rapid thermal annealing (RTA) in nitrogen (or argon)

Subjecting thin oxides formed by RTO to short RT anneals in purified (4.3) nitrogen (or argon) has been claimed (4.3,4.4,4.5,4.7) to improve the electrical properties of the oxides. It has been suggested (4.11)

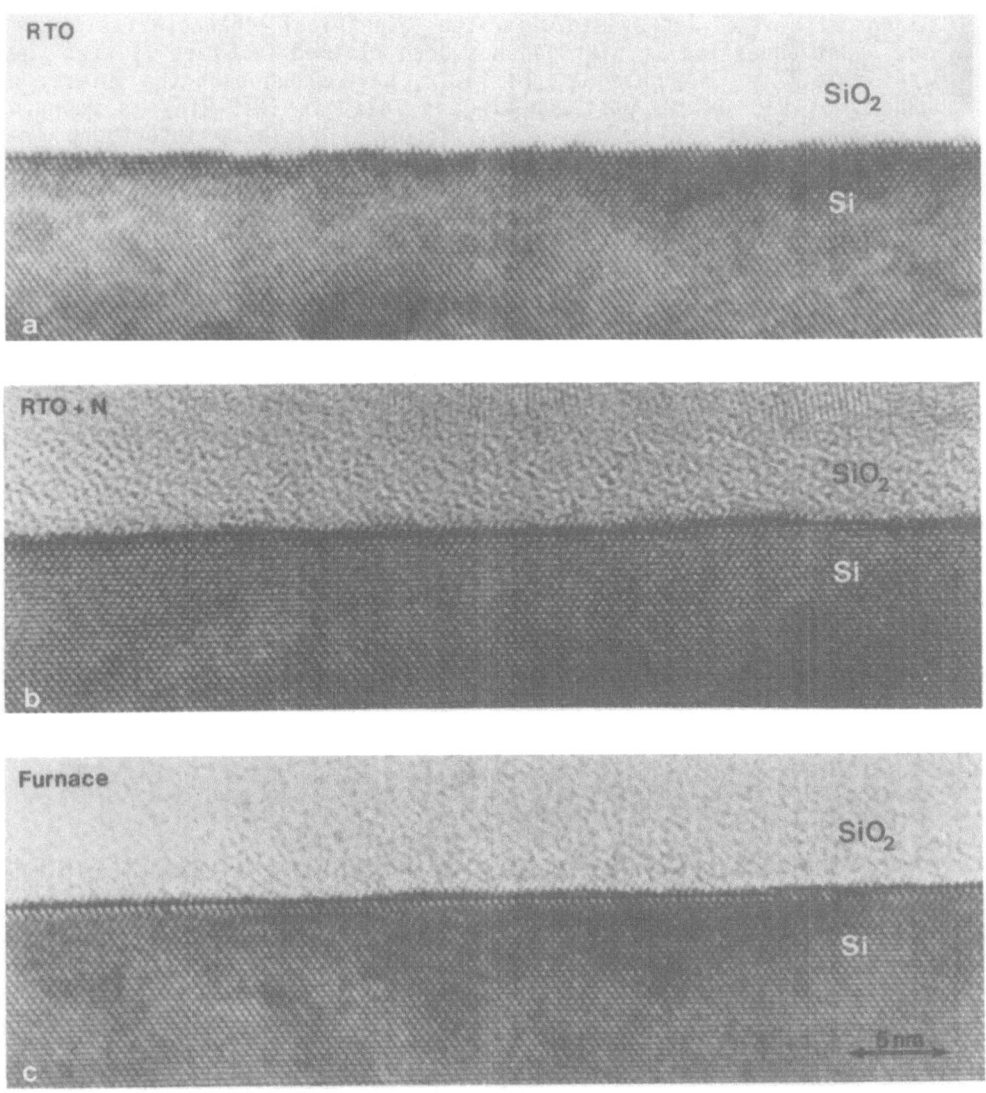

Fig. 11   RTO oxides. Cross-sectional TEM micrographs obtained from oxides on Si substrates formed by a) RTO only; b) RTO followed by RTA in N_2 and c) furnace processing. See text for details.

that RTA changes the local atomic structure of SiO_2, in the bulk and at the SiO_2/Si interface. Small changes in the Si-O-Si bond stretch vibration frequency and the refractive index were interpreted (4.11) as a change in the bond angle of the Si-O-Si structure. Long post-RTO anneals are believed to increase the oxygen deficiency at the SiO_2/Si interface (4.4,4.7,4.9). Additionally during RTA, it is likely (4.4) that water (if present) is removed from the oxide, reducing the number of water-related charge-traps. These charge-traps are thought (4.3,4.9) to be associated with trivalent Si at the SiO_2/Si interface.

In the case of post-RTO anneals in nitrogen, it has been demonstrated (4.2) that initially nitride forms at the surface of the oxide which reduces the number of surface states (4.7). The concomitant reduction in fixed oxide charge (4.7) helps to improve the breakdown voltage (4.4). Prolonged anneals cause a nitrogen enrichment of the SiO_2/Si interface, followed by conversion of the complete film to oxynitride.

Along with the improvement in the electrical characteristics of oxides, post-annealing in nitrogen has been claimed (4.5) to flatten the SiO_2/Si interface. TEM observations (4.12) have shown that the interface roughness of RTO oxides post-annealed in nitrogen (RTO+RTA) is reduced when compared with RTO oxides - see figure 11b. These interfaces are still not as flat as those encountered in furnace oxides - see figure 11c. The interface sharpness of RTO+RTA oxides is also improved - compare figures 11a and 11b. The sharpness of RTO+RTA oxides and furnace oxides is approximately the same - compare figures 11b and 11c. The electrical properties of the oxide subjected to post-annealing, displayed in figure 11b, were not as good as those of the oxide formed by furnace oxidation (figure 11c) although they were better than those of the RTO oxide (figure 11a).

A poor example of an oxide produced by RTO + RTA in nitrogen ($1150^{\circ}C$, 24 sec., O_2 plus $1050^{\circ}C$, 30 sec., N_2) is displayed in figure 12. The roughness of the SiO_2/Si interface is clearly visible. The electrical-breakdown characteristics of this oxide were particularly bad.

Fig. 12   Cross-sectional TEM micrograph of a RTO oxide which exhibited poor electrical characteristics - immediate electrical breakdown during Q_{bd} stepped current stressing (4.12).

c) Post-annealing in oxygen

A short oxygen anneal following post oxidation anneals has been found to reduce hole-trapping and shallow electron trapping (4.4,4.9). It is thought that such an anneal restores the stoichiometry of the silicon dioxide at the originally oxygen deficient interface, reducing the number of traps and thus improving the electrical quality of the oxides.

5. RAPID THERMAL PROCESSING OF SILICIDES

Basically there are two methods of producing silicides as illustrated in figure 13. Either titanium is deposited on silicon or titanium and silicon are simultaneously co-sputtered onto silicon and, in both cases, the structures are annealed forming the silicides. In the following section, the characteristics of the materials produced by two respective methods will firstly be discussed independently and later the characteristics will be compared. Silicides formed by reacting the metal directly with silicon will be discussed first. Titanium disilicide has been chosen as a representative silicide used in the IC-industry.

5.1) Ideal self-aligned process

The ideal self-aligned titanium disilicide process (5.1) is illustrated in figure 14. Titanium is deposited over silicon dioxide and exposed silicon areas. The structure is then annealed at 600 °C in a nitrogen atmosphere. Where titanium is in direct contact with silicon, titanium

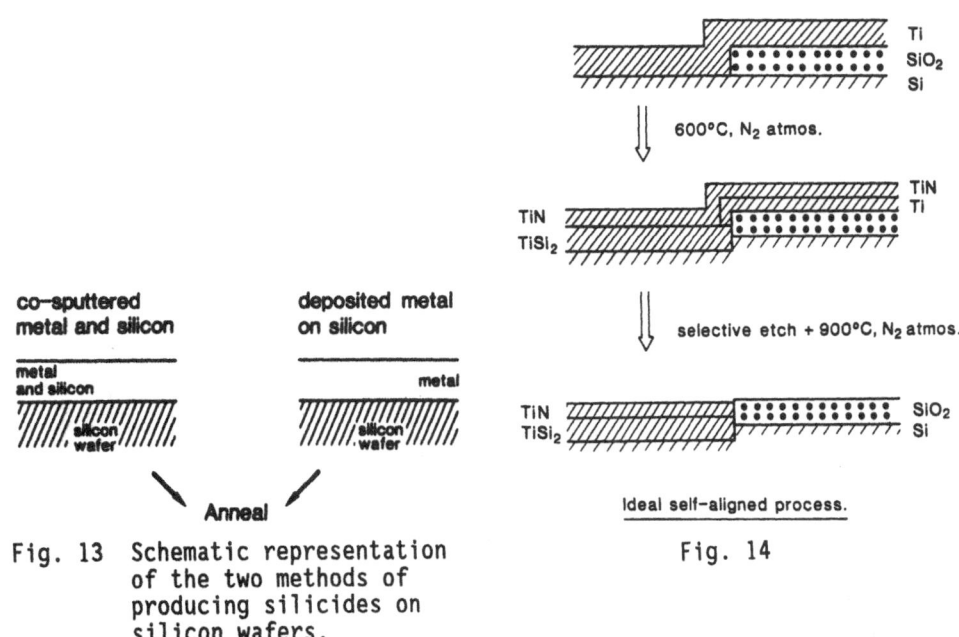

co-sputtered
metal and silicon

deposited metal
on silicon

Fig. 13 Schematic representation
of the two methods of
producing silicides on
silicon wafers.

Fig. 14

disilicide forms. Where titanium is in direct contact with silicon dioxide, no reaction is required. At the exposed surface of the titanium a reaction occurs between the titanium and the nitrogen ambient forming a thin titanium nitride layer over the complete structure as illustrated in the figure. After selective etching to remove the titanium and titanium nitride from silicon dioxide areas, a second anneal at typically 900 °C a nitrogen ambient is carried out.

Why is 600°C chosen for the first anneal? The reason for choosing this temperature can be understood by referring to reference 5.1 (figures 1-4). When 28 nm of titanium reacts with silicon at 600°C, approximately 28 nm of silicide is formed which is considered (5.1) a reasonably thick silicide layer for IC applications. Temperatures below 600°C produce only a few nanometres of silicide. Therefore, temperatures of at least 600 °C are required.

In the case of titanium on silicon dioxide, 28 nm of titanium _in fact_ reacts with silicon dioxide (5.1, 5.2, 5.3) at 600 °C in a nitrogen ambient to form 4 nm of silicide. This silicide is not the disilicide but another titanium rich silicide: Ti_5Si_3. When the same anneal is carried out in argon, 6 nm of Ti_5Si_3 forms. It appears therefore that the nitrogen ambient (forming titanium nitride) suppresses the formation of silicide on silicon dioxide. This is more clearly shown in figure 15- specimens annealed at 900°C in nitrogen and argon. In this figure it can be clearly seen that the formation of a nitride (in this case oxy-nitride) suppresses the thickening of the Ti_5Si_3. (In the case of annealing in an argon ambient, two titanium oxides form above the Ti_5Si_3 layer. Rutherford backscattering spectroscopy and Auger electron spectroscopy confirmed that these are respectively TiO_2 and TiO as shown in the figure). Therefore from the above discussion it can be seen that

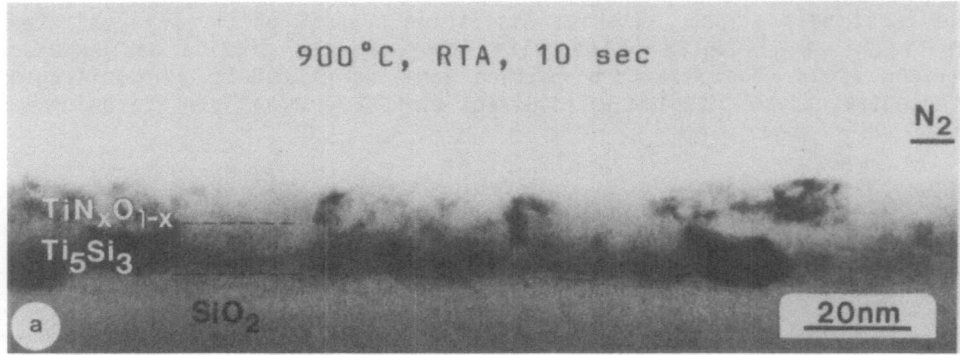

Fig.15   Cross-sectional TEM micrographs displaying the layers formed
during the reaction of Ti with SiO$_2$ at 900°C in a) N$_2$ and b) Ar
ambients.

the first anneal must be carried out at temperatures of about 600 ^{o}C so
that a minimum thickness of silicide forms over silicon dioxide regions
while a reasonable thickness of disilicide is formed on silicon areas.

Why then is the second anneal required at the higher temperature
(typically 800 to 900°C)? Low resistivity C54-phase titanium disilicide
only forms at temperatures above about 700°C. It is this low resistivity
material which is required for IC applications. Therefore in the
formation of titanium disilicide via the self-aligned process, a second
anneal at a temperature around 800°C is needed, after the selective etch,
to form the required low resistivity material. Additionally if the second
anneal is carried out at a sufficiently high temperature (e.g. 900°C), a
thin nitride layer forms on the surface of the disilicide (5.5) as shown
in the lower diagram in figure 14. The presence of a thin titanium
nitride on the surface of titanium disilicide has been shown to act as a
self-aligned diffusion barrier (5.6) between subsequently deposited
aluminium and the underlying titanium disilicide.

Figure 16 shows a titanium disilicide layer that has undergone both
anneals at 600 and 900°C, and the intermediate selective etch. The
thickness of the disilicide is about 27 nm. A thin titanium nitride
layer about 10 nm thick is present on the surface of the disilicide. It
should be noted that the interface between the titanium disilicide and
the substrate is fairly rough (roughness about 10 nm.). Such a roughness

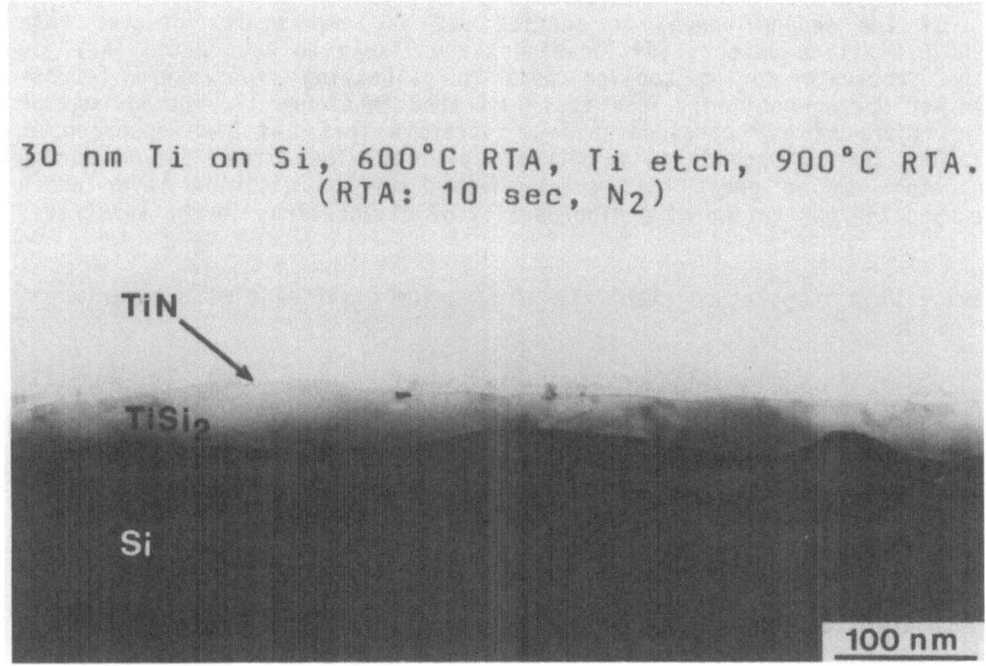

Fig. 16   Cross-sectional TEM micrograph of a titanium disilicide layer
that has undergone both anneals and the intermediate etch, as
described in the text.

Fig. 17   Cross-sectional micrograph of a $TiSi_2$ layer RT processed at
$1100^{\circ}C$ for 10 sec. Note the break in the layer which generates
dislocations in the silicon.

may cause junction "spiking" problems when this layer is used in combination with shallow junctions.

If the second anneal is carried out at temperatures greater than 900°C, lattice defects (dislocations) are likely to be created (5.1) in the substrate during cooling from the annealing temperature (stress formed during cooling). This is illustrated in figure 17. The micrograph in figure 17 was obtained from a substrate that had been annealed at 1100°C for 10 seconds in a nitrogen ambient. The stress formed during cooling was so great that cracks formed in the silicide layer which caused the generation of a high density of dislocations in the substrate.

5.1.1 High temperature stability of titanium disilicide with covering oxide

After the formation of titanium disilicide, oxide is normally deposited over the silicide. This oxide is then frequently subjected to a flow and/or reflow anneal at elevated temperatures. It has been found (5.7) that very bad "thermal-grooving" can occur in the silicide film, leading to physical breaks in the layer, during the anneals associated with flow and/or reflow. Obviously such a discontinuous silicide film is highly undesirable.

A so-called thermal-groove develops between two grains in a layer so that the resultant of the surface tension of each grain and the grain-boundary tension, vanishes along the line of the grain-boundary. Physical breaks in a layer occur by the progressive deepening of the thermal groove during annealing which occurs by the migration of atoms along the surface of the layer away from the groove. More information about thermal grooving can be found in reference 5.7.

Figure 18 displays a silicide film which became discontinuous (5.7) by thermal-grooving. In this example, tetraethoxysilane oxide (TEOS) was deposited over C49-phased titanium disilicide. The resulting structure was then subjected to a 1050°C RTA for 10secs. in nitrogen. The silicide film is clearly discontinuous, as was confirmed by four-point probe resistance measurements. When a TEOS film was deposited over C54-phased titanium disilicide and subjected to the same RT anneal, the silicide remained continuous and thermal grooving was not observed. It appears that the small-grained C49-phased silicide is particularly susceptible to thermal grooving when in contact with an oxide surface layer. Additionally it was found that the thermal grooving that occurs in

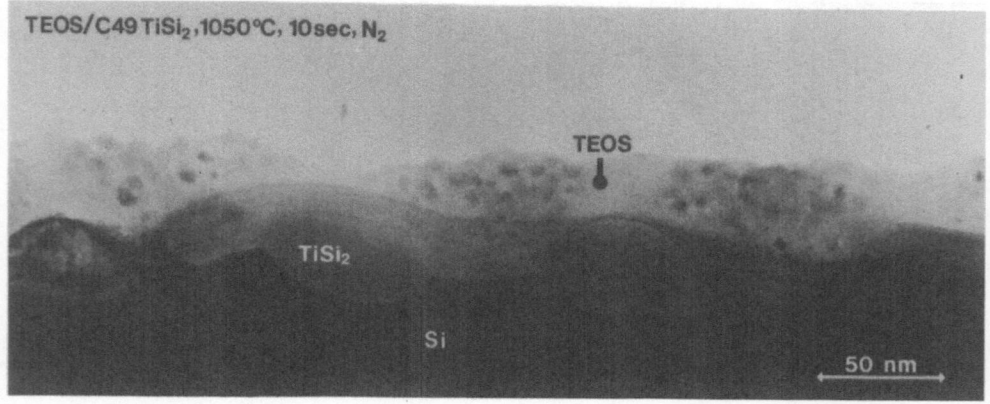

Fig. 18   Discontinuous titanium disilicide layer covered with silicon dioxide (TEOS). Thermal grooving has occurred during the RTA treatment.

titanium disilicide during RTA is suppressed by the presence of a titanium nitride film on the surface of the silicide prior to TEOS deposition.

5.1.2 TiSi$_2$ strap process

In order to locally electrically interconnect junction regions separated by oxide areas, in an integrated circuit, so called titanium disilicide straps (5.8) can be formed over the oxide areas. These straps are produced by blanket-depositing titanium, followed by a deposition of amorphous silicon - see figure 19. After masking, patterning and etching, the structure is reacted during an anneal to form localized titanium disilicide inter-connect, the "straps"- see lower diagram in figure 19. In the case where the amorphous (α) silicon is deposited in an argon plasma sputtering system, frequently imperfect strap structures (5.9) form during subsequent annealing. For the case of strap structures deposited on monocrystalline silicon regions with a Ti/α-Si ratio of less than 0.5 (i.e. additional silicon available above stoichiometry), an additional silicon layer forms on the substrate below the silicide during annealing. The additional silicon available (above stoichiometry) in the α-layer diffuses through the silicide to the silicide/silicon interface during the anneal. Here, epitaxial growth of this extra silicon occurs on the substrate. This extra undoped silicon could cause contact resistance problems between the silicide and the underlying junction.

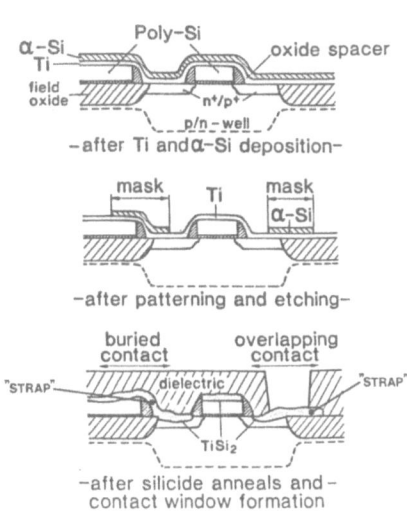

-after Ti and α-Si deposition-

-after patterning and etching-

-after silicide anneals and - contact window formation

Fig. 19 Schematic representation of the process for forming TiSi$_2$ straps over oxide regions using patterned amorphous (α) silicon (5.9).

In the case of straps over silicon dioxide regions where again the Ti/Si ratio is less then 0.5 ,"blisters" are frequently observed on the silicide layer (see figure 20) causing the surface of the silicide layer to be rough. The "blisters" consists of an argon/hydrogen bubble capped by a thin polySi layer. They form during the anneal by diffusion of argon and hydrogen from the amorphous silicon (5.10) to Kirkendall voids that form during the reaction (5.11). Kirkendall voids form at the interface between amorphous silicon and titanium during annealing. The excess silicon (above stoichiometry) crystallizes during the anneal to form a surface capping layer (of polysilicon) over the voids filled with gases. In this way the gases are retained and blisters form on the surface of the silicide. Thus for implementation in the commercial process the

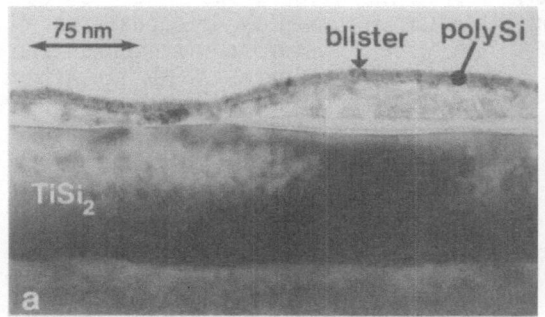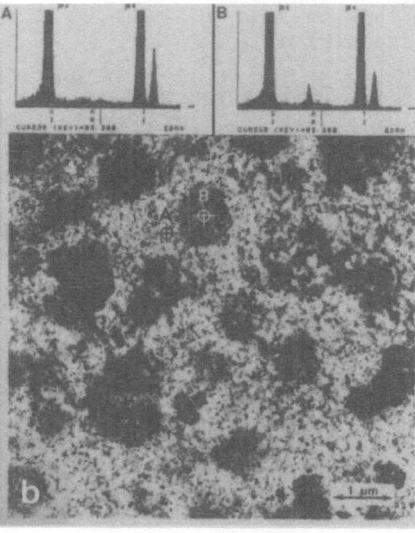

Fig. 20   "Blisters" on TiSi$_2$: a) Cross-section and b) plan-view of "blister". X-ray microanalysis (b) indicates that blisters contain (at least) argon.

atomic ratio Ti/α-Si should be choosen to be between stoichiometric and "slightly silicon deficient" to avoid blister formation. In this way a polysilicon capping layer will not form on silicide over oxide regions and no extra silicon layer will form between the substrate and the silicide layer.

As mentioned earlier there are two methods of producing silicides. Up to this point, the method of reacting titanium directly with silicon has been discussed. Now the method of reacting co-sputtered titanium-silicon on substrate silicon will be discussed.

5.2   Co-sputtered titanium - silicon layers

5.2.1 Grain size of C49 titanium disilicide depends on heating rate

In situ observations in a TEM (figure 21a) of nucleation and growth phenomena in co-sputtered titanium-silicon layers have indicated (5.12) that nucleation has a higher activation energy than growth. This implies that during low heating rates, only a few nuclei form and grow together, thus forming large grains. By obtaining high temperatures almost instantaneously (as can occur by RTP) many nuclei form and grow together, thus forming small grains. These experiments suggest that differing heating rates produce materials with differing grain-sizes. This prediction was, in fact, confirmed by further TEM observations-see figure 21b. Large grain size material is formed during low heating rates and small grain size material is formed during high heating rates (RTP).

More interestingly it was found that the grain size affects the C49-C54 phase transition in the disilicide. Small grain size material was found to transform at a lower temperature than large grain-size material. It is believed that the additional strain energy in the small grain-sized material induces the phase transition to occur at the lower temperature (5.13).

5.2.2 Microstructural defects in annealed co-sputtered layers

Two microstructural features have been observed (5.16, 5.17) in

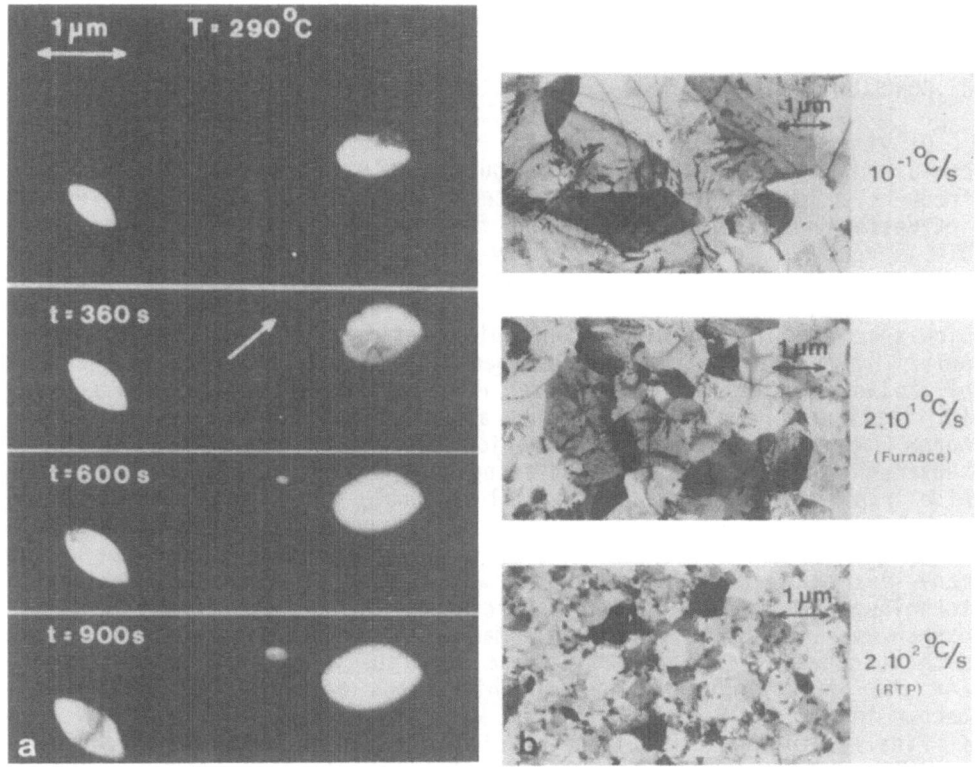

Fig. 21   Plan-view micrographs obtained from co-sputtered Ti-Si layers:
          example micrographs used to determine a) the nucleation and
          growth characteristics of the layers and b) the influence of
          the heating rate of the crystallization anneal on the grain
          size of the layers.

annealed co-sputtered titanium disilicide layers. These are namely
stacking faults and precipitates. The high density of stacking faults
have been found (5.16) to originate from omissions or insertions of
close-packed (060) planes in the C49 titanium disilicide lattice .
Precipitates are only noted in silicon rich (Ti/Si<0.5) material. Many
precipitates are observed in very silicon-rich material whereas a
similar density of small precipitates is noted in slightly silicon-rich
material. Electron diffraction in the TEM has indicated that these
precipitates are comprised of silicon. It is thought that these
precipitates originate from the precipitation of the excess silicon to
yield the stoichiometry of the disilicide compound.

5.3 Comparison of the silicides produced by the two production methods

In the following paragraph the properties of silicides formed by the
two production techniques will be discussed and compared. From the above
it is now understood that annealed co-sputtered material can contain a
high density of stacking faults and precipitates. If the resistivity of
the two types of C49 phased material is compared, it is found (5.18) that
the annealed co-sputtered material has a resistivity of 120 $\mu\Omega$cm whereas
material formed from titanium on silicon has a resistivity of 60 $\mu\Omega$cm.
Comparing the stacking fault density in the two materials is very
revealing. The annealed co-sputtered material has a stacking fault
density of approximately 3×10^6 faults cm^{-1} whereas material formed from
titanium on silicon has a stacking fault density of approximately 2×10^5
faults cm^{-1}. It has been concluded (5.18) that the extremely high

137

stacking fault density in the annealed co-sputtered material causes the large restivity noted in that material. High values of resistivity had been noted in other similar silicides (5.19, 5.20).

6. CONCLUDING REMARKS

Of the many production steps in integrated circuit manufacture that could, in principle, be carried out by RTP, currently only two are frequently executed with this technique. These steps are dopant activation and silicidation. In the future, the number of steps in which RTP is routinely used will certainly rise as device dimensions decrease. In this review, a number of these steps has been discussed. For example, it is highly likely that RTP will be used in the future production of high-speed bipolar circuits containing shallow emitters formed from a polySi layer for the reasons descibed in the text. However, in the case of shallow p^+ junction formation in MOS devices, RTP may not be required for lattice defect removal if pre-amorphization is carried out before boron implantation. Dopant activation would then be achieved during a subsequent high temperature processing step, such as during silicidation. Also rapid thermal oxidation will probably not immediately replace furnace oxidation due to a reluctance by process engineers to accept a technique in which they have relatively little experience. Due to the many years of experience gained while using conventional oxidation techniques, furnace oxides can be produced, at low temperatures in dilute oxygen ambients, in a controllable manner. The properties of such oxides, with thicknesses down to about 8 nm, are _sufficient_ for current needs. For the production of oxides of thicknesses below 8 nm, RTO may well become important.

Finally, in the future manufacturing of advanced integrated circuits, it is highly likely that all (if not most) high temperature (>700°C) process steps after source/drain dopant implantation will have to be executed by RTP. This is likely to occur because small dimension devices demand shallow, accurately known dopant profiles. RTP will ensure an accurate control of the time-period and temperature of each constituent anneal after the implantation. This, in turn, will allow the total "thermal-process" encountered by the dopant to be determined. A sound knowledge of the total "thermal-process" will enable the ultimate dopant profiles to be accurately predicted.

Acknowledgements

I wish to thank M. Geyselaers, R. Hokke, B. Otterloo, H. v. Houtum, R. v. Es, A. Walker and other colleagues at Philips Research Laboratories for their assistance, making their work available and their cooperation.

References

1.1 T.O. Sedgwick, Proc.Symp."Reduced Temperature Processing for VLSI", Eds.G.R. Strinivasan and R. Reif 86-5, Electrochem.Soc., Princeton, NJ 1086, p.49.

1.2 H.C. de Graaf and J.G. de Groot, IEEE Trans. Electron Dev. ED-26, 1771 (1979)

1.3 C.Y. Wong, A.E. Michel, R.D. Isaac, R.H. Kastl and S.R. Mader, J.Appl.Phys. 55, 1131 (1984).

1.4 T. Hara, H. Suzuki and M. Furukawa, Jap.J.Appl.Phys. 23, L452 (1984).

1.5 N.S. Alvi and D.L.Kwong, J.Electrochem.Soc. 133, 2626 (1986)

1.6 J.S. Mercier, L.D. Madsen and I.D. Calder, Mat. Res. Soc.Proc. 52, 251 (1986).

1.7 C.S. Pai, E. Cabreros, S.S. Lau, T.E. Seidel and I. Suni, Appl. Phys.Lett.$\underline{46}$, 652 (1985).

1.8 R.A.M. Wolters and A.J.M. Nellissen, Proc. 4th Int. IEEE VLSI Multilevel Interconnect Conf. 1987, p.351.

1.9 N.S. Alvi, S.M. Tang, R. Kwor and M.R. Fulcher, J. Appl. Phys. $\underline{62}$, 4878 (1987).

2.1 D.K. Sadana, MRS Proc. $\underline{92}$, 319 (1987).

2.2 S.M. Davidson and G.R. Booker, Rad. Ef., $\underline{6}$, 33 (1970).

2.3 T.E. Seidel, D.J. Lischner, C.S. Pai, R.V. Knoell and D.H. Maher, Nucl Inst. and Meth. $\underline{B7}$, 251 (1985).

B^+ - medium doses

2.4 C.C. Ho, R. Kwor, C. Aranjo and Gelpy, MRS Proc. $\underline{52}$, 225 (1986).

2.5 J. Huang and R.J. Jaccodine, MRS Proc. $\underline{52}$, 57 (1986).

2.6 C.R. Peter, J.P. de Souza and C.M. Hasenack, J.Appl.Phys.$\underline{64}$, 2696 (1988).

B^+ High doses

2.7 W.K. Hofker, Philips Res. Reports Suppl. 8 (1975).

2.8 H-Y Liu, P-H Chang, J. Liu and B-Y Mao, MRS Proc. $\underline{92}$, 15 (1987).

2.9 N.E.B. Cowern, K.J. Yallup, D.J. Godfrey, D.G. Hasko, R.A. Mc. Mahon, Ahmed, W.M. Stobbs and D.S. McPhail, MRS Proc. $\underline{52}$, 65 (1986).

2.10 D.G. Hasko, R.A. McMahon, H. Ahmed, W.M. Stobbs and D.J. Godfrey, Inst. Phys. Conf. Series $\underline{76}$, 99 (1985).

Heavy ions

2.11 D.K. Sadana, M.C. Wilson, G.R. Booker and J. Washburn, J. Electrochem. Soc. $\underline{127}$, 1589 (1980).

Heavy ions - medium doses

2.12 D.K. Sadana, J. Washburn and G.R. Booker, Phil. Mag. $\underline{B46}$, 611 (1982).

2.13 J. Narayan, D. Fathy, O.S. Oen and O.W. Holland, J. Vac. Sci. Technol. $\underline{A2}$, 1303 (1984)

Heavy ions - high doses

2.14 Ion Implantation in Semiconductors, J.W. Mayer, L. Eriksson and J.A. Davies (Academic Press, New York), 1970, Chapter 5.

2.15 N.R. Wu, D.K. Sadana and J. Washburn, App. Phys. Lett. $\underline{44}$, 782 (1984).

2.16 D.K. Sadana, W.Maszara, J. Wortman, G.A. Rozgony, and W.K. Chu, J. Electrochem. Soc. $\underline{131}$, 943 (1984).

2.17 D.M. Maher, R.V. Knoell, M.B. Ellington and D.C. Jacobson, Proc. MRS $\underline{52}$, 93 (1986).

2.18 W. Maszara, D.K. Sadana, G.A. Rozgonyi, T. Sands and J.Washburn, Proc. MRS. $\underline{35}$, 277 (1984).

2.19 D. Wouters, D. Avau, P. Mertens and H.E. Maas, MRS Proc. $\underline{52}$, 217 (1986).

2.20 C.C. Ho, R. Kwor, C.Araujo and J. Gelpy, Proc. MRS $\underline{52}$, 225 (1986).

2.21 D.L. Kwong, N.S. Alvi, Y.H. Ku and A.W. Cheung, MRS Proc. $\underline{52}$, 241 (1986).

2.22 Ion Implantation, T.E. Seidel in "VLSI Technology", ed. Sze, McGraw-Hill, 1984.

2.23 A.H. Reader, F.W. Schapink and S. Radelaar, Inst. Phys. Conf. Series $\underline{76}$, 151 (1985).

Si$^+$, Ge$^+$, Sn$^+$ preamorphization

2.24 M. Delfino, D.K. Sadana and A. Morgan, App. Phys. Lett. 49, 575 (1986)
M. Delfino, A.E. Morgan and D.K. Sadana, Nucl. Inst. Meth. B19/20, 363 (1987).

2.25 W. Maszara, D.K. Sadana, G.A. Rozgonyi, T. Sands and J. Washburn, Proc. MRS 35, 277 (1984).

2.26 W.W. Park, M.F. Becker and R.M. Walser, J. Mater. Res. 3, 298 (1988).

2.27 D.K. Sadana, T. Sands, W. Maszara and G.A. Rozgonyi, Inst. Phys. Conf.Series 76, 93 (1985)

2.28 C. Carter, W.Maszara, D.K. Sadana, G.A. Rozgonyi, J. Liu and J. Wortman, Appl. Phys. Lett.44, 459 (1983)

2.29 T. Sands, J. Washburn, E. Myers and D.K. Sadana, Nucl. Inst. & Meth. B7/8, 337 (1985).

Working 1 um CMOS circuit incorporating Ge$^+$ preamorphization

2.30 D. Sharma, S. Goodwin-Johansson, D-S. Wen, C.K. Kim and C.M. Osburn, Electrochem. Soc. (Philadelphia, Pennsylvania) Meeting May 1987.

2.31 D.S. Wen, C.M. Osburn, G.A. Rozgonyi, and P. Smith, Electrochem. Soc. (Philadelphia, Pennsylvania) Meeting May 1987.

3.1 C.Y. Wong, A.E. Michel, R.D. Isaac, R.H. Kastl and S.R. Mader, J.Appl.Phys.55, 1131 (1984).

3.2 W.L.M.J. Josquin, P.R. Boudewijn and Y. Tamminga, Appl. Phys. Letts. 43, 960 (1983).

3.3 G.L. Patton, J.C. Bravman and J.D. Plummer, IEEE trans. ED-33, 1754 (1986).

3.4 J.M.C. Stork, M. Arienzo an C.Y. Wong, IEEE Trans. ED-32, 1766 (1985).

3.5 A.H. Reader, F.W. Schapink and S. Radelaar, in "Poly-micro-crystalline and amorphous Semiconductors"., MRS-Europe Conf.Proc.1984, eds. P.Pinard and S. Kalbitzer, published by Les Edition de Physique, Les Ulis, France, p.253.

3.6 S.R. Wilson, R.B. Gregory, W.M. Paulson, S.J. Krause, J.D. Gressett, A.H. Hamdi, F.D. Mc Daniel and R.G. Downing, J. Electrochem. Soc. 132, 922 (1985).

3.7 H.J. Bohm, H. Wendt, H. Oppolzer, K. Masseli and R. Kassing, J. Appl. Phys. 62, 2784 (1987).

3.8 H.J. Bohm, H. Kabza, T.F. Meister and H. Wendt, Int Electron Devices meeting 1987, published by IEEE, p. 269.

3.9 J.L. Hoyt, E.F. Crabbé, J.F. Gibbonsn and R.F.W. Pease, Appl. Phys.Lett., 50, 751 (1987).
J.L. Hoyt, E.F. Crabbé, J.F. Gibbonsn and R.F.W. Pease, MRS Proc. 92, 47 (1987).

3.10 M. Delfino, J.G. de Groot, K.N. Ritz, P. Maillot and A.E. Morgan, submitted to J. Electrochem. Soc. (Note: capping oxide important to prevent As loss).

3.11 R.H. Reuss and T.P. Bushey, MRS Proc. 92, 221 (1987)

3.12 A. Kermani, F. Van Giesen, S. Litwin, R. Sullivan, T.J. DeBolske and J.L. Crowley, MRS Proc. 92, 227 (1987).

3.13 J.L. Hoyt, E.F. Crabbé, R.F.W. Pease, J.F. Gibbons and A.F. Marshall, J. Electrochem. Soc. 135, 1773 (1988).

3.14 V. Probst, H.J. Bohm, H. Schaber, H. Oppolzer and I. Weitzel, J. Electrochem. 135, 671 (1988).

4.1 J.C. Gelpey, P.O. Stump and R.A. Capodilupo, MRS Proc. 52, 321 (1986).

4.2 J. Nulman, J.P. Krusius and P. Rentteln, MRS Proc. 52, 341 (1986).

4.3 Z.A. Weinberg, J.N. Nguyen, S.A. Cohen and R. Kalish, MRS Proc.52, 327 (1986).
4.4 Z.A. Weinberg, D.R. Young, J.A. Calise, S.A. Cohen, J.C. Deluca, V.R. Deline, Appl. Phys. Lett., 45 (11), 1204 (1984).
4.5 N.Chan. Tung, Y. Caratini and J.L. Buevoz, Mat. Res. Soc. Proc. 92, 147 (1987).
4.6 M.M. Moslehi, MRS Proc. 92, 73 (1987).
4.7 L. Dori, M. Arienzo, Y.C. Sun, T.N. Nguyen, J. Wetzel, Mat.Res.Soc.Proc. 76, 259 (1987).
4.8 J. Nulman, J.P. Krusuis, N. Shah, A. Gat, A. Balduin, J. Vac.Sci. Technol. A 4, 1005-1008 (1986).
4.9 P. Balk, M. Aslam, D.R. Young, Solid State Electronics 27, 709 - 719 (1984).
4.10 M.M. Moslehi, K.C. Saraswat, S.C. Shattss, Appl.Phys.Lett. 47 (10), 1113-1115 (1985).
4.11 J.T. Fitch, G. Lucowsky, MRS Proc. 92, 89 (1987).
4.12 A.J. Walker and A.H. Reader, unpublished.
4.13 C.H. Seager and W.K. Schubert, J.Appl. Phys. 63, 2869 (1988).

5.1 A.E. Morgan, E.K. Broadbent and A.H. Reader, MRS. Proc.52, 279 (1986).
5.2 R. Beyers and R. Sinclair, J. Vac. Sci. Technol.B2, 781 (1984).
5.3 A.E. Morgan, E.K. Broadbent, K.N. Ritz, D.K. Sadana and B.J. Burrow, submitted to J. Appl. phys.
5.4 A.E. Morgan, E.K. Broadbent and D.K. Sadana, Appl. Phys. Lett.49, 1236 (1986).
     J.C. Barbour, A.E.T. Kuiper, M.f.C. Willemsen and A.H. Reader, Appl. Phys. Lett. 50, 953 (1987).
5.5 M.F.C. Willemsen, A.E.T. Kuiper, A.H. Reader, R. Hokke and J.C. Barbour, J. Vac.Sci. Technol. B6, 53(1988).
5.6 M. Delfino, E.K. Broadbent, A.E. Morgan, B.J. Burrow and M.H. Norcott, IEEE Electron Device Lett. EDL-6, 591 (1985).
5.7 H.J.W. van Houtum, I. Menting, A. Moet, M.L.J. Geyselaers and A.H. Reader, to be published.
5.8 S.S. Wong, D.C. Chen, P. Merchant, T.R. Cass, J. Amano and K.Y. Chiu, IEEE Transactions on Electron Devices, ED-34 (3), 587 (1987).
5.9 H.J.W. van Houtum, A.A. Bos, A.G. M. Jonkers and I.J.M.M. Raaymakers, to be submitted to J. Appl. Phys.
5.10 G.C.A.M. Janssen and P.J.J. Wessels, J.Appl. Phys. 62, 2993 (1987).
5.11 I.J.M.M. Raaymakers, A.H. Reader and P.H. Oosting, J.Appl. Phys. 63, 2790 (1988).
5.12 I.J.M.M. Raaymakers, A.H. Reader, H.J.W. van Houtum, J. Appl. Phys. 61, 2527 (1987).
5.13 H.J.W. van Houtum, I.J.M.M. Raaymakers and T.J.M. Menting, J. Appl. Phys. 61, 3116 (1987).
5.14 E.K. Broadbent, A.E. Morgan, B. Coulman, J.W. Huang, and A.E.T. Kuiper, Thin Solid Films, 151, 51 (1987).
5.15 T. Brat, J.C.S. Wei, J. Poole, D. Hodul, N. Parikh and C. Wickersham, MRS Proc., 92, 191 (1987).
5.16 A.H. Reader, I.J. Raaymakers and H.J. van Houtum, Inst. Phys. Conf. Ser. No. 87, 523 (1987).
5.17 R. Beyers and R. Sinclair, J. Appl. Phys. 57, 5240 (1985).
5.18 A.H. Reader, A.H. van Ommen, and H.J.W. van Houtum, MRS Proc. 92, 177 (1987).
5.19 F.M. d'Heurle, F.K. LeGoues, and R. Joshi, Appl.Phys.Lett 48, 332 (1986).
5.20 A.H. van Ommen, A.H. Reader and J.W.C. de Vries, J.Appl.Phys. 64, 3574 (1988).

RAPID THERMAL ANNEALING - THEORY AND PRACTICE

C. Hill, S. Jones and D. Boys

Plessey Research Caswell Ltd
Towcester, Northants, England NN12 8EQ

1.   INTRODUCTION

   Throughout the three decades since the beginnings of the integrated
circuit industry in 1958, heat-treatment has been an essential part of the
processing technology for annealing, oxidation, interfacial reaction, solid
state diffusion and many other physical changes in the silicon and its
overlayers. Despite this central role, heat-treatment has for much of this
time been carried out in basically the same way, in contrast to the radical
changes that have occurred in other areas, e.g. lithography, etching, epitaxy. The industry standard heat treatment is still to load a large batch
of wafers, stacked vertically along a boat, into a resistively heated oven,
and to heat the whole oven to temperature, controlling this by monitoring
the oven temperature. The disadvantage of this technique is that the times
to ramp up and down are necessarily long, both because of the large thermal
masses involved, and because the indirect method of heating the wafer
results in large thermal gradients and plastic deformation (slip) if rapid
heating is attempted. This disadvantage has become significant in recent
years because the reduced thermal processing required by modern small-
geometry processes cannot always be satisfactorily achieved by reducing
temperature: and furnace times are already as short as is feasible.

   The history of searching for a different method of heat treatment goes
back two decades at least, to 1968, when Fairfield & Schwuttke published a
paper on Silicon Diodes made by Laser Irradiation[1]. Between then and 1978,
the first international conference on Laser Processing of Semiconductors[2],
the heating technique almost exclusively explored was that of irradiating
the silicon surface with a very short light pulse (20-50 nanoseconds) from
a pulsed Q switched laser. Between 1978 and 1982, a wide range of other
radiant heating sources were explored[3,4,5], including continuous lasers and
electron beams (scanned or unscanned), lamps (both filament and arc) and
even ion beams, and research continues in all these areas today. However,
once practical application to planar technology was seriously considered,
systems using lamps as the source of radiation emerged as clear favou-
rites[6-10], for reasons discussed later. This paper describes the present
state of theory and practice of lamp-based transient annealing systems for
heat-treatment times in the 1-300 second range. It will be seen in the
theory sections (2,3) that, because the silicon slice is not in thermal
equilibrium with its surroundings, the actual time-temperature

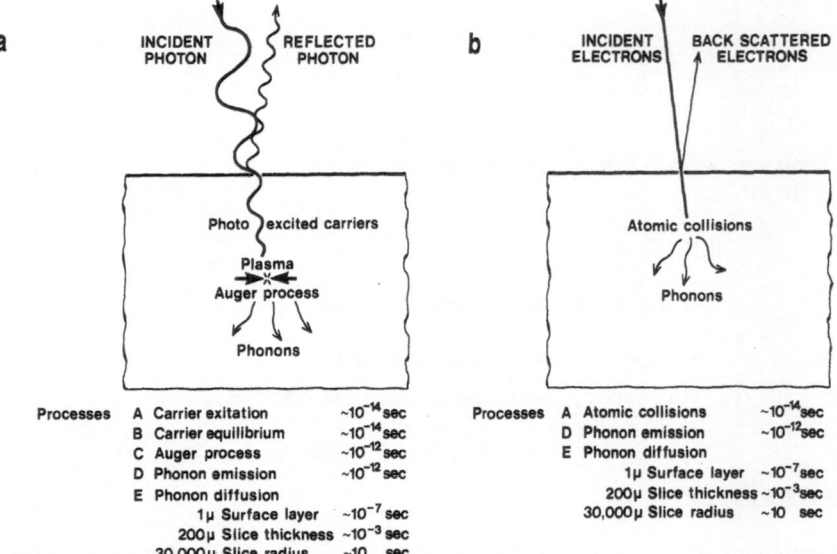

Processes
A Carrier exitation   ~10^{-14} sec
B Carrier equilibrium   ~10^{-14} sec
C Auger process   ~10^{-12} sec
D Phonon emission   ~10^{-12} sec
E Phonon diffusion
   1μ Surface layer   ~10^{-7} sec
   200μ Slice thickness   ~10^{-3} sec
   30,000μ Slice radius   ~10 sec

Processes
A Atomic collisions   ~10^{-14} sec
D Phonon emission   ~10^{-12} sec
E Phonon diffusion
   1μ Surface layer   ~10^{-7} sec
   200μ Slice thickness   ~10^{-3} sec
   30,000μ Slice radius   ~10 sec

Fig 1   Schematic representation of the physical processes involved in transfer of energy from an energy beam to a solid and typical relaxation times for (a) a photon beam (b) an electron beam.

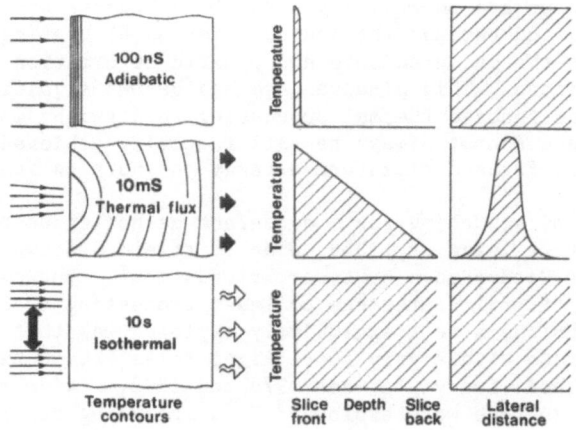

Fig 2   Schematic representation of three characteristic modes of beam processing. From left to right, the parameters illustrated are beam configuration, heating pulse duration, temperature contours in the depth of the wafer, heat flow, and temperature profiles perpendicular and parallel to the wafer surface (Ref 11).

history is a complex function of the optical and thermal parameters of slice and machine. The relative importance of these parameters are discussed in the modelling sections (4,5). Experimental methods developed for reproducible and accurate use of RTA machines, and comparison of experimental data with model predictions (sections 6-8) lead on to description of some fundamental problems and suggested solutions (section 9) and some basic types of machine design which have evolved (section 10).

2.   TRANSFER OF ENERGY IN RADIANT HEATING SYSTEMS

All radiation heating systems transfer energy directly from source to slice by absorption of the energy-carrying beam in the slice surface. The important elements in this transfer are the fraction of the beam that is transmitted through the slice surface, the depth in which this radiation is absorbed (and the process by which this occurs), and the rate at which the resultant thermal energy is transferred throughout the slice by thermal conduction. Approximate time constants for each of these processes are summarised in Fig. 1 for optical and electron beam absorption in silicon. It is evident that even for radiation pulses as short as nanosecond, all the absorption processes are faster still and the rate-controlling step is the transfer of heat by thermal conduction throughout the slice. This gives rise to three characteristic modes of beam processing[11], in which the heating time is such that the thermal diffusion distance is approximately equal to (i) the absorption depth, say 1 micron (ii) half the slice thickness, say 300 microns, (iii) the dimension of a complete circuit across the slice surface, say 11mm. The equivalent heating times are approximately 100ns, 10 milliseconds and 10 seconds, and the thermal distributions to which they give rise are shown in Fig.2.

The different techniques which have been used to heat treat whole wafers in each of these three modes are summarised in Fig.3 Of all these, only (e), (g) and (h) are currently beng used in VLSI development.

The adiabatic mode was found to be unsuitable for VLSI applications because, although the heating is confined to the silicon surface region, only by melting the silicon can materials changes be effected in the 20-500 nanosecond effective heating time. The volume changes that occur on melting and refreezing disrupt other layers (e.g. oxides and nitrides).

The thermal flux mode is still used for controlled melt recrystallization of deposited silicon layers on dielectric into single crystal material, both using laser and electron beams, but the very different equipment puts it outside the scope of this paper : it will not be dealt with further here. There are some solid state processes which can be effected in this time frame (1-10 milliseconds), but present geometries do not demand such short times and the equipment is unnecessarily complex for global heat treatments.

The isothermal mode is the one that has found application in VLSI development and will certainly be an integral part of VLSI processes. The heating times accessible in this regime (5-500 seconds) are a considerable reduction on minimum times available from furnaces (about 2000 seconds) and significant materials changes can be induced in these times (e.g. annealing, oxidation, silicidation, nitridation). Both lamp and electron beam systems (multiscanned to create a uniform heating beam from a primary electron beam much smaller than the slice) have been made available commercially: the optical systems are overwhelmingly preferred.

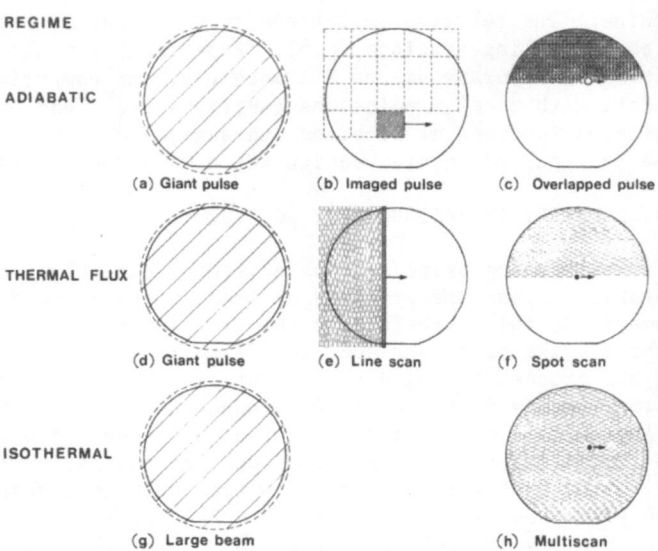

Fig 3   Schematic illustration of the eight possible approaches to beam-
         processing large area wafers. The radiant beam shape and size,
         and its motion, are shown superimposed on a wafer, which can be
         75-150 mm diameter. (Ref 11).

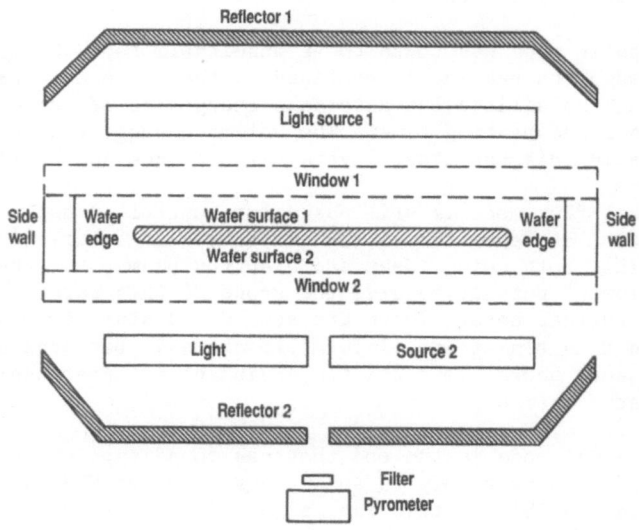

Fig 4   The basic features of a lamp source rapid thermal anneal system,
         shown schematically in vertical section. A silicon wafer (seen
         edge-on) is held (usually on thin quartz pins) in thermal
         isolation inside a cell, containing a controllable gas ambient.
         One (or two) light sources heat the wafer through transparent
         window(s), aided by reflectors. A pyrometer views the wafer back
         through a series of windows and a filter, in order to monitor and
         control wafer temperature.

146

Some of the reasons are:

(1) The power required to maintain a silicon slice at 1200°C, assuming
     that the thermal radiation from the slice is all radiated freely away,
     is about 19 watts per sq.cm., or a total of 34kW for a 6 inch wafer.
     This level of power is more easily and more cheaply achieved control-
     lably using lamps sources rather than electron beams.

(2) Optical heating can be transmitted through gaseous ambients, allowing
     reactive processing and deposition. Electron beams, of necessity,
     must be operated in vacuum.

(3) The individual photon energies derived from lamps (0.1 - 3eV) create
     no damage in the silicon. By contrast, the 20-30keV electrons typi-
     cally used do create lattice damage, though most of this anneals out
     when the beam is switched off and the slice is still hot.

(4) Optical systems can be made very compact. The need to scan the elec-
     tron beams necessitates a minimum column height.

3.   PHYSICS OF RAPID THERMAL ANNEALERS USING LAMP SOURCES

     The basic features of a lamp-source rapid thermal anneal system are
shown schematically in Fig.4. The silicon slice is held in as close an
approximation to thermal isolation as possible inside a chamber, in which
the ambient atmosphere can be controlled. The lamps, with their associated
reflectors, are mounted parallel with the slice. In some systems only the
top lamp and reflector is used. Wherever the chamber walls come between a
lamp source and the slice, they must, of course, be transparent to the lamp
radiation. A pyrometer views the bottom surface of the slice through the
chamber wall, to measure and monitor continuously the slice temperature.
Through a feedback loop, this signal modulates the power to the lamps so
that the temperature-time experience of the slice follows a pre-programmed
schedule.

     The basis for understanding the physics of energy transfer in a rapid
transient annealer is to recognise that it consists of a number of
thermally radiating bodies at different temperatures, each with its characteristic energy spectrum. The lamps may be at 2000-3000°C (filament) or
~6000°C (arc): the slice at temperatures between 20-1400°C: and the chamber
walls between 20 and 600°C depending on system design and slice temperature.

3.1 Emission of Radiation

     The radiation emitted by hot objects has a well-defined spectral
distribution. In the simple case of black bodies, this distribution is a
function only of temperature, and given by Planck's relation:

Power emitted at wavelength $\lambda = \varepsilon_\lambda = \dfrac{c_1 \lambda^{-5}}{\exp\left(\dfrac{c_2}{\lambda T}\right) - 1}$

$$(1)$$

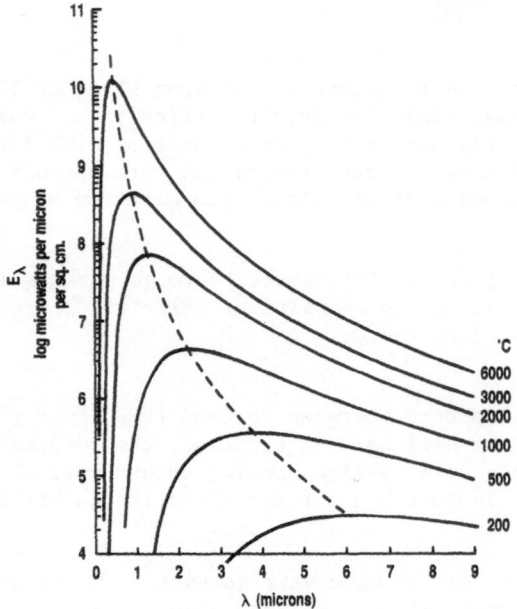

Fig 5   Spectral distribution of the power radiated away from a black
body as a function of black body temperature.   The dotted line
shows how peak power is emitted at progressively shorter wavelengths as temperature increases.

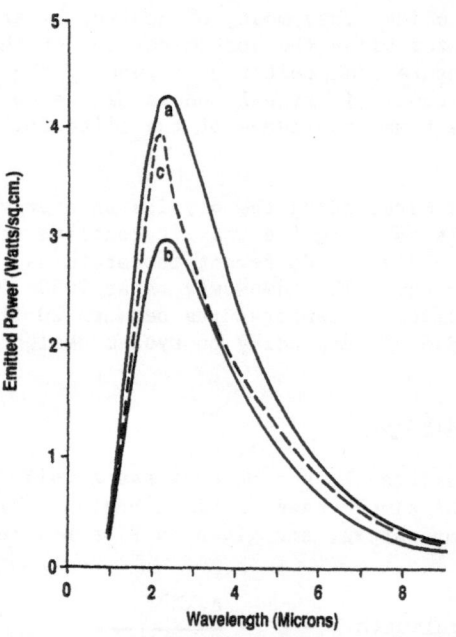

Fig 6   Spectral distribution of thermal radiation at 1000°C from (a) a
black body (b) a grey body of emissivity 0.7 (c) a body with
spectrally-dependent emissivity varying between 0.7 and 0.95 (eg,
silicon coated with a one-micron thick layer of SiO_2).

where ε_λ has units of microwatts cm^{-2} microns^{-1}
when C_1 = 3.7403x10^{10} microwatts microns^4 cm^{-2}
and C_2 = 14384 micron degrees
and λ is in microns
and T is in °K

The power distributions for a black body at various temperatures are plotted in Fig.5. Two important features should be noted. As the temperature of the source increases, the radiant energy output at <u>all</u> wavelengths increases: in addition, the output at shorter wavelengths increases very rapidly with temperature so that the wavelength range containing the bulk of the radiant energy shifts from about 6 microns (200°C) through 4 microns (500°C), 2.2 microns (1000°C), 1.2 microns (2000°C), 0.8 microns (3000°C) to 0.4 microns (6000°C). The total power radiated away from a black body at temperature T is given by the integral under the appropriate curve of Fig.5, numerically equal to that given by Stefan's Law

$$P = 5.7 \times 10^{-12} \ T^4 \ \text{watts/cm}^2 \qquad\qquad (2)$$

where T is in °K

Black-body values of P for a range of temperatures are given in Table 1.

Table 1. Power radiated away from a hot black body at temperatures between 600 and 1400°C, per sq.cm. of exposed plane surface.

600	700	800	900	1000	1100	1200	1300	1400	°C
3.3	5.1	7.6	11	15	20	27	35	45	watts/cm^2

The real components of a system are not, of course, perfect black bodies: they emit less radiation at each wavelength than a black body. The simplest approximation to real emission behaviour is to assume that the emitted radiation is that of a black body, but reduced at each wavelength by a constant factor e, the emissivity. Such a "grey" body has a total emitted power e times that of a black body at the same temperature, as illustrated in Fig.6. At high temperatures, a bare silicon wafer corresponds fairly well to a grey body with emissivity ≈0.7, and thus radiates about 70% of the power emitted by a black body at the same temperature. This grey body approximation is not adequate under all conditions, and in general a wavelength-dependent function, the spectral emissivity, is needed to specify the radiation behaviour at each temperature.

In silicon, deviations from grey-body behaviour occur:

1. When the silicon slice is not completely opaque to the incident radiation. This occurs for photon energies too low to excite carriers across the band gap, when there are also too few free carriers to absorb these low energy photons (in lightly doped silicon below 600°C). The spectral emissivity is then a strong function of both temperature and wavelength[12] as shown in Fig.7a.

2. When the optical properties of the silicon surface are modulated by the presence of overlayers of differing refractive index. As described later, this results in optical interference effects which impose a regular modulation on the intensity of the emitted (and absorbed) radiation, with a characteristic periodicity for each wavelength. Spectral emissivity is then a function of the layer structure and can vary between quite wide

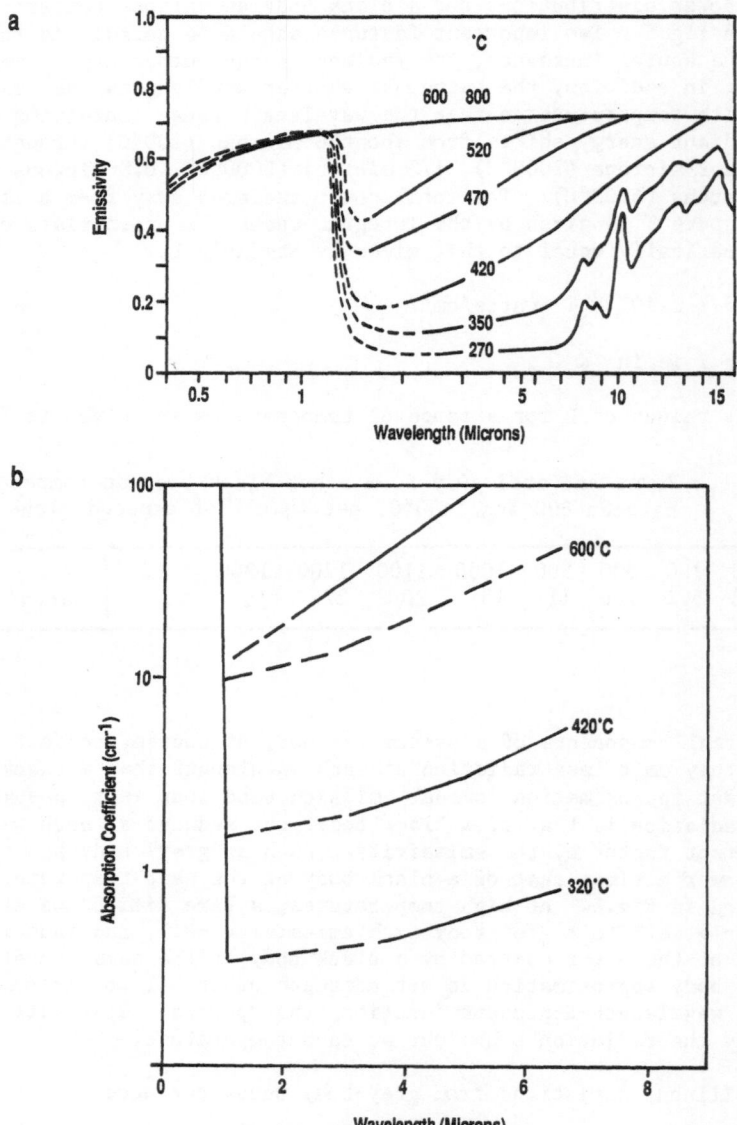

Fig 7 Optical properties of single crystal silicon as a function of
wavelength and temperature. (a) Emissivity of lightly doped
silicon 1.8mm thick (after Sato[12]) (b) Absorption coefficient of
lightly doped silicon[12] as a function of temperature, and of
heavily doped silicon[13] (1 x 10^20 As+ ions/cc) at room
temperature.

limits (0.3 - 0.9). The calculated effect of a 1 micron silicon dioxide layer on the emitted radiation from a silicon slice is shown in Fig.6.

3.2 Absorption of Radiation

Radiation incident on a surface is transmitted through the surface and absorbed. A black body absorbs all the radiation incident upon it. Real bodies reflect a fraction (R) of the radiation, and absorb a fraction (A): it follows that A+R = 1. It can be shown from thermodynamic equilibrium arguments, that when black bodies are in thermal equilibrium established through exchange of radiant energy, the incident radiation P must equal the sum of reflected (RP) and emitted (eP) radiation, and thus 1 = R+e and hence e = 1-R = A. Thus the emissivity and absorptivity of a body are equal, and are related to the reflectivity. Similar arguments establish that the spectral emissivity (e_λ) and absorptivity (a_λ) at any wavelength λ are also equal, and given by : $e_\lambda = 1-R_\lambda = a_\lambda$

An important consequence of this is that while for grey bodies the _average_ emissivity and absorptivity are always equal, for other bodies they may not be. This is because, although the emissivity and absorptivity have the same spectral and temperature dependence, the emitted radiation (e.g. from the silicon slice) and the incident radiation (e.g. from the lamps) have different spectral distributions. The emissivity and absorptivity are then being averaged over different wavelength ranges, and thus will not, in general, have equal average values.

3.3 Thermal Conduction inside the Silicon

Although the mode of thermal processing used in lamp annealers is described as "isothermal", this is only approximately true. Temperature differences can develop across the silicon slice, due to a different balance between emission and absorption in different regions. Common causes are:

1. The different geometrical and radiation situations at the edge as compared with the centre of a wafer.

2. The different energy balances that occur in regions with different types, areas and shapes of patterned overlayers.

3. The small variations in the uniform incident radiation that can be magnified into large temperature differences between adjacent regions because of the strong temperature dependence of both the absorption coefficient[12,13] (Fig.7b) and the thermal conductivity[14] (Table 2).

Deviations from the isothermal situation are largest where the rate of heat input is greatest, i.e. during the ramp-up to steady state temperature.

Table 2. Thermal Conductivity of Silicon and Silicon Dioxide as a Function of Temperature (from Godfrey et al[14]). Units are $Wcm^{-1} deg.K^{-1}$

Temp °C	0	200	400	600	800	1000	1200	1400
Silicon	1.40	0.80	0.55	0.42	0.32	0.27	0.24	0.21
SiO_2	0.014	0.018	0.023	0.028	0.033	0.038	0.043	0.047

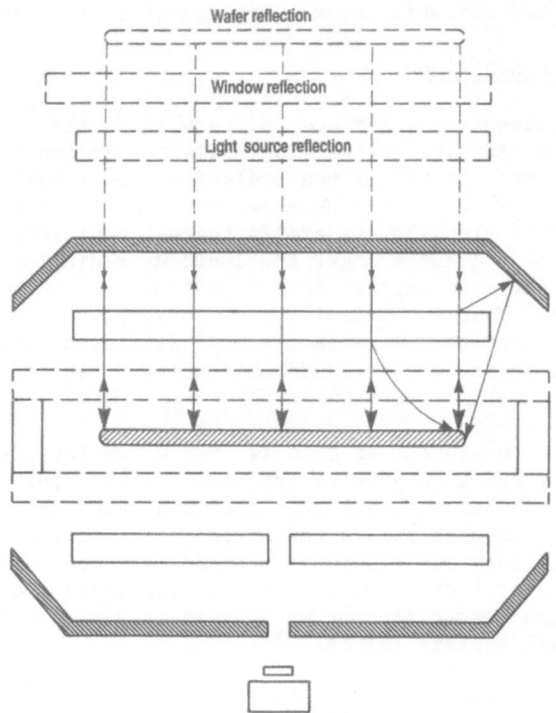

Fig 8   Schematic illustration of the sources of optical radiation in a
         rapid thermal anneal system. The lamps, the wafer, and the walls
         plus the reflections of each of these, act approximately as grey
         bodies of different temperatures producing radiations which all
         contribute to the incident radiation at the wafer.

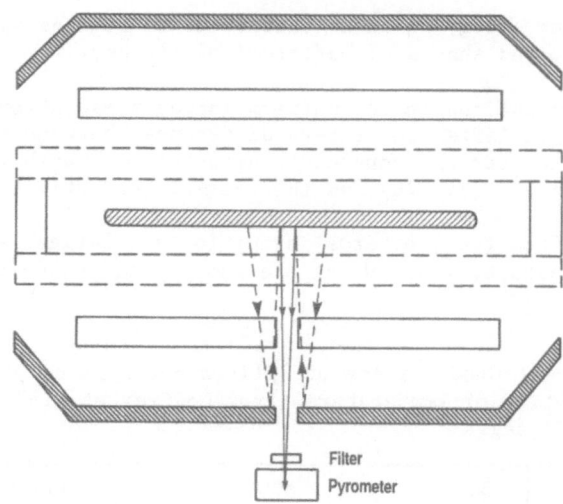

Fig 9   Schematic illustration of the different types of radiation which
         many contribute to the pyrometer reading. As well as direct
         radiation from the wafer, radiation from wafer and lamps may
         enter the pyrometer by reflection in the reflecting walls and the
         wafer.

3.4 Conversion of Absorbed Radiant Energy to Heat

The absorbed radiation interacts with the matrix electrons which then transfer heat energy to the lattice as summarised in Fig.1. This interaction is inefficient if the density of free electrons is low, and the photon energies are lower than the threshold required to excite electrons across the bandgap (wavelengths longer than 1 micron). The bulk absorption coefficient is then a strong function of both wavelength and temperature, in a similar way to that previously described for emissivity. The relationship is shown in Fig.7b.

3.5 Sources of Radiation

The significant sources of radiation in a typical system include the lamps, their reflections, the reflections of the slice itself, and in some systems the chamber walls and their reflections. These are shown schematically in Fig.8. Each of these sources has its own spectral distribution of radiant energy, given by the product of the black body curve and the spectral emissivity at each wavelength. Typical temperatures are 6000°C (arc lamps), 3000°C (tungsten halogen lamps), 600-1200°C (silicon slice), 500°C (quartz walls). At these temperatures, the emissivities of the materials in each case are fairly constant with wavelength in the peak emission and greater than 0.5, so the appropriate black body curves in Fig.5 are a good approximate guide to the relative magnitude and spectral distribution of the majority of the emitted radiation from each source.

3.6 Pyrometry

The slice temperature is almost universally monitored by selecting a band of the emitted radiation for measurement in a pyrometer. It will be evident from the foregoing discussion and from Fig.5, that the total power emitted into a fixed wavelength band from a black body uniquely defines the body temperature. However, it is also evident that for real bodies, the reliability of the pyrometric measurement depends on knowing the average emissivity of the body in the relevant wavelength band. As we have seen, this average value can be a function of slice temperature and the precise nature of the overlayers on the silicon surface. Because of the wavelength dependence of emissivity, there is no general way of measuring absolute slice temperature by pyrometry alone. Some hybrid techniques will be discussed later. An additional problem, shown schematically in Fig. 9, is that the pyrometer may also receive radiation from the lamps, the quartz and the slice by reflection in the slice. The first two effects necessitate automatic corrections to the pyrometer output: the last effect changes the effective emissivity of the slice, and profoundly modifies the effect of overlayers on emissivity, as will be described quantitatively later.

4. MODELLING OF RAPID THERMAL ANNEALING - TEMPERATURE CONTROL

4.1 Introduction

An important issue regarding the operation of rapid thermal anneal machines which requires modelling is that of temperature control. As discussed above, the wafer temperature is controlled by a pyrometer and thus requires accurate knowledge of the emissivity of the region of the wafer viewed by the pyrometer. The sensitivity of this emissivity with respect to layer structure and to pyrometer wavelength can be modelled. This gives a guide to the correct choice of emissivity value and also gives warning of particularly sensitive structures. It is also straightforward to model the expected temperature error due to uncertainty in the emissivity value (which may arise from layer thickness tolerances during previous processing of the wafer).

The model we have developed determines the emissivity of a given layer structure on top of a semi-infinite single crystal silicon substrate for radiation of a given wavelength and angle. For a given pyrometer this information can be used to determine the average value of emissivity across the pyrometer waveband and acceptance angle. In addition, enhancements to the emissivity due to chamber reflectivity (as shown schematically in figure 9), and reductions due to finite wafer thickness can be taken into account.

4.2 Determination of Emissivity

The emissivity, e, can be immediately determined from the reflectance, R, as e = 1-R for the case of zero transmittance (this follows from Kirchoff's Law as demonstrated above). For an opaque body with no surface layers the reflectance is given as:

$$R = \left|n-1\right|^2 / \left|n+1\right|^2 \tag{3}$$

where n is the material refractive index (complex for absorbing material). For silicon in the near infra-red n = 3.43 (negligible absorption) and so R equals 0.3 which gives an emissivity of 0.7. For partially transmitting bodies the emissivity is reduced; the exact equation for this will be presented later.

The determination of the reflectance of a planar multi-layer structure in which the layers are optically flat requires consideration of the interference of radiation multiply reflected from each of the interfaces. This is a common problem in optics and full details can be found in, eg, Born and Wolf[15]. The general N layer problem is shown below, in Fig 10, in which layer ℓ has depth d_ℓ, refractive index n_ℓ (complex to allow for absorption) and propagation vector at angle Θ_ℓ to the normal. The figure shows the case of two orthogonal polarisation states: E and H parallel to the plane of the layer surfaces.

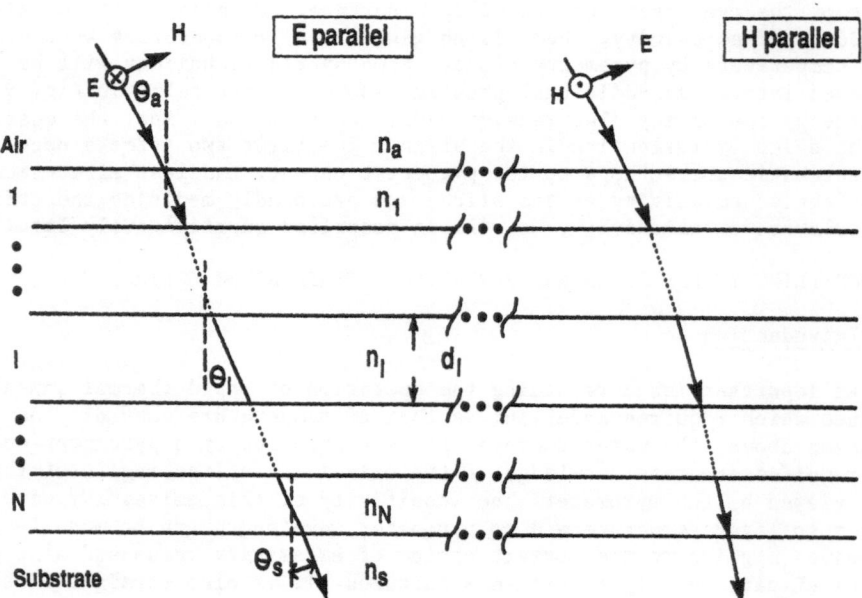

Fig 10 Schematic view of the path of radiation through a planar N layer structure for two different polarisations.

In summary the procedure is to relate the tangential E and H fields at each layer interface to the preceding values by determining the phase change incurred in traversing the layer. This must be done for the two orthogonal polarisation states; for unpolarised radiation these are then averaged. As tangential components of E and H are continuous across a boundary we can eliminate all intermediate values to leave the following matrix equation:

$$
\begin{bmatrix} E \\ Z_o H \end{bmatrix}_{air} = M_1 M_2 \cdots M_N \begin{bmatrix} E \\ Z_o H \end{bmatrix}_{substrate} \tag{4}
$$

where Z_o is the impedance of free space ($= \sqrt{\mu_o/\varepsilon_o}$) and M is a characteristic matrix for each layer. For an arbitrary layer, ℓ, this matrix is:

(a) E parallel

$$
M_E = \begin{bmatrix} \cos(k_o n_\ell d_\ell \cos\Theta_\ell) & \frac{-i}{p} \sin(k_o n_\ell d_\ell \cos\Theta_\ell) \\ -ip \sin(k_o n_\ell d_\ell \cos\Theta_\ell) & \cos(k_o n_\ell d_\ell \cos\Theta_\ell) \end{bmatrix} \tag{5}
$$

(b) H parallel

$$
M_H = \begin{bmatrix} \cos(k_o n_\ell d_\ell \cos\Theta_\ell) & \frac{-i}{q} \sin(k_o n_\ell d_\ell \cos\Theta_\ell) \\ -iq \sin(k_o n_\ell d_\ell \sin\Theta_\ell) & \cos(k_o n_\ell d_\ell \cos\Theta_\ell) \end{bmatrix} \tag{6}
$$

where k_o is the incident wave-vector amplitude ($= 2\pi/\lambda$), $p_\ell = n_\ell \cos\Theta_\ell$, and $q_\ell = n_\ell/\cos\Theta_\ell$.

In order to determine the amplitude reflectance, r (and transmittance, t, into the substrate) assume that the matrix product has been evaluated to yield

$$
M_1 M_2 \cdots M_N = \begin{bmatrix} M_{11} & M_{12} \\ M_{21} & M_{22} \end{bmatrix}
$$

then for the two polarisation states:

(a) E parallel

$$
r_E = \frac{(M_{11} + M_{12} P_s) P_o - (M_{21} + M_{22} P_s)}{(M_{11} + M_{12} P_s) P_o + (M_{21} + M_{22} P_s)} \tag{7}
$$

(b) H parallel

$$
r_H = \frac{(M_{11} + M_{12} q_s) q_o - (M_{21} + M_{22} q_s)}{(M_{11} + M_{12} q_s) q_o + (M_{21} + M_{22} q_s)} \tag{8}
$$

The intensity reflectance R is then determined finally as

$$
R = \tfrac{1}{2} (|r_E|^2 + |r_H|^2) \tag{9}
$$

In order to evaluate the matrix product all that is required is a knowledge of the refractive index at the pyrometer wavelength, and the angle for each layer. The latter can be determined from the incident angle at the surface using Snell's Law.

4.3 Modifications to Emissivity

(a) The emissivity calculated above should be reduced for a finite thickness substrate according to [16]:

$$e(t_w) = \frac{e_o (1 - \exp(-\alpha t_w))}{(1 - \exp(-\alpha t_w)) + e_o \exp(-\alpha t_w)} \tag{10}$$

where t_w is the wafer thickness and α is the absorption coefficient per unit depth ($= 4 \pi \operatorname{Im}(n)/\lambda$). This is important for temperatures less than 700°C and wavelengths longer than the band-gap energy in silicon. This is shown plotted as a function of the thickness absorption coefficient product in figure 11 for undoped silicon.

(b) If reflected radiation can enter the pyrometer the effective emissivity as seen by the pyrometer is increased. The sum of the radiation directly entering the pyrometer plus that multiply-reflected from the chamber wall and wafer surface gives an effective value of [17]:

$$e(R_{ch}) = \frac{e_o}{(1 - R_{ch}) + R_{ch} e_o} \tag{11}$$

where R_{ch} is an effective chamber reflectivity at the pyrometer wavelength which includes a factor to account for the detailed geometry around the pyrometer. The effective emissivity as a function of chamber reflectivity is plotted in figure 12 for a material with emissivity 0.3.

(c) If the pyrometer has a wide waveband then the emissivity determined above as a function of wavelength must be averaged over the pyrometer wavelength window.

4.4 Temperature Sensitivity to Emissivity

The pyrometer is calibrated to convert the detected intensity signal into a temperature: for a given emissivity and wavelength a unique value is obtained as can be seen from the black-body curves of figure 5. The temperature sensitivity to emissivity can be determined from these curves. If the pyrometer emissivity is set at e_o, whereas the actual wafer value is e, then the pyrometer will register a temperature T_o (not equal to the true temperature T) given as:

$$e_o \varepsilon_\lambda(T_o) = e \, \varepsilon_\lambda(T) \tag{12}$$

where $\varepsilon_\lambda(T)$ is the black-body spectrum as given in equation 1, at wavelength λ. This equation cannot be solved analytically for T though to a good approximation the solution is [18]:

$$\frac{1}{T} - \frac{1}{T_o} = \frac{\lambda}{C_2} \ln \frac{e}{e_o} \tag{13}$$

provided that $C_2/\lambda T \gg 1$ which implies $\lambda T \ll 14384 \ \mu K$

For a wide waveband pyrometer, ε_λ must be integrated over the waveband and equation (12) inverted numerically by, eg, the Newton-Raphson method.

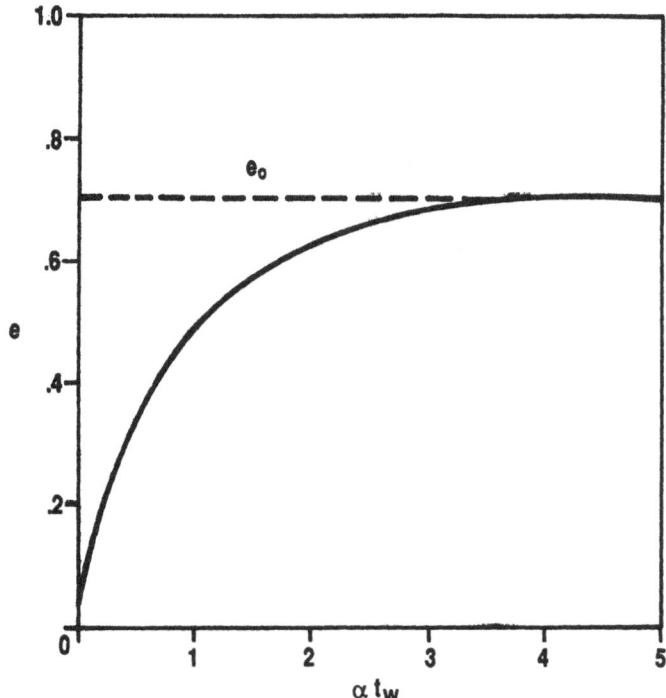

Fig 11 The effective emissivity for a finitely absorbing silicon substrate plotted as a function of the thickness-absorption coefficient product.

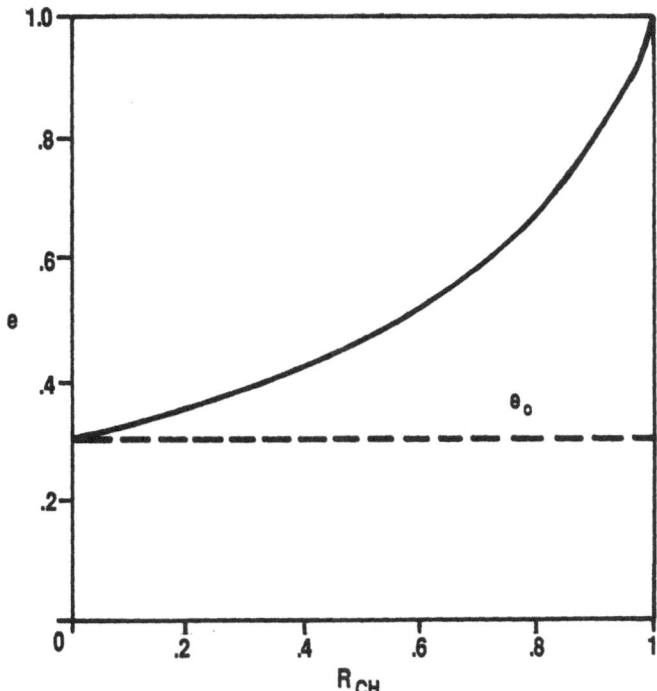

Fig 12 The effective emissivity plotted as a function of chamber reflectivity for a material of emissivity 0.3 in a reflecting chamber.

4.5 Model predictions

As a demonstration of the use of the above equations the emissivity of
a single layer of oxide on top of a silicon substrate is shown in figure 13
as a function of oxide thickness. The pyrometer in this case is a narrow
band centred on 3.5 μm which views the surface at normal incidence with a
narrow acceptance angle. The optical properties chosen were: n = 1.48 for
the oxide; and n = 3.43 + 0.1i for the silicon (corresponding to a heavily
doped substrate). The periodicity of the emissivity corresponds to a thickness change of λ /2n. The results for both a black chamber and a shiny
chamber (70% reflectivity) are shown. It can be seen that the effect of
the reflections are to enhance the emissivity and to compress the relative
variation. Figure 14 shows the same calculation but for a wide-band pyrometer at a longer wavelength (4.8 − 5.6 μm) in which case the oxide becomes
absorbing. It may be expected that the use of a wide band pyrometer will
average out the emissivity and so reduce the overall variation. To some
extent this does occur as can be seen by close comparison of figures 13 and
14.

Having determined the emissivity variation for such a structure we can
proceed to calculate the temperature sensitivity of these structures. From
equation 13 above we expect a sensitivity of approximately 4°C and 6°C per
percentage relative change in emissivity at a temperature of 1000°C for the
pyrometer at 3.5 μm and 4.8 μm, respectively. In order to calculate the
temperature change exactly we assume that the pyrometer emissivity is
correctly tuned to obtain the correct temperature (which we will take as
1000°C) for a bare silicon wafer (ie, set for e =70%) and then determine
what temperature would be obtained if there were an oxide layer on the
silicon. This is shown in figures 15(a) and 15 (b) for the 3.5 μm and
4.8−5.6 μm pyrometers respectively. It can be seen that the shiny chamber
has a reduced sensitivity compared to the black chamber. It can also be
seen that the longer wavelength pyrometer is more temperature sensitive
than the shorter wavelength one. This can be understood by reference to the
black body curves (fig 5). Both pyrometers operate in the tail of the
spectrum: the spectra are increasingly temperature insensitive for
increasing wavelength. Thus emissivity errors further out in the tail of
the distribution result in larger temperature errors.

The model can be used to predict 'critical structures': those which are
particularly sensitive to emissivity, and so should be avoided. An example
is shown in figure 16 for the case of the 3.5 μm pyrometer. The structure
is a two-layer one of polycrystalline silicon on oxide on the silicon
substrate. The oxide thickness is 0.5 μm and acts as a quarter-wave plate
(ie, there is a relative phase of 180 degrees between radiation reflected
off the top and the bottom surface of the oxide). It can be seen that
there is almost an order of magnitude variation in emissivity as the polycrystalline silicon thickness is varied. Any uncertainty in the polycrystalline silicon thickness would thus lead to a large uncertainty in the
choice of emissivity value (and hence very large temperature errors). For
a variation in polycrystalline silicon thickness of ±100Å the emissivity
can vary by up to 6 % in a black chamber around a value of 50%; this would
lead to a temperature uncertainty of ±48°C.

5. MODELLING OF RAPID THERMAL ANNEALING − TEMPERATURE UNIFORMITY

5.1 Introduction

Temperature uniformity of wafers during RTA may not be perfect for three
reasons: the effects of increased radiation from the wafer edges; the effect

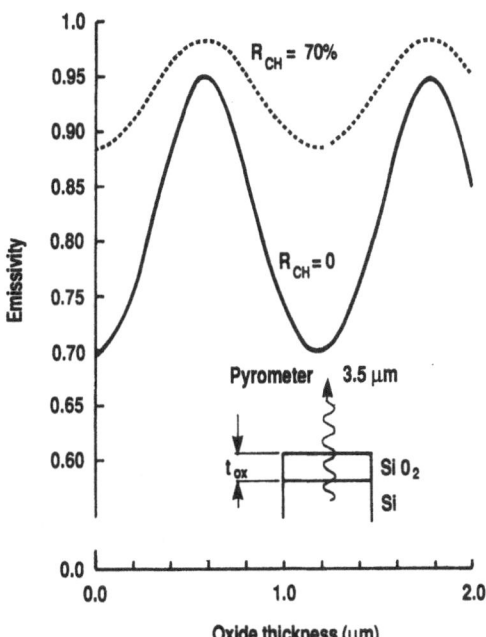

Fig 13 The variation of emissivity with oxide thickness for a planar
oxide layer on a fully absorbing silicon substrate for a single
wavelength pyrometer operating at 3.5µm. The cases of black and
shiny (70% reflecting walls) chambers are shown.

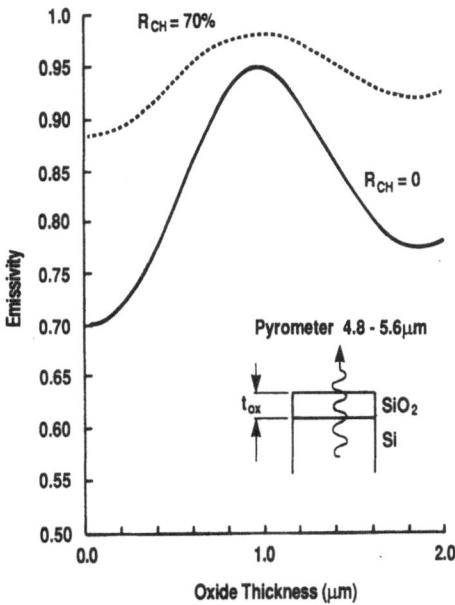

Fig 14 The variation in emissivity with oxide thickness for the structure described in figure 13, but with a wide waveband pyrometer
operating over the range 4.8 - 5.6µm.

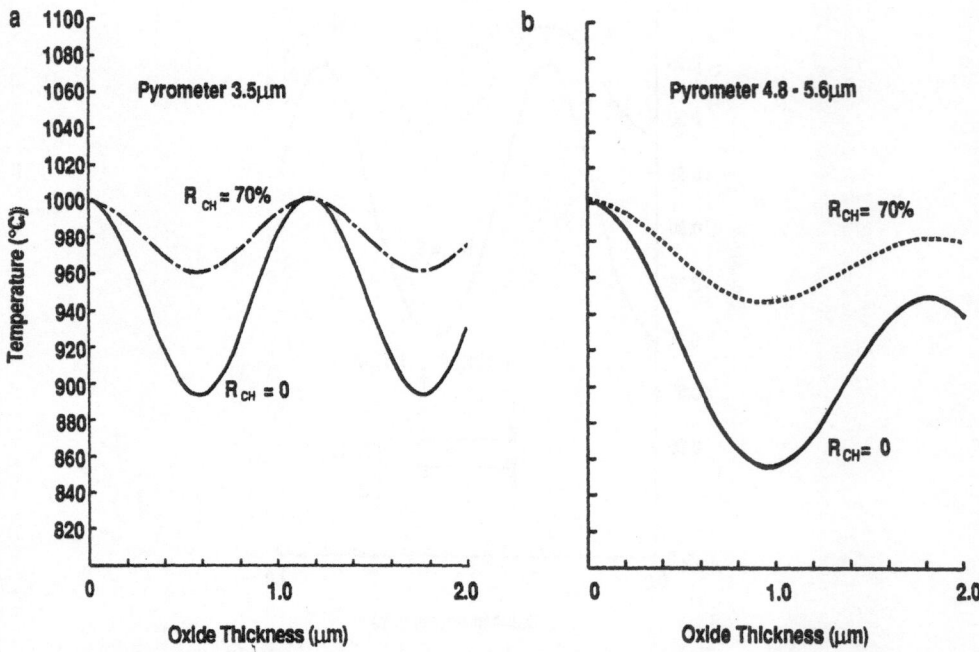

Fig 15 The variation in the temperature achieved by the wafer described in figure 13, assuming that the pyrometer is calibrated to read 1000°C for a blank silicon wafer. The case of (a) a single wavelength pyrometer (3.5µm); and (b) a wide waveband pyrometer (4.8 − 5.6µm) are compared.

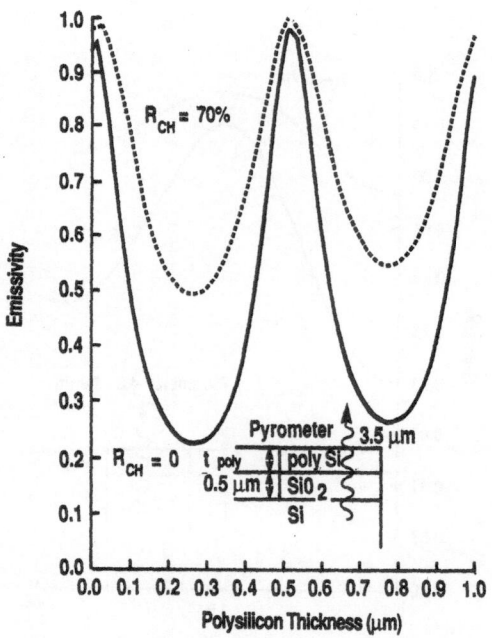

Fig 16 The variation in emissivity with polysilicon layer thickness for a planar structure of poly on 0.5µm oxide on the silicon substrate. The pyrometer is operating at 3.5µm and the cases of black and shiny (70% reflecting) chambers are shown.

of non-uniform incident intensity; and the effect of the differential heating rate of regions of the wafer with different overlayers having different emission and absorption properties. These effects can be modelled by numerically solving the equations governing heat conduction within, and radiation into and out of, a wafer with an arbitrary patterning of overlayers. Such modelling can help to determine the optimum machine conditions (ramp-rates etc) to minimise any temperature non-uniformity.

5.2 Model Details

The model determines the full temperature distribution within a silicon wafer at all stages of the anneal cycle. It can thus predict any significant non-uniformities which could result in wafer slip as well as processing variation. In order to do this the model must solve the heat conduction equation within a 3D wafer subjected to radiative heating, accounting for all possible radiation sources (lamps, reflected radition, hot walls), and radiative losses. The model which has been developed makes the following assumptions and approximations:

(i) The wafer is assumed to be radially symmetric (ie only radial and depth dependencies of the temperature are considered); thus the absorption and emission coefficients of the wafer surface are specified on annular regions as shown in figure 17.

(ii) Each radiation source is spectrally averaged to yield averaged absorption and emission coefficients. The averaging is done over two typical distributions: that of the lamps (black-body at approximately 3000°C); and that of the wafer (1000°C). As an example the averaged emission coefficient for the case of a single layer of oxide on a single crystal substrate is shown in figure 18 as a function of oxide thickness. The spectral emissivity was averaged over a black-body distribution at three temperatures: 900°C, 1200°C and 3000°C. (For simplicity the optical coefficients of the materials were treated as constant over these spectra, and normal incidence was assumed.) For bodies in thermal equilibrium the emissivity is equal to the absorptivity. In the case of RTA the wafer and lamps are not in thermal equilibrium: thus the absorption coefficient should be taken from the curve at the lamp temperature, and the emission coefficient from the curve at the wafer temperature.

(iii) It is assumed that absorption in the wafer occurs at the surfaces. This is accurate at temperatures over 700°C; at lower temperatures there is significant transmission through the wafer. In the latter case the fraction that is absorbed overall is assumed to occur at the surface and heat conduction allows this to diffuse into the bulk.

(iv) The equations are solved on a rectangular finite difference grid as shown in figure 17. The rounded wafer edge is treated by enhancing the radiating area of the nodes at the edge by a geometrical factor of $1/\sin \Theta$ where Θ is the angle of the edge taper.

The equations to be solved by the model are that in the bulk wafer the temperature is governed by the thermal diffusion equation:

$$\frac{\partial T}{\partial t} = \nabla (D (T) \nabla T) \tag{14}$$

where D(T) is the temperature dependent thermal diffusivity equal to K/(ps) with K the thermal conductivity; p the density; and s the specific heat capacity. The temperature dependence of these was taken from Merli[19] and yields values of D(T) :

T (K)	: 273	473	673	873	1073	1273	1473
D(T) mm^2/sec	: 94.7	43.9	26.2	17.7	13.0	11.3	10.1

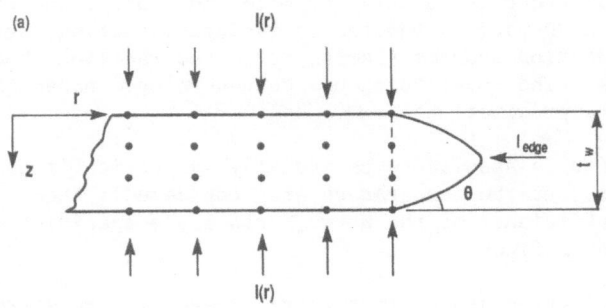

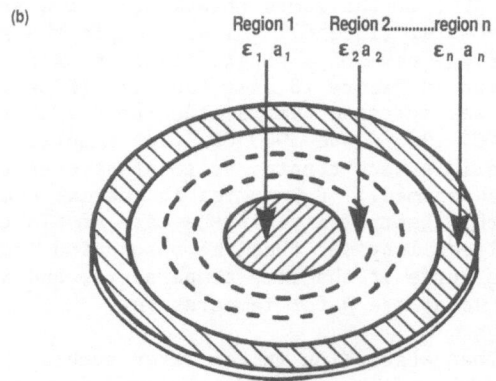

Fig 17 A schematic view of the wafer as modelled: (a) shows a vertical
section through the wafer at the edge showing the rectangular
grid. The intensity field I has a radial dependence, and radiation from the nodes at the edge is enhanced by a geometrical
factor 1/sinΘ where Θ is the taper angle of the edge. Figure (b)
shows how the absorption and emission properties of the surfaces
are modelled; the wafer is assumed to be radially symmetric.

At the surfaces the conduction is balanced with the net radiation into the wafer (where { } indicates upper or lower surfaces):

$$-\frac{K\partial T}{\partial z}\bigg|_{\substack{z=\{o\}\\tw}} = \{\pm\}\ (a\ I(r) + a\ \sigma T_q^4 - e\ \sigma T^4)$$

(15)

where $I(r)$ is the incident radiation field; T_q is the background tempera-ture (of the walls or quartz windows); a and e are the absorption and emission coefficients, respectively; and σ is Stefans constant.

The effects of reflections of the lamp from the chamber walls and the partial absorption by the quartz windows all appear in modifications to e and a following the approach outlined above in section 4.3. The emissivity and absorptivity are modified as in equation 11, with the modification that R_{ch} is replaced by $\gamma^2 R_{ch}$, where γ is the transmission coefficient of the quartz (γ = 0.94 for radiation from the lamps, and γ = 0.6 for radiation from a wafer at 900°C assuming that the quartz absorbs radiation at all wavelengths greater than 4 μm).

The temperature dependence of e and a is modelled by dividing these into two parts: a temperature independent part corresponding to wavelengths less than 1.1 μm; and the temperature dependent part for wavelengths greater than 1.1 μm. The latter is determined by calculating the transmission co-efficient through the wafer as a function of carrier concentration (follow-ing the results of Sato[12]) which is temperature and doping dependent. The relative proportions of these two contributions depends upon the tempera-ture (of the lamp for a and of the wafer for e).

The model can simulate any RTA cycle by specifying the intensity at each stage (ramp-up, steady-state, ramp-down) and solving simultaneously equations 14 and 15.

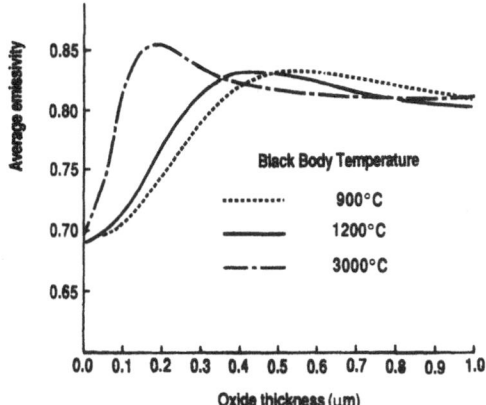

Fig 18 The spectrally averaged emissivity for a varying thickness planar layer of oxide on a silicon substrate for three different black body temperatures.

5.3 Results : Edge effects on temperature uniformity

The model can be employed to study the edge effects on the wafer temperature uniformity during an anneal cycle. Figure 19 shows the radial temperature profile (along the top surface) across a 6 inch wafer at three points of a cycle in which the wafer is ramped to 1100°C and held for 15s before cooling. The points correspond to the ramp-up and steady-state parts of the cycle and the simulation is performed for both the case of a uniform intensity field and one in which the intensity is linearly enhanced to a maximum of 8% over the last 15 mm of the wafer, (as shown in figure 20). The edge was modelled assuming that the angle θ shown in figure 17 is equal to 30 degrees; that the lamp radiation was incident vertically on the wafer; and that there was no reflector at the edge of the wafer.

It can be seen that for the case of a laterally uniform incident intensity the edge is approximately 30°C cooler than the centre of the wafer in the steady state (16 secs). The 8% enhancement produces uniform temperature across the whole wafer to less than 10°C in the steady state. During the ramp-up the edge is actually at a greater temperature than the centre (since it has a larger surface to volume ratio and the incident radiation greatly outweighs that emitted). Figure 20 shows the time temperature profile for this anneal cycle for the case with the optimised intensity. The temperature at a point near the wafer centre and the temperature difference (compared to the centre) at a point near the edge are shown as a function of time. It can be seen that the temperature non-uniformity is transient and occurs in both the ramp-up and the cool down.

The model can be used to optimise the intensity distribution to produce uniform wafer temperature for any system by including specific features such as edge reflectors.

5.4 Results : Surface effects on temperature uniformity

The above example considered a wafer with uniform surfaces and hence uniform absorption and emission properties (apart from the geometrical effects at the edge). In general a wafer subjected to RTA will not have uniform surfaces. Different regions will have different absorption and emission properties. As an example of the consequences of this consider a hypothetical case in which an otherwise blank wafer (absorptivity 70 %) has a narrow 5 mm annular region with low absorptivity (20 %) at a radius of 50 mm. To simplify matters we assume that the emissivity is equal to the absorptivity within each region.

The predicted time-temperature profile for this wafer ramped to a temperature of 1000°C is shown in figure 21 for three different ramp rates (20%, 40% and 80% of total lamp power, 46W/cm^2). The figure shows the temperature at the wafer centre as a function of time and also the temperature difference between the centre of the low absorptivity ring and the wafer centre. It can be seen that significant temperature non-uniformity is incurred (even for the lowest ramp rate) during the ramp-up to temperature. The conduction in the bulk wafer on these time scales is slow even for a distance of 5 mm. Once at the anneal temperature the wafer achieves a steady state uniform temperature (this is a consequence of choosing the emissivity and absorptivity values equal within each region). The modelling predicts that by using slower ramp rates the temperature non-uniformity can be reduced, although not eliminated.

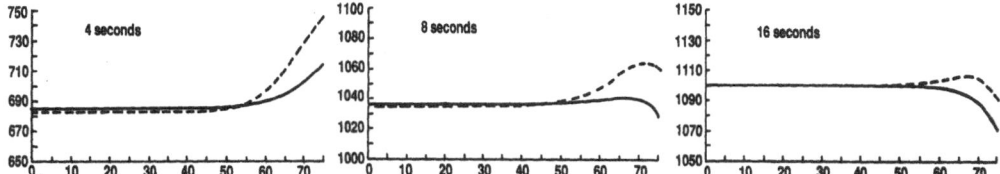

Fig 19 The radial temperature profile across a blank 150mm wafer, at
        three times (4, 8 and 16 seconds) of an anneal in which the wafer
        is ramped from 200 to 1100°C. The solid curve shows the case of
        a laterally uniform incident intensity; the dashed curve has an
        enhanced intensity over the last 15mm as described in the text.

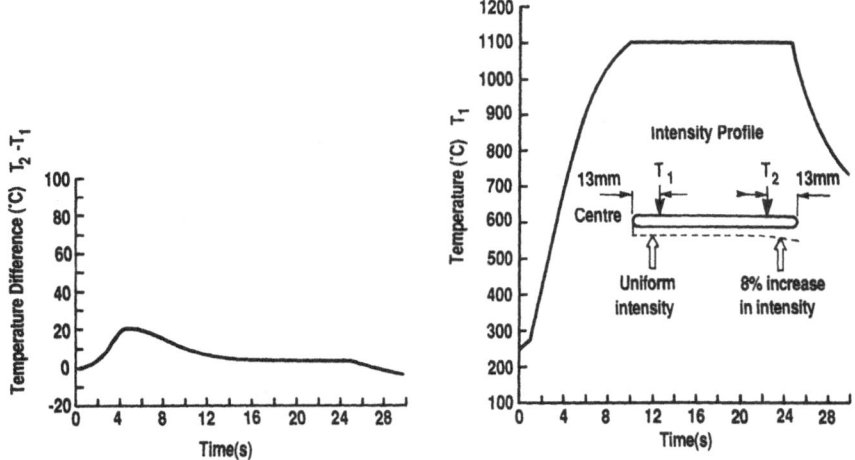

Fig 20 The right hand figure shows the time-temperature profile at a
        point near the wafer centre (T_1) for the anneal cycle described
        in figure 19. The left hand diagram shows the temperature
        difference at a point near the edge relative to the centre.
        Model details are: R_{ch} equal to 80%; γ equal to 70% and 100% for
        radiation from the wafer and lamps, respectively; T_q equal to
        240°C.

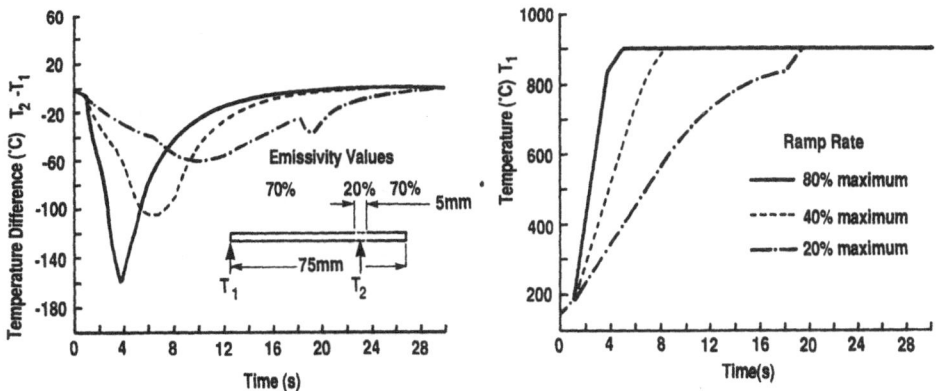

Fig 21 The time-temperature profile for the case of a silicon wafer with
        a 5mm wide overlayer having a lower absorptivity (20%). The
        right hand diagram shows the profile at the wafer centre; and the
        left hand diagram shows the temperature difference at the centre
        of the overlayer. Results for three different ramp rates are
        shown.

165

6.   TEMPERATURE MEASUREMENT

6.1   Direct Measurement

The generally accepted method for measuring the absolute temperature of a silicon wafer is by means of a properly embedded and corrected thermocouple[20],[21]. The thermocouple bead must be in intimate thermal contact with the interior of the silicon over as large an area of the bead as possible, and the whole area of the bead must absorb and reflect radiation in a similar way to the wafer itself. These criteria can be met by laser drilling a hole through the wafer, cleaning and thermally oxidising the slice to line the hole with a thin barrier oxide, and inserting a tightly filling thermocouple bead, subsequently filling the hole top and bottom with cement sprinkled with silicon dust. Such an assembly follows the local silicon temperature faithfully. A thermocouple which is merely touching the silicon, or close to it, does not report the silicon temperature, but comes slowly into equilibrium with the radiation falling on it from the lamps and the slice, as shown by the results in Fig.22. The embedded thermocouple is the primary temperature measurement technique to which all others are referred back: it is, however, very cumbersome and inconvenient for routine monitoring.

The technique almost universally used for temperature monitoring is pyrometry. The pyrometer is usually calibrated by comparing the signal from a hot silicon slice with that of an embedded thermocouple, and adjusting the pyrometer output circuit so that a linear relation between pyrometer signal and temperature is obtained over the working range. This calibration is, of course, only valid for silicon slices with back surfaces of exactly the same emissivity characteristics as the calibration wafer. The calibration can be extended to other types of wafer if the effective emissivity of such wafers <u>under the exact heat-treatment conditions</u> can be accurately measured. This is covered in a later section.

6.2   Indirect measurement

Because it is not practicable to embed a thermocouple into every wafer to be processed, and pyrometry requires an in-situ technique of effective emissivity measurement (as yet not generally available), reliable indirect techniques of temperature measurement are required. A change in the material that is sensitive to temperature can be the basis of such a technique. Suitable material changes in silicon processing are amorphous layer regrowth, silicidation, dopant activation after implant and oxidation. At Caswell, oxidation is used as the main indirect temperature measurement technique, for the temperature range 900-1300°C.

The oxide thickness grown depends on the substrate and its previous surface treatment, the time, the temperature and the ambient gas and its pressure. All these parameters are maintained constant except temperature to establish a master calibration graph of oxide thickness versus temperature. A silicon slice straight from the maker's box, which has been previously calibrated directly with an embedded thermocouple so that pyrometer readings can be used as the temperature measurement, is oxidised for 2 minutes in dry oxygen at a fixed flow rate. The long time at steady-state temperature, as compared with the ramp-up and down times, both ensures the transient effects have insignificant influence on the oxide thickness grown, and yields an oxide thick enough for accurate measurement. A typical calibration graph is shown in Fig.23.

To measure the steady state temperature of any uncharacterised slice, a small square (about 15mm) cut from the reference material is

placed in the centre of the slice, and the calibration conditions
repeated. The oxide thickness grown is measured ellipsometrically and the
temperature read off from a curve like Fig 23. It is found that the oxide
grown on such a superposed square is exactly that grown on the slice on
which it sits, for long oxidation times, as shown in Table 3. The technique is repeatable to about ±2Å, which corresponds to a temperature
uncertainty of between ±1 and 2°C.

Table 3. Oxide thickness grown on lightly doped (100) silicon slice, and on a superposed square of the same
material, after 2 mins. in dry oxygen at various
temperatures. Oxide thicknesses before and after
oxidation are given in Å.

Temperature °C	Slice		Square	
	Initial	Final	Initial	Final
1000	14	89	17	89
1100	14	211	17	210
1150	14	304	17	306
1200	14	397	17	401

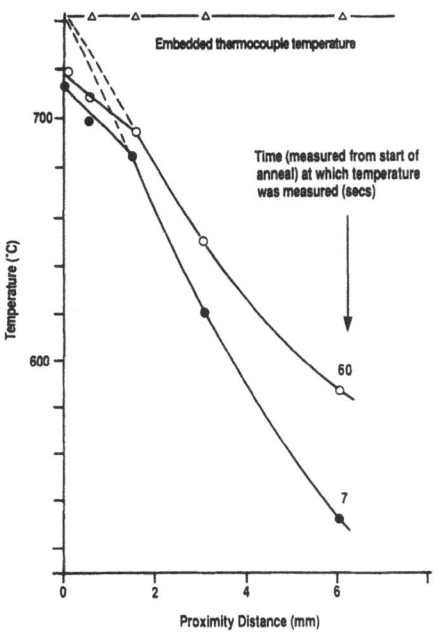

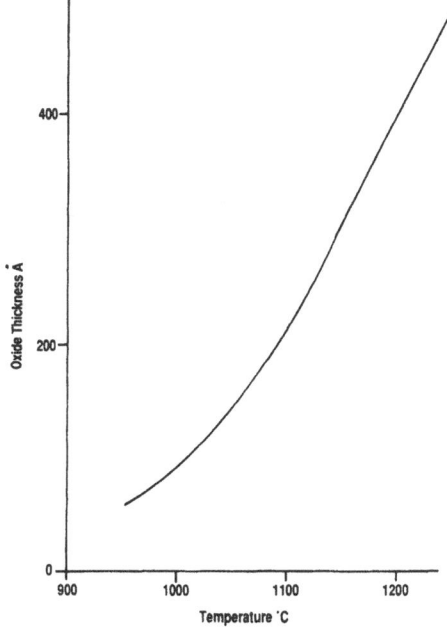

Fig 22 Comparison of experimentally measured tempera-
tures, monitored by a
thermocouple embedded in
a 5 inch silicon wafer,
and a thermocouple in
close proximity to the
wafer at the same radial
position.

Fig 23 Experimentally determined calibration curve
of oxide thickness
versus temperature, for
thermal oxide grown in
dry O_2 on (100) silicon
(5ohm cm p CZ) during a
2 minute RTA.

167

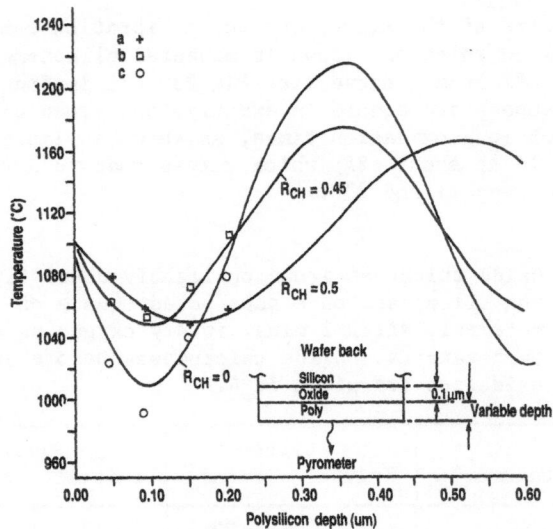

Fig 24 The actual temperature reached by a wafer with a polysilicon
layer of varying thickness on top of 0.1μm of oxide on the wafer
back. Data is given for three machines: (a) pyrometer 4.8 –
5.6μm; (b) and (c) pyrometer 3.5μm. The solid curves are as
predicted from the modelled emissivity with a fitted chamber
reflectance.

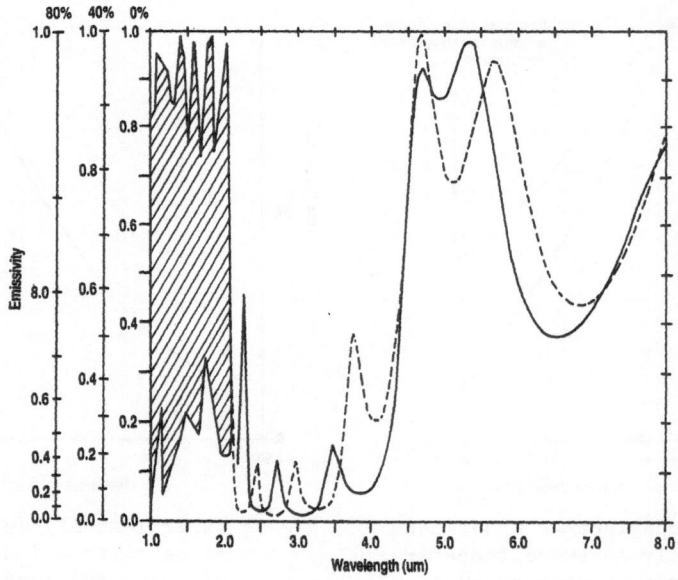

Fig 25 The emissivity versus wavelength for a ten layer structure (comprising poly, oxide and nitride layers) on the silicon wafer
back. The two curves show the effect on the emissivity of varying two of the layer thicknesses over the process tolerance
range. Values of emissivity are given for three chamber reflec-
tances: 0, 40 and 80%.

7. EMISSIVITY MEASUREMENT

In order to use pyrometry to control known temperature-time cycles on
silicon wafers different from those with which the system has been calibrated, the reference signal controlling the pyrometer feedback must be
altered to take account of the differing emissivity of the wafer back. On
most machines, this is achieved by setting an "emissivity" value manually
between 0 and 1, the calibration silicon value being fixed at a convenient
intermediate value, e.g. 0.50. This "emissivity" value (henceforward
referred to as EV) is not identical with the physical emissivity as defined
in Section 3, because the limits of EV, 0 and 1 are not necessarily set to
correspond to silicon slices with zero and 100% emission respectively and
the characteristics of the chamber and the electronics may make the relationship between EV and true emissivity not perfectly linear. This does
not matter, as long as the EV setting is used simply to modify the pyrometer signal in order to match the temperature set to the temperature
achieved. This can be done by controlling the time-temperature cycle with
the pyrometer, monitoring true temperature and adjusting EV until the true
temperature is identical with the temperature set. True temperature can be
measured with an embedded thermocouple, or more conveniently, by an
indirect technique such as the oxidation method.

Quite large variations in emissivity can arise from the interference
effects in multiple layers described in Section 4. If the EV is kept
experimentally at the value corresponding to the bare silicon calibration,
these variations are observed directly as temperature variations, and can
be compared to the predictions of the model described previously. Since
the effective emissivity is affected by the reflectivity of the machine
walls then, unless this latter quantity is known, the comparison of model
with data must involve a fit of the chamber reflectivity.

Figure 24 shows the temperature reached by wafers with a varying
thickness of polysilicon on a fixed 0.1 μm oxide layer on the single crystal substrate. The pyrometer was calibrated so as to reach 1100°C for a
blank silicon wafer; the true temperature reached was measured via the
thickness of oxide thermally grown either on a silicon square in contact
with the wafer, or on the wafer itself, during the anneal. The measurement
were performed on three different machines and the model was used to fit
the chamber reflectivity of each machine. It can be seen that for machines
a and b a very good fit is obtained with reasonable values for the chamber
reflectivity of 45 % and 50 % respectively. The model appears to under-
estimate the magnitude of temperature variation of machine c even if we
assume a black chamber (giving the maximal effect). The shape of the curve
is, however, correctly modelled. The total range of emissivity deduce
from these results (corresponding to maximum and minimum temperature) is:

Machine a (pyrometer at 3.5 μm):   69% − 93%
Machine b (pyrometer at 4.8−5.6 μm): 74% − 90%
Machine c (pyrometer at 3.5 μm):   53% − 87%

The emissivity of a more complicated multilayer structure across the
wavelength range of 0.5 − 8μm is shown in figure 25. A ten layer structure
on the wafer back arises from a bipolar process schedule immediately prior
to the emitter drive-in, including the isolation trenches and two layers of
poly. An uncertainty in two of the layer thicknesses arises from the trench
oxidation step: the two curves in figure 25 show the resulting range in
emissivity. It can be seen that there is a large variation in emissivity
across this wavelength range for both curves individually; thus for a particular pyrometer wavelength careful tuning is required to obtain correct
temperature control. However, comparison of the two curves shows that at
most wavelengths there is significant uncertainty in the emissivity due

to the layer thickness uncertainty. This implies unpredictability in the pyrometer temperature control: ie, wafers from the same batch could reach different temperatures for the same pyrometer setting. This is indeed seen experimentally when the wafer temperature is measured independently of the pyrometer as is indicated in the table below. The pyrometer is at 3.5μm and EV was set to be correctly calibrated for bare silicon to reach a temperature of 1100°C. When the control pyrometer reads the backs of nominally identically processed wafers, the temperatures reached differ by 140°C. When the control pyrometer reads the back of an underlying bare silicon susceptor wafer, the processed wafers both reach temperatures within 4°C of the set temperature. Since the predicted emissivity for the ten layer structure at a wavelength of 3.5μm is much lower than that of bare silicon, we should expect the wafer to reach a temperature greatly in excess of 1100°C. Indeed for even a highly reflecting chamber (80%) the predicted temperature range for the two curves in figure 25 is 1300 – 1800°C. (Note that any uncertainty in other layer thicknesses would have a strong effect on these predictions).

Table 4. Temperatures reached during steady-state 2 min. anneal of two nominally identical wafers at a set temperature of 1100°C and EV setting for bare silicon. (a) Anneal of wafer alone (b) Anneal of a wafer on bare silicon susceptor.

Wafer No.	Details	Wafer Alone	Wafer on Susceptor
31	2 wafers from a batch of identically processed wafers with sequence of layers on back described in text	1245	1100
32		1104	1096

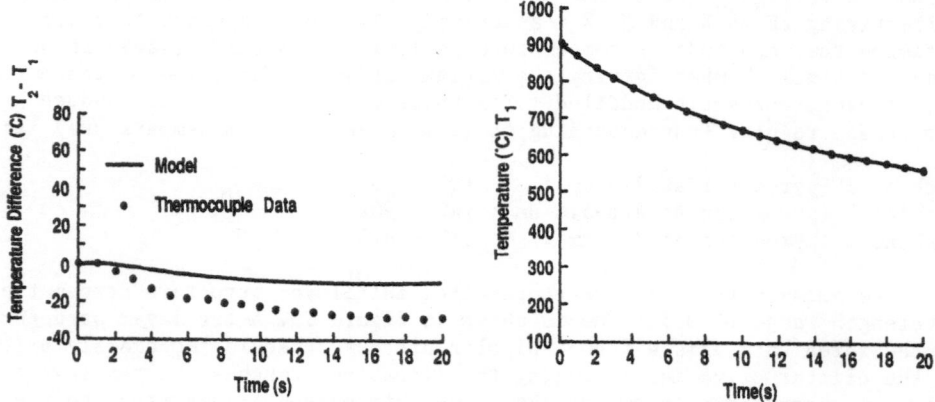

Fig 26 A comparison between the model prediction and measured temperature profiles of the cool down from 900°C of the susceptor mounted test wafer (see text). The right hand figure shows the time-temperature profile for a point near the wafer centre (T_1); and the left hand figure the temperature difference between a point 13mm from the edge and the centre ($T_2 - T_1$).

8. TEMPERATURE NON-UNIFORMITY

8.1 Experimental Measurement

The model discussed previously in section 4 predicted temperature non-uniformity to arise for two different reasons: through machine effects such as the non- uniform incident intensity designed to compensate edge effects; and slice effects such as non-uniform coating layers and the geometrical effects associated with the wafer edge. These have been investigated experimentally by monitoring the temperature at two points of a 6 inch test wafer throughout an RTA cycle via two embedded thermocouples (13 mm from the centre and 13mm from the edge of the wafer). The test wafer was uniformly coated with $0.03\mu m$ gate oxide, $0.46\mu m$ of polysilicon. The top surface had a further uniform layer of $1.0\mu m$ Low Temperature Oxide (LTO), whereas the lower surface had an LTO layer of variable thickness (from approximately $0.1\mu m$ in the centre up to $1.0\mu m$ at the edge; see the inset to Figure 29 for full details). This back surface is then a potential source of temperature non-uniformity. By mounting the wafer on a susceptor (a blank 6 inch wafer is used) this source of temperature non-uniformity could be masked, and so the two different types of temperature non-uniformity could be studied independently.

8.2 Machine Effects

The simplest case for comparison with the model is that of the wafer cooling down. In the case of an uncontrolled cool the lamps are switched off and the wafer cools radiatively. The only other source of radiation is from hot walls or quartz within the chamber. It is expected that the edge of the wafer should cool more rapidly than the centre due to the increased surface area. Figure 26 shows the measured time-temperature profile measured at the centre thermocouple of the susceptor mounted test wafer as it cools down from $900°C$; also shown is the temperature difference between the edge and centre thermocouples as a function of time. The model gives excellent agreement with the data for the centre thermocouple indicating that the combination of emissivity and chamber reflectivity have been correctly chosen. The background temperature was fitted as $240°C$. The experimental data shows that the edge is in fact cooler than the prediction of the model. This may imply that the edge has a greater radiating area than assumed in the model. An alternative explanation is seen by examination of the radial temperature distribution at the various instances of the cool down shown in figure 27. It can be seen that the position of the thermocouple lies on the edge of a steep temperature gradient falling to a value similar to that which is measured. Since the thermocouple itself acts as a sink of heat (not modelled) it may cause the temperature gradient to move further inwards and so lower the temperature at the thermocouple position. In addition, gas flow cooling of the wafer, which is not modelled, may contribute to the temperature non-uniformity.

For the case of ramp-up to temperature the model requires the full information of emissivity and absorptivity across the wafer surfaces, the incident power, the chamber reflectance and the transmission coefficients of the quartz. The emissivity and absorptivity values for the top surface were taken from figure 18 for $1.0\mu m$ of oxide (note this does not take into account the underlying layers, which have only a small effect). For the case of a susceptor mounted wafer the emissivity and absorptivity values for the back surface are those of bare silicon (0.7). The intensity was taken as uniform across the wafer surface with an additional linearly increasing enhancement up to a maximum of 8% over the last 15mm up to the wafer edge. Figure 28 shows a comparison of the measured time-temperature profile and that predicted for the susceptor mounted test wafer. The wafer was heated at an initial rate of 80% maximum lamp power and switched to a

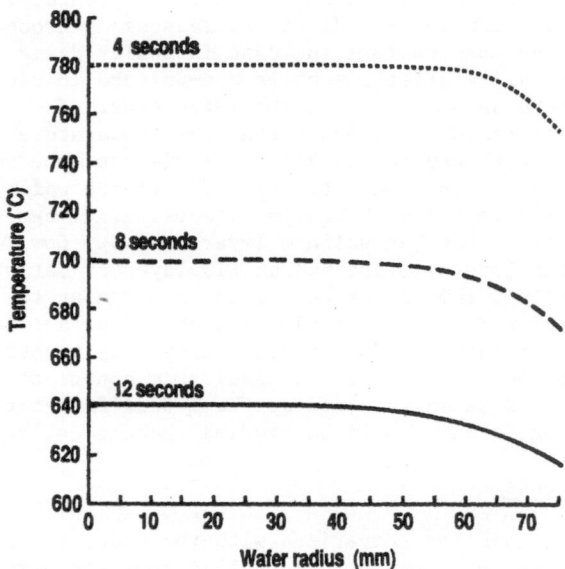

Fig 27 The radial temperature profile at three instants of the cool down
as predicted by the model for the wafer described in figure 26.
The outer thermocouple is located a distance of 13mm from the
edge, coinciding with the edge of the steep temperature gradient.

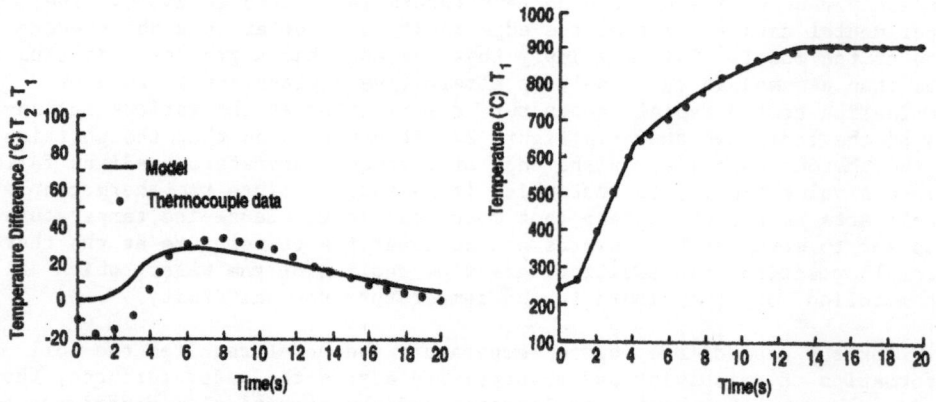

Fig 28 A comparision of the model prediction and the measured time-
temperature profiles for the ramp-up to 900°C at 80% power of the
susceptor mounted test wafer. The right hand figure shows the
temperature at a point near the wafer centre (T_1); the left hand
figure shows the temperature difference between a point 13mm from
the edge and the wafer centre ($T_2 - T_1$).

lower ramp rate at 650°C. The profile is in good agreement with the model prediction for the centre thermocouple. The temperature difference recorded at the edge is also in good overall agreement showing a similar transient of approximately 25-30°C lasting for about 10 seconds. This transient is due to the excess heating at the edge during the ramp-up as discussed in Section 5.

8.3 Slice Effects

If the wafer is not susceptor mounted then the back face of the wafer is exposed to the incident radiation. We then expect an additional contribution to temperature non-uniformity due to the varying absorptivity of the wafer back. The values of emissivity and absorptivity used for the simulation were calculated by assuming the dominant effect to be due to the LTO layer and so ignoring the thin underlying gate oxide. These values are shown in table 5 in which the back LTO profile on the back surface is approximated as a series of steps of uniform thickness.

Table 5 Emissivity (e) and absorptivity (a) values calculated for the back surface of the test water.

Radius (mm)	Oxide thickness(μm)	e	a
0 - 50	0.12	0.71	0.81
50 - 65	0.24	0.76	0.85
65 - 70	0.60	0.83	0.82
70 - 73	0.80	0.82	0.81
73 - 75	1.00	0.81	0.81

A comparison of the measured and predicted profiles for the case of a 40% ramp of the test wafer to 900°C is shown in figure 29. The temperature profiles at the centre thermocouple are in good agreement, but the predicted difference at the edge is not as large as is measured. In addition the model predicts that in the steady state the edge is at a lower temperature

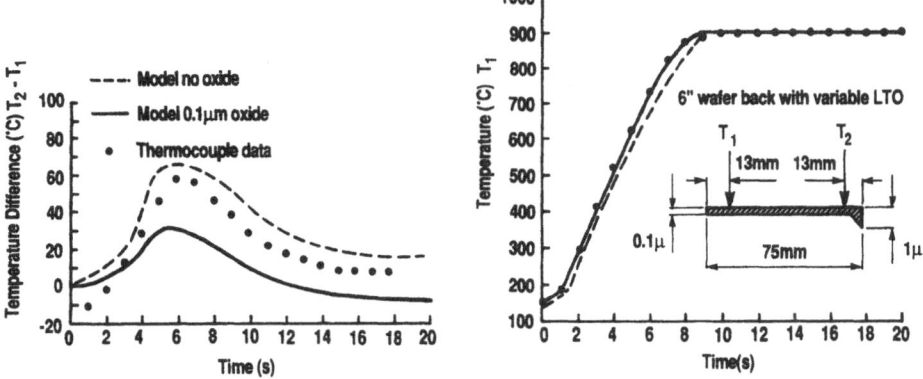

Fig 29 A comparison of predicted and measured temperature profiles for the ramp-up to 900°C at 40% power of the unmounted test wafer. The right hand figure shows the temperature at a point near the wafer centre (T_1); the left hand figure shows the temperature difference between a point near the edge (T_2) and the centre. The solid curves are model predictions corresponding to an oxide thickness of 0.1μ at the wafer centre (as shown); the dashed curves correspond to zero thickness of oxide at the centre.

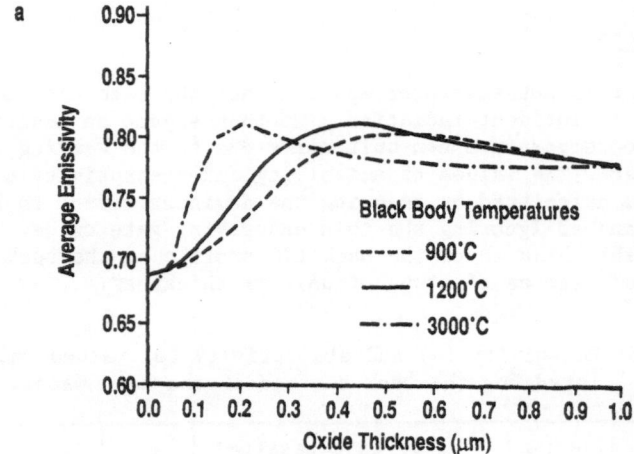

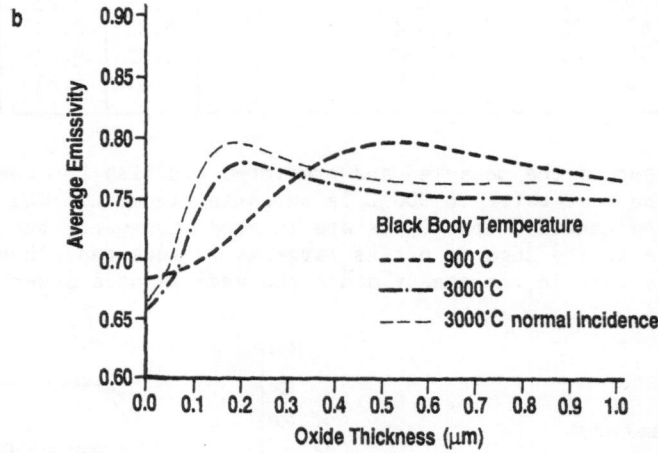

Fig 30 Improved calculations for the spectrally averaged emissivity for
a varying thickness oxide layer on a silicon substrate for
various black body temperatures (cf fig 18)

(a) Spectral averaging is performed with the full wavelength
dependent optical data of the materials and angular averaging
of the radiation emitted.

(b) Additional polysilicon (0.46µ) and gate oxide (0.028µ) layers
are included between the oxide and substrate. Averaging is
performed as in (a) except for the normal incidence curve in
which there is no angular averaging

than the centre whereas the measured results give the opposite behaviour. The reason for a non-uniform steady-state temperature is due to differences in the ratio of emissivity to absorptivity across the wafer, since this determines the net radiation balance at the surface. Regions which have a larger value of e/a will be at lower temperature than regions with lower e/a values. Examining the e and a values in table 5 it can be seen that the values of e/a at oxide thickness 0.12 µm (wafer centre) and 1.0µm (wafer edge) are 0.88 and 1.0 respectively. Thus we should expect the edge to be cooler than the centre which is indeed what the model predicts. The experimental data thus casts doubt upon the e and a values used in the simulation. There are several reasons for uncertainty in the chosen e and a values:

(a) Figure 18 shows that the absorptivity is quite sensitive to oxide thickness around 0.1 µm, thus uncertainty in the exact oxide thickness leads to uncertainty in absorptivity. Indeed, if the centre of the wafer is modelled with zero oxide thickness then the correct type of steady state behaviour is obtained, as is shown by the dashed curve in figure 29.

(b) The exact temperature of the lamp is not known; since this determines the incident intensity spectrum it affects the averaged absorptivity. If the lamp temperature required to hold the wafer at 900°C is lower than 3000°C then the absorptivity curve in figure 18 will move to the right (giving lower absorptivity at 0.1 µm).

(c) The thin underlying gate oxide and polysilicon layers were not included in the calculation of emissivity and absorptivity in figure 18. These layers have a negligible effect on the emissivity since the 900°C spectrum is at too long a wavelength to resolve the oxide layer. There is a small effect on the absorptivity as the lamp spectrum has substantial power in shorter wavelengths. In addition if the calculation of emissivity and absorptivity includes angular averaging (assuming an isotropic radiation distribution) then the resulting values are lower than those obtained at normal incidence.

Figure 30 shows an improved calculation of averaged emissivity (at 900°C) and averaged absorptivity (at 3000°C): averaging over the appropriate black body spectrum was performed using full optical data for silicon[23] and silicon dioxide[24]; angular averaging was performed assuming an isotropic distribution (in the substrate). The angular averaging results in approximately a 2% reduction, and this is the only change for the emissivity determined at 900°C. The improved optical data result in a reduction in the peak of the absorption curve. The effect of the thin gate oxide layer is to lower the absorptivity curve by approximately 2 - 3% relative to the emissivity curve.

In conclusion these simulations show the usefulness of the model in understanding and predicting the magnitude of any temperature non- uniformity during the RTA cycle. An important requirement is accurate knowledge of the emissivity and absorptivity over the spectrum of the lamps and the emitted spectrum of the wafer; this requires values of the optical constants over a wavelength range of approximately 0.5 to 20µm for common processing materials (silicon, polysilicon, silicon dioxide, silicon nitride etc). The model can then be used to study alternative machine conditions to suggest ways of reducing such non-uniformities: these may include altering the ramp rate; using a susceptor; controlling the cool down, etc. The model can also provide a useful input to process models simulating the physical processes occuring during the anneal such as diffusion and oxidation which are sensitive to the exact time-temperature conditions.

9. PROBLEMS AND SOLUTIONS

9.1 Temperature Control

This affects mainly the reproducibility of the process from batch to batch and from slice to slice. Knowledge of the exact temperature-time cycle is essential for modelling the solid state processes occurring in the wafer (e.g. oxidation, diffusion).

CASE 1.   Backs of wafers are identical to the reference base silicon wafer used for pyrometer linearisation and calibration of "emissivity" value (EV) (see section 7).

Solution a

CASE 2.   Backs of wafers are different from that of the reference wafer, but are identical and uniform.

Solution b

CASE 3.   Backs of wafers are different from the reference wafer, are uniform but are not identical.

Solution c or d

CASE 4.   Backs of wafers are different from the reference wafer and are not uniform across the wafer back.
Solution d

CASE 5.   Backs of wafers change emissivity during the heat cycle, but do so uniformly across the wafer back.
Solution c or d

CASE 6.   Backs of wafers change emissivity during the heat cycle, but do so non-uniformly across the wafer back.
Solution d

9.2 Temperature Uniformity

This affects mainly the uniformity of thermal processing across an individual wafer, and therefore the matching of components and circuit characteristics. The most serious effect of poor temperature uniformity is thermal gradients which cause differential thermal expansion and plastic deformation (slip).

CASE 1.   Radial non-uniformity in steady state caused by additional heat loss at slice edges.
Solution d, e or f

CASE 2.   Radial non-uniformity during ramp-up and cool-down caused by different surface/volume ratio at slice edges.
Solution d, f or g

CASE 3.   Lateral non-uniformity in steady-stage caused by non-uniform emissivity and absorptivity across the wafer surface.
Solution d

CASE 4.   Lateral non-uniformity during ramp-up and cool-down, caused by non-uniform emissivity and absorptivity across the wafer surface.
Solution d or g

Solution a.   Manual setting of EV setting to bare silicon reference value.

Solution b.   Manual setting of appropriate EV setting to achieve true temperature = set temperature, after measurement of EV setting using one slice.

Solution c.   Automatic measurement of appropriate EV setting and real-time feedback to pyrometer during the heat-cycle.

Solution d.   Support of wafer on susceptor of known EV.

Solution e.   Energy loss supplied by additional lamp radiation around wafer rim.

Solution f.   Energy loss supplied by reflection of wafer radiation back into wafer edge. This can be achieved either by optical reflection from a cold reflecting surface close to the wafer edge, or approximated by provision of a hot surface in the same position: in the latter case, the surface must follow the same time-temperature cycle as the wafer.

Solution g.   Temperature gradients minimised by allowing time for solid state thermal conduction, during slow ramp-up and cool-down.

10. PRACTICAL MACHINE DESIGNS

Three types of machine design which have proved practical are illustrated schematically in Fig 31. Examples of all these are available commercially. Design A is built around a quartz processing chamber sealed with a cooled metal door. Transmission of both lamp and pyrometer radiation is thus through quartz, and because of the poor thermal conduction of quartz the whole chamber can become quite warm (400 - 600°C) during processing. Advantages are (i) contamination control - the quartz cell can be periodically etched to semiconductor processing standards and (ii) corrosive ambients can be used during processing eg, HCl gas. (iii) Since the lamp reflectors are external to the cell, they can be selected, without compromise, for high reflectivity: this reduces the effective emissivity variation caused by coatings on the slice, as discussed in Section 4. Disadvantages are mainly in engineering complexity associated with accurate pyrometry. The pyrometer signal is affected by its passage through the quartz wall, which begins to absorb strongly at wavelengths greater than 4 microns, and at progressively shorter wavelength as temperature increases above room temperature. The reflecting chamber, with lamps on the same side of the slice as the pyrometer, also means that considerable attention has to be paid to minimising and compensating for lamp radiation at the pyrometer wavelength which may enter the pyrometer aperture.

Design B is built around a watercooled metal chamber, sealed with a cooled quartz plate. Transmission from the lamps is again through quartz, but the pyrometer radiation is transmitted through an aperture in the metal wall which is cooled and thus need not be sealed by a quartz window. Advantages of this design are (i) easier fabrication of the metal cell (ii) option of using the quartz plate as an active optical element in controlling radiation uniformity (iii) much easier measurement of temperature by pyrometry. This latter advantage arises from shielding the

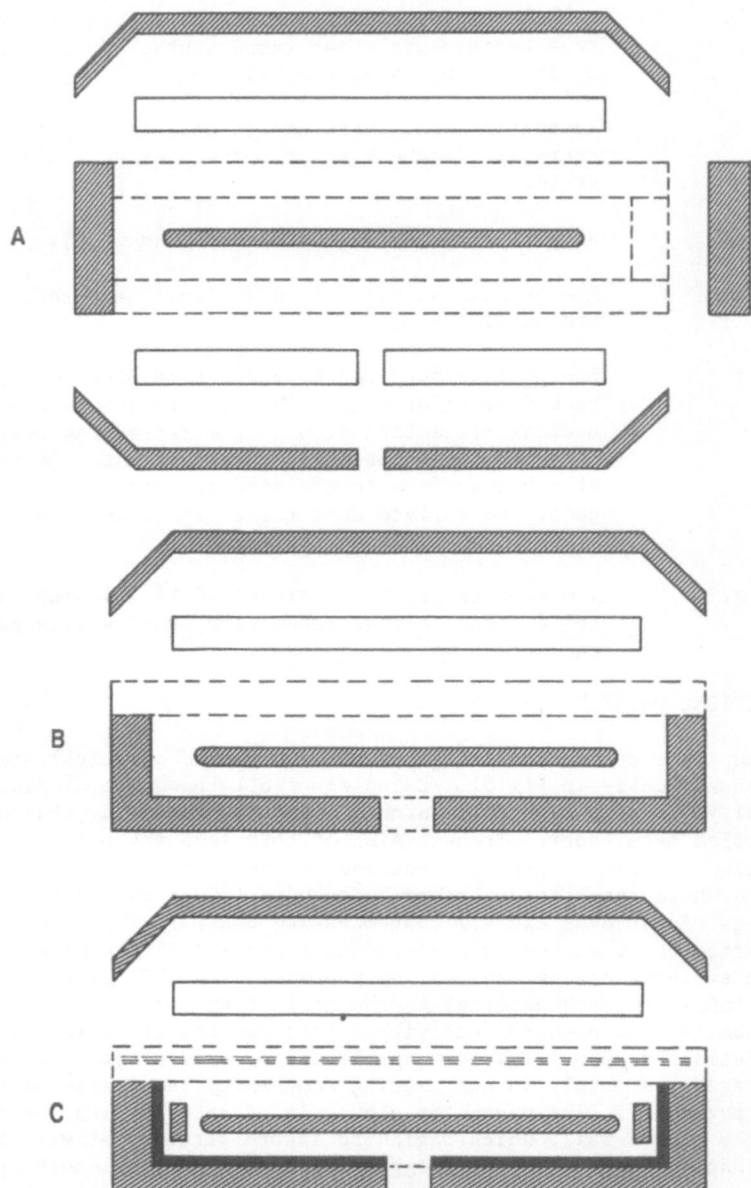

Fig 31 Schematic representations of three designs of RTA machine:
(A) Quartz chamber, halogen lamps above and below wafer, high
reflectivity external reflectors; (B) Metal chamber, moderately
reflective, arc lamp backed by highly reflective external reflector above, transparent quartz plate between lamps and wafer,
acting both as gas seal and optical element; (C) Metal chamber
with arc lamp, non-reflecting walls, reflecting ring around wafer
periphery.

pyrometer from lamp radiation with the silicon slice, and from being able to choose the pyrometer radiation band unconstrained by the absorption properties of quartz. The choice is generally for long wavelengths (> 5 micron) which have the advantages of having no interfering signal from the lamps (stopped by the quartz window), and being nearer to the peak of radiation from the slice when heat treated in the lower temperature range (400 - 600°C). Disadvantages of the design are non-compatibility with corrosive gases, difficulty of high-quality cleaning, and less reflective chamber walls.

Design C is similar to that of B, but is shown with the additional features of non-reflecting walls and a reflecting ring around the periphery of the silicon slice. This is one engineering solution to the problem of obtaining uniform temperature across a wafer, combining a uniform radiation field across the wafer surface, with a self-balancing radiation field around the wafer edge which exactly compensates for edge losses of wafer radiation. This approach has been shown to be very effective in avoiding slip due to thermal gradients[22]. Disadvantages are that differential emissivity effects are maximum in a non-reflecting chamber, and also a much higher power lamp source must be used since less of the lamp radiation reaches the slice.

This summary is necessarily incomplete, since the incorporation of engineering solutions based on the physics described in this paper is a very active area in machine design. All of the solutions described in Section 9 are expected to become generally available over the next two years, and no doubt the current development of RTA machines capable of deposition and etching will require and give rise to new solutions.

11. ACKNOWLEDGEMENTS

The authors thank Vic Browne (Caswell), Arnon Gat and Jain Nulman (AG), Tim Stulz and Ahmed Kermani (Peak) for useful discussions and permission to present some results of collaborative experiments; also Julian Black (Eaton) for information on the ROA400 and 500 systems. The work was partially funded by the Alvey Directorate through the Procurement Executive MOCVD and thanks are due to them and the Plessey Company for permission to publish.

12. REFERENCES

[1] J.M.Fairfield and G.H.Schwuttke Solid State Electronics 11 1175 (1968).

[2] Laser-Solid Interactions and Laser Processing - 1978 (S.D.Ferris, H.J.Leamy, J.M.Poate eds) American Inst.Phys.(1979)

[3] Laser and Electron-Beam Processing of Materials (C.W.White and P.S.Peercy eds) Academic Press N.Y. (1980)

[4] Laser & Electron-Beam Solid Interactions and Materials Processing (J.F.Gibbons, L.D.Hess, T.W.Sigmon eds). Mat.Res.Soc. Proc 1 North Holland (1981).

[5] Laser and Electron-Beam Interactions with Solids (B.R.Appleton & G.K.Celler eds) Mat.Res.Soc. Proc 4 North Holland (1982).

[6] Energy Beam - Solid Interactions and Transient Thermal Processing (J.C.C.Fan, N.M.Johnson eds) North Holland (1984).

[7] Rapid Thermal Processing (T.O.Sedgwick, T.E.Seidel, B.Y.Tsaur) Mat.Res.Soc. Proc. 52 MRS. Pittsburgh (1986).

[8] Rapid Thermal Processing of Electronic Materials (S.R.Wilson, R.A.Powell, D.E.Davies eds) Mat.Res.Soc. Proc.92 MRS. Pittsburgh (1987).

[9] Rapid Thermal Processing of Electronic Materials and Devices Extended Abstracts 63-92 Electrochemical Society Meeting Atlanta May 15-20 (1988).

[10] Y.Miyai, K.Yoneda, H.Oishi, H.Uetida, Morio Inone, J.Electrochem.Soc. 135 No.1 150 (1988).

[11] C.Hill in 'Laser Annealing of Semiconductors' (J.Poate, J.W.Mayer eds) 479 Academic Press NY (1982)

[12] T.Sato Japan J.Appl.Phys. 6(3) 339 (1967)

[13] C.Hill unpublished data

[14] D.J.Godfrey, A.C.Hill and C.Hill J.Electrochem.Soc. 128 (8) 1798 (1981)

[15] M.Born and E.Wolf 'Principles of Optics' 6th edition Pergamon Press (1981)

[16] T.S.Moss 'Optical Properties of Semiconductors' Butterworth, London (1980)

[17] D.W.Pettibone, J.R.Suarez and A.Gat Ref.7 p.209 (1986)

[18] R.E.Sheets Ref.7 p.191 (1986)

[19] P.G.Merli Optik 56(3) 205 (1980)

[20] S.A.Cohen, T.O.Sedgwick and J.L.Speidell Ref.6 p.321 (1984)

[21] J.L. Hoyt and J.F. Gibbons Ref.7 p.15 (1986)

[22] J.Blake, J.C.Gelpey, J.F.Moquin, J.Schlueter, R.Capodilupo Ref.8 p. 265 (1987)

[23] D.E.Aspnes, 'Optical Functions of intrinsic Si', in: 'Properties of Silicon', EMIS data review 4, INSPEC, The Institution of Electrical Engineers, London (1988).

[24] H.R.Philipp, 'Silicon Oxides : optical functions', (ibid).

RAPID THERMAL PROCESS INTEGRATION

Iain D. Calder

Advanced Prototyping Laboratory
Northern Telecom Electronics
P.O. Box 3511, Station C
Ottawa, Ontario, Canada, K1Y 4H7

INTRODUCTION

The capability of rapid thermal processing (RTP) to perform many of the thermal steps required for semiconductor device fabrication has been investigated and demonstrated thoroughly over the past few years and many excellent reviews of these techniques are now available[1,2]. Hundreds of papers have been published dealing with isolated RTP process techniques such as oxidation/nitridation[3,4], junction annealing[5-7], thermal silicidation[8,9], densification and reflow of deposited oxides[10,11], and contact sintering[12]. These techniques are now being implemented in prototyping and manufacturing lines, often replacing several different conventional furnace processing steps. Therefore integration of RTP into a full device process must be studied. The effects of RTP fall into two general categories: for the circuit designer, optimized circuit performance and reliability are required; for the process designer, the need is for a simple, robust, manufacturable process in which interactions between different steps do not adversely affect the properties of existing materials and structures. These interactions are particularly significant for thermal steps, since the thermal budget of a complete device process is cumulative.

The next section briefly describes a generic state-of-the-art CMOS process, with emphasis on the thermal treatments that may be replaced by RTP steps. This summary will provide a basis for later discussions of interactions between process steps. The third section addresses the relations between RTP processing and final device characteristics. The fourth section looks at interactions between process steps and the compromises that must be made to overcome these problems. Finally, the fifth section explores some opportunities that arise in process integration because of the flexibility and capability of RTP.

CMOS DEVICE PROCESSING

Simple CMOS processes are described in a number of standard textbooks[13-15] and have been developed with several process enhancements[16] to the state-of-the-art generic process flow shown in Figure 1. This double level metal twin well process uses advanced features such as epitaxial silicon, a lightly doped drain (LDD), self-aligned silicidation (Salicidation) of source, drain and polysilicon, titanium nitride local interconnect and (optional) selectively deposited tungsten contact and via plugs. Since the use of reflowed glass (such as borophosphosilicate glass, or BPSG) for the passivation dielectric is still widespread, I shall discuss this process technique as well. There are 12 photoengraving steps (including the scratch protection, not shown), which would increase in number with the addition of more polysilicon or metal layers, for example.

There are a total of 26 thermal steps in this process, of which 11 are some form of chemical vapour deposition (CVD), 7 are oxidations, 6 are anneal or reflow steps, and 2 are sintering steps (for silicide formation and final metal sinter). RTP can replace most of these steps, except for the long anneals such as well drives and field oxidations, although the anneal/reflow procedures are generally finding their way into manufacturing first. Metal sintering and growth of thin oxides are also reasonably mature, while rapid thermal deposition is, for the most part, still in the exploratory stages of development[17,18]. Some of the process interactions to be discussed later are clear by looking at Figure 1. For example, the implanted source/drain junctions may be subjected to three high temperature steps: junction anneal, silicide anneal, and BPSG reflow; the thermal budget is cumulative through these steps. On the other hand, BPSG reflow takes place after silicide formation, so the reflow temperature is limited to avoid damage to the $TiSi_2$.

DEVICE/CIRCUIT OPERATION CONSIDERATIONS

A circuit designer has two major criteria he wants to see met by semiconductor integrated circuits--performance and reliability. Optimization of circuit performance does not necessarily arise from conditions that optimize each process step individually. A time/temperature process window must be selected that provides the best compromise satisfying each criterion. The greatest advantage can be obtained by the use of response surface methodology, a statistical technique for experimental design and analysis[19].

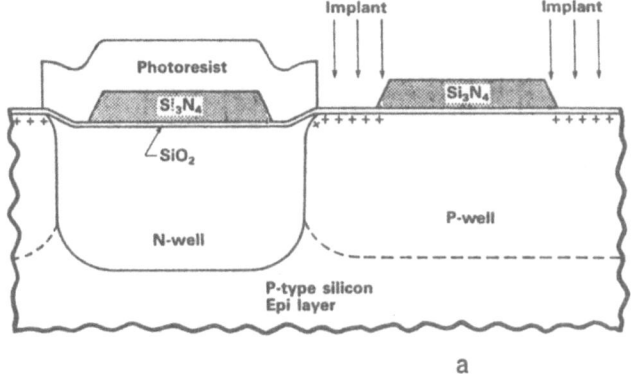

Implant   Implant

Photoresist

Si₃N₄   Si₃N₄

SiO₂

N-well   P-well

P-type silicon
Epi layer

a

First pad oxidation
First LPCVD nitride deposition
N–well PE
First nitride etch
N–well implant
Resist strip
Alignment oxidation
First converted nitride etch
First wet nitride strip
P–well implant
Well drive
Oxide strip
Second pad oxidation
Second LPCVD nitride deposition
Device well PE
Second nitride etch
P–guard Pe
Field implant

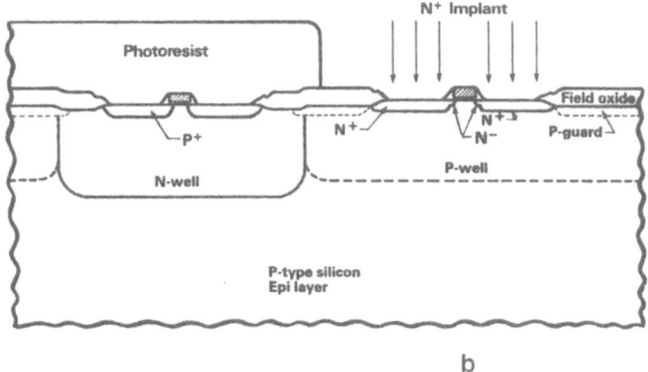

N⁺ Implant

Photoresist

Field oxide

P⁺   N⁺   N⁺,   P-guard
     N⁻

N-well   P-well

P-type silicon
Epi layer

b

Resist strip
Field oxidation
Second converted nitride etch
Second wet nitride strip
Wet pre–gate oxidation
Gate oxidation
VT adjust implant
Poly deposit
Poly doping implant
Poly anneal
Poly PE
Poly etch
Poly oxidation
P+ PE
P+ implant
Photoresist removal
N– LDD implant
CVD oxide deposition
Sidewall spacer formation
N+ PE
N+ implant
N+ anneal

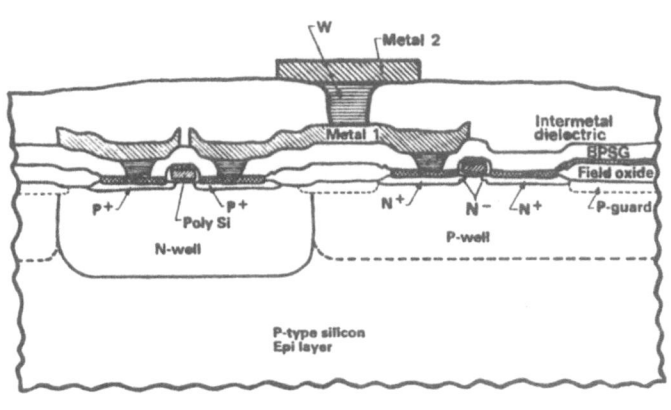

W   Metal 2

Metal 1   Intermetal
dielectric

BPSG

Field oxide

P⁺   P⁺   N⁺   N⁻   N⁺   P-guard

Poly Si

N-well   P-well

P-type silicon
Epi layer

c

BPSG deposit & densification
Contact PE
Contact dry etch
Contact reflow
Tungsten contact plug deposition
First interconnect metal deposit
First interconnect metal deposit
First metal PE
First metal dry etch
Ti deposition
Silicide formation
Local interconnect
Unreacted Ti strip
Planarizing oxide deposit
Planarizing etch
Intermetal dielectric deposition
Via etch
Tungsten via plug deposition
Second interconnect metal
   deposition
Second metal PE
Second metal etch
Sinter

Fig. 1. Process flow for a generic state-of-the-art double level metal CMOS
process. There are 26 thermal steps in total, including BPSG
reflow, which is actually unnecessary when tungsten contact plugs
are used.

Junction Annealing

One set of criteria that has been published for choosing an optimum
junction anneal cycle is the combination of minimal junction diffusion (Figure
2) with elimination of extended defects caused by ion implantation[1,5] (Figure
3). Then a narrow process window opens up at high temperatures and short
times (Figure 4), since the activation energy for dopant diffusion is greater
than for defect removal. However it is more valuable to look at the
optimization of junction annealing from the customer's (circuit designer's)
viewpoint, i.e. by aiming at high performance and reliability. Shallow
junctions are desirable for improved circuit speed, while the elimination of
end-of-range defects should reduce junction leakage and improve reliability.
It is certainly true that the shallow junctions obtained from rapid thermal
annealing (RTA) can improve circuit speed, as demonstrated in Figure 5,
although this result is not obvious, a priori. The reduced diffusion obtained
by RTA, compared to furnace annealing, actually results in a longer effective
channel, which would decrease speed. However this effect is more than offset
by the reduced junction capacitance obtained with a shallow junction, and is
further reduced when the LDD is present. The improvement due to RTA becomes
greater still at submicron geometries. A third implication of shallow
junctions for device characteristics is the potentially greater drain leakage
current because of the more abrupt junction. Therefore it may be more
desirable to use device leakage and circuit performance, or at least junction

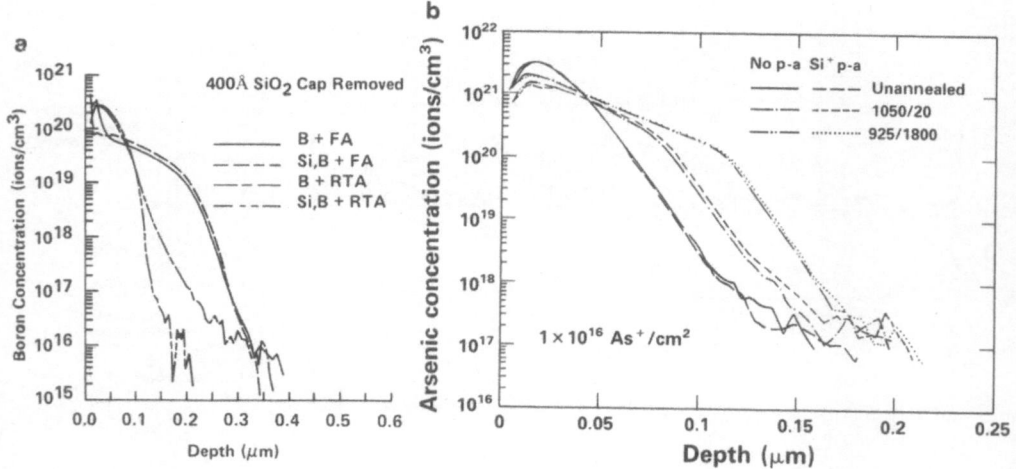

Fig. 2. SIMS profiles of implants of (a) 3×10^{15} boron/cm^2 @ 10 keV and (b)
1×10^{16} arsenic/cm^2 @ 50 keV showing diffusion of dopant after
various annealing cycles. Boron requires both pre-amorphization
(with 2×10^{15} Si$^+$/cm^2 @ 50 keV) and RTA to achieve a shallow
junction, while pre-amorphization is unnecessary with arsenic.
From References 6 and 7.

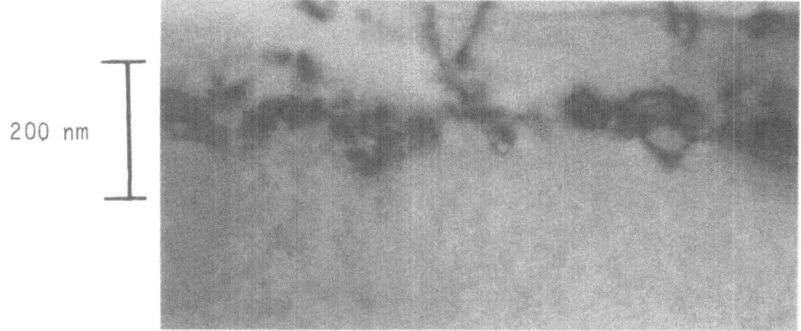

200 nm

Fig. 3. Transmission electron micrograph showing end-of-range defects after annealing of a junction, pre-amorphized with $2x10^{15}$ Si^+/cm^2 @ 60 keV, and implanted with $1x10^{16}$ As^+/cm^2 @ 50 keV. RTA was carried out at 1050°C for 20s.

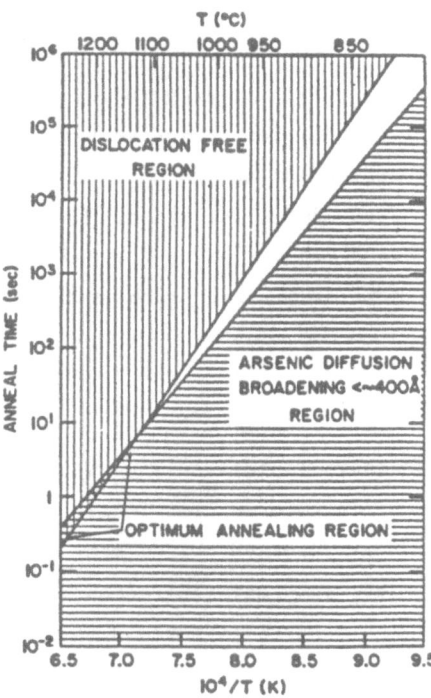

Fig. 4. Process design for n^+p junction formation after As^+ implantation using criteria of removal of all end-of-range defects, and limitation of dopant diffusion to $\leq$400 Å. Then a process window opens up at short times and high temperatures. From Reference 1.

depth minimization, as criteria for optimizing the process. Figure 6 gives an example using several set of data[7,20-23] for RTA of arsenic implanted n^+p junctions. When maximum leakage current (arbitrarily set at 50 na/cm^2 in this case) is used as a criterion rather than defect removal, the allowable process window shifts to low temperatures at long times. A compromise cycle of

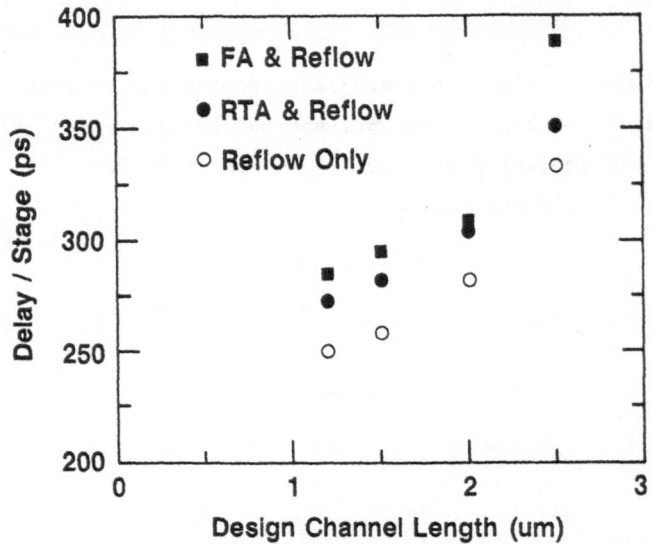

Fig. 5. Variation of ring oscillator stage delay for a CMOS circuit as a function of FET channel length and heat treatment. The furnace annealing was carried out at 925°C for 30 min and followed by a reflow at 1050°C for 20s. The RTA of the junction was also carried out at 1050°C for 20s. The highest speed is obtained by omitting the junction anneal step, so that the reflow step performs both functions.

1050°C/20s provides good characteristics along with reasonable throughput. A similar examination can be carried out for implanted p^+n junctions, but a single unified experiment is required, since many more factors, such as dose, implanted species (B^+ or BF_2^+), preamorphization, and silicidation, have an effect on p^+n junctions and results gathered from disparate sources[22,24-29] are not particularly consistent. Results from a single experiment are shown in Figure 7. Here a specification for the junction yields a process window; for example, a junction depth of $\leq 0.3\mu m$ and a sheet resistance of $\leq 60\Omega/\square$ defines a process window between 1000°C and 1060°C for a 20s anneal. There-fore our selected window of 1050°C for 20s is still valid. No process window exists for furnace annealing.

186

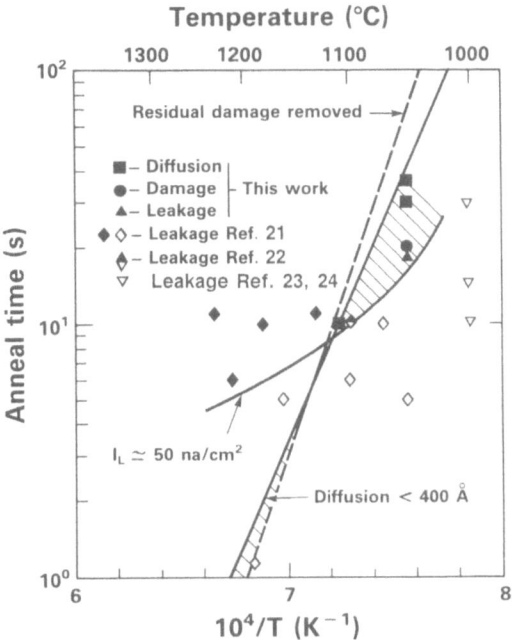

Fig. 6. Process design for n+p junction formation after As+ implantation using criteria of limited dopant diffusion ($\leq$400 Å) and limited diode leakage ($\leq$50 na/cm^2). Open symbols indicate I_L > 50 na/cm^2. The resulting process window is at long times and low temperatures. From Reference 7. Compare with Figure 4.

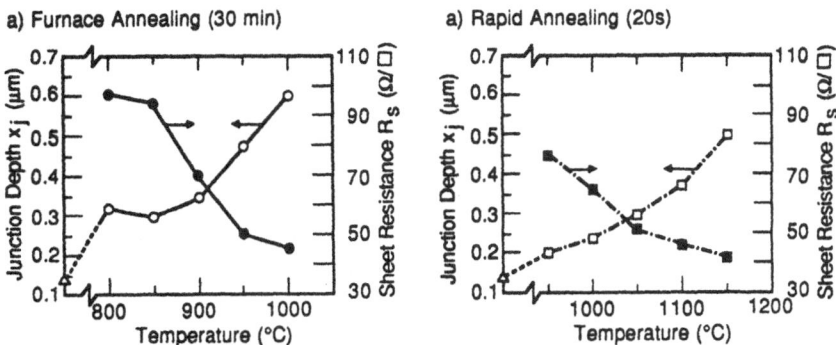

Fig. 7. Optimization of p+n junction characteristics for an implantation of 3×10^{15} B+/cm^2 @ 10 keV into silicon preamorphized with 2×10^{15} Si+/cm^2 @ 60 keV. A process window is selected by specifying maximum junction depth and maximum sheet resistance. From Reference 10.

As: 2.0E15/cm², 70 keV

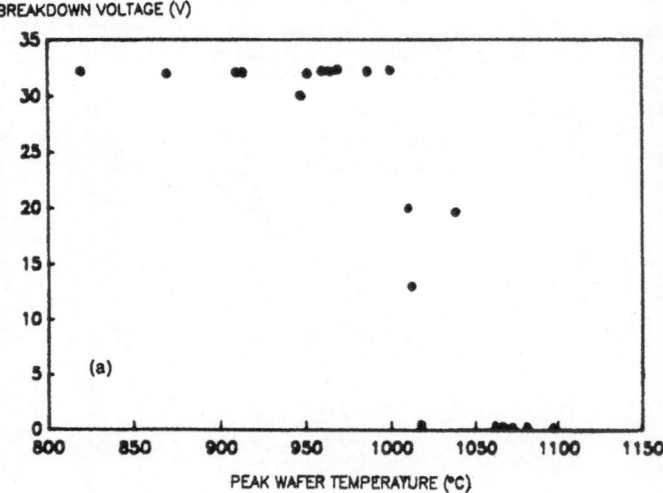

BF₂: 2.5E15/cm², 50 keV

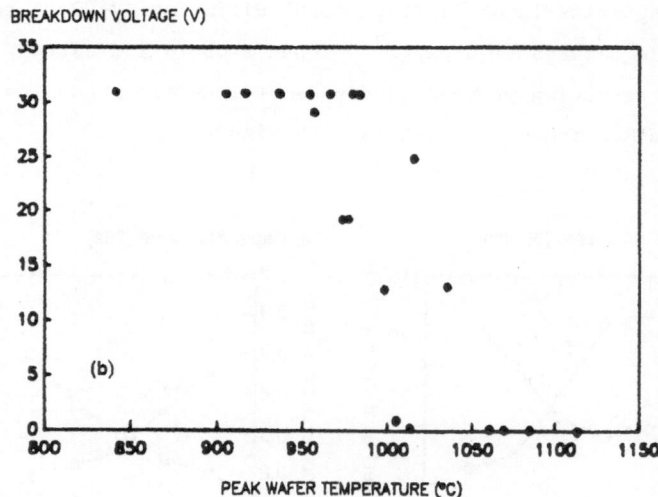

Fig. 8. Breakdown voltage for a 300 Å thermal oxide as a function of RTA temperature. The 2250 Å polysilicon plate was doped with phosphorus and, after patterning, implanted with (a) As⁺ or (b) BF₂⁺, as indicated. From Reference 30.

Oxide Degradation

Concerns about the effects of RTP on thin (gate) oxide integrity were first raised in 1986 when Finn and Coe[30] demonstrated degraded breakdown voltages for cycles in excess of 950°C (see Figure 8). However, more recent work[31] has shown that RTP at temperatures up to 1100°C is not harmful, unless preceded by reactive ion etching (RIE); then an furnace anneal is required to remove the RIE damage. This problem has been solved[32] by annealing in an

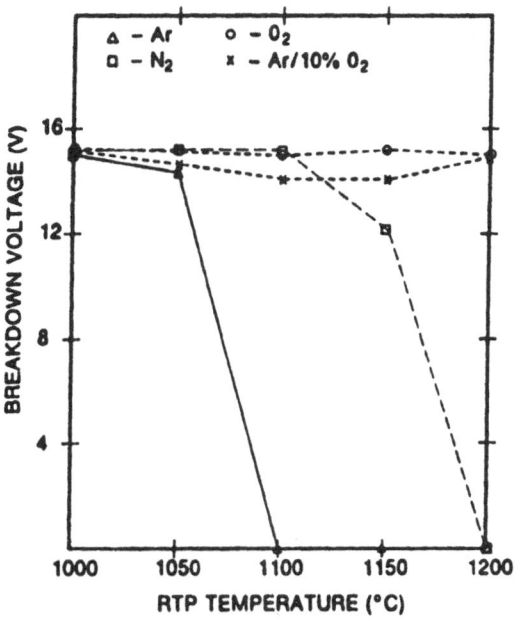

Fig. 9. Breakdown voltage for nominally 150 Å thermal oxide as a function of RTA temperature for various annealing ambients. An oxidizing ambient prevents breakdown up to at least 1200°C. From Reference 32.

oxidizing ambient prior to RTP, or by including a small amount of oxygen in the RTP ambient, as seen in Figure 9. As little as 1-2% O_2 will prevent instability in the Si/SiO_2 interface that can result in the formation or volatile SiO and degradation of the oxide integrity[33]. Similar results, with no breakdown voltage degradation for 80Å oxides annealed at 1100°C for 10s or for 150Å oxides annealed at 1125°C for 10s, were obtained without the deliberate use of oxygen[34,35]; in these cases it is likely that the use of only a short purge allowed some trace O_2 to remain in the annealing chamber.

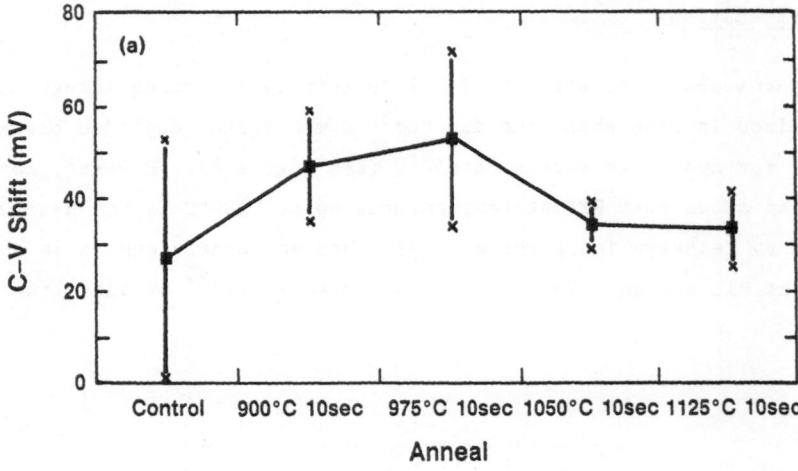

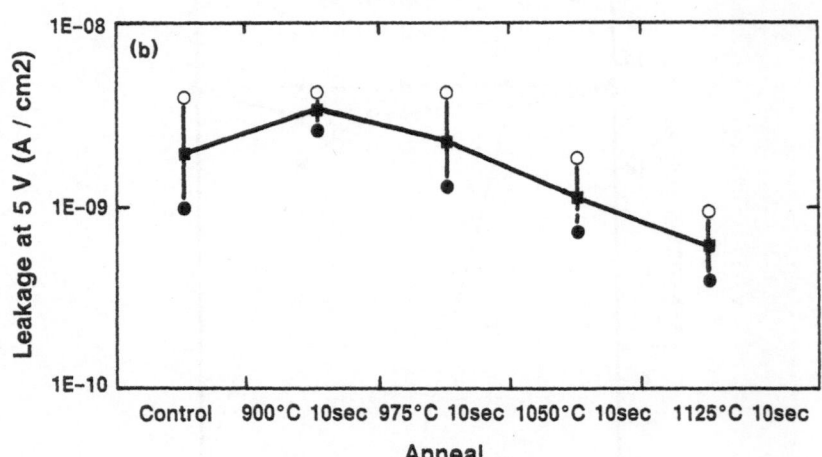

Fig. 10. Effect of RTP on (a) threshold voltage and (b) leakage for 150 Å
thermal oxide. No degradation up to at least 1125°C has occurred;
in fact some improvement in oxide leakage is observed. From
Reference 35.

Several studies of charge trapping, especially by the group centered at
the University of Texas[34,36-37], have also shown little or no degradation for
RTP temperatures up to 1100°C. Electron trapping is unaffected or reduced by
RTP, while hole trapping is somewhat enhanced by high temperature/long time
RTP[36,37]. However, annealing for long times, once more in an oxidizing
ambient, has been shown to reduce hole trapping as well as electron
trapping[38]. Measurements by Cosway and Hodel[35] have shown no increase in
threshold shift or oxide leakage after an anneal of 1125°C for 10s (Figure
10). In addition, accelerated stress conditions show a greater degradation of

a   b

Fig. 11. **Scanning electron micrographs of hillock formation on Al-1%Si after (a) furnace annealing at 400°C for 30 min or (b) RTP at 450°C for 20s. Hillock formation is suppressed by RTP. From V.Q. Ho, private communication.**

threshold voltage in RTP annealed devices, than for conventional furnace annealing[38]. This more pronounced hot electron degradation may be understood because of the more abrupt junctions formed by RTP.

<u>Metal Reliability</u>

Problems coming under the heading of metal reliability include hillock formation, electromigration, and poor metal step coverage over underlying oxide. The sintering of aluminum alloy metallization is accompanied by an increase in grain size and, frequently, the formation of hillocks or protrusions on the metal surface that can contribute to electromigration induced failure mechanisms. Hillock formation is suppressed by the use of multilayer metals and by rapid thermal sintering, as seen in Figures 11 and 12a. However, it has been found that a subsequent thermal treatment in a furnace will result in hillock formation after all[39]. Such a thermal cycle may be unforeseen if it occurs during the packaging process, for example. This problem can be prevented by using a high sintering temperature, as shown in Figure 12b. Note that temperatures in this study appear too high for aluminum processing, possibly because of some temperature calibration problem, common with RTP systems. More work should be done to clarify this effect of sintering on hillock formation.

One of the most useful applications of RTP, along with junction annealing and silicide formation, is fusion of doped glass films in order to improve the

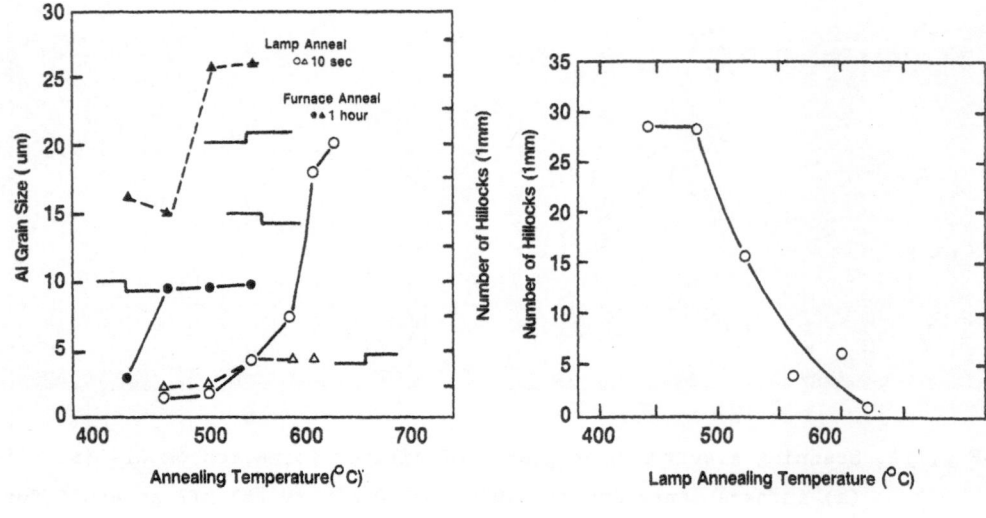

a

b

Fig. 12. Effect of a subsequent thermal treatment of hillock formation after rapid thermal sintering. The triangular symbols in (a) show the reduction in the density of hillocks per mm of interconnect after RTP compared to FA, confirming the results in Figure 12. However, as seen in (b), a subsequent FA at 440°C for 1 hour increases the hillock density 15X after a 440°C RTP, but if the rapid sinter is at a high temperature, then the hillock density is not affected by a subsequent heat treatment. Temperature calibra- tion may have erred on the high side in these experiments, since aluminum cannot normally be subjected to a temperature of 600°C. From Reference 39.

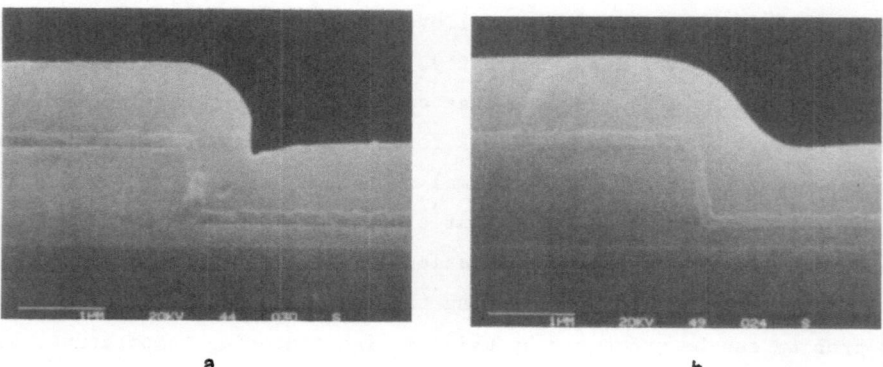

a

b

Fig. 13. Scanning electron micrographs illustrating flow of borophosphosilicate glass of composition 4 w/o (weight percent) B and 4 w/o P deposited by LPCVD over a polysilicon coated oxide step. Results are shown for (a) the as-deposited profile, where the step is slightly reentrant (beyond vertical) so that metal step coverage will be compromised, and (b) RTP at 1050°C for 20s. From Reference 10.

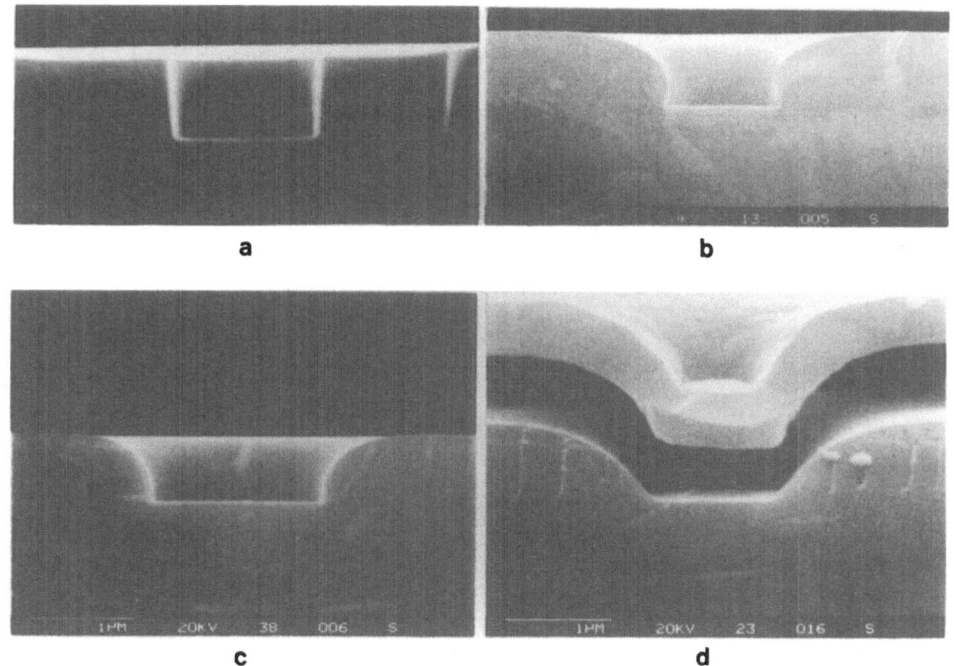

Fig. 14. Scanning electron micrographs showing reflow of small (1 μm) contact holes after (a) no anneal, (b) 1050°C for 20s resulting in a reentrant profile, (c) 1050°C for 20s for a 1.5 μm hole (no reentrancy), and (d) 1100°C for 30s, resulting in good metal step coverage (shown). From Reference 47.

step coverage of overlying metal layers. Flow and reflow are defined to be thermal surface smoothing steps occurring respectively before and after the cutting of contact hole openings, while fusion is the generic term for both processes. Rapid isothermal fusion (RIF) has been used to improve the surface morphology of phosphosilicate glass (PSG)[40-44] and borophosphosilicate glass (BPSG)[10,45]. Most experiments have studied the effect of RIF on glass step coverage over an underlying step (Figure 13) or at an edge etched in the glass to simulate a contact opening. Few results have yet been published on the reflow of small geometry contact holes cut in glass layers. There is a proximity effect that causes closely spaced holes to exhibit reentrant profiles (i.e. an overhang of the hole by the glass edge), even after RIF[46]. Typical results[47] are shown in Figure 14. Levy and Nassau[48] have used a simple argument to explain these results. Three considerations define the model. The minimum surface energy (in two dimensions) occurs for a circular cross-section; the cross-sectional area of the region between two spaces is assumed to be conserved; and the experimental observation that the contact point does not move during reflow is included (Figure 15). Then the final

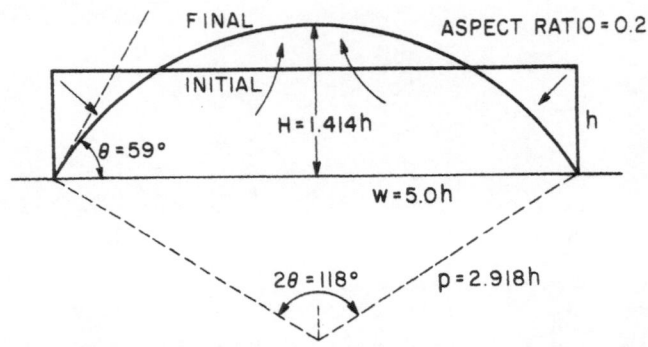

Fig. 15. Phenomenological model for glass reflow, assuming minimization of
surface area (in two dimensions here) and immobility of the contact
points during heat treatment. Initial and final profiles are shown
for a glass line before and after reflow. The cross-section
remains constant. From Reference 46.

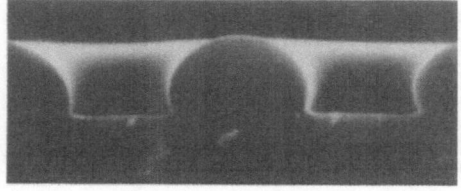

a

b

Fig. 16. Scanning electron micrographs showing reflow of contact holes with
(a) high aspect ratio, and (b) small aspect ratio and two different
separations, confirming the model illustrated in Figure 15. The
reflow angle is improved for low aspect ratio and widely separated
contacts. From Reference 46.

equilibrium contact angle is completely determined. For small lines, a
droplet-like reentrant profile results, while for wide lines (widely spaced
contacts), reflow is never carried on long enough to obtain a circular
cross-section--only edge rounding occurs (Figure 16). Therefore reflow
provides an important limitation on design rules, for closest approach of
neighboring contact openings.

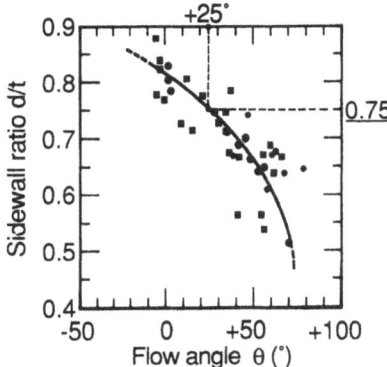

Fig. 17. Sidewall thinning as a result of BPSG reflow. Data is shown for
rapid reflow, furnace reflow, LPCVD BPSG, and atmospheric pressure
CVD BPSG. All results fall on a universal curve, allowing a
compromise to be chosen between sidewall thinning (e.g., 75% of
film thickness) and flow angle (e.g., 30°). From Reference 10.

More fusion is obtained for more severe thermal cycles, but then dopant
diffusion is excessive, since the flow or reflow cycles take place after
source/drain doping. Therefore a compromise must be made based on results
such as those shown in Figures 17 and 18. The acceptable flow angle must be
balanced against the junction characteristics and the sidewall thinning that
occurs with overly long or high temperature flow cycles. Figure 17 illustrates that the relation between flow contact angle and sidewall ratio is a
universal curve for a given material, whether rapid or furnace fusion is
employed[10]. If larger contact angles are required for a process, then fusion
in a steam ambient will produce more flow for a given cycle, widening the

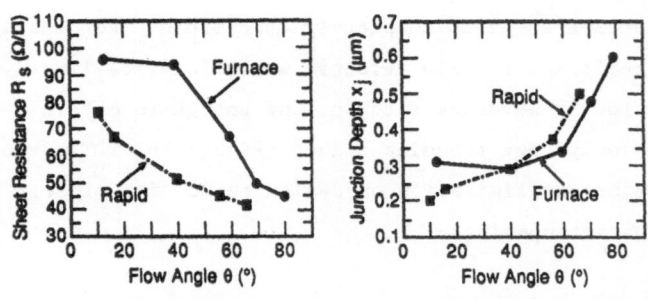

Fig. 18. Relationship between flow angle and junction characteristics for 4
w/o B and 4 w/o P BPSG and a junction formed by implantation of
3×10^{15} B^+/cm^2 @ 10 keV into silicon pre-amorphized by an implant of
2×10^{15} Si^+/cm^2 @ 60 keV. If process specifications require, for
example, a flow angle of 30°, then the sheet resistance is 57 $\Omega/\square$,
and the junction depth is 0.26 μm, within our example
specification for the junction. From Reference 10.

acceptable process window by reducing the required temperature by
25-50°C[10,43].

Densification is a process step required immediately after CVD of doped
glass to stabilize the PSG or BPSG film in order to maintain the impurity
content and to enable reproducible etching. Recent work has shown that
densification also greatly affects the reflow properties of BPSG, even though
the typical densification temperature is 200-300°C lower than the reflow
temperature[11]. It can be seen from Figure 20 that there is a threshold
densification temperature of 700°C required to enable reflow to occur, even at
1100°C. We might speculate that the composition of the corner of the contact
opening has been modified by the reactive ion etching used to form it, and
some redistribution of the impurities is needed to allow the onset of reflow.

The results of the considerations discussed in this section should be,
typically, summarized in a process window chart such as that given in Figure
20, where we find that a reasonable compromise cycle to combine junction
annealing and glass reflow would be 1050°C for 20s.

PROCESS INTERACTION PROBLEMS

In this section we look at problems with RTP which, while they certainly
will affect final device and circuit performance, have more direct effect on

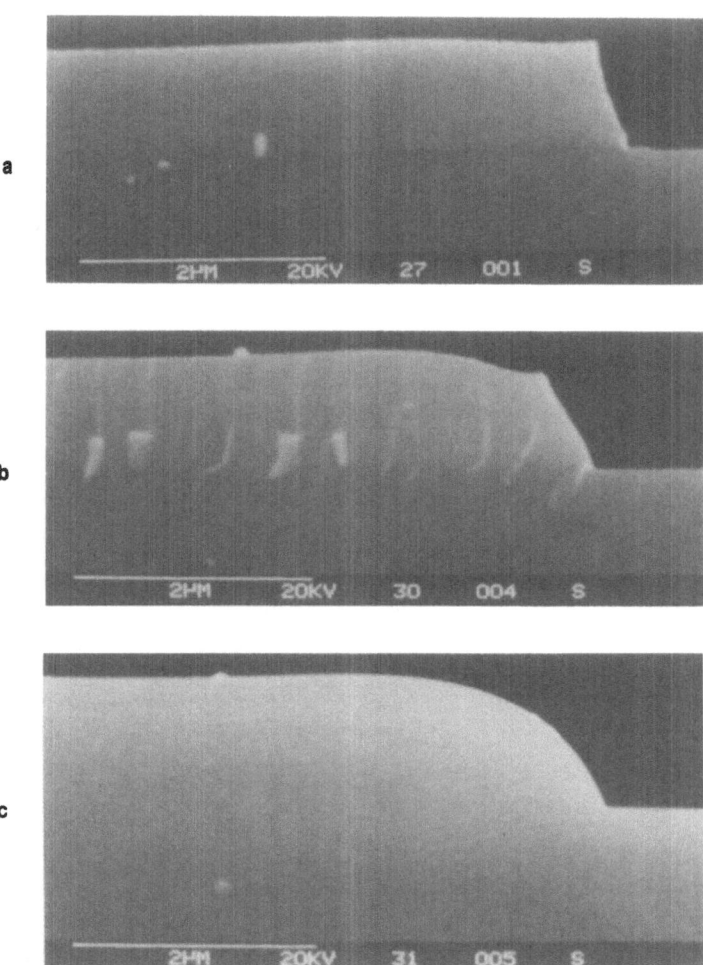

Fig. 19. Scanning electron micrographs showing the effect of densification on reflow, for 4 w/o B and 4 w/o P BPSG and (a) no densification, (b) 600°C/20s densification, and (c) 700°C/20s, all followed by a reflow cycle of 1100°C for 20s. Only after a densification of at least 700°C for 20s will the reflow proceed properly, even at the process temperature of 1100°C. From Reference 11.

other aspects of the process flow. As discussed by Sedgwick[1], an optimized cycle is one which best balances desirable effects against undesirable effects. If two competing effects are both thermally activated, than high temperatures are required when the desired process has the higher activation energy, and low temperatures are needed in the opposite situation. Figures 4, 6, and 20 illustrate examples of optimization.

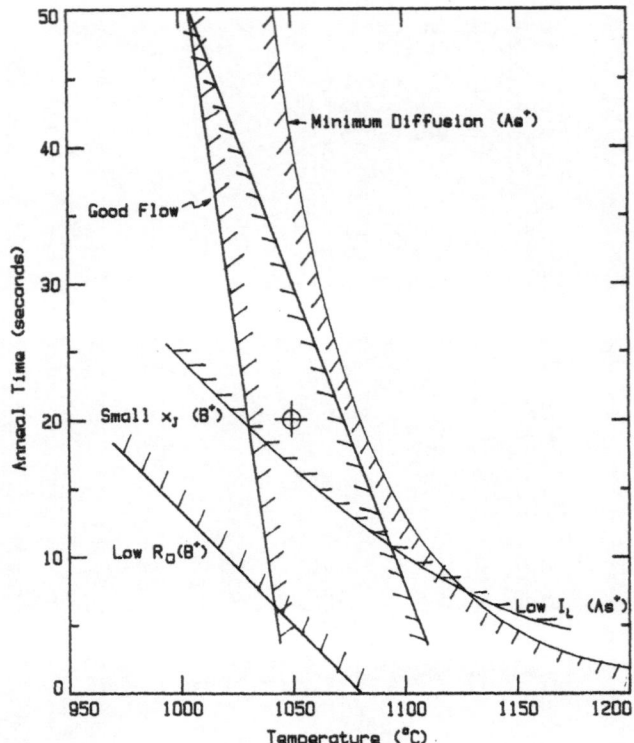

Fig. 20. Summary of process design including the requirements for acceptable
As$^+$ and B$^+$ diffusion, diode leakage (n$^+$p), sheet resistance (p$^+$n),
and flow angle. The hatched sides of the lines indicate acceptable
processing conditions. Then an acceptable compromise thermal cycle
is 1050°C for 20s (shown as the target symbol).

Dopant Diffusion

A study of Figure 1 reveals that there are four high temperature steps
following source/drain implantation, excluding silicide and metal sinter
steps. Silicide anneal and glass densification do not require excessive
temperatures; the remaining steps of junction anneal and glass reflow are the
most significant and will together result in more diffusion than expected from
one step alone. A simple solution would appear to be the elimination of the
anneal step so that the reflow process serves two functions. However, as
discussed in detail later, titanium does not react well with silicon containing a high surface concentration of unactivated dopant, so the junction
anneal is still necessary. Alternatively, the glass reflow step can be
removed if more sophisticated processing is used to selectively deposit
tungsten plugs[48] or cut contact holes with sloped walls[49]; in the last
analysis this procedure may be necessary in any case since titanium silicide
does not stand up well to processing temperatures in excess of ~950°C.

In many cases, if extremely shallow junctions are not required for a process, consecutive diffusion steps are acceptable since the total increase in junction depth is less than the sum of the results of two single processes operating on as-implanted material. This interaction occurs because of the transient enhanced diffusion observed for short times or low temperatures[1,6,50], especially for boron implantation, as seen in Figures 21-23a. Transient diffusion is much reduced but not entirely eliminated for silicon that has been pre-amorphized by a Si[+] or Ge[+] implant. Two different mechanisms have been suggested to explain these results. Fair[51] attributes the enhanced diffusion to the generation of point defects by a dislocation

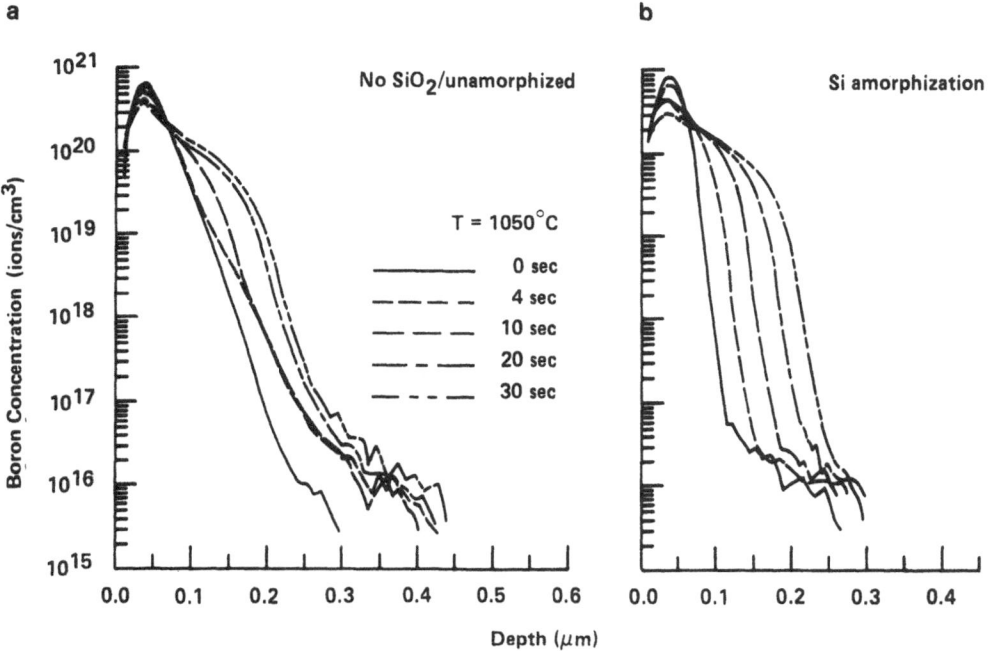

Fig. 21. Dopant profiles for p[+]n junctions formed by implantation of 3×10^{15} B[+]/cm^2 @ 10 keV and RTA, (a) with pre-amorphization of 2×10^{15} Si[+]/cm^2 @ 50 keV and (b) without pre-amorphization. The junction is not only shallower with pre-amorphization; there is no clear evidence of the transient diffusion seen in (a) for the profile after 4 seconds, while the tail is unchanged after further processing up to 10 seconds. From Reference 6.

network that forms within the region of high damage at the peak in concentration of the implanted ions; the concentration of these point defects decreases with time as annealing takes place, and consequently the enhancement of diffusion outside the peak region also decays. In this model all impurities

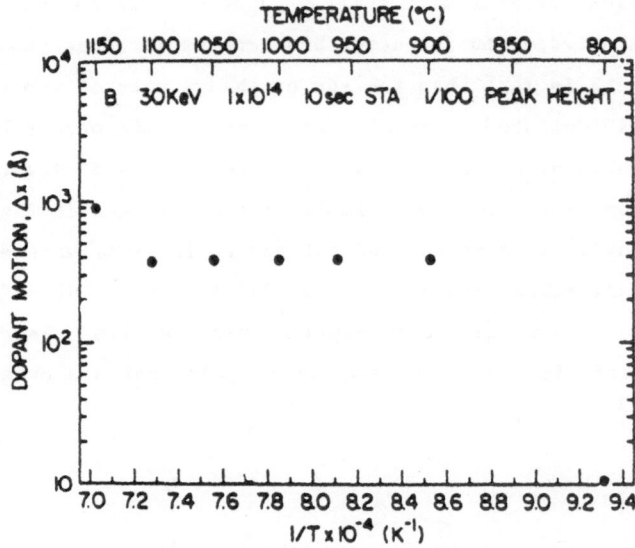

Fig. 22. Transient diffusion of boron, showing dopant movement after various
RTA cycles. There is negligible diffusion at 800°C, but by 900°C
motion of ~400 Å is observed, which does not increase up to 1100°C.
From Reference 53.

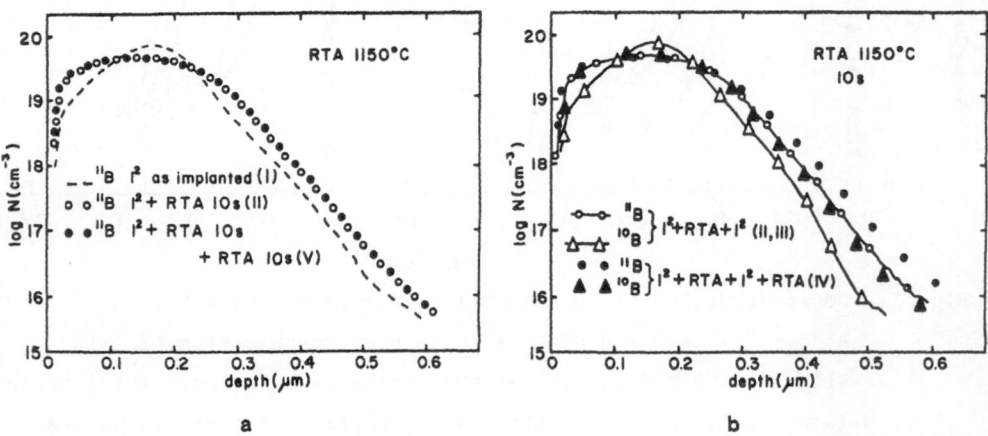

Fig. 23. (a) Transient diffusion of boron (implanted with a dose of 1×10^{15}
ions/cm^2 @ 50 keV) is observed after RTA at 1150°C for 10s, but a
second identical anneal results in negligible further diffusion.
(b) Investigation of the mechanism of boron transient diffusion.
Two isotopes of boron are used; ^{11}B is implanted first and
annealed, resulting in transient diffusion; subsequently ^{10}B is
implanted with the same energy and dose and followed by a second
identical RTA. Transient diffusion of the ^{11}B is observed after
each RTA step. From Reference 55.

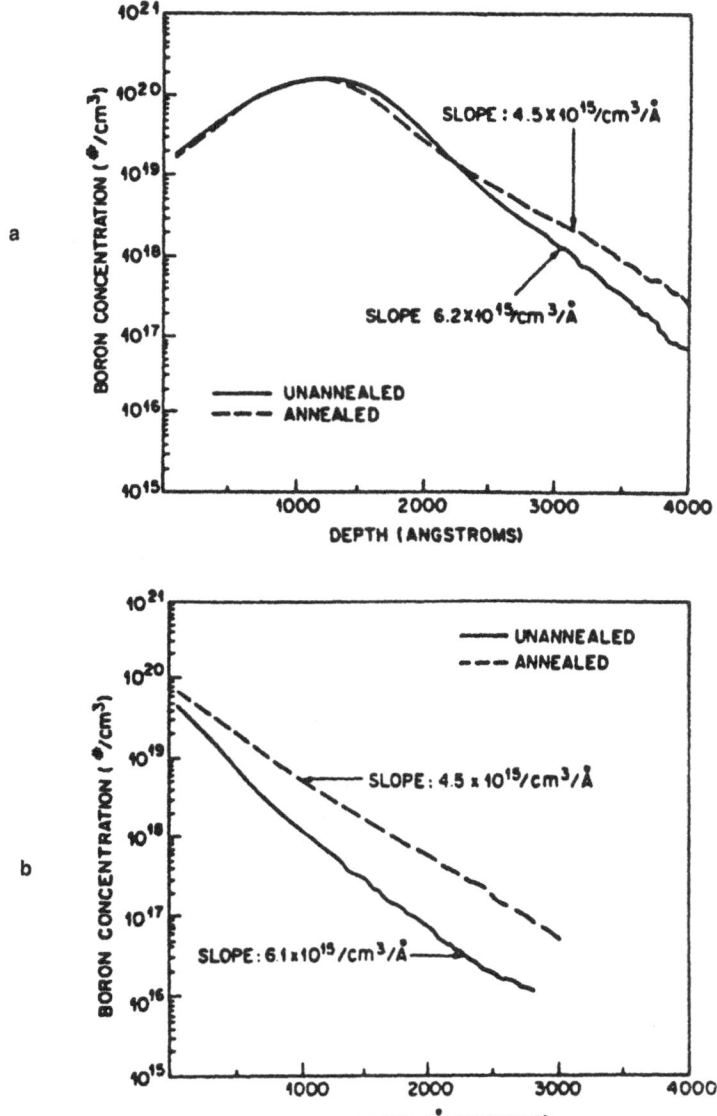

Fig. 24. Investigation of the mechanism of transient boron diffusion. Boron
was implanted at a dose of 2×10^{15} ions/cm^2 and an energy of 40 keV.
Then RTA of 1100°C for 10s was carried out, on samples that (a)
were otherwise untreated, or that (b) had had ~2000 Å of silicon
etched off to remove the high damage high dose region that might
act as a defect generation source. The absolute concentrations are
not easily comparable, but the identical slopes of the profiles in
the two situations suggest that the implant peak does not affect
transient diffusion. From Reference 54.

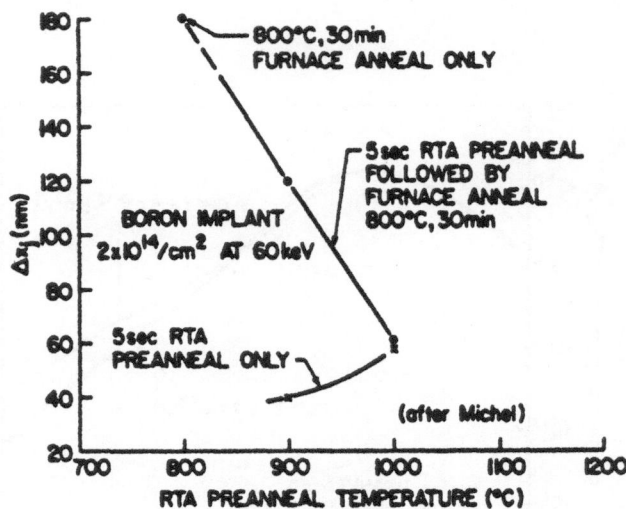

Fig. 25. Reduction in boron transient diffusion as a function of an RTA
pre-anneal, for an 800°C/30 min furnace anneal. A 1000°C/5s
pre-anneal reduces junction movement from 180 nm to only 60 nm,
showing that transient diffusion is greater for furnace annealing
than for RTA. From Reference 1 with data taken from Reference 53.

outside of the high damage region experience an increased mobility. Within
the heavily damaged region the dislocations act as sinks for point defects and
no diffusion enhancement occurs. An alternative approach, proposed by
Morehead[52] and elaborated by Michel[53], assumes that only a fraction of
impurity ions experience enhanced mobility; this fraction diffuses via an
interstitialcy mechanism. The enhanced diffusion coefficient decays with time
as the dopant is activated. Experimental attempts to differentiate between
these mechanisms have not been conclusive, although they tend to support the
"two fluid" type models. When the high damage region at the peak of the
implant is stripped before annealing[54], diffusion rates in the tail region are
unchanged (Figure 24), contradicting the model in which defects generated at
the implantation peak enhance diffusion. On the other hand, in experiments in
which ^{11}B implantation and annealing is followed by ^{10}B implantation and a
second anneal[55], enhanced diffusion of the ^{11}B is observed for each anneal
(Figure 23b), suggesting that implantation damage is responsible for the
observed effects, and impurities that are interstitial or in the mobile
fraction in general, which should be made substitutional after the first
anneal, are not a factor.

It is interesting to note that transient enhanced diffusion is an effect
that is not confined only to RTP processes. In fact it is somewhat suppressed

by RTP, but is very pronounced in low temperature furnace annealing processes[1,53]. Figure 25 shows the reduction in junction movement obtained by
employing an RTP pre-anneal before an 800°C/30 min furnace anneal. The
dramatic improvement in junction depth is attributed to removal of implantation induced defects, a process that takes place much more readily at the
high temperatures of RTP, since it has a large activation energy of about 4.7
eV. The enhanced diffusion with RTP can also be reduced with a pre-anneal
step which performs the solid phase epitaxial regrowth at a temperature of
800°C or less, so that little diffusion occurs (see Figures 22 and 23).

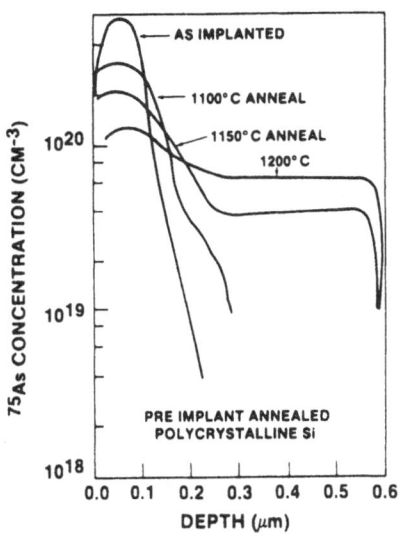

Fig. 26. Diffusion of dopant in polycrystalline silicon after a pre-anneal
        at a peak temperature of 1250°C for ~30s and implantation of 5×10^{15}
        As^+/cm^2, followed by RTA at the peak temperatures shown. From
        Reference 57.

    Diffusion of dopants in polysilicon is also reduced by RTP. For
conventional processing, the only concern is whether any dopant will penetrate
the thin gate oxide and poison the FET channel region; the high grain boundary
enhanced diffusion rate ensures uniform doping within the polysilicon.
However, with RTP, diffusion is reduced so much that uniform doping may not be
obtained under normal RTP conditions, especially if the grain size in the
polycrystalline silicon has been increased by a pre-anneal step before
doping[56,57], through a poly oxidation step, for example. If the dopant
concentration at the poly/oxide interface falls much below the density of

states at the conduction band edge ($2-3 \times 10^{19}$ cm^{-3}), then the threshold voltage of a MOSFET, which tracks the polysilicon surface potential, will be increased. Figure 26 shows a SIMS profile for an implant of 5×10^{15} ions/cm^2 of As$^+$ into 0.58 μm of polysilicon after RTP at peak temperatures of 1100, 1150 or 1200°C, with and without a pre-anneal. If the polysilicon has undergone a pre-implant anneal, then an RTA of at least 1150°C is needed to diffuse the arsenic throughout the thickness of the gate. At these temperatures, out-diffusion of dopant may be another concern. Since the peak temperatures in these experiments were only reached for times of less than 1 second, lower temperatures for longer times might result in acceptable diffusion, but in general the use of implanted polysilicon will set a lower limit on RTP annealing cycles.

Mechanical Distortion

Any single wafer processing system has the potential of introducing mechanical distortion into the silicon, because of non-uniform temperature distributions across the wafer. Undesirable effects of this type include wafer warpage, slip, and pattern shifts. Careful design of RTP equipment has fairly well eliminated the warpage problem and has even improved wafer flatness by stress relief, as long as temperatures are maintained at 1150°C or less[58]. Slip-free processing has also been obtained within the same temperature limits, even up to 1200°C for 150 mm wafers[59,60]. The incidence of slip increases for large diameter wafers and the maximum permissible temperature decreases[59] due to an increase in stress across the wafer[61].
Temperature non-uniformities across a wafer can result in another form of mechanical distortion known as pattern shift. This effect may occur in particular during the reflow step, which takes place after the opening of contact holes, when stresses in the wafer may be relieved so that the contact holes effectively move during this step and prevent proper alignment of the next mask level (first metal). Figures 27 shows an example of the shift of

contact openings, measured by an electrical technique, after reflow of BPSG. Shifts of over 0.3 μm were observed, far too large to be acceptable for stepper patterning, and were only corrected by improving the uniformity of the heating in the RTP system.

Titanium Reactions

TiSi$_2$ makes an excellent contact metallization because of its low sheet resistance, its low contact resistance to silicon, its stability at fairly high temperatures, its capability of thermally forming a self-aligned silicide (SALICIDE) structure[9,62-66] (Figure 28), and its reactivity with nitrogen to

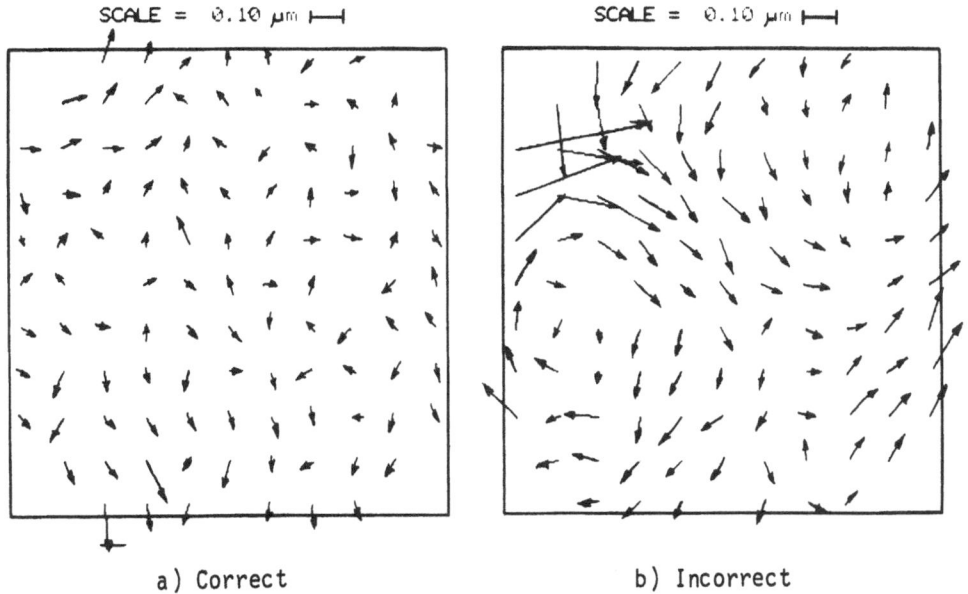

a) Correct   b) Incorrect

Fig. 27. Pattern shift due to temperature non-uniformities during RTP. A
good die is shown in (a) and an unacceptable die in (b), where the
shifts of up to 0.3 μm are well beyond the pattern tolerance.

form TiN barrier metal[8,9]. However this same reactivity can also be a problem. Titanium reacts easily with oxygen and requires a lengthy purge of a
furnace to form $TiSi_2$ rather than TiO_2[67]. This problem is solved by RTP
because the fast, high temperature process sequence favors silicide formation
as long as the process is carried out in an inert gas like N_2 or Ar and the
chamber is sealed against leakage from the atmosphere. Reactivity between Ti
and O_2 then to a certain extent becomes an asset since the Ti will reduce any
residual SiO_2 remaining at the Ti/Si interface, so that a dip in dilute HF is
usually sufficient to prepare the surface for silicidation[9], although an in
situ sputter clean also works well[65].

From the phase diagram[68] shown in Figure 29, it is seen that Ti and SiO_2
cannot co-exist in equilibrium; therefore they will react to form titanium
oxides and titanium silicide (likely Ti_5Si_3) on SiO_2. However, reduction of
SiO_2 by Ti also takes place over field, sidewall spacer, or passivation oxide,
forming TiO[69,70], or a mixture of TiN_xO_y (when processed in N_2), TiO_x, and
$TiSi_y$, which may be difficult to remove in a selective etch. The reaction
occurs at temperatures above 450°C, but is not a serious problem until
temperatures greater than 600°C are reached (Figure 30). Another problem
arises for high temperature silicidation. Since Si diffuses into Ti during
the silicidation reaction, bridging of $TiSi_2$ over oxide (especially the
sidewall spacer) occurs at temperatures above ~650°C[65], as seen in Figure 31.

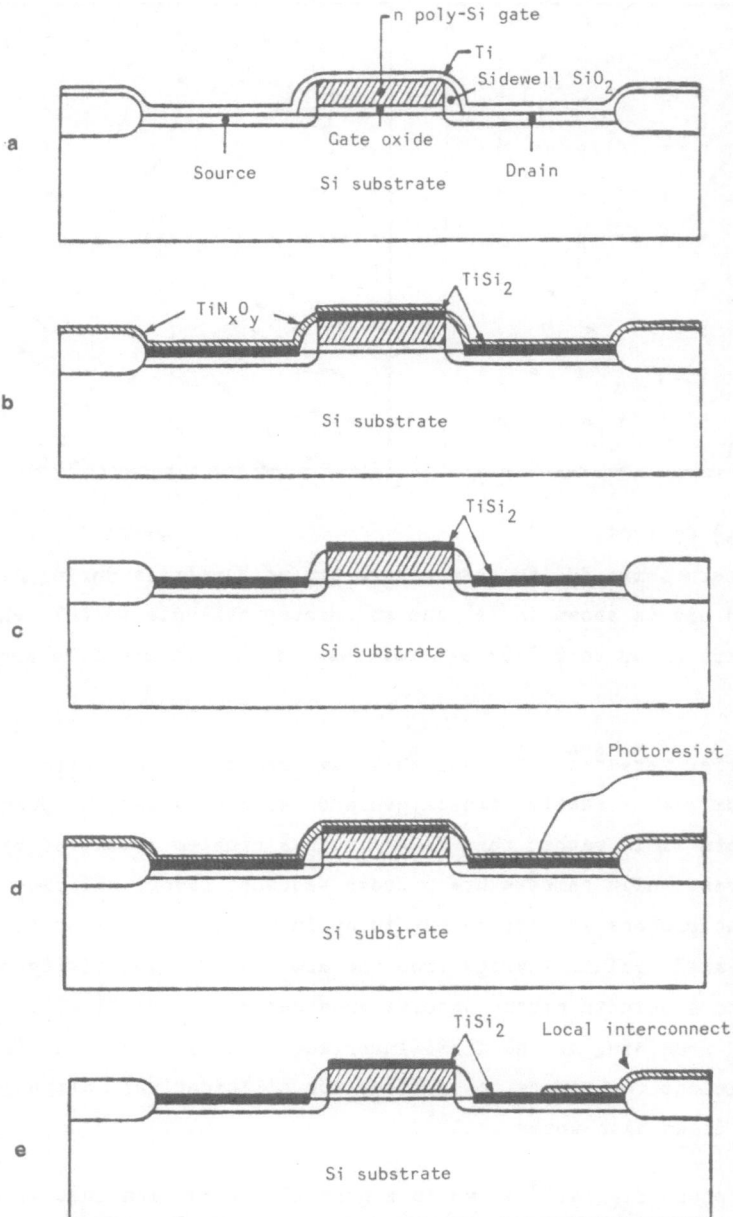

Fig. 28. Typical SALICIDE process. Metal (Ti) is deposited in (a), and
sintered (at ~600°C) in (b) to form $TiSi_2$ over Si and Ti plus a
TiN_xO_y/Ti mixture ("unreacted Ti") over SiO_2. Then the unreacted
Ti is stripped off in (c), leaving selective silicide. A further
high temperature anneal (800-850°C) is then used with $TiSi_2$ to form
the low resistance C54 phase. Additional TiN can be formed
optionally by a second Ti deposition and a long anneal in N_2. A
TiN local interconnect can be formed by partial masking during the
Ti strip, shown in (d) and (e). See References 65 and 71.

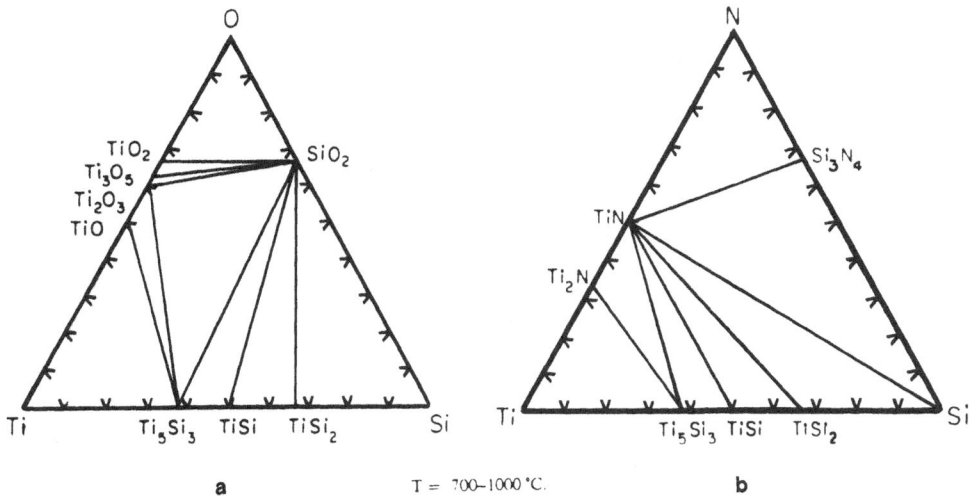

a   T = 700–1000 °C.   b

Fig. 29. Three component phase diagrams for (a) Ti-Si-O and (b) Ti-Si-N.
Solid lines link those phases that can co-exist. For example in
(a), $TiSi_2$ and SiO_2 co-exist, but Ti and SiO_2 do not, so they will
form other phases, typically a mixture of TiO, Ti_5O_3 and SiO_2.
From References 68 and 82.

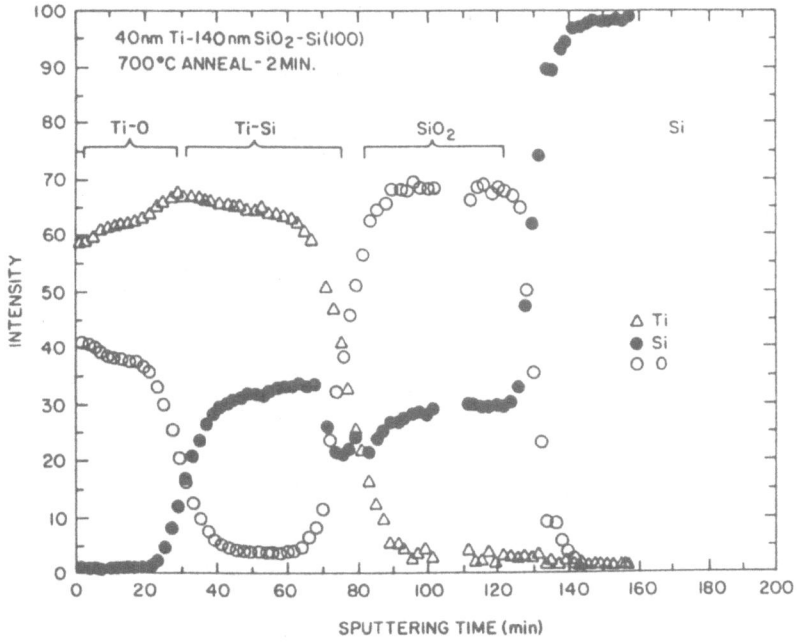

Fig. 30. Auger profile showing the composition of 40 nm Ti on SiO_2 after RTP
at 700°C for 120s and the conversion into TiO_x and Ti_5Si_3, as
predicted by Figure 29(a). From Reference 69.

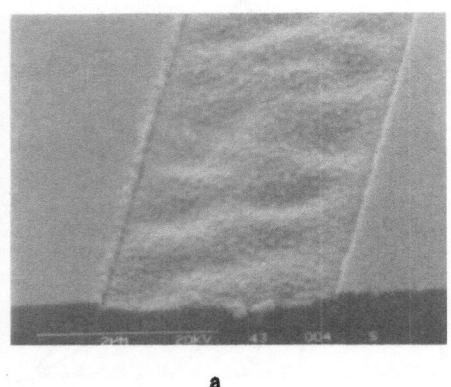
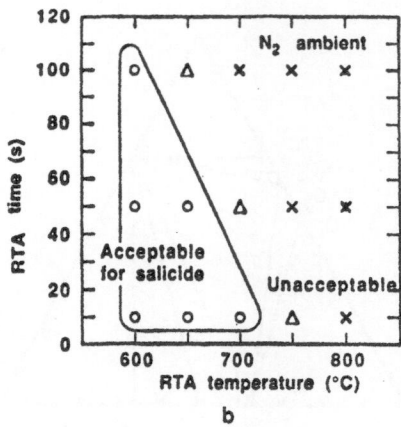

a

b

Fig. 31. (a) Scanning electron micrograph of lateral silicidation after RTP
at 700°C for 20s of 500 Å of Ti over a 1000 Å SiO_2 step. The
silicide has crept up the oxide and over onto the surface. A
suitable process window for negligible lateral silicidation is
shown in (b). From Reference 65.

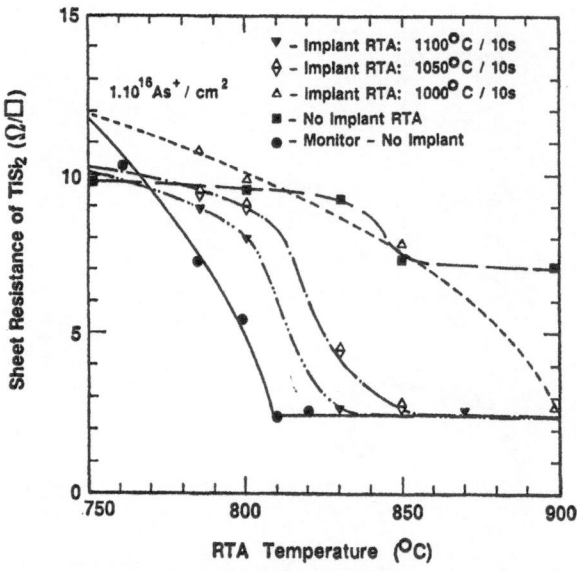

Fig. 32. Retardation of silicidation by high surface concentrations of As^+.
The $TiSi_2$ formation is complete when the sheet resistance reaches
the high temperature value of 2.4 $\Omega/\Box$. For an unannealed implant,
silicidation is not complete up to at least 900°C, and even after
an 1100°C/10s junction anneal, the silicide anneal must be 25°C
higher than for the unimplanted case. From Reference 76.

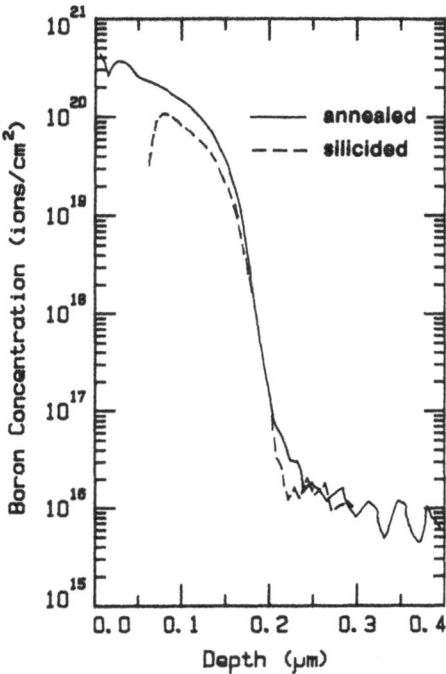

Fig. 33. Boron loss due to the formation of 800 Å of TiSi$_2$ at 800°C/30s.
The concentration at the silicide/silicon interface is decreased by
a factor of approximately 2 after silicidation, but little change
is seen in the tail, implying that most boron is lost through
out-diffusion. The implant was 3×10^{15} B$^+$/cm^2 @ 10 keV into a
silicon substrate pre-amorphized with 2×10^{15} Si$^+$/cm^2 @ 60 keV and
subsequently annealed at 1050°C/20s. From Reference 76.

Therefore titanium Salicidation must be carried out in a two step process,
with a low temperature sinter preceding the selective etch, which is then
followed by a higher temperature anneal step to form the low resistance (C54)
phase of TiSi$_2$. This phase is stable at relatively high temperatures, up to
920°C for 30 minutes[71], but many process integration engineers may feel that
that is too close for comfort when contemplating a BPSG reflow cycle. Good
devices can be made with a 1050°C/20s cycle after Salicidation[65], but long
term reliability is still an open question.

The reaction of Ti with nitrogen to form TiN is covered in the section on
Process Integration Opportunities.

Silicidation of Doped Silicon

The thermal process steps of junction annealing and silicide formation are
intimately related. If dopant atoms, especially arsenic, remain unactivated
in the silicon, then titanium silicidation is retarded[72-76]. Conversely,

silicidation may cause significant redistribution of the dopant because of segregation at the silicide/silicon interface[72,74-77].

Figure 32 illustrates the effect of unactivated arsenic[76] (implanted at 50keV) on the formation of $TiSi_2$. For low to medium doses, $5x10^{14}$ and $1x10^{15}$ As^+/cm^2, the final silicide phase is actually reached at a lower temperature than for unimplanted material, because of the ease of reacting arsenic with amorphous silicon. However, for doses above $5x10^{15}$ As^+/cm^2, silicidation is retarded and silicide does not form at all for a dose of $1x10^{16}$; in fact peeling of the films is observed. For this high dose, annealing of the junction at 1050°C or more is required to permit silicidation at temperatures of 850°C or less. A similar but less pronounced effect is observed for BF_2^+ doping[72]. The effect decreases for longer or higher temperature anneals implying that the key parameter determining the reaction is the surface concentration of dopant[76]. It has also been found that silicidation is retarded by knock-on oxygen atoms present in the silicon after dopant implantation through a masking oxide[73]. Therefore in a device process using thermal titanium silicide, implantation should be carried out into bare silicon, and annealing should be carried out before silicidation.

There is not significant segregation of boron at the silicide/ silicon interface[72,76,78], but boron does out-diffuse rapidly through $TiSi_2$ by grain boundary diffusion and evaporates at the surface (Figure 33), although that loss can be prevented by the presence of TiN diffusion barrier[78]. Arsenic exhibits a similar behavior, with a diffusion coefficient estimated[78] to be ~10^{-11} cm^2/s at 800°C in $TiSi_2$. However RTP conditions can be chosen to obtain low resistance silicide with minimal dopant loss[76,77], as shown in Figure 34. Dopant redistribution does then place an even more severe restriction on post-silicidation processing temperatures than does reaction with SiO_2, effectively precluding the use of a reflow step.

Other silicides exhibit very different dopant segregation properties from those of $TiSi_2$. Little or no outdiffuson of As^+ is observed after $CoSi_2$ formation[79,80], but silicidation temperatures should be kept below 800°C for good contacts; similar results are observed for boron doped junctions[80,81]. Noble metal silicides like PtSi and Pd_2Si show very different behavior from the refractory metal silicides, since they segregate, or "snowplow" dopant into the silicon to build up a high interface concentration. This phenomenon will be explored further in the next section.

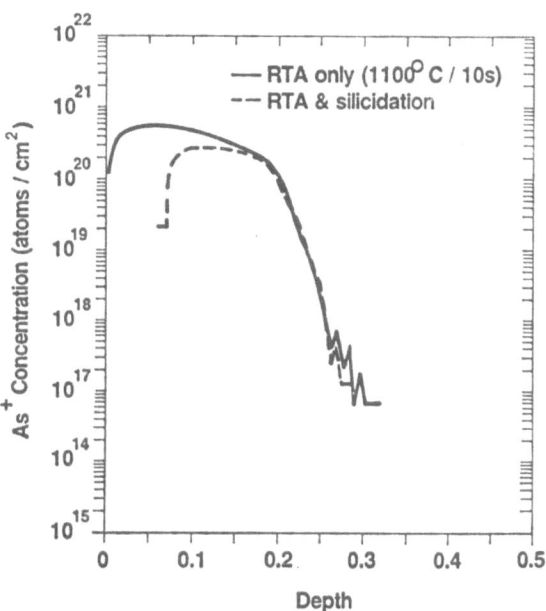

Fig. 34. As in Figure 33, but for a substrate doped with 1×10^{16} As$^+$/cm^2 @ 50 keV and annealed at 1100°C/10s.

PROCESS INTEGRATION OPPORTUNITIES WITH RTP

The most immediately apparent opportunity that arises from the use of RTP is the greater flexibility in selection of thermal cycles. Certainly RTP enables a very shallow junction, if called for by the process design and device specifications. In many cases, however, the minimum attainable junction depth is not necessary; therefore more thermal cycles are possible with RTP than with a furnace. Additional useful oxidation steps can be added, for example, to clean surfaces, oxidize polysilicon, perform re-ox or pre-oxidation steps, and so forth, without requiring very clever process design to limit dopant diffusion. On the other hand, two or more process steps can be combined into one, just as with furnace annealing. The example case of simultaneous implantation annealing and glass reflow has already been discussed.

More interesting are the entirely new processes that are possible with RTP. Many of these arise in conjunction with silicidation, because of the ease of carrying out this procedure with RTP. Therefore the remainder of this paper will look at opportunities originating with the formation of silicides.

Formation of Titanium Nitride

The results obtained from a titanium silicidation step depend on the reaction ambient since titanium reacts not only with silicon, but with nitrogen[82] and oxygen[68] as well. Annealing in N_2 or NH_3 yields $TiSi_2$ with a TiN overlayer[9,65,83-88] (Figure 35a) which acts as a very good diffusion barrier[83,84]. There is usually some oxygen mixed with the "TiN" film, if only from reduction of native oxide on the underlying silicon, so the film is actually a TiN_xO_y mixture[9,65,86] (see Figure 35b), but the barrier properties seem unaffected. The reaction of titanium with nitrogen is energetically favored over that with silicon, but it proceeds much more slowly; therefore only a thin TiN film forms after silicidation in N_2. A slightly thicker, more stable film can be formed in NH_3, but the structure is qualitatively similar[87,88]. A thick TiN_xO_y is formed over the field oxide. The TiN film is removed by the selective etch used to form a Salicide structure, but if only silicided contacts are being formed, then the TiN can remain over the entire wafer to be patterned along with the (Al alloy) interconnect layer. In a Salicide structure, a second nitridation step is required to form TiN (Figure 28); if the cycle is long enough (1100°C/90s), the exposed silicide is entirely converted to TiN[84,86].

A minor increase in the process complexity results in the use of TiN to provide, essentially, self-aligned contacts and a new interconnect material. Over exposed oxide, such as the field or passivation oxide, RTP forms mostly a TiN_xO_y alloy with good conductive properties ($\rho = 30-50$ $\mu\Omega$-cm), yielding a sheet resistance of 3-5 $\Omega/\square$ for a 1000 Å thick film. Since this film is provided essentially for free during the Salicidation process, it can be used as an interconnect simply by patterning a photoresist layer before selective etching of the "unreacted" titanium (Figure 28). The resulting self-aligned local interconnect is particularly useful in memory devices where it acts as a sort of buried contact, permitting the fabrication of high density static random access memories (SRAMs)[89].

Simultaneous Formation of Junctions and Silicides

If defect removal is not a requirement, then the typical anneal cycle used to form $TiSi_2$ is sufficient to activate dopant in silicon. Extended defects will not be formed in the silicon if the damage region due to ion implantation is confined to a metal layer, and then dopant is subsequently diffused into the substrate. Dopant can be implanted either into the deposited Ti before silicide formation[90], or into the $TiSi_2$ formed after the sinter step[91-93]. Similar experiments have been carried out by implantation into $TaSi_2$[94] and

CoSi$_2$[95]. This technique may be especially useful with CoSi$_2$, since the large consumption of silicon during silicide formation is not a factor when the junction is automatically formed below the silicide/silicon interface. Figure 36 shows the dopant profile obtained after implanting 1×10^{15} As+/cm^2 into 82 nm of TiSi$_2$, followed by an anneal at 900°C for 90s. The profile was measured after stripping off the silicide layer, to avoid confusion at the interface. A sheet resistance of 100 $\Omega/\square$ and diode leakage of <0.1 pa/cm^2 were obtained without residual damage. The surface concentration of 6×10^{19} As$^+$/cm^2 is barely sufficient for a low contact resistance but a higher implant dose of arsenic

a

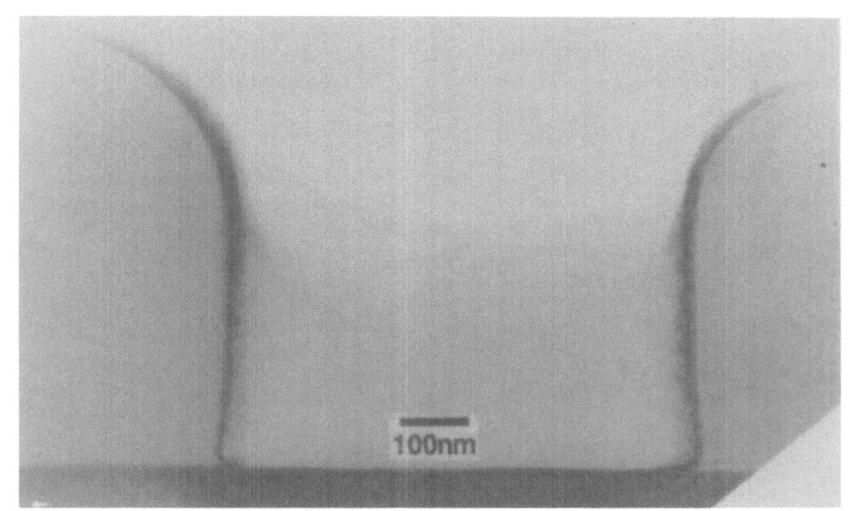

b

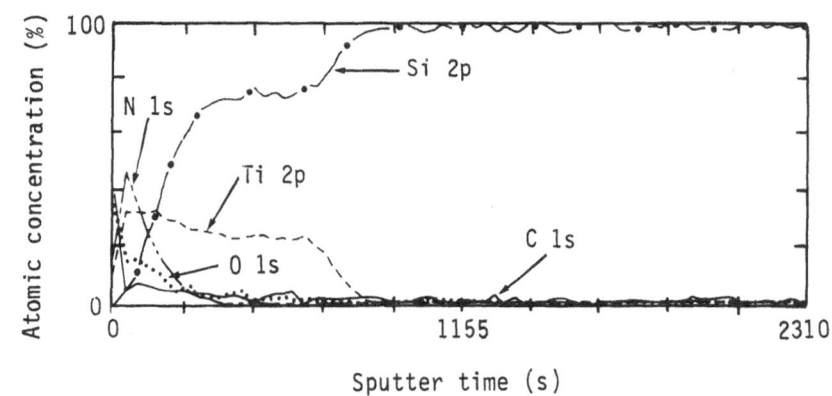

Fig. 35. (a) Transmission electron micrograph of a silicided contract hole, formed by sputtering of 500 Å Ti followed by sintering at 750°C in N$_2$. On the silicon surface TiSi$_2$ is formed with a thin TiN surface layer, while on the oxide surfaces, TiN alone is formed; (b) ESCA profile of the composition of the TiN/TiSi$_2$ bilayer on silicon shown in (a). From V.Q. Ho, unpublished.

should solve that problem. Ditchek et al.[95] also found that the leakage of
p^+n junctions formed without gettering decreased from 1 A/cm^2 to 1 na/cm^2 when
implantation into CoSi$_2$ was used. Caution should be exercised in using this
technique at higher temperature (~1100°C) since Ti-As compounds may form and
degrade the junction quality[93].

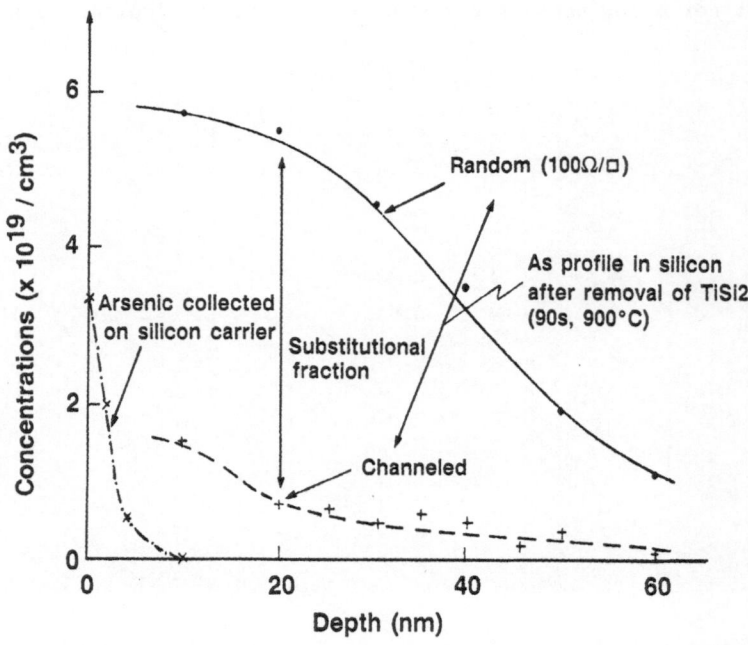

Fig. 36. Rutherford backscattering/ion channelling analysis of the diffusion
of arsenic from an implanted TiSi$_2$ source to form a shallow defect
free n$^+$p junction. The implant was 1x10^{15} As$^+$/cm^2 into 82 nm of
silicide, followed by a diffusion/ anneal of 900°C/90s. From
Reference 92.

Moderately Doped Junctions

Short channel MOSFETs are now often made with a lightly doped drain LDD
structure (Figure 37a), used to decrease the electric field at the drain/
channel interface and reduce hot electron effects[96]. Formation of this
structure complicates the device process because an additional implantation is
required for the PMOS transistors and for the NMOS transistors. The high
dose implantation is still required to give a high surface concentration and
ensure a low contact resistance.

This process can be simplified by forming a noble metal silicide which preferentially segregates dopant into the silicon and builds up a high surface concentration, thereby lowering the contact resistance without the need of a high dose implant. The process is illustrated in Figure 37b. This moderately doped drain (MDD), or "silicide on lightly implanted drain" (SOLID)[97] structure has been formed with PtSi[98,99] or Pd$_2$Si[99,100]. An example of the dopant profile before and after silicidation is shown in Figure 38. The snowplowing of arsenic ahead of the forming silicide results in a concentration significantly greater than the implanted concentration in the silicon, so that doses of 1-5x10^{14} should be sufficient to form a good contact while reducing hot electron effects. Conversely, the use of PtSi results in a very low contact resistance with more conventional doses[98], down to 7.5x10^{-8} Ω-cm^2. The junction must be annealed before metal deposition since the noble metal silicides cannot be processed at temperatures much above 600°C. High dopant concentrations used to form MDD or SOLID structures. Much less snowplowing occurs for boron, but neither hot electron nor contact problems are so significant for p$^+$n junctions.

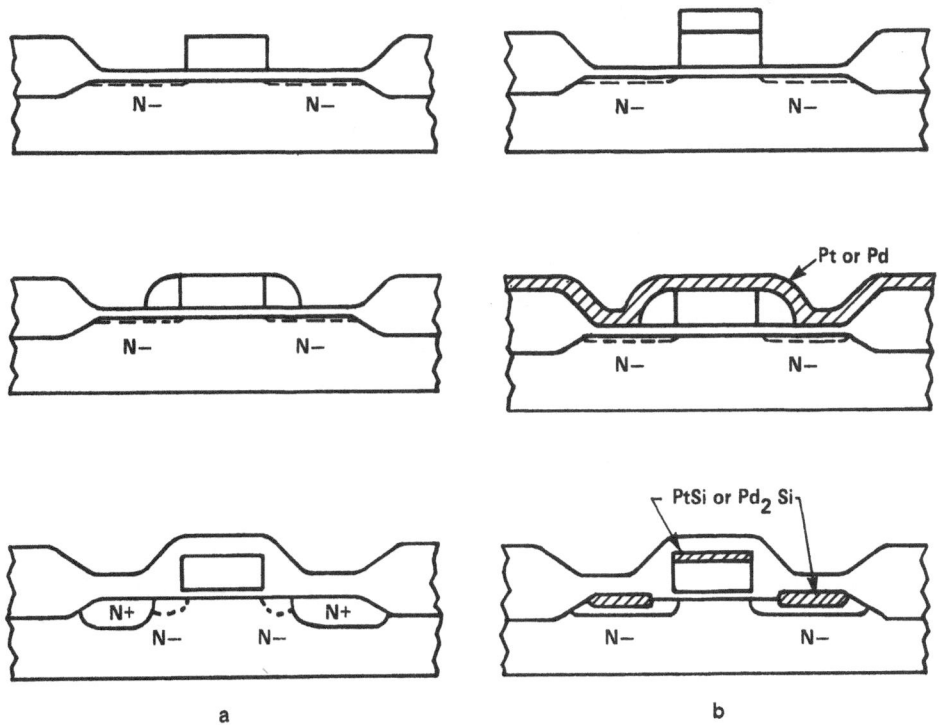

a   b

Fig. 37. (a) Lightly doped drain process. The N$^-$ implant is typically 1-5x10^{13} ions/cm^2, while the N$^+$ implant is a conventional high dose; (b) Moderately doped drain process. Only one implant of 1-5x10^{14} ions/cm^2 is required because of the snowplowing of dopant by the platinum or palladium silicidation process.

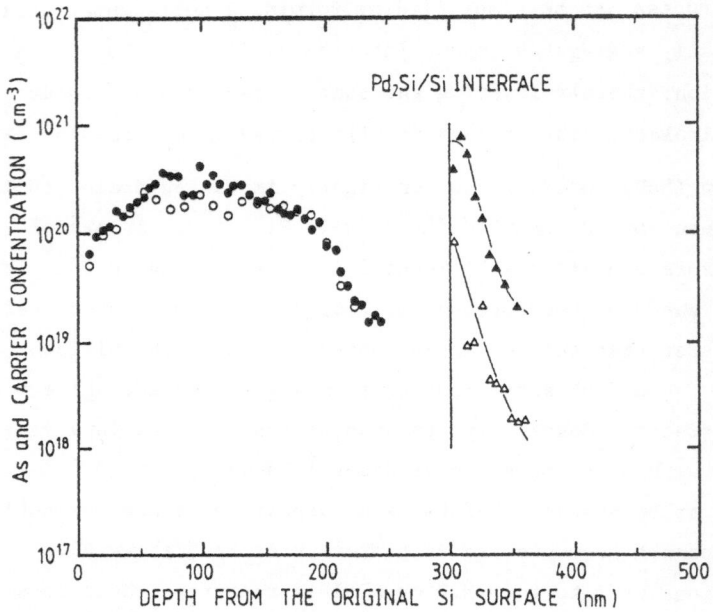

Fig. 38. Snowplowing of arsenic after thermal silicidation of Pd on silicon. A very high interface concentration (closed symbols) of As[+] is obtained, although the activated con- centration (open symbols) is lower than desired. The implant was 5×10^{15} As[+]/cm^2 @ 140 keV followed by a 900°C/ 30 min anneal. The silicide was formed from a 600 nm Pd film annealed at 250°C/9.75 hr. From Reference 99.

Defect Engineering

Removal of extended defects created by end-of-range ion implantation damage is desirable in order to reduce junction leakage currents, although as discussed earlier, a process should be engineered around reduced leakage, not around defect removal per se. Defects do not need to be removed if dopant diffusion is sufficient to extend the junction into the silicon beyond the damaged region so that any residual dislocation damage is fully contained within the high conductivity drain and does not extend into the depletion region. Junction leakage has been found to be reduced by more than three orders of magnitude by this means[101]. For self-amorphizing implants such as As[+] or Sb[+], the anneal cycle must simply be long enough to produce sufficient diffusion. However for B[+] or BF$_2$[+], shallow junctions are often obtained by using a pre-amorphizing implant of Si[+] [6,102-104] or Ge[+] [105,106]. Then the depth of the pre-amorphizing implant must be carefully matched to the dopant implant depth so that the junction will be beyond the damage region after diffusion[6] (see Figure 39). End-of-range defects are much more easily eliminated for As[+] implants if no pre-amorphization has been used (Figure 40).

TEM cross-section

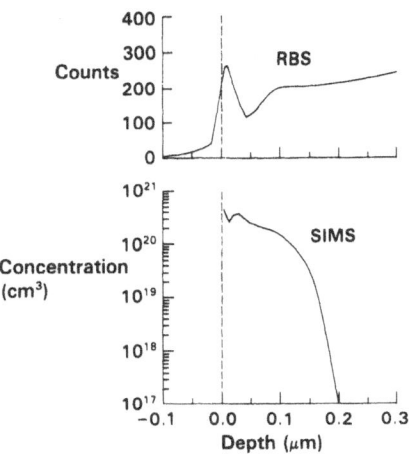

Fig. 39. Disabling of end-of-range defects by diffusing the dopant
sufficiently to place the junction edge and the depletion region
beyond any residual defects. From top to bottom are illustrated a
TEM cross-section, a Rutherford backscattering ion channelling
profile, and a SIMS profile, all on the scale depth scale. The TEM
and RBS result show a damage area well within the junction depth of
0.22 μm (where the doping concentration reaches the background of
$1x10^{15}$ ions/cm^2). From Reference 6.

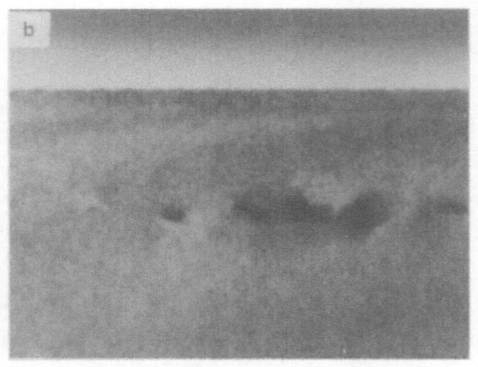

Fig. 40. Effect of pre-amorphization ($3x10^{15}$ Si/cm^2 @ 60 keV) on
end-of-range defects remaining after implantation of $1x10^{16}$ As$^+$/cm^2
@ 50 keV and RTA at 1050°C for 20s. Without pre- amorphization (a)
all residual defects are removed, but with pre-amorphization (b)
they remain. From Reference 7.

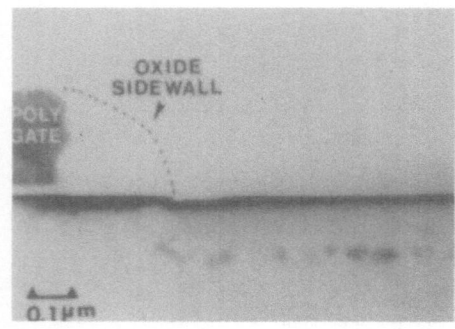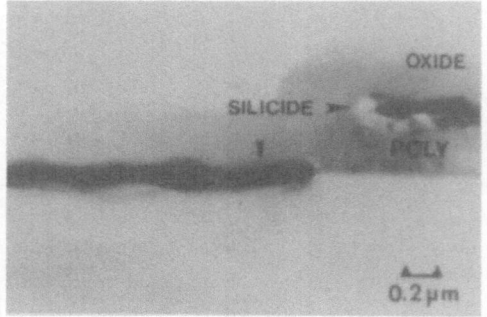

a   b

Fig. 41. Elimination of end-of-range defects by enhanced dislocation climb
during silicidation. The structures in both (a) and (b) were
formed by pre-amorphization with $1x10^{15}$ Ge$^+$/cm^2 @ 85 keV,
implantation of $1x10^{15}$ B$^+$/cm^2 @ 10 keV and an RTA of 1050°C/10s.
End-of-range defects from the Ge$^+$ are visible in (a), but in (b),
where a TiSi$_2$ Salicide was formed by a two-step anneal of 650°C/10s
and 850°C/10s, all extended defects have been removed. Since TiSi$_2$
forms by silicon diffusion, vacancies are created in the substrate
which enhance the rate of dislocation climb. From Reference 107.

It has recently been found that defects can in fact be more readily removed during the titanium silicidation process[107]. Since $TiSi_2$ forms by diffusion of Si into Ti, vacancies are created in the silicon substrate that enhance the climb rate of dislocations towards the surface. A comparison of residual damage for a 900°C/30s cycle with and without the presence of Ti on the surface is shown in Figure 41. The greater-than-equilibrium concentration of vacancies should also enhance (or retard) dopant diffusion. Further work is needed to explore this phenomenon.

SUMMARY

RTP affords an additional degree of flexibility in process design because the reduced thermal budget permits additional thermal steps to be included, and because a wider range of processes are possible. These processes compete and interact with one another, but they also offer new opportunities in process control through shallow junction formation, local interconnect fabrication, and defect engineering. Several competing effects will determine the optimum cycle for a given application. Temperature and time are limited by dopant diffusion, device performance, chemical reactions (especially with Ti), degradation of oxide and silicide, and mechanical distortion. On the other hand, thermal treatments must be hot or long enough to decrease junction leakage, remove damage, diffuse dopant beyond a damaged region, and properly reflow glass. Careful design of an entire device process as a whole is required to satisfy these sometimes conflicting requirements.

REFERENCES

1.   T.O. Sedgwick, Mat. Res. Soc. Symp. Proc. 92, 3 (1987).

2.   R.T. Fulks, ibid, p. 249.

3.   J. Nulman, J.P. Krusius, and A. Gat, IEEE Electron Dev. Lett. EDL-6, 205 (1985).

4.   M.M. Moslehi, Mat. Res. Soc. Symp. Proc. 92 73 (1987).

5.   T.E. Seidel, C.S. Pai, D.J. Lischner, D.M. Maher, R.V. Knoell, J.S. Williams, B.R. Penumalli, and D.C. Jacobson, Mat. Res. Soc. Symp. Proc. 35, 329 (1985).

6.   I.D. Calder, H.M. Naguib, D. Houghton, and F.R. Shepherd, ibid, p. 353.

7.   A.A. Naem and I.D. Calder, J. Appl. Phys. 62, 596 (1987).

8. S.P. Murarka, "Silicides for VLSI Application" (Academic Press, New York 1983).

9. Y.H. Ku, S.K. Lee, E. Louis, D.K. Shih, and D.L. Kwong, Mat. Res. Soc. Symp. Proc. 92, 155 (1987).

10. J.S. Mercier, Solid State Technol. 30-7, 85 (1987); and J.S. Mercier, L.D. Madsen, and I.D. Calder, Mat. Res. Soc. Symp. Proc. 52, 251 (1986).

11. L.D. Madsen and J.S. Mercier, to be published in Proc. Fourth Can. Semi. Technol. Conf. (1988).

12. M.L. Reed, B. Fishbein, and J.D. Plummer, Appl. Phys. Lett. 47, 40 (1985).

13. D.B. Scott, K.-L. Chen, and R.D. Davies, chapter 10 in "VLSI Handbook", ed. by N.G. Einspruch, (Academic Press, Orlando, 1985).

14. L.C. Parillo, chapter 11 in "VLSI Technology", ed. by S.M. Sze, (McGraw-Hill, New York, 1983).

15. I. Brodie and J.J. Muray, "The Physics of Microfabrication", (Plenum, New York, 1982).

16. R.A. Chapman, R.A. Haken, D.A. Bell, C.C. Wei, R.H. Havemann, T.E Tang, T.C. Holloway, and R.J. Gale, IEDM Tech. Digest 87, 362 (1987).

17. J.F. Gibbons, S. Reynolds, C. Gronet, D. Vook, C. King, W. Opyd, S. Wilson, C. Nauka, G. Reid, and R. Hull, Mat. Res. Soc. Symp. Proc. 92, 281 (1987).

18. M.M. Moslehi, K.C. Saraswat, and S.C. Shatas, ibid, p. 295.

19. G.E.P. Box and N.R. Draper, "Empirical Model-Building and Response Surfaces", (John Wiley, New York, 1987).

20. A. Kamgar, W. Fichtner, T.T. Sheng, and D.C. Jacobson, Appl. Phys. Lett. 45, 754 (1984).

21. A.L. Butler, D.J. Foster, and A.J. Pickering, Mat. Res. Soc. Symp. Proc. 71, 417 (1986).

22. S.K. Lee, D.K. Shih, Y.H. Ku, E. Louis, and D.L. Kwong, Proc. SPIE 797, 20 (1987).

23.   D.L. Kwong, N.S. Alvi, Y.H. Ku, and A.W. Cheung, Mat. Res. Soc. Symp. Proc. 52, 241 (1986).

24.   J.B. Lasky, J. Appl. Phys. 54, 6009 (1983).

25.   I.-W. Wu, R.T. Fulks, and J.C. Mikkelsen, Jr., J. Appl. Phys. 60, 2422 (1986).

26.   M.E. Lunnon, J.T. Chen, and J.E. Baker, J. Electrochem. Soc. 132, 2473 (1985); Appl. Phys. Lett. 45, 1059 (1984); and Appl. Phys. Lett. 46, 35 (1985).

27.   H. Mikoshiba, H. Abiko, and M. Kanamori, Jpn. J. Appl. Phys. 25, L631 (1986).

28.   S. Solmi, E. Landi, and P. Negrini, IEEE Electron Dev. Lett. EDL-5, 359 (1984).

29.   E.K. Broadbent, M. Delfino, A.E. Morgan, D.K. Sadana, and P. Maillot, IEEE Electron Dev. Lett. EDL-8, 318 (1987).

30.   M.A. Finn and M.E. Coe, Electrochem. Soc. Ext. Abstr. 85-2, 374 (1985).

31.   S.B. Felch, D.T. Hodul, and M. Salimian, Mat. Res. Soc. Symp. Proc. 92, 235 (1987).

32.   N.E. McGruer and R.A. Oikari, IEEE Trans. Electron Dev. ED-33, 929 (1986).

33.   J. Nulman, private communication.

34.   S.K. Lee, D.L. Kwong, and N.S. Alvi, J. Appl. Phys. 60 3360 (1986),

35.   R.G. Cosway and M.W. Hodel, J. Electrochem. Soc. 135, 533 (1988).

36.   S.K. Lee, D.K. Shih, D.L. Kwong, N.S. Alvi, N.R. Wu, and H.S. Lee, Mat. Res. Soc. Symp. Proc. 71, 449 (1986).

37.   S.K. Lee, D.K. Shih, Y.H. Ku, E. Louis, and D.L. Kwong, Proc. SPIE 797, 20 (1987).

38.   Z.A. Weinberg, D.R. Young, J.A. Calise, S.A. Cohen, J.C. DeLuca, and V.R. Deline, Appl. Phys. Lett. 45, 1204 (1984).

39.   S. Onishi, K. Nishizawa, and K. Sakiyama, Proc. Spring Mtg. Jpn. Appl. Phys. Soc., p 484 (1987).

40.	T. Hara, H. Suzuki, and M. Furukawa, Jpn. J. Appl. Phys. <u>23</u>, L452 (1984).

41.	J.S. Mercier, I.D. Calder, R.P. Beerkens, and H.M. Naguib, J. Electrochem Soc. <u>132</u> 2432 (1985).

42.	N.S. Alvi and D.L. Kwong, J. Electrochem. Soc. <u>133</u>, 2626 (1986).

43.	N. Shah, J. McVittie, N. Sharif, J. Nulman, and A. Gat, Mat. Res. Soc. Symp. Proc. <u>52</u>, 233 (1986).

44.	I. Barsony. H. Anzai, and J.-I. Nishizawa, J. Electrochem. Soc. <u>133</u>, 157 (1986).

45.	J.R. Gigante, J.M. Geneczko, and R.N. Ghoshtagore, Electrochem. Soc. Ext. Abstr. <u>85-2</u>, 382 (1985).

46.	R.A. Levy and K. Nassau, J. Electrochem. Soc. <u>133</u>, 1417 (1986).

47.	T. Abraham and I. Wylie, to be published (1988).

48.	H. Kotani, T. Tsutsumi, J. Komori, and S. Nagao, IEDM Tech. Dig. 87, p. 217 (1987).

49.	G. Jolly, unpublished.

50.	S.J. Pennycook and R.J. Culbertson, Proc. SPIE <u>797</u>, (1987); and Mat. Res. Soc. Symp. Proc. <u>74</u>, 379 (1987).

51.	R.B. Fair, J. Vac. Sci. Technol. <u>A4</u>, 926 (1986).

52.	F.F. Morehead and R.F. Lever, Appl. Phys. Lett. <u>48</u>, 151 (1986); and Mat. Res. Soc. Symp. Proc. <u>52</u>, 49 (1986).

53.	A.E. Michel, Mat. Res. Soc. Symp. Proc. <u>52</u>, 3 (1986).

54.	L.C. Hopkins, T.E. Seidel, J.S. Williams, and J.C. Bean, J. Electrochem. Soc. <u>132</u>, 2035 (1985).

55.	K. Cho, M. Numan, T.G. -Finstead, W.K. Chu, J. Liu, and J.J. Wortman, Appl. Phys. Lett. <u>47</u>, 1321 (1985).

56.	S.R. Wilson, R.B. Gregory, W.M. Paulson, S.J. Krause, J.D. Gressett, A.H. Hamdi, F.D. McDaniel, and R.G. Downing, J. Electrochem. Soc. <u>132</u>, 922 (1985).

57. S.R. Wilson, R.B. Gregory, W.M. Paulson, S.J. Krause, J.A. Leavitt, L.C. McIntyre, Jr., J.L. Seerveld, and P. Stoss, Appl. Phys. Lett. 49, 660 (1986).

58. M. Current and A. Yee, Solid State Technol. 26-10, 197 (1983).

59. J. Blake, J.C. Gelpey, J.F. Moquin, J. Schlueter, and R. Capodilupo, Mat. Res. Soc. Symp. Proc. 92, 265 (1987).

60. A. Gat, private communication.

61. G. Bentini, L. Correra, and C. Donolato, J. Appl. Phys. 56, 2922 (1984).

62. K. Tsukamoto, T. Okamoto, M. Shimizu, T. Matsukawa, and H. Nakata, IEDM Tech. Dig. 84, p. 130 (1984).

63. D. Pramanik, M. Deal, A.N. Saxena, and O.K.T. Wu, Semi. Int. 8-5, 94 (1985).

64. C.Y. Ting and S.S. Iyer, Proc 1985 VLSI Multilevel Interconnet Conf., p. 307 (1985).

65. V.Q. Ho and D. Poulin, J. Vac. Sci. Technol. A5, 1396 (1987).

66. D.B. Scott, R.A. Chapman, C.-C. Wie, S.S. Mahant-Shetti, R.A. Haken and T.C. Holloway, IEEE Trans. Electron Dev. ED-34, 562 (1987).

67. K.L. Wang, T.C. Holloway, R.F. Pinizzotto, Z.P. Socczak, W.R. Hunter, and A.F. Tasch, Jr., IEEE Trans. Electron Dev. ED-29, 547 (1982).

68. R. Beyers, J. Appl. Phys. 56, 147 (1984).

69. L.J. Brillson, M.L. Slade, H.W. Richter, H. Vander Plas and R.T. Fulks, Appl. Phys. Lett. 47, 1080 (1985).

70. M. Natan, Mat. Res. Soc. Symp. Proc. 74, 679 (1987).

71. M. Delfino, A.E. Morgan, E.K. Broadbent, P. Maillot and D.K. Sadana, J. Appl. Phys. 62, 1882 (1987).

72. T.P. Chow, W. Katz and G. Smith, Appl. Phys. Lett. 46, 41 (1985).

73. H. Matsui, H. Ohtsuki, M. Ino and S. Ushio, Mat. Res. Soc. Symp. Proc. 54, 769 (1986).

74. T. Okamoto, K. Tsukamoto, M. Shimizu and T. Matsukawa, J. Appl. Phys. 61, 4530 (1987).

75. S.W. Sun, F. Pintchovski, P.J. Tobin and R.L. Hance, Mat. Res. Soc. Symp. Proc. 92, 165 (1987).

76. A. Naem, private communication.

77. A.A. Pasa, J.P. de Souza, I.J.R. Baumvol and F.L. Freire, Jr., J. Appl. Phys. 61, 1228 (1987).

78. S.W. Sun, F. Pintchovski, P.J. Tobin and R.L. Hance, Mat. Res. Soc. Symp. Proc. 92, 165 (1987).

79. M. Tabasky, E.S. Bulat, B.M. Ditchek, M.A. Sullivan and S. Shatas, Mat. Res. Soc. Symp. Proc. 52, 271 (1986) and IEEE Trans. Electron Dev. ED-34, 548 (1987).

80. L. Van den Hove, R. Wolters, K. Maex, R.F. De Keersmaecker and G.J. Declerck, IEEE Trans. Electron Dev. ED-34, 554 (1987).

81. A.E. Morgan, E.K. Broadbent, M. Delfino, B. Colman and D.K. Sadana, J. Electrochem. Soc. 134, 925 (1987).

82. R. Beyers, R. Sinclair and M.E. Thomas, J. Vac. Sci. Technol. B2, 781 (1984).

83. M. Delfino, E.K. Broadbent, A.E. Morgan, B.J. Burrow and M.H. Norcott, IEEE Electron Dev. Lett. EDL-6, 591 (1985).

84. P.J. Rosser and G.J. Tomkins, Mat. Res. Soc. Symp. Proc. 35, 457 (1985).

85. H. Kaneko, M. Koyanagi, S. Shimizu, Y. Kubota and S. Kishino, IEDM Tech. Dig. 85, p. 208 (1985).

86. A.E. Morgan, E.K. Broadbent and A.H. Reader, Mat. Res. Soc. Symp. Proc. 52, 279 (1986).

87. B. Cohen and J. Nulman, Mat. Res. Soc. Symp. Proc. 92, 171 (1987).

88. T. Okamoto, M. Shimizu, A. Ohsaki, Y. Mashiko, K. Tsuokamoto, T. Matsukawa and S. Nagao, J. Appl. Phys. 62 4465 (1987).

89. T. Tang, C.-C. Wei, R. Haken, T. Holloway, C.-F. Wan and M. Douglas, IEDM Tech. Dig. 85, p. 590 (1985).

90. M. Horiuchi and K. Yamaguchi, IEEE Trans. Electron Dev. ED-33, 260 (1986).

91. D.L. Kwong, Y.H. Ku, S.K. Lee, E. Louis, N.S. Alvi and P. Chu, J. Appl.
     Phys. 61, 5084 (1987); Y.H. Ku, S.K. Lee, E. Louis, D.K. Shih and D.L.
     Kwong, Mat. Res. Soc. Symp. Proc. 92, 155 (1987); and Y.H. Ku, S.K. Lee,
     D.K. Shih, D.L. Kwong, C.-O. Lee and J.R. Yeargain, Proc. SPIE 797, 61
     (1987).

92. D.X. Cao, H.B. Harrison and G.K. Reeves, Mat. Res. Soc. Symp. Proc. 100,
     737 (1987).

93. V. Probst, H. Schaber, P. Lippens, L. Van den Hove and
     R. De Keersmaecker, Appl. Phys. Lett. 52, 1803 (1988).

94. H. Gierisch, F. Neppl, E. Frenzel, P. Eichinger and K. Heibar, Mat. Res.
     Soc. Symp. Proc. 71, 183 (1986).

95. B.M. Ditchek, M. Tabasky and E.S. Bulat, Mat. Res. Soc. Symp. Proc. 92,
     199 (1987).

96. P.J. Tsang, S. Ogura, W.W. Walker, J.F. Shepard and D.L. Critchlow, IEEE
     Trans. Electron Dev. ED-29, 590 (1982).

97. M. Horiuchi and K. Yamaguchi, Solid-State Electron. 28, 465 (1985).

98. H.-C.W. Huang, R. Cook, D.R. Campbell, P. Ronsheim, W. Rausch and B.
     Cunningham, J. Appl. Phys. 63, 1111 (1988).

99. I. Ohdomari, K. Konuma, M. Takano, T. Chikyow, H. Kawarda, J. Nakanishi
     and T. Ueno, Mat. Res. Soc. Symp. Proc. 54, 63 (1986).

100. N.S. Alvi, D.L. Kwong, C.G. Hopkins and S.G. Bauman, Appl. Phys. Lett.
     48, 1433 (1986).

101. H. Ishiwara and S. Horita, Jpn. J. Appl. Phys. 24, 568 (1985).

102. T.E. Seidel, IEEE Electron Dev. Lett. ED-4, 353 (1983).

103. C. Carter, W. Maszara, D.K. Sadana, G.A. Rozgonyi, J. Liu and
     J. Wortman, Appl. Phys. Lett. 44, 459 (1984).

104. R.A. Powell, J. Appl. Phys. 56, 2837 (1984).

105. D.K. Sadana, E. Myers, J. Liu, T. Finstead and G.A. Rozgonyi, Mat. Res.
     Soc. Symp. Proc. 23, 303 (1983).

106. A.C. Ajmera and G.A. Rozgonyi, Appl. Phys. Lett. 49, 1269 (1986).

107. D.S. Wen, P.L. Smith, C.M. Osburn and G.A. Rozgonyi, Appl. Phys. Lett.
51, 1182 (1987).

INTRODUCTION TO DIRECT WRITING OF INTEGRATED CIRCUIT

Geoffroy Auvert

Centre National d'Etudes des Télécommunications

BP.98, 38243 Meylan, France

INTRODUCTION

Integrated circuit fabrication steps employing localized-laser writing involves maskless and sequential processing of circuit layout. These steps are fundamentally different from those used in optical lithography but may be used for some process steps in device fabrication. The laser direct writing technique, sometimes called laser pantography, allows a rapid and discretionary interconnection network drawing of prefabricated gate-arrays and is suitable for either development of prototype circuits or filling low-volume custom orders of integrated circuits. It appears to be the most practical near-term application of this technology. Another application of the laser direct writing technique is the discretionary fabrication of high performance devices.

LASER PANTOGRAPHY PROCEDURE

Laser-induced surface modification is achieved either by driving a chemical reaction between the chip surface and surrounding gaseous molecules or by modifying the electrical properties of a deposited film.

a) Using gas reactions

After placing a circuit in a chemical chamber, a circuit image is produced on a video screen by adjusting the circuit position in the focal plane of the microscope (Fig. 1)[1]. The laser beam is turned on at low laser power in order to visualize the spot position on the circuit. The laser beam is then turned off. When no leakage is detected, the chamber is filled with the reacting gas at a constant pressure. The laser beam power is turned to

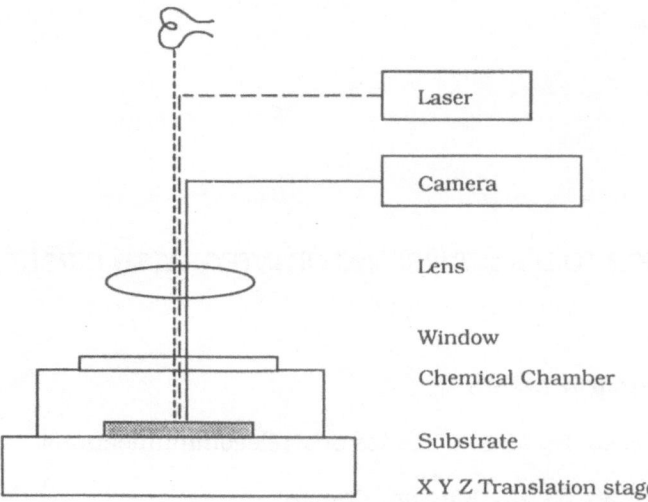

Fig. 1. General schematic shows a laser pantography system. All elements can be computer controlled. The optical beam is used to visualize the circuit, the laser beam is focussed on the circuit and the camera forms on a video screen an image of both the laser spot and the circuit. The circuit is locally heated up by the laser and decomposition of gaseous molecules occurs on the circuit.

an appropriate laser power and the laser spot is displaced on the circuit surface in order to draw a line. Next, the chamber is purged with an inert gas and an other gas can be introduced to trigger another reaction. Finally, the circuit is taken out from the chamber and electrical tests can be performed. The main advantage of gaseous processes stems from the possibility of drawing different lines without circuit manipulation .

b) Using solid phase reactions

After deposition of a solid over the entire circuit surface, the wafer is placed under the laser pantography equipment [2]. No chemical chamber is needed. The irradiation is performed as explained above. The irradiation induces a chemical reaction in the deposited solid film which is locally transformed into a conductive material. The circuit is removed from the equipment and the non transformed solid is chemically etched. If another layer is required, this procedure is performed a second time with a different deposited solid film and modified under different laser conditions. The main advantage of this process is to operate in ambient air .

In the following, we will focus our attention on the laser direct writing technique using gaseous reactions.

If a molecule is decomposed by the laser beam in the gaseous phase or on the surface, the distance between the excitation point and the deposition point represents the resolution of the laser pantography technique. This distance depends mainly on the mean free path of molecules. For an optical absorption in the gaseous phase, the mean free path L depends on the gas pressure P according to the following formula:

$$L = 1 / \left[2^{1/2} n \sigma \right]$$

where n is the number of molecules per volume unit and σ is the molecule surface. The mean free path decreases with increasing gas pressure. Therefore, if a photolytic regime in the gas phase occurs, a high pressure has to be used to increase the resolution.

The mean free path in the adsorbed layer does not depend on the gas pressure (at a pressure above a few Torr) due to saturation of the surface by the adsorbed species and the resolution depends strongly on the decomposition mechanism. A better resolution is obtained in the etching of materials where the limiting process is the desorption of chemically adsorbed species which are strongly fixed to the surface.

Another very important point which limits the resolution of the laser pantography technique is correlated with the beam size. In a photolytic regime, reaction occurs where the beam is, and the resolution is equal to the beam diameter. Therefore, the resolution depends directly on the quality of the optical modification of the laser beam in the optical path. The formula giving the smallest beam diameter behind a focussing lens of focal length f, is:

$$F = \frac{\lambda z}{\pi f}$$

where z is the beam diameter at the lens plane and λ is the laser beam wavelength. According to this formula, the laser pantography resolution is increased by using larger optics and shorter wavelengths. In practice, a resolution of less than one micron can be reached using equipment with appropriate optics.

For a pyrolytic reaction, the resolution is lower than in a photolytic reaction and is determined by the heat profile which will be studied in the next section.

LASER INDUCED TEMPERATURE

a) Measurement

In a pyrolytic reaction, the laser beam is locally absorbed and the beam energy is transformed into heat. The irradiated surface increases in temperature and when equilibrium is reached, a constant surface temperature profile is established. The chemical reaction takes place in this hot area (Fig.2). For a thermally conducting material, the stationary regime is reached faster than in an insulating medium. However, in the former case, a lower laser beam power is necessary to reach the same equilibrium temperature. Typical transient times are of a few microseconds when irradiating a silicon substrate, whereas several milliseconds are necessary in a quartz substrate (see at Fig.3). The equilibrium temperature can be calculated by analytical or digital methods but as the simulation of a multi-layer substrate is difficult, measurement of the local temperature is often preferable. This is achieved by measuring the laser power inducing a phase change in the irradiated substrate [3]. It thus becomes possible to compare different substrates and establish a calibration curve. For example, the laser power needed to melt a silicon coating on a quartz substrate is described in Fig.4a. This curve can be compared to a similar one achieved on a silicon substrate (Fig.4b). According to these curves, the

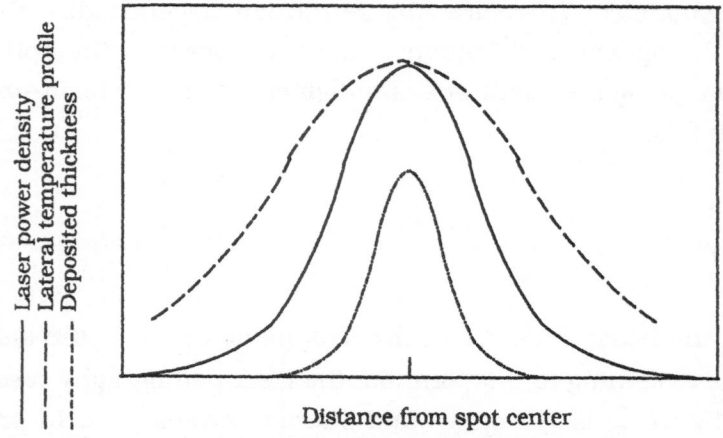

Fig. 2. Lateral laser induced temperature profile using a laser beam power of 1 W, a spot radius of 100 μm, a quartz substrate of 1 mm thick and a calculated deposited thickness with a thermally activated decomposition process. The heat profile is twice the laser beam whereas the deposition is only a half of it.

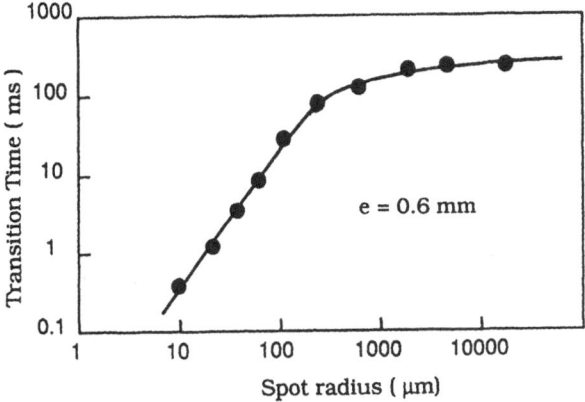

Fig. 3. Characteristic temperature rise time versus laser spot radius. Substrate is silicon coated quartz plate of thicknesses 0.001 and 0.6 mm respectively. A 3D heat flow is calculated for a small spot size and a 1D for a large size.

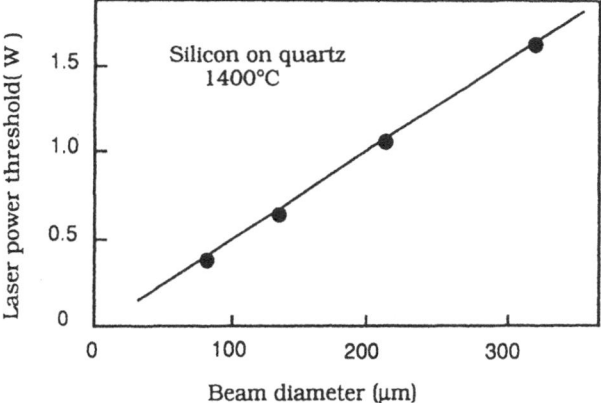

Fig. 4a. Experimental dependence of the laser beam power needed to melt a silicon coated quartz substrate on the laser spot size.

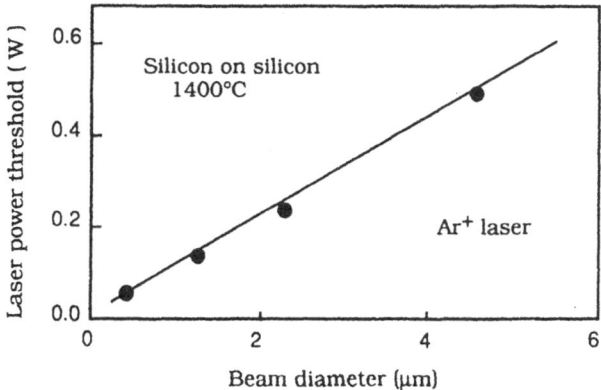

Fig. 4b. Experimental dependence of the argon beam power needed to reach the melting point of a virgin silicon substrate on the laser beam spot size. The spot size depends on the objective used to focus the beam and is determined by measuring the width of the irradiated area at a laser power twice the threshold corresponding to the melting point.

silicon melting temperature is reached by using half of a Watt with a spot diameter of 100 μm on the quartz substrate and 5 μm on the silicon substrate. This result shows the importance of local heat conductivity on laser-induced temperature.

Several other experiments have been performed to measure the laser-induced temperature. Historically, these measurements were made to evaluate the laser-induced annealing temperature in laser annealing experiments. T.T.Kodas has deposited a thermocouple of 10 μm on a silicon wafer to directly measure the laser-induced temperature obtained by irradiating a silicon substrate with an argon laser focussed down to 20μm in diameter [4]. His main finding is a strong influence of the thermocouple geometry on the laser-induced temperature. This influence is due to the very high thermal conductivity of materials used in the thermocouple. In fact, in nearly all experiments, the nature of the thermal probe limits the possibility of measuring the local value of the laser-induced temperature.

Following A.Compaan [5], some authors have measured the value of the laser-induced temperature using the stokes to antistokes ratio of the Raman effect [6]. This Raman technique is currently limited at laser spot diameters higher than 100 μm in order to have an observed area big enough for the measurement of a constant temperature. G.D.Pazionis [7] has avoided this difficulty by limiting the Raman emitting area to a surface of 4 μm in diameter. Using this local probe, he has measured the laser induced temperature on a quartz substrate coated with a thin silicon layer heated by an argon laser. The laser-induced temperature is found to be proportional to the laser power used. This result is important and will be extensively used to evaluate the laser-induced temperature when using a quartz substrate.

The linear dependence of laser-induced temperature versus laser power can be used in addition to a reference measurement. F.Shaapur [3] has shown that melting of NaCl or KCl thin films deposited on the irradiated substrate gives the ratio between surface temperature increase and laser power. The author has used the same technique by melting silicon or germanium films deposited on a quartz substrate. These materials melt at 1400°C and 900°C respectively and give a good accuracy for indirect temperature evaluation in laser-induced chemistry for pyrolytic and photolytic reactions.

Another technique used to measure the temperature, is the optical pyrometer technique. This type of measurement is very classical and its main limitation in spot diameter results from the minimum size of the observed area which is limited to around 100 μm [8,9]. This technique is used by D.Bauerle to study laser-induced chemical reactions [10] which can be transposed into a laser-pantography process. The pyrometer technique is classically limited to temperatures above 800°C and cannot be used for most photolytic experiments in which the local temperature is far below 500°C.

D.Bauerle has evaluated the time-dependent laser-induced temperature during deposition of nickel on a silicon substrate [11]. On a silicon substrate, the dependence of the laser power necessary to reach the silicon melting point during scanning experiments is shown in Fig.5 [12]. The non linearity of these curves comes from the temperature dependence of the silicon heat conductivity .

b) Calculation of laser-induced temperature

Theoretical evaluation of the laser induced temperature has been performed firstly by an analytical method and then by a numerical one. Each method uses a different way to solve the same heat flow equation:

$$\text{div} (K \text{ grad } T) = \rho \ C_v \frac{\delta T}{\delta t}$$

Where K is the heat conductivity, T the temperature, r and Cv are the specific mass and thermal capacity respectively.

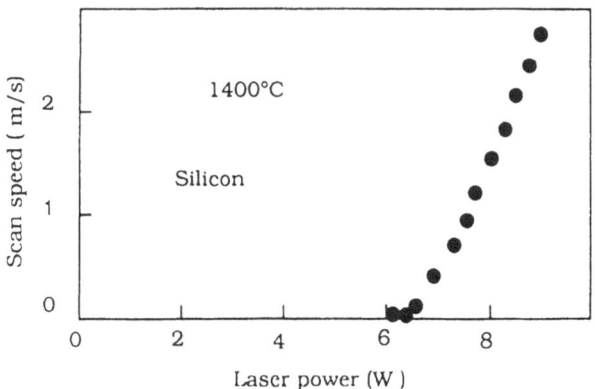

Fig. 5. For each scan speed, the cw argon laser power needed to induce melting in virgin silicon is plotted. The substrate is at room temperature. The beam spot size is 9 μm.

In addition to the heat flow equation, boundary conditions have to be introduced and the laser irradiated surface is taken as follows:

$$Q(x,y) = \frac{P(1-R)}{2\pi r^2} \exp\left(-\frac{(x-vt)^2 + y^2}{2\pi r^2}\right)$$

where Q is the laser power density, P the total laser power, R the surface reflectivity; the laser spot of radius r moves along the x direction at velocity v.

The analytical solution to this equation, introduced by Lax [13] has been extensively developed in order to take into account the specificity of the substrate geometry and substrate physical properties [14,15,16]. Since then, it has been improved by A.Maruani and Y.I.Nissim to introduce the transient regime [17].

Numerical calculations have been developed by D.Tonneau in the case of silicon-coated quartz substrates [18,19]. Fig.6 shows the dependence of the laser induced temperature on the laser power. The influence of the thermal properties of the silicon coating is detected for a spot size below 20 μm. In Fig.6, at constant laser power, the laser induced temperature depends on the spot size. The heat flow occurs in three dimensions for small spot sizes and in only one for spot sizes larger than three times the substrate thickness. For small diameters, a non linear dependence is calculated due to the non linearity of the thermal parameters of the silicon coating.

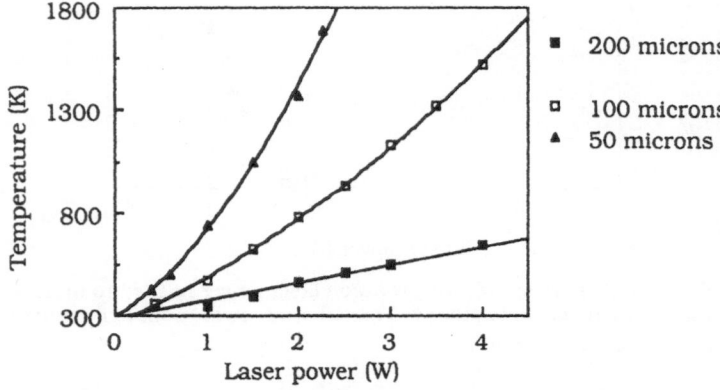

Fig. 6. Calculated laser induced temperature versus laser power on a quartz substrate for three spot sizes: (▲) 50, (□) 100 and (■) 200 μm.

In the laser direct writing technique, it is necessary to control the temperature profile and the temperature stability of the hot area in order to draw an interconnection line with a constant shape all along the line. In practice, by scanning the laser spot on an integrated circuit, several underlayer configurations are encountered and the local temperature can completely change during the spot displacement. Another cause of a temperature change comes from the local reflectivity. An aluminum line reflects nearly 90% of a visible laser beam and only 40% on a silicon surface. Therefore, temperature control is difficult in laser direct writing equipment.

LASER WAVELENGTH AND LASER POWER

The laser pantography can be used to write connections on integrated circuits. The laser beam power has to be high enough to locally modify the equilibrium of the gas-surface interface. For a pyrolytic reaction, the local temperature must increase to around 500°C or more. The corresponding laser beam power, sufficient to draw a 1 μm wide line, is around 1 Watt. Furthermore, in order to control the process, a continuous wave laser may be preferable to a pulsed laser which induces very fast temperature changes. In some cases, this fast rising in temperature is useful when a rapid temperature gradient is needed or when the material has some thermal instabilities.

From the point of view of a thermally activated phenomenon, the limiting process is more the number of molecules reaching the hot surface rather than the laser beam power. Consequently, a constant temperature, regulated with a cw laser beam, will induce a higher deposition rate than a pulsed laser inducing a high reacting temperature during only a short time and corresponding to a small average deposition rate. This means that a significant deposit cannot be accomplished during only one laser pulse of a few nanoseconds. Note that the same idea can be used for a photolytic regime.

From a wavelength point of view, pyrolytic and photolytic regimes cannot always be induced at the same wavelength. Pyrolytic processes are induced at wavelengths in the absorption domain of the substrate. Photolytic decomposition of the gaseous molecules occurs either in the ultraviolet domain or in the infrared one when vibrational states can be populated with several infrared photons. From this point of view, UV cw lasers are very interesting for photolytic processes. In practice, most laboratories studying laser pantography use either cw lasers emitting in the visible range (for pyrolytic processes) or pulsed lasers emitting in the far UV (for photolytic processes).

MAIN DIFFICULTIES IN LASER DIRECT WRITING

One important difficulty of laser pantography is the presence of two gas-solid irradiated interfaces in the optical path. The first one is on the circuit where interconnections have to be built or etched. The second is the chemical chamber window through which the laser beam passes. In the case of a photolytic regime, the reaction occurs simultaneously on the circuit and on the window. The window deposit modifies the total transmittance of the window and unless the deposited material is transparent to the laser beam, the chemical process stops and the window has to be cleaned before the next experiment [20].

In the case of a pyrolytic reaction, the laser induced temperature at the window level must be lower than at the circuit level. This is obtained by placing the window as far as possible from the circuit at a position where the laser beam diameter is more than ten times larger than on the circuit surface.

DEPOSITION AND ETCHING RATES

For small volume reactions as in laser pantography, the nature and kinetics of the surface and near-surface processes can be very different from those of conventional large-area chemical processes [21,22,23]. One consequence of immediate practical importance is that the rates of many reactions are orders of magnitude faster than those of the corresponding large area reaction. Typically, the deposition rate is expressed in µm/s as unit of speed for laser assisted processes, whereas nm/mn is the most commonly used unit for conventional large area processes. As far as we know, this difference stems from the fact that in large-area processes, the gas phase is at high temperature and some homogeneous chemical reactions may limit the temperature level. For example, above 700°C, a silicon powder is formed in a silane atmosphere and no adherent film can be deposited on a substrate. For the laser pantography technique, the gas phase is at room temperature and no homogeneous decomposition can occur, except in the laser irradiated area where the temperature can be as high as the substrate melting temperature. For example, silicon can be deposited at 1400°C by laser pantography with a deposition rate of 3μm/s under a silane pressure of 20 Torr [24].

From the deposition rate point of view, laser assisted processes are excellent in comparison with conventional chemical vapor deposition processes. Unfortunately, the number of wafers coming off a large-area process line is higher than those coming out from the laser-pantography technique. The laser technique is a serial process in which connection lines are drawn one by one. In this case, using a very high deposition rate of about one millimeter per second, an integrated circuit with 4 meters of connections takes more than one hour to write. During the same time, 10000 chips can be made by using large area processes. This comparison is interesting to determine a realistic application domain for the laser pantography technique. To make prototypes or unique circuits, laser pantography able to write a chip in less than one day, will be the best technique. As the market for custom or prototype circuits is rapidly expanding, the laser pantography technique seems to have a wide future application.

PROCESS FOR SILICON MICROELECTRONICS

Laser pantography in silicon microelectronics can be divided into three topics:
-modification of one or two connections in a finished integrated circuit,

-writing all metallic connections on a gate array,

-building all transistors and connections.

These three topics will probably be developed in this order due to the increasing complexity of the corresponding industrial equipment. In fact, the first topic can almost be made by hand, the second has to be fully automatic but with only one elementary process, whereas the third needs a very complex process. Let us examine these three topics in more detail.

a) Modification of interconnection network

The process is described in Figs. 7a and 7b. Two kinds of processes are possible depending on the laser induced temperature.The first step is a local insulation of certain conducting connections (generally aluminum) by deposition of an insulator (Fig. 7a). The second step consists of deposition of metal lines as new connections between two aluminum lines and above the deposited insulating dots of the crossing area.

For a high temperature process, one of the step is at too high a temperature for the materials used in the integrated circuit. Therefore, this process must include an extra step which consists of a local replacement of the low melting temperature material by a high melting temperature one (Fig. 7b). This can be achieved by an etching reaction followed by deposition of a refractory metal. In silicon microelectronics, aluminum has the lowest melting point and must be replaced by nickel or tungsten in order to locally deposit an insulator such as silicon nitride which can be produced only at temperatures higher than the melting point of aluminum [1].

b) Connections in gate arrays

In silicon gate arrays, all transistors are already available on the starting wafer and the necessary via holes are open. According to Fig. 8, if only one metallic layer has to be drawn, all interconnections can be written

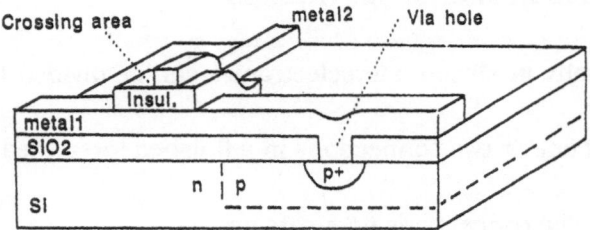

Fig. 7a. 3D view of a two level crossing line obtained by using a low temperature interconnecting process. The local deposition of an insulator is performed at temperatures lower than the melting point of metal 1 which is currently aluminum.

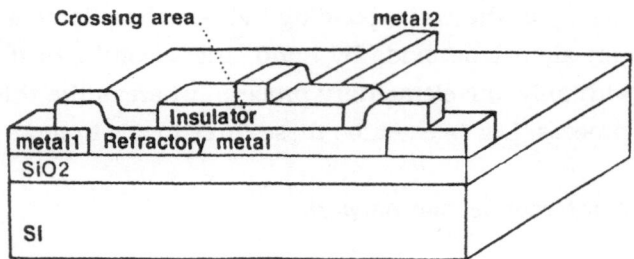

Fig. 7b. In a high temperature process, the aluminum line (metal 1) must be changed into a nickel line in the crossing area. The refractory metal must have a low deposition temperature and a low contact resistance with the existing aluminum line in order to have a reparation process for integrated circuit.

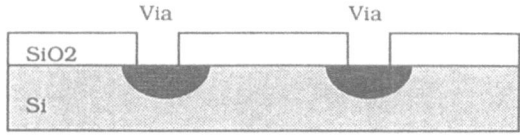

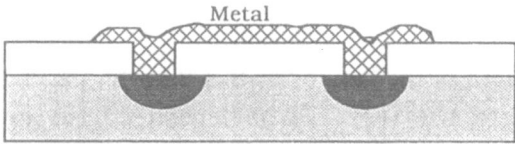

Fig. 8. Schematic view of a gate array circuit before (top) and after (below) connections using laser pantography equipment. Via holes are opened and a conductive material is locally deposited in order to connect at least two of them.

by a single scan. If two interconnection levels are necessary, deposition of an insulator has to be performed. This can be achieved in two different manners. By using a localized process, deposition is performed only where lines cross each other. Another method is proposed for commercial equipment in which a low temperature insulating layer is deposited all over the circuit and vias are opened by using a laser etching process.

c) Making chips

To make a complete chip, the procedure is very similar to that of a conventional process since there are exactly the same number of steps. The advantage of the laser technique is to avoid the use of all lithography steps. On the other hand, the drawback is that transistors are drawn in a serial process and not a parallel one as in the conventional technique.

To date, only a few attempts have been made to build transistors [25,26].

LASER INDUCED CHEMICAL REACTIONS

The first deposited atoms were silicon atoms coming from silane molecules decomposed using an infrared laser beam. Since this time, numerous molecules have been decomposed using various laser beams. The decomposition is either a photolytic phenomenon or a pyrolytic one. Focussing our attention on decompositions with high enough rates and obtained using cw lasers, the following materials seem to be suitable for laser pantography:

a) Metal deposition

Cadmium deposition

Using a wavelength of 676 nm or in the range of 337-514 nm given by cw high power lasers, cadmium can be deposited from dimethylcadmium $Cd(CH3)2$ [27]. The deposit nucleates via a photolytic effect followed by a pyrolytic growth. An important induction time is observed. For laser pantography, nucleation time must be as small as possible and can be experimentally obtained with UV light. Unfortunately, due to this photolytic effect, a delocalization of the deposit is detected and this UV wavelength is not useful for laser direct writing. These observations correspond to very general characteristics of the laser pantography and have been nearly always found when metalorganic molecules are used to deposit metals.

Tin deposition

An argon ion laser was used to decompose $SnCl4$ on the surface of a semiconductor [28]. It was found during these experiments that tin was buried in the semiconductor. The chlorine atoms etch the substrate and tin is deposited at the same place. Therefore, chlorine atoms are transferred from tin atoms to substrate atoms. The laser induced chemical reaction occurring on the semiconductor surface is the following:

$$Si + SnCl4 \rightarrow SiCl4 + Sn$$

in which Si and Sn are in solid phase and chlorine compounds are volatile. Consequently, in a complete process of laser pantography, these kinds of carrier molecules containing chlorine atoms have to be avoided.

Copper deposition

Copperchelate such as $Cu(HFAcAc)2$, has been commonly chosen as a precursor for copper depositions [29,30,31]. This solid compound has a vapor pressure of about 5-10 Torr at room temperature. As this pressure is small, the deposition rate is also small. The gas pressure can be increased by heating all the experimental apparatus at temperatures above 50°C.

The decomposition mechanism of the precursor is found to be pyrolytic in the adsorbed layer.The laser induced temperature threshold for the deposition of copper lines was estimated to be around 300°C. A thermal desorption of the precursor from the surface is the commonly used explanation for the volcano-shaped deposit obtained at high laser power. This desorption problem prevents high deposition rates being reached by increasing the laser induced temperature. The deposition rate of electrically suitable copper is lower than 0.1 µm/s and the maximum writing speed has been found to be around 0.1 mm/s. This pyrolytic decomposition has been performed by using an argon ion laser with a power density of above 105 W/cm2.

Platinum deposition

Decomposition of platinumacetylacetonate, Pt(HFAcAc)2, at saturated vapor pressure from the solid phase at 0.3 Torr gives good electrical properties at a speed of 10 µm/s using 150 mW of an argon laser emitting in the visible domain (458 nm). This result is mentioned in several publications made by H. van der Berg [26,32]. The author has studied the dependence of the deposition rate on the laser power used. At low laser power, no deposition is observed. At moderate power, a pyrolytic decomposition is possible. Under high power, the deposition rate is limited by the mass transport of the platinum carrier molecules onto the hot zone.

The dependence of the vapor pressure used in these experiments, on temperature is shown in Fig.9. This curve shows that at room temperature

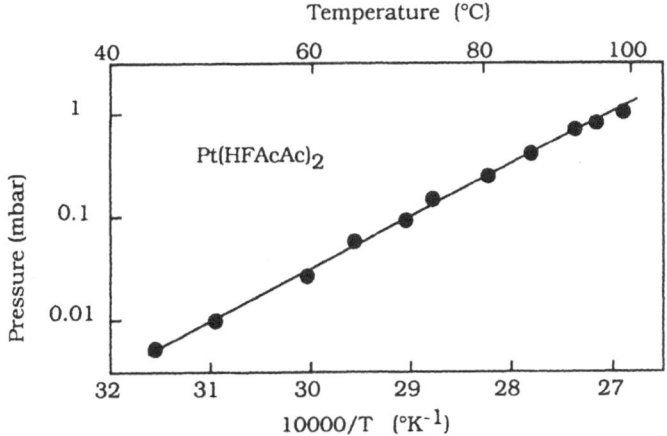

Fig. 9. The vapor pressure of Pt(HFAcAc)2 versus the reciprocal temperature. According to Fig: 12, a theoretical deposition rate for platinum at 4 µm/s can be achieved by heating up the experimental setup at 90°C.

the partial pressure is too small to expect high writing speeds. Therefore, as a general result for this sort of compound, a good writing speed can be achieved by heating the entire experimental set-up.

Tungsten deposition

Tungsten carrying molecules are WF6 and W(CO)6. The feasibility of tungsten line writing has been demonstrated for both molecules [33,34,35,36]. The first molecule (WF6) can be decomposed according to the following reactions:

$$WF6 + 3 H2 \longrightarrow W + 6 HF$$

or $$2 WF6 + 3 Si \longrightarrow 2 W + 3 SiF4$$

HF and SiF4 are volatile molecules and tungsten deposition occurs. The first reaction is not selective and tungsten can be deposited on various substrates. The second reaction is selective on silicon surface (no deposition occurs on SiO2) and stops when the deposited tungsten thickness reaches around 100 nm. This thickness is too small to expect a low connection resistance in an integrated circuit. Only the second reaction is useful to increase the deposited thickness. The first reaction has been carefully studied by Y.S.Liu [33] and the second one by D.Bauerle [34]. A deposition rate of around 20 µm/s is achieved. This value is very high in comparison with other deposition rates. It has also been found that the deposition rate depends on the partial pressure of WF6   [35]. In our experimental apparatus, by irradiations of a silicon quartz substrate using an argon laser and a mixture of H2+WF6, tungsten dots were deposited. Fig. 10 shows the dependence of the deposition rate

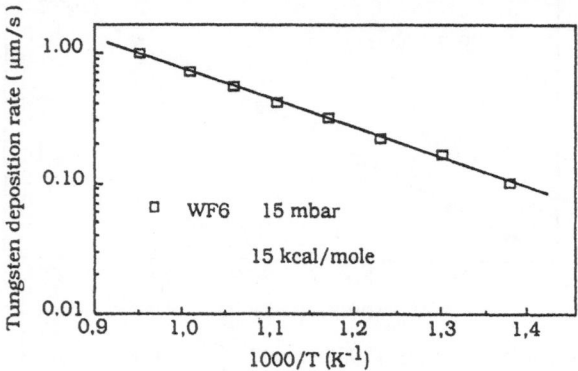

Fig. 10. Arrhenius diagram of tungsten deposition rate. Tungsten fluorine pressure is 11 Torr mixed with 220 Torr of hydrogen. Deposition rate of 1 µm/s is achieved at temperatures around 800°C.

on the laser induced temperature. A deposition rate of 1 µm/s is achieved at around 800°C indicating that 20 µm/s can be achieved only at very high temperature.

The second carrier of tungsten atoms belongs to the carbonyl family. The decomposition reaction is:

W(CO)6 --> W + 6 CO

CO is volatile and the deposition rate has been studied by D.J.Ehrlich [36]. The deposition rate is proportional to the UV laser power used. The decomposition mechanism is interpreted as a photolytic process without gas phase decomposition. This point is interesting for laser pantography. Unfortunately, the deposition rate is lower than 0.4 nm/s which is too slow for most applications. In our experimental set-up, by studying pyrolysis of W(CO)6 it was found that decomposition occurs at temperatures higher than 700°C by using laser radiations in the visible domain. The deposit has a low reflectivity indicating the presence of carbon atoms [37].

Aluminum deposition

In integrated circuit technology, aluminum is the most widely used interconnecting material due to its very low resistivity. Therefore, all laboratories have tried to deposit this metal. By using infrared or visible laser beam, no decomposition of available metalorganic molecules containing aluminum has been found. These molecules are thermally decomposed at high temperature but the deposition of aluminum carbide is always obtained.

Recently, the feasibility of aluminum direct writing lines has been proved [38,39]. Deposition occurs by using a frequency doubled argon ion laser beam at 257 nm. A photolytic effect is at the origin of the decomposition of either trimethylaluminum or dimethymaluminum. The deposition rate is small and a direct writing speed of 10 µm/s seems to be possible. This value is small for laser pantography applications but as the resistivity of aluminum is so small, there will probably be no other possibility.

Nickel deposition

Nickel is a high conducting metal which is not used in silicon microelectronics because it is impossible to etch by the conventional plasma technique based on halogenated gases. In laser pantography, as no etching of the deposited metal is required, deposition of nickel can be interesting.

The nickel carrying molecule is $Ni(CO)_4$. The vapor pressure at room temperature is around 500 mbar. This pressure is sufficiently high to expect a high deposition rate if pressure dependent kinetics occurs. The decomposition reaction of nickel tetracarbonyl is:

$$Ni(CO)_4 \longrightarrow Ni + 4\,CO$$

CO is a volatile compound and leaves the heated zone. By using a CO_2 laser emitting at 10.6 μm, at the beginning of the reaction, a high deposition rate is measured. Due to the high reflectivity of nickel in the infrared, the local temperature then drops as does the deposition rate [24,40]. Consequently, to study nickel deposition kinetics, it is necessary to proceed in two steps to be able to measure a constant deposition rate. The first step consists in depositing a thin nickel film of around 300 nm. In the second step, nickel is deposited at high laser power on the thin film. A deposition rate of 5 μm can be obtained at 400°C under a pressure of 20 Torr.

By using a laser emitting in the visible range, D.Bauerle has studied the dependence of nickel deposition rate and line width on laser scan speed and laser power [41]. A width of 12 μm was performed with around 0.8 μm in thickness by using a writing speed higher than 100 μm/s. This result is very interesting for laser pantography which needs both very high scanning speeds and film thicknesses of around one micron. As the deposition rate does not depend linearly on the laser power used, a thermal decomposition process is assumed for both laser experiments.

By using our experimental set-up, nickel carbonyl has been decomposed [42]. It has been found that at 10.6 μm, 0.5 μm and 0.33 μm the decomposition of nickel carbonyl is thermally activated with an activation energy of 12 kcal/mole (Fig. 11). By studying the deposition rate at various pressures, it can be proved that decomposition occurs in the adsorbed layer (Fig. 12) and that a mass transport limited growth rate can occur at high temperature and pressure below 0.7 mbar. The decomposition yield of molecules impinging on the hot surface is evaluated to be around 60%.

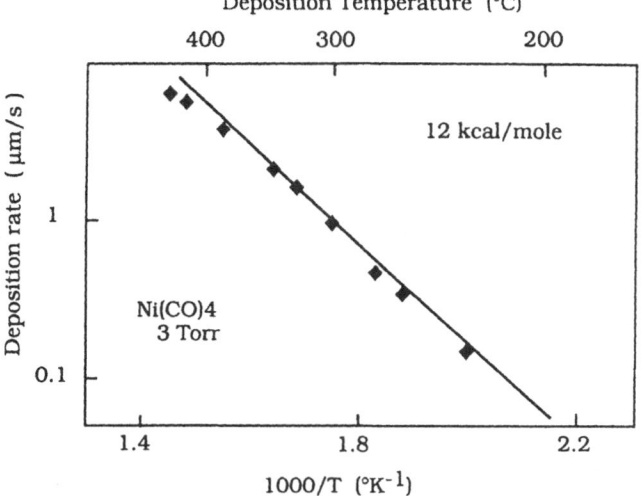

Fig. 11. Arrhenius diagram of the deposition rate of nickel dots on silicon coated quartz substrate. The decomposition process is thermally activated and does not depend on the laser wavelength used.

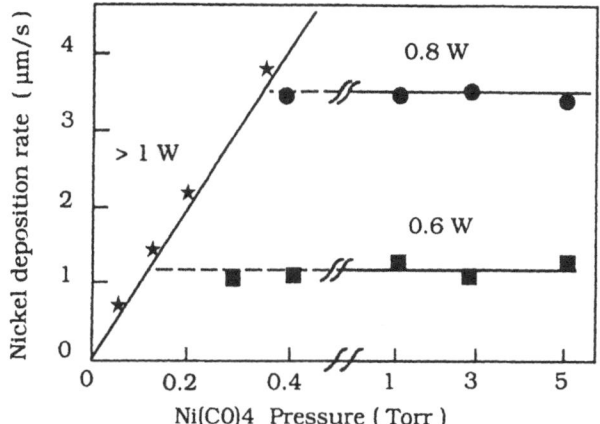

Fig. 12. Effect of Ni(CO)4 pressure on deposition rate. The output laser power was: 0.6 W (■), 0.8 W (●) and higher than 1 W (★). If the laser power is low enough and the gas pressure high enough, the deposition rate does not depend on the pressure and the decomposition occurs in the adsorbed layer. At low pressure, and high laser power, the rate is limited by the number of impinging molecules.

b) Semiconductor deposition

Silicon deposition has been extensively studied whereas for germanium and gallium arsenide only the feasibility has been demonstrated. Deposition of gallium arsenide is very difficult because the two carrier molecules (Ga(CH3)3 and As(CH3)3) are not decomposed at the same temperature [43].

245

By using cw lasers, deposition of silicon is obtained by decomposition of SiCl4, SiH4 or Si2H6. By using SiCl4 as a precursor, deposition of silicon has been measured in Ref. 44. It has been found that the deposition rate increases with laser power. The deposition rate is small and local deposition of a continuous film is achieved after an irradiation time longer than 60 s. The small rate can be explained by a competition between a deposition and an etching process due to the liberated chlorine molecules. The competitive reactions are as follows:

$$Si + SiCl4 \longrightarrow 2\,Si + 2\,Cl2$$

$$Si + SiCl4 \longrightarrow 2\,SiCl2 \qquad \text{(volatile)}$$

In order to avoid chlorine etching, the most widely used precursor to deposit silicon is silane gas. The feasibility of laser induced deposition was proved in 1977 by using a CO2 laser [45]. Thus, for the laser pantography technique, a demonstration of very small lines has been performed [46,47]. Line widths as small as 0.2 μm have been drawn and explained by a thermally activated decomposition process of silane molecules. As expected, the deposited thickness decreases with increasing laser scanning speed according to a constant deposition rate [48].

In order to check the thermally activated process, we have measured the silicon deposition rate using silane and disilane. In Fig. 13, the silicon

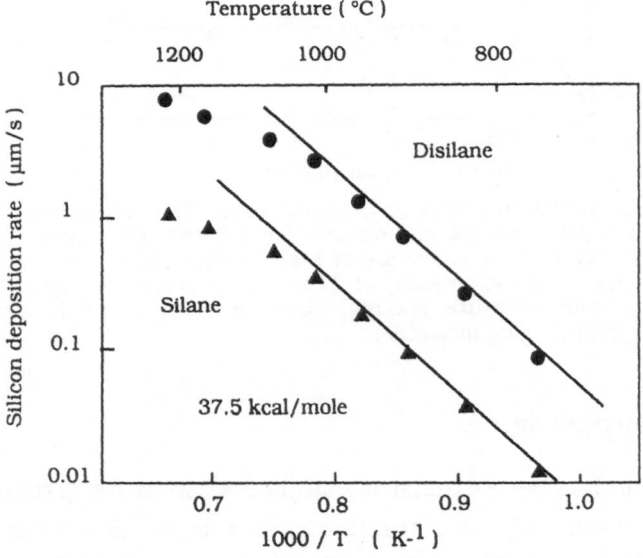

Fig. 13. Arrhenius diagram corresponding to the deposition of silicon dots using an argon laser at power ranging from 1.2 to 1.7 Watt on a silicon coated quartz substrate. Silicon is deposited from (▲) silane at 10 Torr or (●) disilane at 1 Torr.

deposition rate is plotted versus the reciprocal laser induced temperature [49]. A deposition rate of 10 µm/s is measured at a pressure of 25 Torr and at a laser induced temperature of around 1200°C. At a given temperature, disilane molecules decompose seven times faster than silane molecules. It has been proved in Ref. 49 that a photolytic effect is necessary for silane and disilane decomposition at temperatures below 900°C. In the high temperature domain, a collisional process occurs and no photolytic effect is detected. The same deposition rate can therefore be obtained by using a CO_2 laser emitting in the infrared [50]. A collisional process means that an adsorbed molecule can be decomposed during a collision with an impinging molecule from the gas phase [51]. Physical characteristics for a collisional decomposition are both a temperature threshold for the deposition and above the threshold a deposition rate proportional to the gas pressure.

c) Insulating materials deposition

Between two levels of interconnections, an insulating film must be deposited. The most straightforward approach would consist in locally depositing the insulator by using the laser direct writing technique. Unfortunately, localized deposition of insulators is not yet fully understood [52,24].

In silicon microelectronics, insulating films are silicon dioxide and silicon nitride. Silicon dioxide is deposited from a mixture of silane with O_2, N_2O or a similar gas. Silicon nitride comes from a silane ammonia mixture [53]. In our experimental apparatus, deposition kinetics of the silane ammonia mixture has been determined. Fig. 14 shows the deposition rate of silicon nitride versus the reciprocal temperature for two laser wavelengths. From this figure, a photolytic effect is observed. It has been found that the photolytic effect comes from silane decomposition. Therefore, formation of a silicon nitride film starts with deposition of silicon followed by a nitridation. This nitridation is not complete at low nitride pressure whereas at high pressure no deposition occurs. This is interpreted by the formation of volatile molecules when ammonia is in excess. At intermediate pressure, the deposition rate of silicon nitride is nearly equal to the silicon deposition rate by using 2.4 eV photons. For infrared photons, the deposition rate is far below and is not very useful for laser pantography.

d) Etching of insulating films

The insulating film can be deposited on the first level of interconnects using either the conventional chemical vapor deposition method or the light assisted one [53]. Via holes are then opened by laser-assisted etching before laser writing of the second interconnection level.

Localized etching of silicon dioxide has been studied using a CO2 laser [54]. The etching gas is silane and it has been found that an etching reaction occurs for silane pressure below 1 Torr and for a laser induced temperature above 1000°C (see Fig. 15). The etching mechanism proposed is a decomposition of silane molecule followed by a chemical reaction between the formed silicon atom and the oxide layer according to the reaction:

$$Si + SiO2 \longrightarrow 2\ SiO$$
in which the reaction product is volatile at temperatures above 1000°C.

According to Fig. 15, the etching rate is proportional to the silane pressure and is around 0.15 µm/s at 1400°C and 0.7 Torr of silane. This temperature is too high to open a via hole above an aluminum line and further studies are necessary in order to etch the encapsulating layer of an integrated circuit containing aluminum connections.

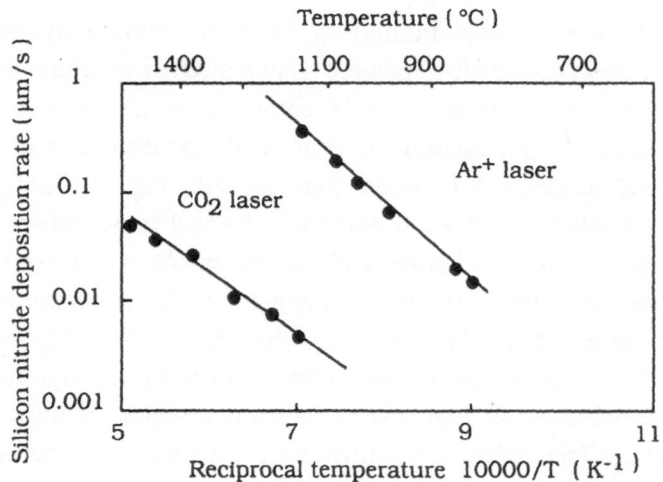

Fig. 14. Deposition rate of silicon nitride versus the reciprocal temperature using either an argon laser or a CO2 laser. Evidence of a photolytic effect can be seen but the limiting step is rather a pyrolytic decomposition rather than a photolytic one. For the argon laser, a silicon coated quartz substrate is used. For the CO2 laser, a virgin quartz substrate is irradiated.

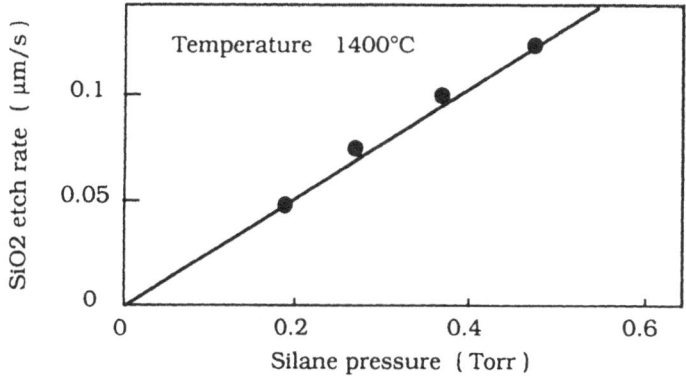

Fig.15. Etching rate of silica versus silane pressure at a laser-induced surface temperature of 1400°C. A CO2 laser is used to locally heat up the silica substrate in which hole depths are measured with a mechanical stylus.

CONCLUSION

Fundamental processes for building new interconnections in an integrated circuit are now available: metal deposition, insulator deposition and insulator etching. Decomposition mechanism of the carrier molecules seems to be understood and high writing speed can be expect in order to draw gate arrays.

BIBLIOGRAPHY

1) G. Auvert, Y. Pauleau, D. Tonneau, in "Emerging Technologies for In-Situ Processing", edited by D.J. Ehrlich and V.T. Nguyen, NATO ASI Series, Serie E : Applied Sciences, Vol. 139, 1988, p. 201.
2) M.E. Gross, G.J. Fisanick, P.K. Gallagher, K.J. Schnoes, M.D. Fennell, Appl. Phys. Lett., 47(9), 1985, p.923.
- G.J. Fisanick, J.B. Hopkins, M.E. Gross, M.D. Fennell, K.J. Schnoes, Appl. Phys. Lett., 46(12), 1985, p. 1184.
- G.J. Fisanick, M.E. Gross, J.B. Hopkins, M.D. Fennell, K.J. Schnoes, A. Katzir, J. Appl. Phys., 57(4), 1985, p. 1139.
3) F.Shaapur, S.D.Allen, Appl. Phys. Lett. 50, 12, 1987, p.723.
4) T.T.Kodas, T.H.Baum, P.B.Comita J. Appl. Phys. 61 (8), 15, 1987, p.2749.
5) H.W.Lo, A.Compaan, J. Appl. Phys. 51, 3, 1980, p.1565.
6) D.Kirilov, J.L.Merz, Mat. Res. Soc. Proceeding, V.17, 1983, p.95.
7) G.D.Pazionis, H.Tang, L.Ge, I.P.Herman, Mat. Res. Soc. Symp. Proceeding, V.101, 1981, p.113.
8) T.O.Sedgwick, Appl. Phys. Lett. 39, 3, 1981, p.254.
9) M.A.Bosch, R.A.Lemons, Phys. Rev. Let. 47, 16, 1981, p.1151.
10) G.Leyendecker, D.Bauerle, Appl. Phys. Lett. 39, 11, 1981, p.921.
11) K.Piglmayer, J.Doppelbauer, D.Bauerle, Mat. Res. Soc. Symp. Proceeding, V.29, 1984, p.47.

12) F.Ferrieu, G.Auvert, J. Appl. Phys. 54, 5, 1983, p.2646.

13) M.Lax, Appl. Phys. Lett. 33, 1978, p.786.

- M. Lax, J. Appl. Phys., 48, 1977, p. 3919.

14) S.A. Kokorowski, G.L. Olson, L.D. Hess, in "Laser and Electron-Beam Solid interactions and Materials Processing", edited by J.F. Gibbons, L.D. Hess and T.W. Sigmon, Mat. Res. Soc. Symp. Proc., Vol. 1, 1981, p. 139.

15) H.E.Cline, T.R.Anthony, J. Appl. Phys. 48, 9, 1977, p.3895.

16) Y.I.Nissim, A.Lietola, R.B.Gold, J.F.Gibbons, J. Appl. Phys. 51, 1, 1980, p.274.

- R.B. Gold, J.F. Gibbons, J. Appl. Phys., 51(2), 1980, p. 1256.

17) A.Maruani, Y.I.Nissim, F.Bonnouvrier, D.Paquet, Mat. Res. Soc. Proceeding, V.13, 1983, p.123.

18) D.Tonneau, G.Auvert, European Mat. Res. Soc. Proceeding Vol. 15, 1987, p.169.

19) D.Tonneau, G.Auvert, Mat. Res. Soc. Proceeding, Vol. 101, 1987, p.131.

20) Y. Pauleau, D. Tonneau, G. Auvert, in "Laser Processing and Diagnostics" edited by D. Bäuerle, Springer Series in Chemical Physics, Vol. 39, 1984, p. 215.

21) R.J.Von Gutfeld, R.E.Acosta, L.T.Romankin, I.B.M. J. Res. Dev. 26, 136, (1982).

22) D.J.Erlich, J.Y.Tsao in "Laser diagnostics and photochemical processing for semiconductor devices" Ed. by R.M.Osgood, S.R.J.Brueck, H.R.Schlossberg North-holland, New York, 1983, p. 3.

23) J.P.Herman, R.A.Hyde, B.M.McWilliams, A.H.Weisberg, L.L.Wood,in "Laser diagnostics and photochemical processing for semiconductor devices" Ed. by R.M.Osgood, S.R.J.Brueck, H.R.Schlossberg North-holland, New York, 1983, p.3.

24) G. Auvert, D. Tonneau, Y. Pauleau, in "Laser Processing and Diagnostics II", edited by D. Bäuerle, K.L. Kompa and L. Laude, les Editions de Physique, European-Mat. Res. Soc. Symp. Proc., Vol. 11, 1986, p. 109.

25) B.M. Mc Williams, I.P. Herman, F. Mitlitsky, R.A. Hyde, L.L. Wood, Appl. Phys. Lett., 43(10), 1983, p. 946.

26) D. Braichotte, H. van den Bergh, International Conference on Lasers, 1985, p. 688.

27) Y.Rytz-Froidevaux, R.P.Salathe, H.H.Gilgen, H.P.Weber, Appl. Phys. A 27, 1982, p.133.

28) J.Tokuda, M.Takai, K.gamo, S.Namba, Mat. Res. Soc. Symp. Proceeding, V.101, 1987, p.261.

29) C.R.Moylan, T.H.Baum, C.R.Jones, Appl. Phys. A 40, 1986, p.1.

30) F.A.Houle, C.R.Jones, T.Baum, C.Pico, C.A.Kovac, Appl. Phys. Lett. 46, 2, 1985, p.204.

31) D.Braichotte, K.Ernst, R.Monot, J.M.Philippoz, M.Qiu, H. van der Berg, Mat. Res. Soc. Symp. Proceeding V.58, 1985, p.879.

32) D.Braichotte, H.van der Bergh, In "Laser processing and diagnostics" Ed. D.Bäuerlé, Springer Series in Chem. Phys. Berlin, 39, 1984, p.183.

33) Y.S.Liu, C.P.Yakymyshyn, H.R.Philipp, H.S.Cole, L.M.Levinson, J. Vac. Sci. Technol. B 3, 5, 1985, p.1441.

34) G.Q.Zhang, T.SzorenYi, D.Bauerle, J. Appl. Phys. 62, 2, 1987, p.673.

35) S.D.Allen, A.B.Tringubo, J. Appl. Phys. 54, 3, 1983, p.1641.

36) D.J.Ehrlich, R.M.Osgood, T.F.Deutsch, J. Electrochem. Soc. Sept.1981, p.2040.

37) D.Tonneau, G.Auvert, Y.Pauleau, Proceedings of the 7th European Conf. on CVD, Ed. M.Ducarroir, L. van den Bulke, C.Bernard, Les Editions de Physique, Paris, 1989.

38) J.Flicstein, J.E.Bouree, J.F.Bresse, A.M.Pougnet, Mat. Res. Soc. Symp. Proceeding, V.101, 1987, p.49.

39) T.Cacouris, G.Scelsi, R.Beach, R.M.Osgood, C.L.E.O. May 1987, p.289.

40) S.D.Allen, J. Appl. Phys. 52, 11, 1981, p.6501.

41) W.Krauter, D.Bauerle, F.Finberger, Appl. Phys. A 31, 1983, p.13.

42) D.Tonneau, Y.Pauleau, G.Auvert, J. Appl. Phys., 67, 10-1, 1988, p.5189.

43) D.Braichotte, H. van der Berg, Frujahrstagung der Schweiz, Phys. Gesellschaft, V.59, 1986, p.1014.

44) V.Baranauskas, C.I.Z.Mammana, R.E.Klinger, J.F.Greene, Appl. Phys. Lett. 36, 11, 1980, p.930.

45) C.P.Christensen, K.M.Lakin, Appl. Phys. Lett. 32, 4, 1978, p.254.

46) A.Ishizu, Y Inoue, T. Nishimura, Y. Akasaka, H. Miki, Japanese Journal of Applied Physics, V.25, N.12,(1986) p.1830.

47) J.G.Black, D.J.Ehrlich,M.Rotschild, S.P.Doran, J.H.C.Sedlacek, J. Vac. Sci. Technol. B 5,1, 1987, p.419.
    S.Leppavuori, J.Lenkkeri, J.Levoska, NATO Adv. Res. Worshop Proceeding, Ed. D.J.Ehrlich, V.T.Nguyen, (1988), p.265.

48) D.Bauerle, P.Irsigler, G.Leyendecker, H.Noll, D.Wagner, Appl. Phys. Lett. 40, 9, 1982, p.819.

49) G.Auvert, D.Tonneau, Y.Pauleau Appl. Phys. Lett. 52, (13), 1988, p.1062.

50) D.Tonneau, G.Auvert, Y.Pauleau, Thin Solid Films, 155, 1987, p.75

51) J.D.Beckerle, Q.Y.Yang, A.D.Johnson, S.T.Ceyer, J. Chem. Phys. Jun 87, p.7236.

52) S.Szikora, W.Krauter, D.Bäuerle, Matr. Lett., V. 2, 4A, 1984, p.263.

53) G.Auvert, Y.Pauleau, D.Tonneau, Mat. Res. Soc. Proceeding, V.101, 1987, p.125.

54) Y.I.Nissim, C.Licoppe, J.M.Moison, J.L.Regolini, D.Bensahel, G.Auvert, In Trends in quantum Electronics, SPIE proceeding, V.1033, (1989), in press.

55) D.Tonneau, Y.Pauleau, G.Auvert, J. Appl. Phys. (Submitted 11/1988).

ION BEAM ASSISTED PROCESSES

Gaetano Foti

Dipartimento di Fisica
Università di Catania
Corso Italia, 57
95125 Catania, Italy

INTRODUCTION

Ion beam assisted processes represent an emerging technology used to change the surface properties of materials at low temperature, thus avoiding diffusion in the bulk.

The main effects induced by ion beam can be divided into two groups: one, where the rate of some process is enhanced by ion irradiation, like ion induced solid phase epitaxy,[1,2,3] ion assisted etching, radiation enhanced diffusion,[4,5] ion induced grain growth,[6,7] and ion beam mixing;[8] another, where the materials are modified in a permanent way, like cross-linking in polymers,[9] phase transitions in oxides, and ion beam assisted deposition.

The energy release from the ion to the target atoms in atomic collisions is the key parameter in understanding ion assisted phenomena in semiconductors and metals, while for insulators and polymers the electronic excitations play a role. In the following, ion assisted regrowth of ion implanted or deposited amorphous layers will be discussed along with possible applications in microelectronic devices, such as high current ion implantation and grain growth.

DEFECT AND COLLISION CASCADE

Before beginning the discussion on the ion assisted regrowth of amorphous silicon, I would like to say a few words about the mechanism of thermal annealing of silicon.

Recrystallization of amorphous silicon requires the transport of a silicon atom to lattice position in the region near the internal interface. The process must therefore contain one step where a vacant lattice site is formed near the interface and migration of an atom to fill the vacant site occurs.

In Fig. 1, two possible mechanisms are shown. In Fig. 1A, a thermally activated transport of an atom occurs in the disordered region to a lattice site (v) in the same region. In Fig. 1B, an atom in the crystalline region is moved to a lattice site (v) in the disordered region; the produced vacancy diffuses in the crystal and is annihilated by an atom of the disordered region. Both processes require the formation of a vacant site lattice near the interface and of an atomic migration across the interface.

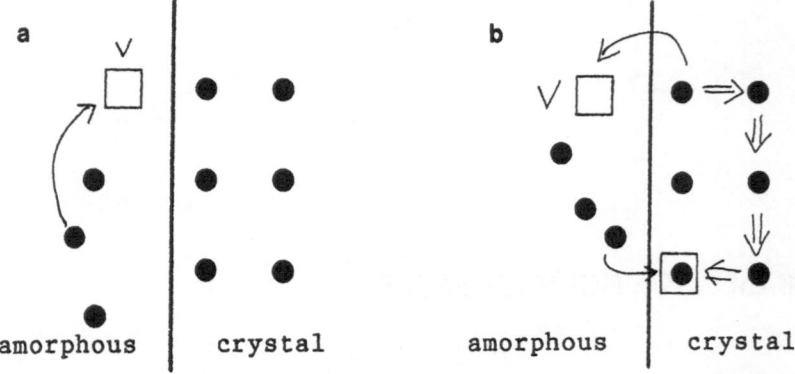

Fig. 1.   Epitaxial regrowth through defect motion in the disordered region (a) and at the interface (b).

Unfortunately, the activation energy for atom mobility is not very different in the amorphous and crystalline phases of silicon, so it is very hard to distinguish between the two processes with temperature dependence measurements. The atomic jump rate can be described by a free-energy diagram during the amorphous to crystalline transition as shown in Fig. 2.

The net rate of atomic flux (a → c) at the interface is

$$\frac{dn}{dt} = fn_\alpha v_o e^{-\Delta G'/KT} - fn_c v_o e^{-(\Delta G'+\Delta G)/KT} ,$$

where f is the fraction of available sites for exchange of atoms or accommodation factor; $n_\alpha \cong n_c$ are the interfacial number densities and v_o is the lattice vibration frequency; ΔG, the difference of free-energy between the amorphous and crystalline phases (for silicon, $\Delta G = 0.1$ ev / atoms) and $\Delta G'$, the free-energy change for an atomic jump across the interface.

The interface velocity v is given by

$$v = \lambda f v_o e^{-\Delta G'/KT} ,$$

where λ is the atomic jump distance across the interface. The accommodation factor in the pre-exponential determines the magnitude of the interface velocity and it is correlated with the chance that an atom can find a position on the amorphous-crystal interface where it can attach itself. Different crystalline orientations present entirely different types of sites, so that the accommodation factor for the motion of atoms from the amorphous towards the crystal varies with the orientation. The less close-packed the plane, the easier it is for atoms to attach themselves to the crystalline interface. This concept is well illustrated in Fig. 3 where the atomic arrangement on (110) and (100) surfaces are reported for a diamond-type structure.[10]

The free-space in the surface available for the accommodation of an atom as it joins the crystal is larger for the less close-packed planes. As a result of this difference, the regrowth velocity along the (100) orientation will be higher compared to the (110) orientation. This effect has been measured for thermal regrowth where the ratio v(100)/v(110) is approximately a factor of four.

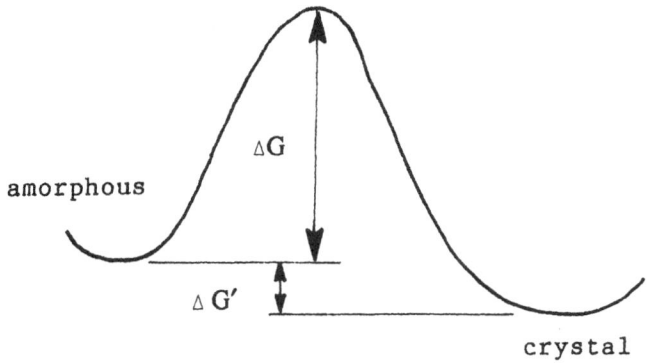

Fig. 2.   Schematic free-energy diagram between amorphous and crystalline phases.

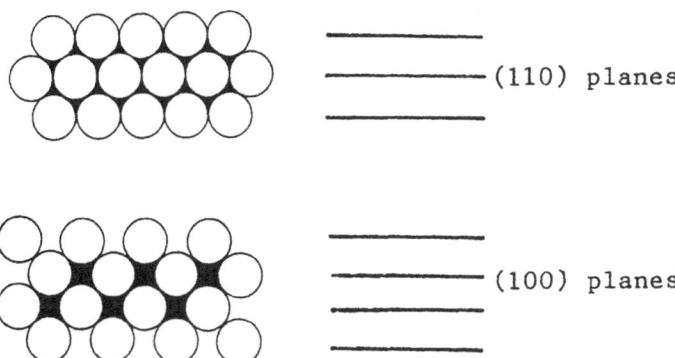

Fig. 3.   Closed-packed planes in diamond-type crystal.

The kinetics of crystal growth from the amorphous phase have been widely investigated. The main conclusions from those investigations are that it is a first order phase transition, as shown in Fig. 2, and G', interpreted as the activation energy of 2.68 eV, is involved in the breaking of bonds (defect formation) and defect diffusion.

Ion irradiation of semiconductors produces a high concentration of vacancies and interstitials through creation of Frenkel pairs well above the equilibrium concentration. If the matrix temperature is sufficiently high for both defect species to be mobile, the excess of point defects can be annihilated through two mechanisms: the recombination of vacancies and interstitials, or the annihilation of both defect types at sinks at the internal boundaries. Of course, the recrystallization process leads with the last reaction as shown in Fig. 4a, where a Frenkel pair is created in the crystal, or as in Fig. 4b, where interstitial atoms (i) are injected by collision and migrate to the vacant site near the interface, either directly or via Frenkel pairs.

For dilute collision cascade (low nuclear stopping power values), the defect production by ion irradiation is well described in terms of formation of isolated defects, either in the amorphous or crystalline regions. For a dense collision cascade (high nuclear stopping power values), the spatial distribution of point defects is not uniform

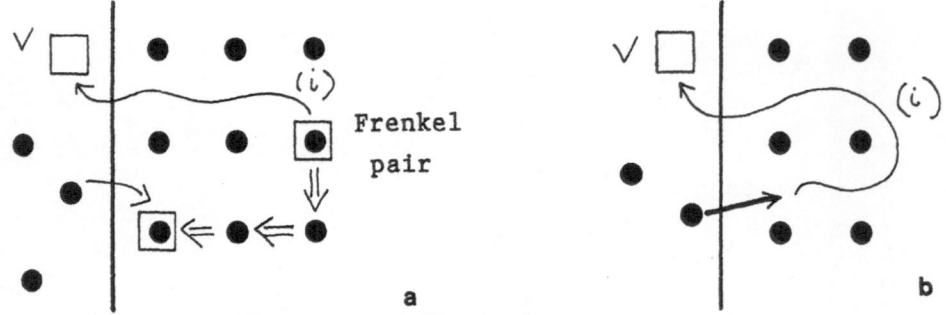

Fig. 4.   Defect motion assisted by Frenkel pairs.

in the matrix and the strong gradient around the ion track cannot be neglected. A simple sequence of events occurs in the solid as a result of the ion bombardment of a crystalline solid. In a very short time (10^{-13} sec), the energy goes from the ion to the target atoms and an energy spike surrounding the ion track leaves a highly disordered region of many broken bonds and displaced atoms (interstitial atoms and vacant lattice sites).

The cross-section of the original disordered core can be evaluated by knowing the energy release rate through atomic collision in the target

$$\sigma = S_n / E_d n_o \,,$$

where S_n is the elastic (or nuclear) energy loss per unit path length (eV/cm); E_d, the energy to displace a target lattice atom (for semiconductor as Ge or Si, $E_d = 14$ ev) and n_o is the atomic number density of the target (atoms / cm^3). Typical values for σ are $10^{-13} - 10^{-15}$ cm^2 and $r_o = 10$–40 Å, the radial dimension of the disordered core. For a better description of atomic displacement distribution within each individual cascade, the recoiling atoms created along the primary ion's path can be evaluated by using a Monte Carlo simulation, as shown in Fig. 5.[11]

Displaced atoms form new bonds and they may stabilize in a "defect" within the collision cascade as a divacancy or with a more complex configuration. Such "defects" escape via thermal diffusion from the disordered core and diffuse in the material to reach the internal interface to drive the recrystallization. To give a temperature range where the proposed mechanism will be effective, we adopt the same model to explain the damage behavior of silicon during ion bombardment. Around each ion track, the disordered core (Fig. 6) of radius r_o is regarded as an amorphous phase at low density from which defects have escaped.

When the diffusion length (δr) is very low compared to the dimension (r_o), the disordered core is quenched as an amorphous zone in the crystal silicon. However, when $\delta r \geq r_o$ the "defects" produced by the incoming ions are not trapped in a stable configuration, but are available to diffuse in the target.

A simple estimate for δr gives

$$\delta r = 2\sqrt{D_v \tau} \,,$$

where D_v is the temperature dependent diffusion coefficient characterizing the process and τ is the "escape" time which is of the order of 10^{-9} sec. The escape time

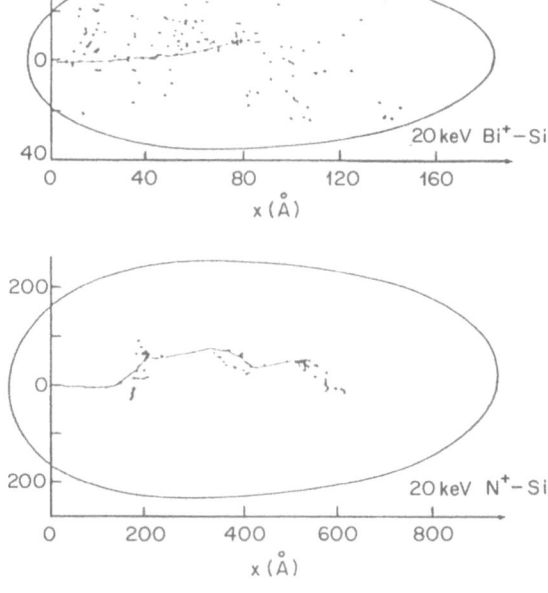

Fig. 5.   Collision cascades for 20 KeV Bi$^+$ and 20 KeV N$^+$ ions in silicon.

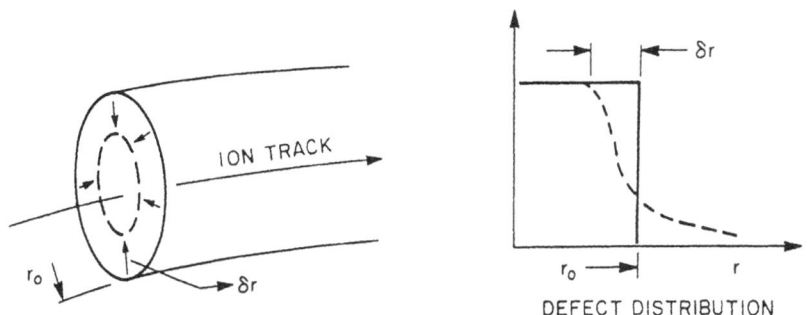

Fig. 6.   Schematic defect distribution around the ion track.

determined by radiation damage production experiments is about four orders of magnitude larger compared to the time for atomic vibration allowing the "defects" to migrate up to large atomic distance.

By comparing $\delta r \cong r_o$, we arrive at a temperature range (300°C-600°C) where the defects induced by ion bombardment are mobile and have an activation energy of $E_{df} = 0.12eV$.[12] In the same temperature range, it has been observed that defect annealing or solid phase epitaxy growth can be induced by irradiation with energetic ions.[2]

In order to complete the discussion, the number of atoms transferred in a collision cascade (ion track displacement) to the matrix will be estimated. Using a modified Kinchin and Pease formula proposed by Sugmund, the number of atoms per ion displaced in a collision cascade is

$$N(E) = 0.84 \, v \, / \, E_d \, ,$$

where ν is the fraction of the ion energy transferred to target atoms in elastic collision and E_d is the displacement threshold already mentioned. For 2 MeV Ar ions, the value $N(E)$ near the surface ($\sim 1\mu m$) is 4 atoms/ion in silicon.

Another interesting quantity is the production rate of Frenkel pairs, given as

$$K = \phi \, N(E) / R \, ,$$

where ϕ is the ion current density ($10^{11} - 10^{13}$ ions / cm^2 sec) and R is the ion range. The ratio $N(E)/R$ is almost constant and for low energy ions equal to

$$N(E) / R \cong 0.5 \, \rho \, \frac{Z_1^{2/3}}{1 + \dfrac{M_2}{M_1}} \, ,$$

where ρ is the target density; Z_1, the atomic number of the ion; and M_1 and M_2 the ion and target mass number, respectively. The knowledge of the production rate of Frenkel pairs allows the average distribution of mobile point defects under ion irradiation to be determined.

If only one defect controls the ion beam assisted process, because, for example, interstitials are fast diffusers, the rate equation in a simplified form is

$$\frac{\partial n_d}{\partial t} = K + D_d \, \frac{\partial n_d^2}{\partial \chi^2} - K_d n_d \, ,$$

where n_d is the defect concentration, D_d is the diffusion coefficient and K_d is the depletion rate. When the diffusion is not important, because the rate is limited by some local bond rearrangement, the equation becomes

$$\frac{\partial n_d}{\partial t} = K - K_d n_d \, ,$$

and the temperature dependence of the process will be included in the term $K_d = K_d^o \exp \left(- E_a / KT \right)$.

ION BEAM ASSISTED EPITAXY REGROWTH

The experimental investigations have demonstrated that the amorphous to crystal transition is not a consequence of macroscopic beam heating of the sample but results more directly from ion-atom interactions in the material. Low temperature ($T = 228^\circ C$) annealing of amorphous germanium (110) has been already observed by measuring "in situ" in the secondary electron emission during 40 KeV Ge ion, at the ion fluence 2×10^{15} ions / cm^2 as shown in Fig. 7.

The sharp transition between the amorphous and crystalline phase is due to the experimental technique being sensitive to only the last 50Å near the surface.[13] Recently, SPEG has been measured in 1000Å thick amorphous layers during 0.6 - 3 MeV Ne$^+$ ion bombardment of silicon. These high-energy ions penetrate to depths of a few microns in the target with a uniform energy deposition rate in the amorphous layer of 1.5 - 5 eV / Å of nuclear energy loss. In this near surface region, the dominant loss of energy is in electronic excitation (~ 100 eV/Å).

Experimental data are reported in Fig. 8, where RBS spectra in the channeling condition have been recorded for samples irradiated with 1.5 MeV Ne$^+$ ions at 318°C with equal ion fluence steps (3×10^{16} / cm^2). From the analysis of the RBS spectra,

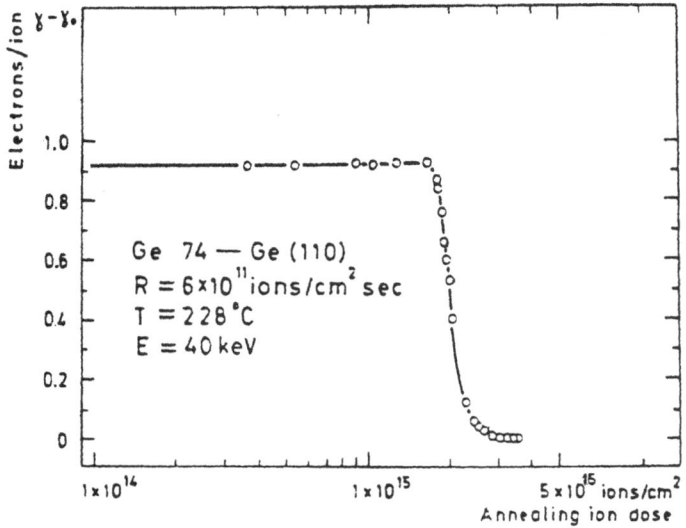

Fig. 7.   Secondary electron emission in Ge bombarded with 40 KeV Ge$^+$ vs. ion fluence.

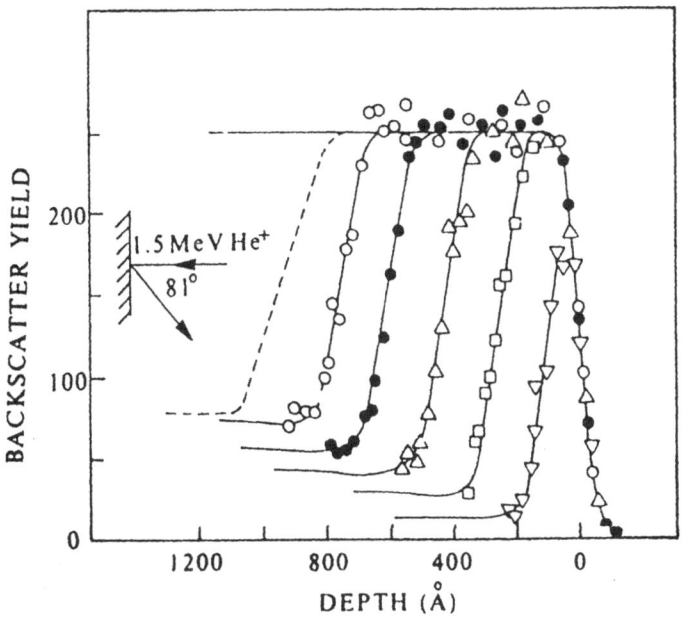

Fig. 8.   Epitaxial regrowth assisted MeV Ne$^+$ beams detected by RBS-channeling spectra along (100) silicon with 1.5 MeV He$^+$.

we get the temperature dependence of beam-induced regrowth, from the Arrhenius plot given in Fig. 9, for different neon-ion energies.

Two distinct beam-induced regrowth regimes are evident: at temperatures below 400°C, the regrowth rate is linear with the ion fluence, and at temperatures above 400°C, the behavior is more complex. In the low temperature regime, ion beam induced epitaxy enhances, by several orders of magnitude, the rate of regrowth when compared to the thermal data and exhibits a single activation energy of 0.24 eV. This

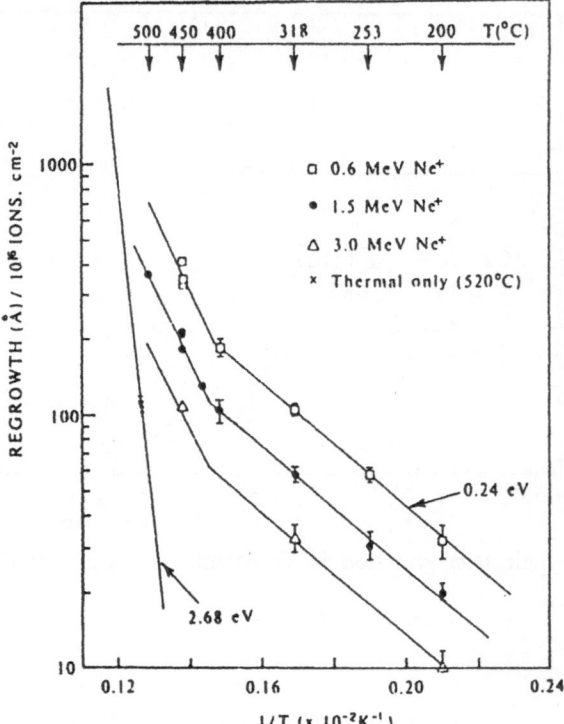

Fig. 9. Arrhenius plot of temperature dependence of regrowth assisted with MeV Ne⁺ beam on (100) silicon.

value is very low compared with the activation energy for thermal regrowth and leads to the conclusion that most of the activation energy measured under pure thermal growth conditions is required for the formation of vacant sites, and only a small fraction for migration of such defects.

The free-energy diagram for atomic jumping through the interface during ion bombardment is modified, as follows:

thermal, $\Delta G' = \Delta G_F + \Delta G_m \cong 2.7$ eV

ion assisted regrowth, $\Delta G' \cong \Delta G_m \cong 0.2$ eV,

where the original $\Delta G'$ is reduced to ΔG_m for defect migration. The regrowth rate observed for Si(100) at 300°C is 100 Å / 10^{16} ions cm^{-2} for 0.6 MeV Ne⁺ irradiation ($S_n = 4.8$ eV/Å).

The cross-section of the disordered region for Ne⁺ ions is about $\sigma \cong 10^{-15}$ cm², which corresponds to a fluence of 10^{15} ions/cm² to uniformly cover a planar interface. If we assume that defects controlling damage in silicon are the same as those controlling regrowth, the value 10^{15} / cm² corresponds to a monolayer step which gives 20 Å / 10^{16} ions cm^{-2} comparable (together with the activation energy) to the measured growth rate of 100Å / 10^{16} ions cm^{-2}. Such values correspond to 2.08×10^{-21}

nm/KeV cm^{-3} in agreement with other data for 600 KeV Kr^{++} − Si(100) as reported in Fig. 10, where the influence of the dopant is also shown for a layer pre-amorphized with Si or P ions.[14]

The adopted ion fluence of P$^+$ implant is 4×10^{15} / cm^2 which corresponds to a concentration of 2×10^{20} / cm^3 at the amorphous-crystal interface. The P-implanted layer has a regrowth rate three times greater than the pure amorphous silicon, but the activation energy is almost unchanged. The thickness, x, (Fig. 11) of the regrown layer increases linearly with v(E) · Fluence.

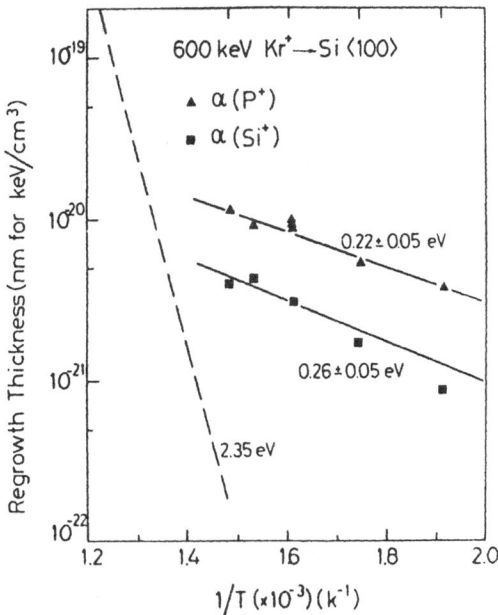

Fig. 10.   Arrhenius plot of temperature dependence of regrowth assisted with 600 KeV Kr$^+$ beam on (100) doped silicon with phosphorus.

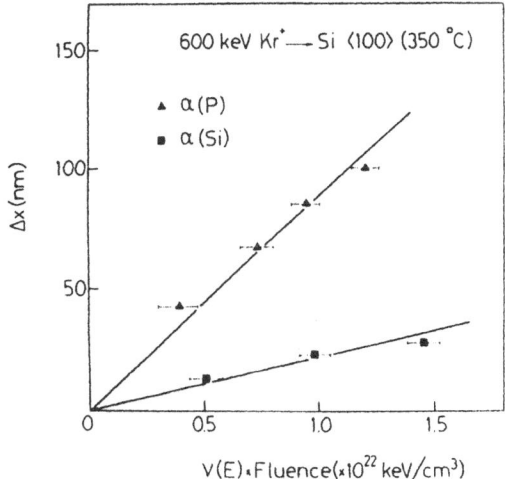

Fig. 11.   Regrown thickness (Δx) vs. deposited energy (KeV/cm^3) where v(E) is the energy deposited into elastic collisions.

In Fig. 12, the behavior in the amorphous layer (Ar^+, Ge^+, P^+, As^+), when the self-ion implanted impurity has been changed, is shown. Here the experimental data are distributed into two sets: one for P and As with high regrowth rate, and the other for Ar and Ge with unchanged rate compared to the self-ion implanted amorphous silicon (Si^+). A weak concentration dependence on growth rate is observed by changing the impurity content of phosphorus from 10^{20}/cm^3 to 3×10^{20}/cm^3, the growth rate increasing only by 25%. Finally, we have to mention that the ion assisted regrowth is faster for less close-packed directions, such as the (100) orientation compared with the (110) or (111) orientations by about a factor of two. This is much less than in the thermal regrowth case, where the variations were a factor four and twenty, respectively.

A systematic comparison between thermal and ion beam assisted regrowth of amorphous layers obtained by ion implantation shows a similar behavior in terms of impurity (or orientation) dependence. The only difference being that in the ion beam assisted case the dependence is weaker. For amorphous layers deposited onto silicon single crystals by CVD or electron gun evaporation, the ion beam assisted regrowth exhibits the unique feature that the layer can be converted to single crystal. This is impossible through thermal annealing.

After deposition of the CVD amorphous layer without any special cleaning procedure, the sample is irradiated with an arsenic beam to extend the amorphous layer about 200Å deeper into the single crystal silicon. The arsenic implantation does not dilute or dissolve the native oxide, as observed by TEM analysis,[15] but it is important to bind the layer to the silicon without any macroscopic cracks or voids at the interface and to remove any crystallization nucleus in the CVD film. A subsequent thermal anneal without any ion beam irradiation, up to 800°C for one hour, causes the epitaxial crystallization of the underlying ion implanted amorphous layer, but the thermal regrowth stops at the initial interface between the CVD deposited layer and the silicon substrate. This high temperature annealing of the CVD layer yields a polycrystalline film as evidenced by electron diffraction. The same sample, without any thermal treatment, exposed to a 600 KeV Kr^{++} ion beam at 450°C shows a very nice epitaxial

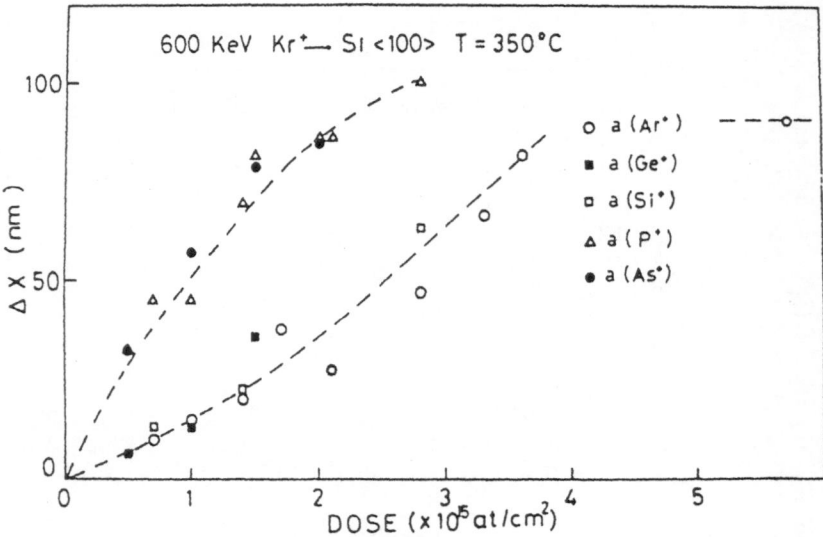

Fig. 12.   Regrown thickness (Δx) vs. ion fluence for ion implanted silicon with several ions.

regrowth for the entire amorphous layer, including the CVD films as shown in Fig. 13.

The regrown thickness, as a function of ion dose, is reported in Fig. 14 for three substrate temperatures; a linear trend is detected only for 450°C Kr$^+$ radiation. The regrowth rate in the amorphous CVD layer is almost a factor of two lower compared to the rate measured in the ion implanted amorphous layer.

The presence of oxygen in the CVD film - crystal interface may explain the observed difference in the regrowth rates because some of the energy deposited by the Kr ion beam breaks the Si-O bonds producing some spreading of the initial SiO$_2$ layer. However, there is always a chemical driving force to reform a stable silicon dioxide, which slows down the regrowth by reducing the mobility of defects in the irradiated sample. The interaction between oxygen and defects during the ion beam irradiation can be indicated by more accurate measurements in the regrown layer as a function of ion fluence, as shown in Fig. 15. Again, the deposited CVD films (800Å) have been ion implanted (1000Å) after deposition.

After the ion assisted regrowth of the first 200Å, the current interface is stopped at the original interface for a well defined amount of ion fluence (incubation fluence ϕ_i). Then, the interface again moves toward the surface with a different growth rate. The ϕ_i-values plotted as functions of temperature give an activation energy of about 0.44 eV for the regrowth, and are very similar to the values observed in radiation-enhanced diffusion in silicon.[16]

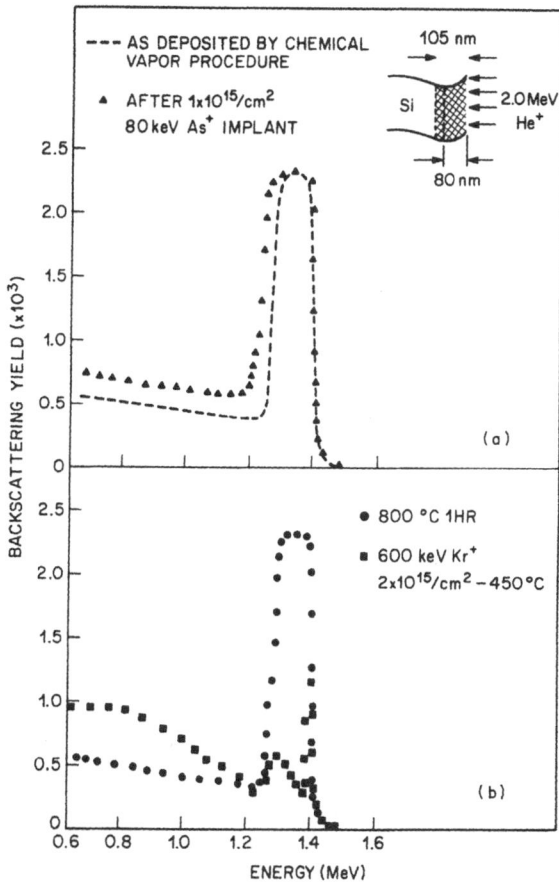

Fig. 13. RBS-Channeling spectra with 2 MeV He$^+$ for CVD deposited films.

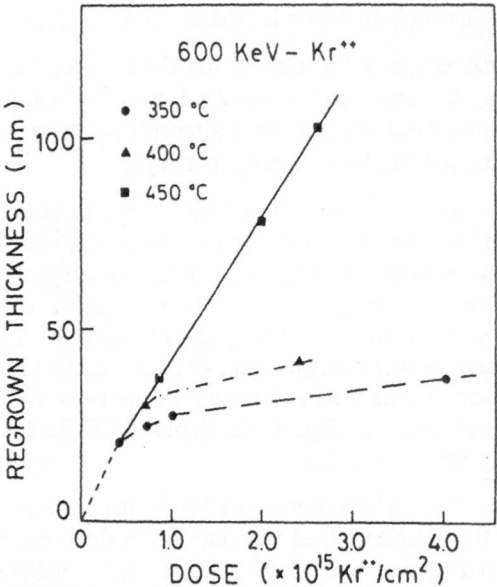

Fig. 14.   RBS-Channeling spectra with 2 MeV He$^+$ for CVD deposited and implanted films after thermal treatment (800°C-1hr) and ion beam irradiation with 600 KeV Kr$^+$ (450°C).

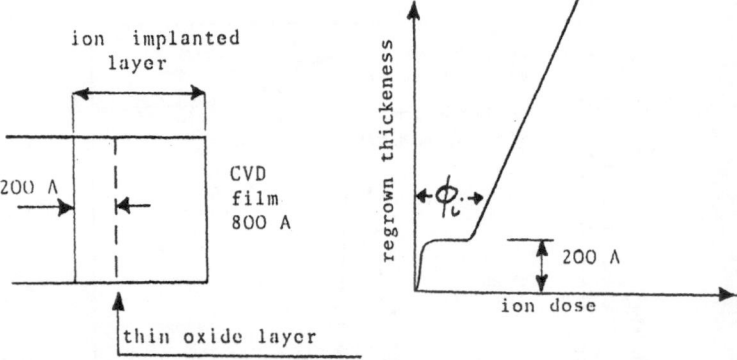

Fig. 15.   Regrown thickness vs. ion fluence for CVD deposited films under ion beam irradiation.

HIGH CURRENT IMPLANT

High-dose implants in the $10^{15} - 10^{16}$ ions/cm^2 range are widely used in microelectronics device fabrication for the drain/source doping of metal oxide semiconductor field-effect devices, or for the emittance of bipolar transistors. Moreover, implanted layers are now used as a predeposition step for diffusion process. To increase the output in many factories, high current implanters, with mA beams and 10^3 watts power intensity, are routinely running.

The large energy release from the ion beam to the wafer during the few minutes of implantation produces a fast temperature rise whose value depends on thermal contact between silicon wafer and the sample holder. As shown in Fig. 16, 400°C degrees can be easily reached where the calculated temperature-dose relationship is given for irradiation with 120 KeV phosphorus and a beam current of 1 mA swept electrostatically over the entire wafer (average density of 9 uA/cm^2). The whole wafer is at the same temperature and a slight difference in magnitude occurs by varying

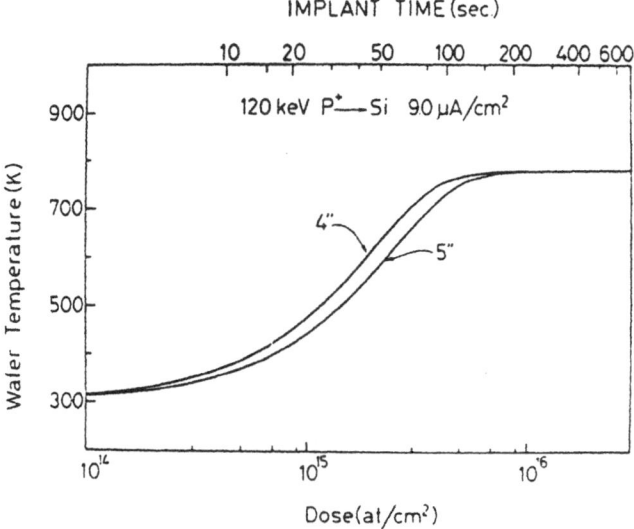

Fig. 16.   Calculated temperature-dose relationship for high current implantation.

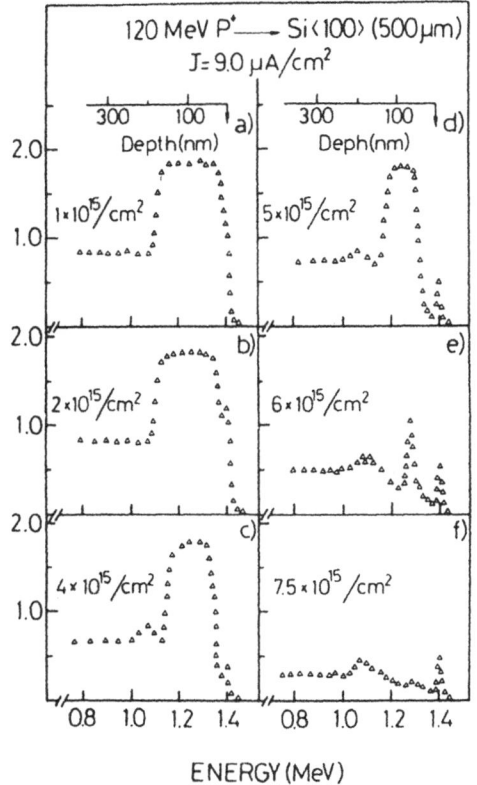

Fig. 17.   RBS-Channeling spectra with 2 MeV He$^+$ for ion implanted silicon with 120 KeV P$^+$ at high current and different ion fluences.

the dimension from 4-in. to 5-in. The implants at high current and dose usually produce extended defects, such as dislocations or microtwins, which vary noticeably from one region to another of the same wafer by slight changes in the implantation conditions, or thermal contact with the sample holder.

For the effects of different ion doses on a 5-in. Silicon (100) wafer (600 μm thick), the results are given in Fig. 17. At low values, an amorphous layer (200 nm thick) forms in the crystal. By increasing the ion dose,[17] the layer shrinks from the surface and the deep interface. Damage is done to the silicon at high dose, when no amorphous layer is present in the sample. The sequence of the experimental data indicates that the annealing takes place only after a certain dose which corresponds to a temperature rise of about 200°C, and it is governed by epitaxial regrowth of two interfaces surrounding the buried amorphous layer. The residual disorder in silicon is due mainly to extended defects (dislocations) which require very high temperature annealing.

The emitter of a bipolar transistor requires a sharp depth profile of the doping impurity to obtain high injection efficiency. One of the conventional approaches in reaching this goal is to implant at high doses a polycrystalline layer deposited on silicon and to use it as an infinite source for doping.

An interesting phenomenon of grain growth has been observed in the polycrystalline layer during the ion bombardment. As was observed for ion beam-induced epitaxy, the grain-boundary motion is proportional to the energy deposited in elastic collisions in the solid. This has been observed in semiconductors (Ge and Si) and metals (Ni, Au, Pt). The rate of grain growth with a constant ion flux at 600°C for 100 nm thick silicon exhibits a variation with time proportional to the square root of time. For metal films, a similar behavior is observed but at very low temperature (23°C).

The activation energy for the temperature dependence of ion beam enhanced grain growth rate for Ge is 0.15 eV. This is lower when compared to the expected value for vacancy migration in Ge. Similar behavior has been exhibited in a polycrystalline silicon layer where an activation energy below 0.1 eV has been found. Both results suggest that self-interstitials, or "defects", migrate essentially athermally during ion irradiation, as does radiation damage production in semiconductors.[6]

Up to now we have been looking at the growth of an amorphous layer on a crystal toward the surface. Another interesting possibility is the lateral growth solid-phase epitaxy (LSPE) of silicon induced by an ion beam. A sample, prepared as shown in Fig. 18, with an amorphous layer in contact laterally with a (100) Si can be ion irradiated with a scanning focused ion beam (FIB) to grow along the (110) direction.[18]

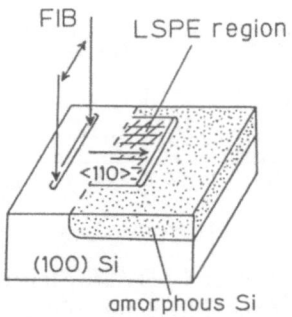

Fig. 18.   Lateral growth assisted with focused ion beam of amorphous layer on (100) silicon substrate.

The crystallization phenomenon can be controlled in three dimensions with the ion energy in depth, and by scanning in X-Y. Because the lateral dimension of the focused beam is below one micron, submicron crystalline islands can be obtained on the silicon surface. The LSPE region extends more than 30 μm from the seed; however, the crystallinity gets worse as the distance from the seed increases, going from crystal to polycrystal-like morphology as evidenced by the etch pit density test.

CONCLUSIONS

Ion implantation has became one of the most widely used processing techniques in the semiconductor industry. High output and small dimensions are the main constraints behind device fabrication causing any process to become more intricate and complex.

High dose rate implantation, ion beam assisted phenomena, and ion beam effect in photoresistant material drove us to understand the basic mechanisms involved in the ion-target interaction and defect behavior under irradiation.

References

1. J. W. Williams, R. G. Elliman, W. L. Brown and T. E. Seidel, *Dominant influence of beam-induced solid phase epitaxial crystallization of amorphous silicon*, Phys. Rev. Lett., **55**, n. 14, 1482 (1985).

2. B. Svensson, J. Linnros and G. Holmen, *Ion beam assisted regrowth*, Nucl. Instr. Methods, **209-210**, 755 (1983).

3. M. Servidori, S. Cannavo, G. Gerla and La Ferla, *Damage created by high current density implants of phosphorus into (100) and (111) silicon wafer*, Appl. Phys., **A44**, 213 (1987).

4. F. Priolo, J. M. Poate, D. C. Jacobson, J. Linnros and J. L. Batstone, *Radiation enhanced diffusion of Au in amorphous Si*, Appl. Phys. Lett., **53**, 1213 (1988).

5. B. Soder, J. Roth and W. Moller, *Anisotropy of ion beam induced self diffusion in pyrolitic graphite*, Phys. Rev. **B, 37**, n. 2, 815 (1988).

6. H. A. Atwater, C. V. Thompson and H. I. Smith, *Interface limited grain boundary motion during ion bombardment*, Phys. Rev. Lett., **60**, n. 2, 112 (1988).

7. J. C. Liu, M. Nastasi and J. W. Mayer, *Ion irradiation induced grain growth in Pd polycrystalline thin film*, J. Appl. Phys., **62**, n. 2, 423 (1987).

8. G. C. Farlow, B. R. Appleton, L. A. Boatner, C. J. McHargue, C. W. White, G. L. Clark and J. E. E. Baglin, in "Ion Beam Processes in Advanced Electronic Material Device Technology", edited by B. R. Appleton, F. H. Eisen and T. W. Sigmon, MRS Symp. Proc., **45**, 137 (1985).

9. L. Calcagno, G. Foti, A. Licciardello and O. Puglisi, *Ion chain interaction in KeV ion beam irradiated polystyrene*, Appl. Phys. Lett., **51**, n. 12, 907 (1987).

10. J. Narayan, *Interface structure during solid phase epitaxial growth in ion implanted semiconductors and a crystallization model*, J. Appl. Phys., **53**, n. 12, 8607 (1982).

11. J. A. Davies, in "Surface Modification and Alloying by Laser, Ion and Electron Beams", Edited by J. M. Poate, G. Foti and D. C. Jacobson, NATO Conf. Series, 8, 189 (1983).

12. F. F. Morehead, Jr. and B. Crowder, *A model for the formation of amorphous Si by ion bombardment*, Rad. Effects, **6**, 27 (1970).

13. G. Holmen, S. Peterstrom and A. Buren, *Radiation enhanced annealing of radiation damage in Ge*, Rad. Effects, **24**, 45 (1975).

14. A. La Ferla, E. Rimini, C. Spinella and G. Ferla, in "Fundamentals of beam-solid interaction and transient thermal processing", Edited by M. J. Aziz, L. E. Rehn and B. Stritzker, MRS Symp. Proc., **100**, 50 (1988).

15. A. La Ferla, E. Rimini and G. Ferla, *Ion induced epitaxial growth of chemical vapor deposited Si layers*, Appl. Phys. Lett., **52**, n. 9, 712 (1988).

16. F. Priolo, A. La Ferla, R. C. Spinella, E. Rimini, G. Ferla and F. Baroetto, *Influence of a thin interfacial oxide layer on the ion beam assisted epitaxial crystallization of deposited Si*, submitted to Appl. Phys. Lett., (1988).

17. S. Cannavo, A. La Ferla, E. Rimini, G. Ferla and L. Gandolfi, *Ion beam annealing during high current density implants of phosphorus into silicon*, J. Appl. Phys., **59**, n. 12, 4038 (1986).

18. T. Kanayama, H. Tanque and M. Komuro, *Laser solid phase epitaxy of Si induced by focused ion beam*, Jap. J. of Appl. Phys., **26**, n. 2, L84 (1987).

MICROMETALLIZATION TECHNOLOGIES

J.M. Martínez-Duart and J.M. Albella

Dept. Física Aplicada (UAM) and Inst, Ciencia de Materiales (CSIC)
Universidad Autónoma, C-12. Cantoblanco, 28049-Madrid, Spain

1. INTRODUCTION

As the semiconductor technology advances towards Ultra Large Scale Integration (ULSI), with device feature in the micron or submicron range, new metallization materials and processes are needed. Specifically, some of the requirements of the materials for ULSI metallization are low resistivity, low contact resistance, fine-line patternability, resistance to electromigration, strong adherence and good step coverage and conformality. Sometimes all of these requirements cannot be optimized simultaneously and one is forced to adopt a compromise.

In order to fabricate an integrated circuit one must make electrical contacts to the devices and connect the individual devices among themselves constituting all these operations the field of metallization. At present, the microelectronic industry has developed a set of standard procedures for metallization of integrated circuits with feature size as low as 1 μm. However, the micrometallization technologies are not as well developed and an intense research effort is continuously in progress. In this paper we summarize the metallization technologies which are already consolidated and also discuss some of the results obtained in different laboratories and which are not yet incorporated as routine fabrication procedures.

This review begins (Section 2) with a short description of metallization technologies, paying special attention to chemical vapor deposition (CVD), including tungsten CVD which allows the possibility of selective deposition. In Section 3 the ohmic contacts to silicon shallow junctions are reviewed. The replacement of the traditional aluminum contacts by several alloys, silicides and refractory metals is treated as well as the salicide (self-aligned silicide) technology. The case of gate contacts is discussed in Section 4 including the case of the silicides deposited on top of polysilicon (polycides) and that of refractory metals. In Section 5 the diffusion barriers are treated. These barriers are used to reduce the diffusion of atoms caused by concentration gradients across multilayered structures. Next we focus our attention in Section 6 on the interconnects since the surface area of an ULSI chip is dominated by them. In this context, the problem of electromigration and corrosion of the interconnects is of paramount significance in the reliability of ULSI devices. Finally we treat multilevel interconnections one of the most demanding technologies in ULSI, including planarization processes and via filling techniques.

2. METALLIZATION TECHNIQUES

Metallization techniques are needed to make ohmic contacts, gate contacts and interconnections in integrated circuits. The two techniques most widely used have been evaporation and sputtering but in recent times chemical vapor deposition (CVD) has also been commercially available. Aluminum, which was at the beginning the most employed material, presents problems of electromigration and step coverage and is now being replaced by other materials such as silicides and refractory metals.

2.1. Evaporation

Evaporation is performed in a high vacuum, i.e., at pressures under 5×10^{-7} torr so that the mean free path of the molecules is much larger than the chamber dimensions. The evaporation is usually performed at a high rate so that a little amount of residual gasses is adsorbed during the film deposition. In order to reduce as much as possible contamination from pump oils and at the same time obtain a high pumping speed the diffusion pumps are being replaced by cryogenic pumps.

There are several methods for evaporating metals. The simplest one uses a resistance heated wire made of a high melting point and low vapor pressure metal, such as tungsten, which is resistively heated. The wire is usually shaped in the form of a helix placing in its interior the metal to be evaporated. Metals like aluminum, gold and palladium can be easily deposited by this method. This technique of evaporation is very simple and inexpensive but has several drawbacks, among them: i) possibility of contamination from the heater; ii) small charges which make difficult to evaporate relatively thick films; iii) refractory metals cannot be deposited by this technique; and iv) difficulty in preserving a given alloy composition in the deposited film.

Electron beam evaporation is based on a flux of electrons which are accelerated by a difference of potential of 10 KV typically and made impinge against the target to be evaporated. Upon the collision the kinetic energy gained by the electrons is transformed into heat. The electron beam can be electromagnetically scanned over the source to get deposited films of uniform thickness since otherwise cavities are formed in the source. The large sources that the electron beam systems use ensures that relatively thick films can be deposited. With various sources multilayer films can be deposited without opening the system to the atmosphere or form alloys by coevaporation. However, electron beam evaporation presents some drawbacks like the depositon of metal droplets if excesive power is applied to the source. Another problem that arises is that for voltages of the order of 10 KV x-rays are generated by the electron beam. The x-rays produced cause the creation of trapped charges in the gate oxide of MOS transistors which can only be removed by an annealing treatment.

Evaporation of metals can also be accomplished by placing the material in a crucible which is heated by radio frequency induction. The crucible is usually made of boron nitride because it reacts very little with metals such as aluminum. One advantage of this method over electron beam evaporation is that no x-rays are generated.

2.2. Sputtering

Sputtering is probably the technique most widely used at present in the deposition of metallic films for integrated circuits. In this technique noble gas ions like Ar^+ are accelerated and made to collide against the target or cathode. The energy of this collision is converted into phonon excitation of the crystal lattice. As a consequence some of the atoms from the surface of the target are dislodged and transported as a vapor to the substrate where they condense. The atoms leaving the target have an average energy of a few electronvolts which is one or two orders of magnitude higher than that of the evaporated atoms.

Some of the advantageous characteristics of sputtering are the following: i) the composition of sputtered alloys is very similar to that of the target; ii) refractory metals like W, Ta and Mo can be easily deposited by sputtering, contrary to what happens in evaporation; iii) the sputtering is usually carried out from large area targets, thus improving the step coverage and the thickness uniformity of the films in relation to evaporation; iv) the adhesion of the deposited films to the substrate is usually good; and v) there are much less x-ray generation than in electron-beam evaporating systems. The requirement of coverage of vertical walls is very important. Some authors have observed that the step coverage can be improved by applying a voltage bias to the substrate since the ion bombardment thus produced increases the surface mobility of the atoms arriving to the substrate. For the same reason, heating of the substrate also enhances the step coverage, specially in low melting materials like aluminum.

The simplest form of sputtering is of the diode type in which the gass discharge takes place between the target or cathode and the substrate or anode. In this configuration, the substrate is part of the electrical circuit and gets heated by the strong electron bombardment. For this reason and specially because of the limited rate of deposition, the diode sputtering is of no practical technical use. If a high deposition rate is needed, then magnetron sputtering must be employed. In this kind of sputtering a magnetic field constrains the movement of the electrons to helical paths, thus enhancing the probability of ionizing collisions. Since the electrons are confined in their trajectories, they do not reach the substrate, which remains at essentially room temperature.

2.3. Chemical Vapor Deposition (CVD)

During the last years, chemical vapor deposition of metals has found a wide acceptance and several commercial systems are at present available. CVD of metals presents several advantages like producing films with excellent step coverage properties. Chemical vapor deposition carried out at low pressures (LPCVD) allows the simultaneous metallization of many substrates and the equipment required is relatively simple as shown in Fig. 1. Another very important characteristic of CVD of metals like tungsten is that the films can be selectively deposited, under certain conditions, i.e., the deposition proceeds in surfaces which are reactive such as silicon and is prevented in passivating surfaces such as silicon oxide.

The technique of CVD is based in the use of volatile compounds, like halides or organometallics, that contain the atoms to be deposited. The compounds are made to react or decompose inside a vessel heated by a furnace. The steps involved in the CVD deposition are the transport of the reacting gasses into the chamber, chemisorption, dissociation and surface migration of the molecules, the chemical reaction itself, and finally the desorption of the by-products and their transportation out of the reaction chamber[1]. Any of these stages can limit the process rate. At first the CVD processes were carried out at atmospheric pressure, but at present most systems operate at low pressures of the order of 1 torr or lower. At these low pressures the uniformity of the deposited films is much better and in addition the wafers can be placed closer to each other, thus increasing the throughput of the system. This is because at low pressures the diffusivity of the reacting gasses is largely increased and the rate limiting step is the surface reaction itself.

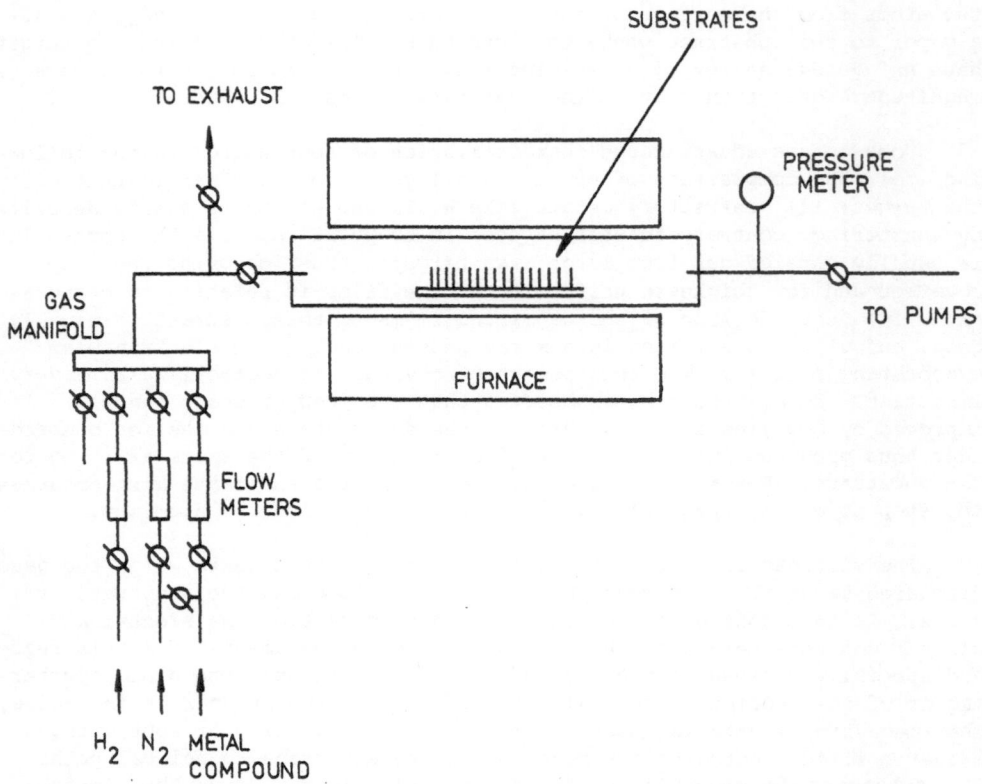

Fig. 1.   Simplified scheme of a low-pressure CVD
          system.

The CVD system shown in Fig. 1 is called a "hot wall" system because
the whole quartz reactor is inside a furnace and is uniformly heated leading
to a high reactor throughput. In the "cold wall" reactors the only zone
which is heated, usually by rf induction heating, is the wafers themselves
and the plate on which they sit, while the walls of the chamber are cooled
by water. In these reactors the film is not deposited on the walls and
therefore there is less utilization of the reactants. The films deposited
do not present a very uniform thickness and the throughput of the systems
is lower than for the hot wall reactors.

Although several metals can be deposited by CVD, the deposition of
tungsten is the most extended one[2,3]. Tungsten is starting to find many
applications as interconnection material, planarized vias, gate electrodes,
diffusion barriers, etc. Tungsten can be deposited from chlorides and
organometallic sources, but the most extended technique utilizes tungsten
hexafluoride[4]. Plasma enhanced CVD of tungsten has also been tried but
produces high resistivity tungsten films that have to be further annealed
at high temperatures. The deposition of tungsten by CVD from WF_6 can proceed by its reduction either by silicon or by hydrogen according to the
reactions.

$$2WF_6 + 3Si \quad \rightarrow \quad 2W + 3SiF_4$$

or

$$WF_6 + 3H_2 \quad \rightarrow \quad W + 6HF$$

respectively.

The selective deposition of tungsten is initiated by the first reaction[2] at temperatures close to 300^0C since the silicon surfaces provide good nucleation sites when they are very clean. This reaction is self-limiting because the W itself acts as a diffusion barrier between the silicon and the WF_6 gass, stopping the reaction when a thickness of about 200 $\overset{\bullet}{A}$ is reached. Then, the reaction proceeds by the hydrogen reduction of WF_6, the thickness of the film being proportional to the square root of the hydrogen partial pressure. This reaction is also selective due to the dissociation of hydrogen molecules by the tungsten surface. On the other hand, tungsten does not nucleate on the silicon oxide unless there are nucleation sites produced by the impurities or if the temperature is high. Very often encroachment and wormholes appear as a consequence of the reaction between Si and WF_6 which extends the deposited W under the SiO_2.

The selectivity of the CVD tungsten deposition is continuously lost as a consequence of increasing partial pressures of the reactants, deposition temperature, deposition time and the relation between the Si and the SiO_2 areas[5]. In the selective or blanket CVD the tungsten gets deposited in all hot surfaces including the silicon oxide. The blanket deposition is employed when the tungsten plays the role of a regular metallization layer which has to be latter patterned. This kind of deposition is very similar to the selective one but carried out at higher temperatures, between 400 and 500^0C. As it is characteristic of the CVD technique the step coverage and the filling of submicron contacts are excellent[6].

3. OHMIC CONTACTS

The advances in the packing density of ULSI demand new metallization materials and techniques requiring metal contacts to shallow junctions, both in bipolar and MOS transistors for ULSI[7]. In the MOS transistors with submicron channel lengths the parasitic resistance of the shallow source and drain and the high contact resistance due to their small size are becoming very deleterious for the proper performance of the devices. The scaling rules makes the contact resistance increase to values between 10 and 100 Ω when the contact size becomes smaller than 1 μm x 1 μm[8]. Kotani et al[9] have studied the contact resistance R_C as a function of the contact size observing that R_C dramatically increases for sizes below 1 μm. At the same time the depth of the junctions of the transistors should be smaller than 0.2 μm. For all these reasons the traditional use of aluminum for contacts to silicon is being replaced by several alloys, silicides and refractory metals.

The ohmic contacts for ULSI should verify several requirements, some of them being that they should be stable to both thermomigration and electromigration effects[10]. In addition the specific resistivity should be lower than about 100 Ω-μm^2. Thermomigration produces the penetration of the metal contact into the junction of a device (Fig. 2) by crossing the metal-semiconductor interfaces, producing the phenomenon known as spiking or pitting. Simultaneously the silicon mixes with the metal (f.i. aluminum). Thermomigration is specially hazardous in ULSI devices in which the junctions are as shallow as 0.1 μm. Contact electromigration is a similar phenomenon but the atoms are transported because of the momentum of the charge carriers and therefore the atoms move in the same direction as the electrons. As a result voids are finally formed close to the metal-semiconductor interface, making the contact to fail. Electromigration is an specially grave problem in ULSI bipolar transistors where the current densities at the contacts are very large. Therefore it is necessary that the metallurgical properties of the contacts should remain stable during the oper-

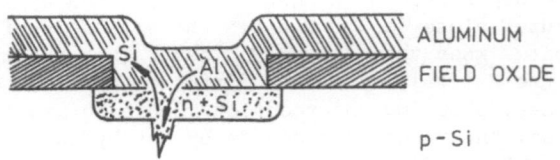

ALUMINUM

FIELD OXIDE

p-Si

Fig. 2. Short-circuited shallow junction caused by aluminum
spiking. Thermomigration produces the penetration
of the aluminum down to the junction.

ation of the devices for long periods of time. For this purpose several
annealing steps are implemented during the fabrication cycle.

3.1. Aluminum Contacts

At the beginning aluminum was the most widely used metal for contacts
in bipolar and MOS transistors integrated circuits. The advantages of aluminum resided in the fact that it has a high conductivity and forms low
resistance ohmic contacts to p^+ and n^+ silicon (the contact resistance is
about $10^2 \, \Omega \mu m^2$). The excellent contact properties to p-silicon are due
to the fact that aluminum is an acceptor impurity in silicon. The ohmic
contact to n^+ silicon is explained by the fact that the silicon surface
is so much doped that the carrier transport across the junction is by tunnelling. Another advantage of aluminum is that adheres well to silicon
and silicon oxide and it can also be used as interconnect material.

After depositing the aluminum onto the silicon substrate, an annealing
is carried out at temperatures of about 500^0C during some 30 minutes in an
inert atmosphere. During this sintering process the aluminum reacts with
the native oxide in silicon. This native oxide has a thickness of only
6 Å for a freshly etched Si surface but can be about 20 Å thick after one
day in contact with the ambient. Aluminum reacts with silicon oxide according to

$$4Al \; + \; 3SiO_2 \quad \rightarrow \quad 2Al_2O_3 + 3Si$$

and this allows to make an intimate contact between the metal and the
semiconductor. The Al_2O_3 molecules produced are thereafter dissolved
in the metal. In addition this reaction is self-limiting ensuring that the
aluminum interconnects on the silicon oxide remain stable.

Pitting of aluminum contacts to silicon has been observed frequently
producing junction spiking in shallow contacts (Fig. 2). This is because
the silicon diffuses into the aluminum until the solid solubility limit
is reached at the annealing temperature. The voids left by the silicon
are replaced by aluminum atoms producing a short between the base and the
emiter in shallow junctions of bipolar transistors.

3.2. Al-Si Alloy Contacts and Other Alloys

The utilization of Al-Si alloys was proposed almost two decades ago
to alleviate the problem of junction pitting. The solid solubility of
silicon in aluminum at the contact sintering temperature is close to 1%

in weight. The silicon atoms migrate in the aluminum over distances as much as 20 μm because the activation energy for silicon diffusion in aluminum is only 0.95 eV, which is lower than the activation energy for self diffusion of aluminum. Studies of Al-Si contacts to silicon carried out by Faith et al[11] have shown that the best values of the contact resistance were almost independent of the Si content in the Al-Si alloy. Therefore one should use the minimum amount of Si to reach the solubility limit so that the alloy itself has the lowest resistivity since, as one should expect, the resistivity of the alloy increases with the Si content.

The use of Al-Cu-Si alloys improves the contact properties as much as the Al-Si alloys and has the additional advantage that the copper doped aluminun interconnects are much more resistant to electromigration than the aluminum interconnects. It has been observed that a small addition of copper inhibits void and hillock formation in aluminum lines, substantially improving the mean time to failure. This is because the hillocks are formed by the relief of the compressive stresses that arise in large grain aluminum lines. By adding copper atoms the grain size is substantially reduced. The use of the Al-Cu-Si alloy with Cu concentrations between 1 and 4 % in contacts has been fairly extended, but problems arise when the contact size reaches the submicron range because the contact resistance can be as much as 100 Ω or higher, thus degrading the device performance.

In addition to the problems already mentioned, the Al alloy films are difficult to produce by CVD methods[12]. In fact, the resulting Al films deposited by CVD show a large surface roughness which introduces new difficulties in the subsequent photolithographic processes. Also, the Al alloy films deposited by evaporation or sputtering present a rather poor conformal step coverage. If the Al alloy films are deposited by electron-beam evaporation the properties of the Si-SiO$_2$ interface can be altered by induced radiation. Therefore it is frequently necessary to subject the samples to additional annealing steps at temperatures which are high for the alloy aluminum contacts to keep their integrity.

3.3. Silicide Contacts

Metal silicides, specially noble metal silicides like PtSi and Pd$_2$Si, are specially suitable to make shallow contacts to silicon and do not lead to junction spiking, as it happens with the Al-Si contacts. To form the silicide contacts the metal is deposited over the silicon windows opened through the silicon oxide by chemical etching. After the sintering to form the silicide on the silicon surface, the metal can be etched from the oxide surface by a selective etchant that removes the metal without attacking the silicide or the SiO$_2$.

Perhaps one of the most important achievements in the IC technology is the salicide (self-aligned silicide) process which serves to reduce the number of masking steps and avoids many of the drawbacks caused by the transfer of patterns due to the difference between the mask dimensions and the final devices. In this process a metal layer is deposited over a MOS structure and is made to react with the exposed Si areas of the source/drain and poly-Si in the gate to form the silicide. The process is illustrated in Fig. 3. In the salicide technology the polysilicon gate is patterned before the silicide formation. An oxide layer is deposited by CVD and anisotropically etched down leaving insulating sidewall spacers besides the poly-Si gate. These spacers have two purposes; they serve as masks during the ion implantation of source/drain areas and also prevent the gate and source/drain regions from being electrically short-circuited. The metal to form the salicide is then deposited and reacted with the silicon. The unreacted metal is etched away by using a selective etch

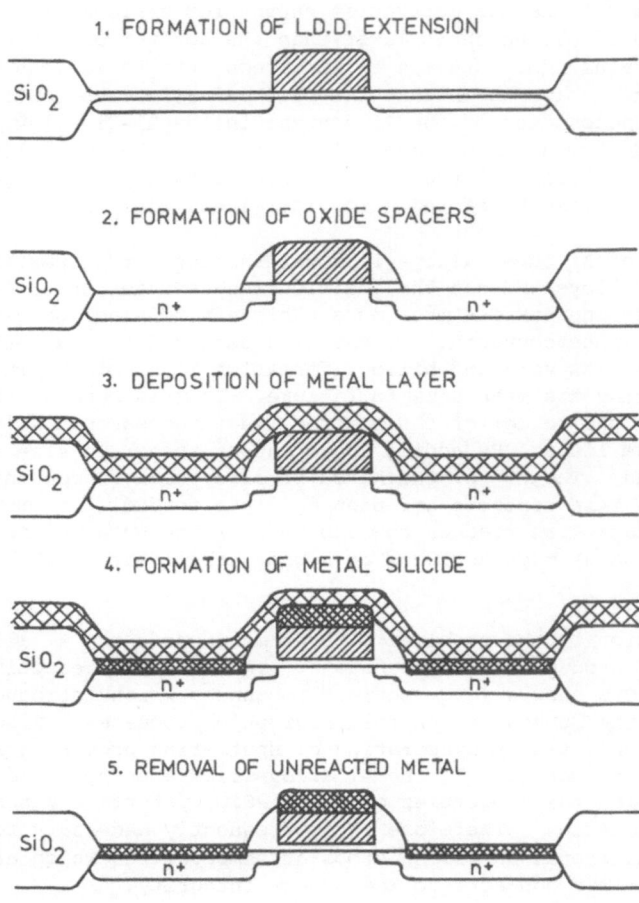

Fig. 3. Salicide processing sequence (from ref. 5).

that does not react with the silicide. Finally, the device is glass passivated and contacts are made. With this technology a lower resistance between the silicide and the silicon is achieved because the area of the interface is much larger than in the case of a conventional metal-silicon contact structure.

a) Noble-metal silicides

The contact pitting mentioned above for the Al-Si contacts can be avoided by using silicides. The silicides most often used are the near-noble silicides ($PtSi$, Pd_2Si, $NiSi_2$) and the refractory metal silicides ($TiSi$, $TaSi_2$, $MoSi_2$, WSi_2). The near-noble silicides are usually formed at low temperatures (below 500^0C) while the refractory ones need temperatures above 600^0C. The silicides for contacts are either formed as a result of an in situ compound formation between the metal and the silicon or from a sintering reaction after codeposition of the metal and silicon in appropriate proportions. The excellent properties of the silicides for contact formation are due to several reasons, among them their metallic

nature, low stresses in contact to silicon and good oxidation properties
to create passivation layers.

Platinum silicide contacts have been very much used for the formation
of ohmic contacts to silicon[13]. The silicide is formed by evaporating
platinum films with a thickness between 200-500 Å onto silicon substrates
and annealing at temperatures between 500 and 600°C in an ambient of H_2,
N_2 or in forming gas. Previous to the deposition is very important to clean
the silicon surface, by in situ sputter-cleaning for instance. During the
annealing, the platinum first reacts with silicon to form the unstable Pt_2Si
at temperatures of about 200°C by the interdiffusion of platinum and silicon[14]. After the platinum is consumed the reaction progresses to form
the PtSi phase at the Pt_2Si/Si interface until the Pt_2Si is consumed[15].
Even minute amounts of oxygen during the evaporation of platinum partially
inhibits the formation of the platinum silicide[16]. Also the existance of
an oxide layer, about 50 Å or thicker, in the contact windows localizes
the silicon-platinum reactions at the discontinuities of the layer, resulting in rough and bad defined contacts. However if the conditions of clean-
liness are satisfied one can produce circuits in the micron range on arsenic-
implanted silicon surfaces with contact specific resistance as low as
$0.5 \times 10^{-7} \Omega \, cm^2$. The resulting resistances are very low because the silicide contacts are intimate contacts. But other factors also contribute,
one of them being the so-called snow plow effect consisting in an enhancement of the dopant concentration at the silicide-silicon interface due to
the lower solubility of the dopants in the silicide than in silicon. This
results in a high electric field at the interface which facilitates carrier
tunnelling across it.

The studies of palladium as an ohmic contact to silicon are much more
scarce than for platinum. Because of the large affinity that Pd has for
hydrogen the sintering is usually performed in argon or nitrogen atmospheres. The reaction between silicon and palladiun at temperatures between
225-275°C produces Pd_2Si the activation energy being 1.1 eV[17], although
other phases such as Pd_9Si_2, Pd_3Si, Pd_9Si_4 and PdSi are known to exist[18].
Palladium, as in the case of platinum, also exhibits arsenic snow plow
effects during the silicide formation resulting in good ohmic contacts to
n^+ silicon. The contact to p-type silicon is also good due to the low
barrier height of the contact.

b) Refractory-metal silicides

Notorius aspects of the refractory metal silicides are their low electrical resistivity, their stability under heat treatments and their very
low electromigration properties. Among them the most popular are $TiSi_2$,
$TaSi_2$, WSi_2 and $MoSi_2$.

The titanium silicide is the one with the lowest resistivity and its
use is not only restricted to the formation of ohmic contacts but also as
interconnect and gate contact material. The specific contact resistance
to arsenic doped junctions[19] is as low as 100 $\Omega \, \mu m^2$. When aluminum is used
as interconnect material in Si/$TiSi_2$/Al structures there might be mutual
diffusion of silicon and aluminum which can be avoided by the use of TiN
diffusion barriers[20] (Section 5). $TiSi_2$ is the most accepted material for
the self-aligned silicidation technologies[21,22]. This technique improves
the sheet resistance of the contacts and allows the possibility of higher
packing density in integrated circuits. The contact and shallow junction
characteristics in submicron CMOS with self-aligned $TiSi_2$ has been studied
in detail[23] for different titanium thickness and shallow source-drain profiles. The titanium consumes a layer of silicon as well as its dopants if
there are no snowplow effects, lowering the doping concentration at the
silicide/silicon interface. Therefore in order to maintain a low contact

resistance the titanium layer should be very thin. Fig. 4 shows the measured $TiSi_2/p^+Si$ contact resistance as a function of implanted boron dose for a contact with an evaporated titanium layer of 350 Å. The contact resistance decreases with increasing boron dose until a dose of about $2 \times 10^{15}/cm^2$. Higher doses will produce a deeper junction due to concentration-enhanced diffusion effects, but will not result in a significantly lower contact resistance.

Tantalum silicide can also be used to make contacts to shallow diffusion regions in silicon[24,25]. In addition a thin $TaSi_x$ layer under the aluminum metallization acts as a barrier against aluminum and silicon diffusion at the contacts. Neppl et al found that the tantalum silicide with the specific composition $TaSi_{1.85}$ was the most stable silicide for its use in contacts for integrated circuits[24]. As standard metallization they also preferred the alloy Al-1.2% Si-0.15% Ti over the more common Al because they found that titanium enhanced the electromigration resistance of the interconnects by one order of magnitude. Putting an intermediate tantalum silicide ($TaSi_{1.85}$) layer between the Al-Si-Ti metallization and n^+ type silicon improves the contact resistance R_c by a least a factor of three making also R_c less dependent on the contact width. The contact improvement was attributed to the lower barrier height and to the prevention of formation of p-Si interface layers. It has also been proved that the $TaSi_{1.85}$ diffusion barrier retards the junction leakage failures and the thermomigration of silicon by a least one order of magnitude[24]. A very interesting recent development is that the tantalun silicide can be selectively deposited by CVD[25] (Section 2.3). This technique has the advantage over the normally used self-aligned technique of consuming no silicon and therefore is very appropiate for shallow junctions in ULSI.

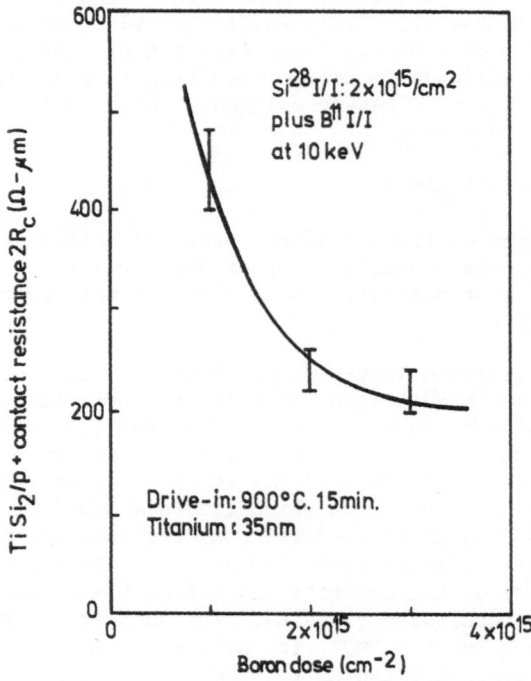

Fig. 4. $TiSi_2/p^+Si$ contact resistance as a function of implanted boron dose (from ref. 23).

c) Noble-refractory metal silicides

A method of making shallow silicide contacts and their diffusion barriers simultaneously has been reported consisting in the reaction of an alloy of noble metals and refractory metals to silicon[26,27]. As we have seen the noble metal silicides form at about 300^0C mainly by the diffusion of the metal atoms while the refractory metals form the silicides at temperatures above 600^0C by silicon out diffusion. Therefore the reaction leads to a two layer structure. The inner layer formed by the noble metal silicide makes the contact to the Si surface while the refractory metal or its silicide, depending on the temperature, constitutes the second or diffusion layer. As an example consider the alloy Pd_2W_8 with a thickness of 1000 Å deposited on silicon. If the system is made to react about 500^0C, Pd_2Si would form consuming no more than 100 Å of Si from the substrate, remaining the W as a diffusion barrier.

3.4. Refactory Metal Contacts

Refractory metals were not considered until recently because they have a higher electrical resistivity than aluminum and show difficulties in making intimate contacts to silicon because of their low reactivity with SiO_2. On the other hand, refractory metal contacts to silicon are much more stable at high temperatures than the Si-Al contact. Some of the refractory metals, such as tungsten, can be selectively deposited on silicon by CVD techniques as we have seen in Section 2.3.

The contact properties of tungsten to silicon have been studied by several authors. S. Swirhun et al[28] measured the contact resistances of Al/W/Si contact structures, with the tungsten being selectively deposited, as a function of the surface doping concentration. The specific contact resistance of tungsten to n^+Si is about a factor of 20 lower than that of aluminum for phosphorus surface concentrations of $10^{20}cm^{-3}$. However the specific contact resistance of W to p+ Si is of the same order of magnitude than that of aluminum. Recently, Kotani et al[9] have studied the contact resistance to p+ and n+ diffusion layers for selectively deposited tungsten as a function of the contact size and have shown that for submicron contact sizes the resistance is much lower than in the case of Al-Si contacts. These authors deposited the W selectively by the reaction of WF_6 with SiF_4 and to demonstrate that this technology is compatible with ULSI processes they applied it to the production of CMOS 1 Mbit DRAM's.

As we shall see later tungsten-titanium alloys have been tested as diffusion barriers and also as direct contacts to silicon. The alloy is ussualy deposited by sputtering[29,30] in order to get a higher Ti concentration than that dictated by thermodynamics . The excess Ti serves to reduce the silicon native oxide thus providing an intimate contact. However direct silicidation of Ti should be avoided to impede the outdiffusion of silicon and hence an increase in contact resistance. Kim et al have observed a contact resistance minimum at an annealing temperature around 625^0C, a temperature sufficient to dissolve the native oxide on silicon and not high enough for the Ti and W reaction with silicon which would increase the sheet resistance and contact resistance[29].

Molybdenum is another refractory metal candidate to make direct contacts to silicon because of its good adherence, low electromigration properties and lack of reaction with silicon at temperatures below 700^0C thus avoiding the risks of contact pitting. This contact has been studied by Mochizuki et al[31] on n+Si substrates without specifying the dopant concentration. The values reported for the specific contact resistance were very low, between 70 and 100 $\Omega\mu m^2$, but this value increases markedly for sintering treatments above 700^0C probably due to the formation of molybdenum silicide. Cohen et al[10] also investigated the contact of molybdenum to n-and p-type implanted Si substrates with a concentration of about $10^{20} cm^{-3}$. The sintering was performed at temperatures below 700^0C to avoid the formation of the silicide. In all cases the contact specific resistance was below 100 $\Omega \mu m^2$ showing that the Mo contacts can be appropriate for use in ULSI technology.

4. GATE CONTACTS

The first MOS transistors made in the 1960s used aluminum as the gate electrode because at the time aluminum was employed in the metallizations for the more popular bipolar transistors. Some years later the aluminum gate was replaced by polysilicon in order to decrease the threshold voltage of MOS transistors and specially because the polysilicon can withstand the high temperature processes needed for the fabrication of integrated circuits. Other advantages are: i) the polysilicon is chemically stable in contact to silicon and to silicon oxide; ii) the poly-Si allows the possibility of self-alignment since the deposition of aluminum has to be performed after the high-temperature processes have been implemented; iii) the breakdown properties of the SiO_2 do not degrade in contact with the polysilicon; and iv) the polysilicon forms ohmic contacts to metals.

4.1. Polysilicon Gates

Polysilicon films are usually deposited by the pyrolysis of silane in a low-pressure reactor at temperatures between 600 and 650^0C. It can also be deposited by the decomposition of silicon tetrachloride at about 900^0C. The electrical conductivity of the poly-Si films is increased by doping with gases such as phosphine, arsine or diborane[32]. The resistivity of the undoped polysilicon films ($\sim 500 \Omega cm$) is reduced by the doping down to about 0.01 Ω cm. The dopant species influence strongly the deposition rate. The diborane (B_2H_6) produces a large increase in the deposition rate, but in contrast the PH_3 and AsH_3 greatly reduce the deposition rate[33].

The polysilicon can be doped by diffusion and ion implantation. Doping by phosphorus is usually carried out at temperatures close to 1000^0C, with excess dopant segregating at the grain boundaries. Silicon doped by P implantation shows a resistivity about one order of magnitude higher than diffused silicon, even for heavy implantations with about 10^{20} phosphorus atoms per cm^3. The polysilicon deposited below 575^0C is amorphous while that deposited around 625^0C presents a columnar structure with (110) preferred orientation[34]. Grain growth and further crystallization take place under heat treatments at about 700^0C.

4.2 Refractory Metal Silicide Gates

One of the major disadvantages of polysilicon as contact and interconnect material is its high resistivity. As a consequence, when the lateral dimensions of the devices are reduced the sheet resistance of the polysilicon interconnect lines becomes too large. The length of these lines influences the RC time constant of the propagating signal. Fig. 5 shows the RC time constant for poly-Si on SiO_2 which is too high when high speeds are needed. Therefore it becomes necessary to substitute the polysilicon for other materials such as the silicides of refractory metals ($TaSi_2$, WSi_2, $TiSi_2$). For instance, if poly-Si is replaced by $TaSi_2$ about one order of magnitude improvement in the RC time constant is obtained (Fig. 5). From the point of view of gate contact and interconnect material, the titanium silicide is very convenient because it presents a very low resistivity and a Schottky barrier height of 0.6 eV to n-type Si.

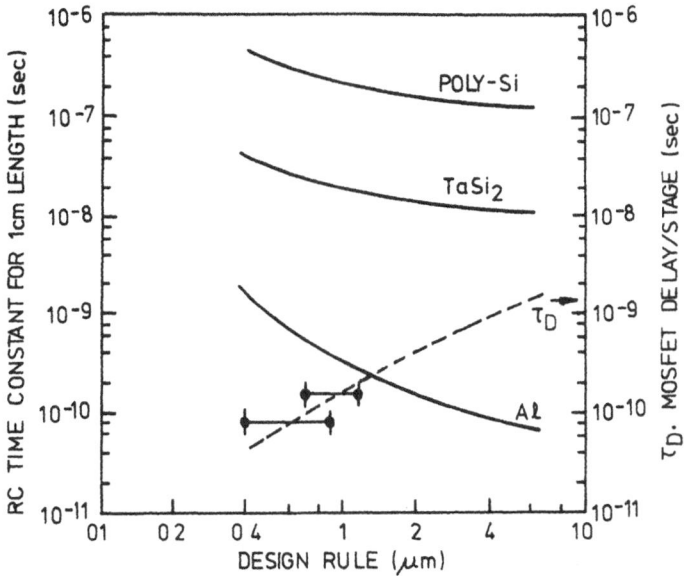

Fig. 5. RC time constant of poly-Si, TaSi$_2$/poly Si
and Al deposited on SiO$_2$. The MOSFET delay
per stage is also shown (from ref. 35).

The application of refractory silicides to gate metallizations and
interconnects has been the subject of numerous review articles and a
book[35-38]. The refractory silicides can be formed by a variety of techniques: (i) by sintering a thin metal film deposited on silicon, as it was
done with the noble metal silicides; (ii) sputtering of a composite silicide target; (iii) cosputtering from a metal and a silicon target; (iv)
coevaporation from two targets as before; and (v) chemical vapor deposition
by the reduction of halide compounds such as fluorides or chlorides. Probably the most frequently used technique is cosputtering from metal and silicon
targets because the silicide produced is purer than in the case of sputtering
from a compound target and also allows the control of the stoichiometry by
changing the deposition rate of each element. When the silicide is formed
by a codeposition technique such as cosputtering it is observed that the
resistivity of the as-deposited material increases with the silicon to metal
ratio as can be appreciated in Fig. 6(A) for TaSi$_x$[39]. The resistivity of
the as-deposited silicides is fairly large and in order to diminish it an
annealing step around 800-900^{0}C is needed as shown in Fig. 6(B). For
tantalum silicide a minimum in the sheet resistance is associated to the
complete transformation of the film into hexagonal TaSi$_2$ structure as a
consequence of the annealing treatment.

Lately the CVD technique is increasingly used for silicide formation
because of the good step coverage properties and the very low oxygen contents of the deposited films. The starting gases are silane and the fluorides or chlorides of the corresponding metals. The reactions to produce
titanium and tungsten silicides are respectively:

$$TiCl_4 + 2\ SiH_4 \rightarrow TiSi_2 + 4\ HCl + 2\ H_2$$

$$WF_6 + 2\ SiH_4 \rightarrow WSi_2 + 6\ HF + H_2$$

These reactions are performed at elevated temperatures. An important aspect
of the titanium silicide is that in addition of having the lowest resistiv-

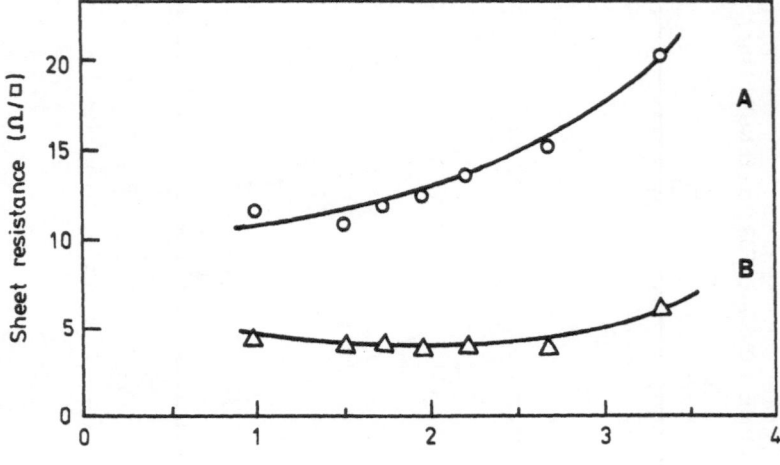

Fig. 6. Sheet resistance of Si-Ta films vs ratio
of Si/Ta atoms for as-deposited films
(curve A) and after annealing at 950°C
(curve B) (from ref. 39).

ity, it has recently been demonstrated that it can be deposited selectively,
i.e., it only deposits on the exposed silicon or polysilicon surfaces but
not on the neighboring silicon oxide[40].

To take advantage of both the excellent properties of the Si-SiO$_2$
interface and the high electrical conductivity of the silicides, a new
structure called "polycide" has been developed in which a polysilicon pad
is deposited between the SiO$_2$ and the silicide layer. In this structure
evidently the sheet resistance of the interconnect lines is controlled by
the silicide film[41].

The implementation of silicides for gate contacts has resulted in
different silicide gate structures. In the simplest one the poly-Si gate
is substituted by the silicide but the trouble with this structure is that
when the surface is oxidized for passivation the silicide film is partially
consumed leading to reliability problems and a high sheet resistance. Therefore, the polycide structure described above is the most commonly used.
Another gate structure uses a refractory metal, usually molybdenum which
is passivated by growing a silicide layer by reacting it with silane[42].
Finally in the salicide structure the poly-Si is reacted with a metal to
form a silicide overlayer. The advantage of this self-aligned structure
is that the silicidation of the shallow junctions of source and drain and
the polysilicon gate proceed simultaneously.

If the refractory metal silicides are to be used in the fabrication
of integrated circuits the necesary processing steps have to be developed.
Proper patternability by plasma etching techniques in CF$_4$/O$_2$ mixtures and
NF$_3$ has been demonstrated[43]. Also the effect of ion implantation has to
be considered since the gate silicide is often used as an implantation
mask. As the technological problems are being solved, short-channel devices
have been fabricated[31,41]. In this respect methods have been described
for the fabrication of submicron self-aligned TiSi$_2$ CMOS. The produced
silicided transistors show less degradation and contact resistance than the
non silicided transistors . In order to check the process stability of

the silicides, a magnitude as the sheet resistance is used. Fig. 7 shows
this parameter as a function of typical processing steeps demonstrating
the stability of the silicides.

A most desirable property of the transition metal silicides is the
possibility of forming a good quality oxide for passivation and electrical
insulation in multilevel metallizations[35]. Practically all the research
has been centered on the thermal oxidation of silicides usually in an steam
atmosphere[45]. In this kind of oxidation a layer of SiO_2 grows on top of the
silicide layer. The silicon atoms are mostly supplied by the silicon substrate, and not by the silicide, specially when the oxidation is carried out
at temperatures above 900^0C. If this supply is limited, as in the case of
the polycide straucture, the integrity of the structure might hamper rapidly.
Many of these unwanted effects can be avoided by anodic oxidation at low
temperatures. The anodization can be carried out either in a liquid electro-
lyte[46] or in a plasma[47]. In the case of $TaSi_2$ the oxide consits of a mixture
of Ta_2O_5 and SiO_2 with a high-surface enrichment of Ta_2O_5 and therefore
there is no need of a large supply of silicon. The oxide films produced
show good quality insulating properties as determined by capacitance, electrical conductivity and dielectric strength measurements.

4.3. Metal Gates

With the scaling down of integrated ciruits, the gate resistance of
the silicide contacts in MOS devices with access times of the order of
nanoseconds may be too large and one is forced to use metal gates. This
is because the gate resistance raises rapidly as the feature size diminishes
down to 1 μm for one megabit memories.

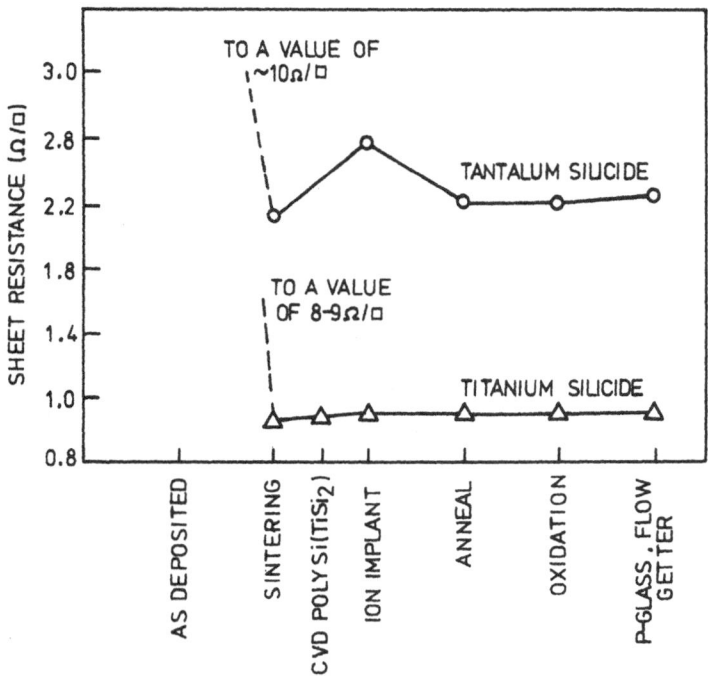

Fig. 7. Sheet resistance of Ti and Ta silicides on
poly-Si as a function of typical MOS processing
steps (from ref. 35).

As a metal, aluminum presents many drawbacks because of its vulnerability to high process temperatures and to electromigration. Of all the metals, only tungsten and molybdenum are the ones that have been seriously considered as gates for submicron technology. These refractory metals show adequate properties for MOS device fabrication, such as adhesion to silicon oxide and fine-line patternability. The main advantage of using refractory metal gates over silicide gates resides in the better electrical conductivity of the former[48]. Another advantages of W and Mo is that they react little with SiO_2 and do not present special problems related to oxide degradation, breakdown, instabilities in the threshold voltage, etc. In addition W and Mo show a good electromigration resistance, can be appropriately deposited by CVD techniques and can also be used for ohmic contacts to silicon. However one problem presented by the refractory metals is that they cannot be subjected to high temperatures in oxidizing ambients because their oxides are volatile.

W and Mo allow to use the self-aligned gate process if some precautions are taken to stop the implanted ions through the thickness of the gate. If the films are not thick enough they can be nitrided to avoid the channeling of the dopant As^+ for the n-channel technology. The W films are usually covered by a phosphosilicate glass (PSG) of about 500 Å thick to impede the channeling. Another possibility consists of growing a thin anodic film on W[49].

One megabit DRAM memories have been fabricated with tungsten[50] and molybdenum gates, the technologies employed being compatible with the conventional polysilicon processes. The technique for preventing channeling consisted in the deposition of a thin PSG film on W[50]. The MOS structures fabricated with refractory metal gates show a fixed surface charge density, a surface state density, a concentration of mobile ions in the gate and breakdown voltages of the oxide similar, if not better, than for the conventional poly Si gate structures.

5. BARRIER LAYERS

When a metal film is deposited on top of a substrate or on another metal layer there is a reaction to a more o lesser extent between both adjacent materials. This reaction usually takes place in the form of a diffusion of one of the metals into the other to give, in some cases, a new compound. The diffusion rate depends on many factors, not only on the nature of the metals and their temperature, but also on the aggregation state (crystallinity, presence of defects, grain size, etc). This makes the film deposition parameters, namely, substrate temperature, residual atmosphere, type of sputtering, presence of bias, etc to strongly influence the diffusion characteristics of the metal overlayers. Therefore, the presence of a diffusion barrier is an essential requirement where metallization layers in device processing intermix with the underlaying substrate.

5.1. Types of Barriers

According to the scheme proposed by Nicolet[51], the barriers can be classified in three kinds: i) "Passive barriers", characterized by their complete chemical inertness against the adjacent layers as well as negligible mutual solubility and diffusivity for the contacting materials (Fig. 8a). Compound materials such as metallic carbides and nitrides constitute the best approaches to the passive barriers due to their low resistivity, high melting point and hardness, good chemical stability and, above all, compatibility with the most common processes for the VLSI devices. ii) In a lower degree of stability are considered the so-called "sacrificial barriers" which are based in a uniform and well controlled reaction rate with

a	b	c
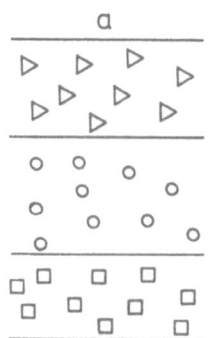	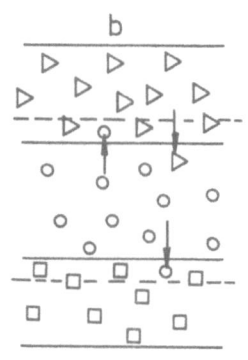	

Fig. 8.   Scheme of the different diffusion barriers
according to Nicolet[51]   : a) Passive barrier;
b) Sacrificial barrier (the broken lines
indicate the limits of the barrier after
reacting); c) Stuffed barrier.

the adjacent materials (Fig. 8b). A typical example involves the use of
a layer of Ti or Cr between the Al and the silicide in Al/Silicide/Si
structures, since the reaction rate of Ti and Cr for both the Al and the
underlying silicide has been well characterized[52,53]. iii)   To minimize
the paths of rapid diffusion along grain boundaries and defects of a barrier
film, a practical approach consists in "stuffing" or "plugging" these easy
paths. The "stuffed barrier", a concept coined by Nicolet[51], is based on
this effect (Fig. 8c). Oxygen and nitrogen, sometimes   added unintention-
ally as impurities from the residual atmosphere during film deposition,
seem to play an important role as blocking agents of diffusion through
polycrystalline films. This is the case of Mo in the metallization system
Ti/Mo/Au, in which the controlled addition of nitrogen improves noticeably
the barrier properties of Mo, probably by the formation of a MoN coating
on the Mo grains. A similar process is thought to occur in the popular
W-Ti barriers frequently used in gold metallizations.

5.2. Barrier Layers in Microelectronic Applications

From the point of view of microelectronic applications, the barrier
layers should obey very stringent demands. Among them we should mention:
i) the transport rate of the materials A and B separated by the barrier X
across the barrier itself should be kept at a minimum;   ii) the chemical
stability of the barrier X with both layers A and B should   be high, so
that the barrier can survive all the expected temperature cycles, and iii)
apart from these essential characteristics, the barrier should show compatibility with the manufacturing process of electronic devices. This requires
additional characteristics for the barrier layers such as good adherence to
the materials A and B, low contact resistivity with both A and B layers,
high electrical conductivity, resistance to thermal shock and mechanical
stresses, etchability, and patternability[54]. Note that some of these
properties cannot be met simultaneously, for instance, no reaction of the
barrier X with the materials A and B   means a poor adherence. Therefore,
one is forced to adopt a compromise.

One should emphasize the importance of the grain boundaries and other
defects in the control of the diffusion processes in thin films. Nowadays,
there is strong evidence that the enhanced atomic diffusion in thin films

comes about along the easy paths formed by grain boundaries and other structural defects. Moreover, it has been demonstrated that the grain size of polycrystalline films exerts also a strong influence in the transport properties of diffusing species. Larger grain sizes, i.e. smaller number of grain boundaries, significantly decrease the rate of diffusion with respect to the smaller grain sizes. Here again, the deposition procedure strongly affects the barrier properties of the films. In addition, it has been experimentally observed a trend for metals with highest melting points to give smaller grain sizes in films when deposited on substrates kept at relatively low temperatures[51]. However, it is also true that during annealing treatments the smaller grains grow faster that the larger ones. These empirical rules, along with a good knowledge of the physico-chemical characteristics of the materials (reactivity and diffusivity data, mostly) may give a good insight into the proper selection of a barrier film of the desired characteristics.

In IC applications, the barrier layers are utilized in different areas of the circuit, namely in ohmic contacts, n+p diodes, Schottky diodes, local interconnects, gate contacts, etc. Several types of barriers have been investigated to stop the reaction of Al in the Al/silicide/Si multilayer structure, among them, different metals such as Ti, Cr or W, their alloys (Ti-W for instance) and metal compounds, mainly nitrides and carbides, like TiN, TaN, etc. In fact, titanium forms the base of many of these barriers probably as a consequence of its chemical affinity for oxygen. As a result it adheres well to both Si and SiO_2 and can dissolve the silicon native oxide during annealing treatments. When added to tungsten, the Ti enhances notably its barrier characteristics by providing a thin layer of titanium oxide which prevents the W of further oxidation.

Metal compounds and particularly titanium nitride have been also extensively studied due to their excellent properties as barrier layers[55-57]. TiN can be deposited in a wide range of stoichiometries, although the lowest resistivity value is found for the ratio 1:1. The deposition of TiN is normally carried out by evaporation or sputtering in a nitrogen reactive atmosphere which allows the sequential deposition of Ti and TiN in the same run. Ting and Wittmer[53] have analyzed the Ti based contacts in detail and reached the conclusion that, in terms of stability, the TiN barriers are able to withstand the highest temperature in annealing treatments (550^0C), followed by the double barrier Ti/W (500^0C) and its alloy Ti-W (450^0C). Pure titanium has poor barrier characteristics since it reacts with Al at 400^0C, although is a good barrier for Si. In contrast, from the point of view of contact resistivity, the barriers they evaluated are in the following sequence of increasing values: Al/W/Ti/Si, Al/Ti/Si, Al/(Ti-W)/PtSi/Si, Al/(Ti-W)/Si and Al/TiN/TiSi$_2$/Si.

One of the most popular metallizations is the multilayer structure formed by Ti/Pt/Au used in the so-called "beam lead technology"[58]. In this technology the tri-layer is deposited on top of ohmic contacs formed by PtSi on heavely doped silicon, so that the complete structure is Si/PtSi/Ti/Pt/Au. In this scheme, the function of the Ti is to give adherence to the silicide beneath of it. Besides, Pt is introduced to separate the unstable couple formed by Ti and Au in contact to each other. Based on this straucture there are many variations to improve the barrier characteristics of the whole system. One obvious version is the introduction of a layer of TiN between Ti and Pt to take advantage of the good barrier properties of this compound.

TiN has been successufully tested as metal contact for low barrier diodes made on n-Si[59]. The height of the Schottky barrier is 0.49 eV which compares fairly well with the barrier of Ti, 0.5 eV, and TiSi$_2$, 0.6 eV. The value of 0.49 eV is near midgap, so that TiN can be also used

286

for low barrier diodes when deposited on p-type silicon.

6. INTERCONNECTIONS

The interconnection lines connect the different devices in an integrated circuit. In the submicron technologies, the interconnections may take over two thirds of the chip area and the reliability of the ever more complex integrated circuits depend many times on them. The most important failure mode of interconnections is caused by electromigration and to a lesser extent by corrosion and stress induced nucleation of vacancies in voids[60].

With the shrinking of ULSI device dimensions, the scaling factor results in interconnection lines with cross sections which are reduced by the square of the scale factor, giving rise to a current density j that can be as much as 10^5 A cm^{-2} for MOS transistors and 10^6 A cm^{-2} for bipolar transistors. The relatively large Joule heating generated by this current density has to be dissipated through the substrate and the passivating layer. Therefore any defect in the interconnection may lead to runaway thermal conditions[8].

6.1. Single-level Interconnections

Several metal layer configurations have been tried as interconnections. The simplest ones are formed by a single layer of aluminum or by the Al-Si alloy which avoids contact pitting as mentioned before. Addition of copper to the AlSi significantly improves the electromigration porperties of the interconnections. Other interconnections take advantage of the good corrosion and conductive properties of gold. However gold films show a very poor adhesion to silicon dioxide and therefore gold is incorporated in either bilayer configurations: Ti:W/Au, Mo/Au or in trilayer ones such as Ti/Pd/Au. In these cases, the Ti:W, Mo and Ti provide a good adherence to SiO_2 and also serve as barrier layers.

Electromigration in thin interconnections is the most important failure mode, thereby strongly influencing the reliability of microelectronic devices. Electromigration consists in a displacement of the atoms in a conductive material as a consequence of a direct momentum transfer from the conduction electrons in the direction of their motion. The resulting net atomic flux is proportional to the current density and to the effective charge of the migrating atoms; it also depends on the atomic diffusivity and temperature. For the electromigration to induce failures in interconnection lines there should be regions with a divergence in the atomic flux which is often caused by defects in the microstructure and temperature gradients. The transport of mass causes removal of material in some locations, which generates voids, and accumulation of material in other locations. Therefore, open circuits or cracks in the area where the material is removed, can frequently occur, as well as short circuits caused by touching of two lines where the material builds up.

The electromigration failure rates are characterized by the so called median time to failure (MTF) which is the time necessary to reach a 50% failure in a set of interconnections. The failure times usually obey a lognormal distribution yielding an straight line when the logarithm of the individual failure times is plotted in probability paper as a function of the cumulative percent of fails in the whole population as shown in Fig. 9 for Al-4% Cu interconnections with a current density of 2.5x10^6 A cm^{-2} and a temperature of 548 K[61]. The MTF is read directly from the graph of Fig. 9 as the time at which 50% of the samples have failed. As criterium of failure, there are several, but one of them consists in passing a current

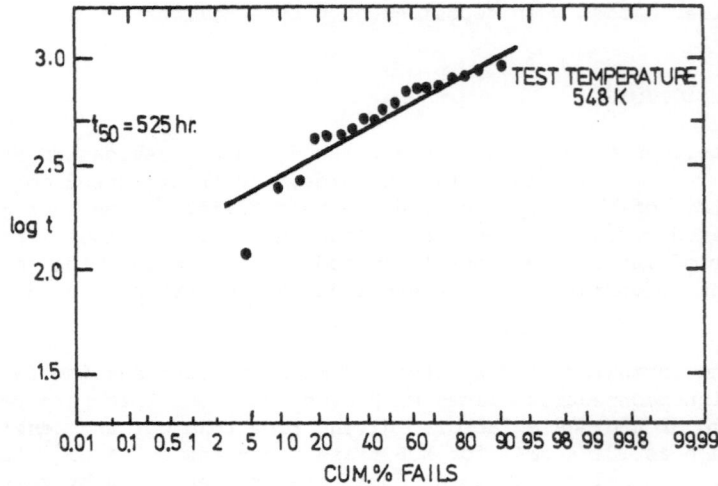

Fig. 9. MTF as a function of cumulative percent of
samples that failed for Al-4%Cu interconnections
stressed at 2.5×10^6 A cm^{-2} (from ref. 61).

through the interconnection and monitor the voltage across two points until
the line opens or its resistance increases to a predetermined value.

Empirically, the equation relating the MFT to the current density J
and temperature T is of the form

$$\text{MTF} = \frac{F \text{ (geometry)}}{J^n} \exp\left(\frac{E}{kT}\right)$$

where F is a factor which depends on the geometry of the line, in particular its thickness and width, k is Boltmann's constant, E is the activation
energy and n is the exponent for current acceleration which is found to be
about two[62]. The activation energy at a given current density is obtained
from an Arrhenius plot of MTF vs temperature, taking into account that T
is the temperature of the line, i.e., the oven temperature plus temperature
rise due to Joule heating. The activation energy gives a magnitude which
depends on the cumulative effects of defects in the film, composition,
temperature gradient, passive coatings, etc. In order to perform the
reliability proofs in not too long time spans, the experiments are carried
out with currents and temperatures higher than the operating ones under
normal conditions and assuming that the failure mechanism under the stressed
and operating conditions are similar. This condition is not always ful-
filled and one has to be careful in the interpretation of the results from
accelerated tests.

The first interconnection lines were made from aluminum deposited by
evaporation or sputtering. The aluminum was soon substituted by Al-Si
alloys to avoid contact pitting. The problem with the Al-Si alloy is that
it is easily prone to electromigration failures with the formation of voids
and hillocks. Sometimes, the voids in Al-Si interconnections of about 2
or 3 microns wide are caused by high stresses provoked by passivation films
or the substrate which are relieved by the formation of voids[60]. The
values of the electromigration activation energy measured for aluminum and
its alloys varies from 0.4 to 0.8 eV[63]. Therefore, the atomic migration
takes place by grain boundary diffusion, which is the predominant mechanism

of diffusion for electromigration in fine lines. This conclusion is reached because the self diffusion of Al in the bulk crystalline material is
characterized by an activation energy of 1.4 eV. Shima et al[64] have found
that the MTF's of Al-Si% lines, deposited by magnetron sputtering, with
grain sizes of about 2 μm are larger by one order of magnitude than those
lines with grain sizes of about 0.5 μm. However, one has to be careful
with these results because they are in contradiction with the results obtained by other investigators[61].

The aluminum films for interconnections are being substituted by alloys
of aluminum with small amounts of such elements as Cu, Ni and Mg. In
particular the Al-Cu-Si alloy is found to improve both the contact pitting
and the electromigration properties. The introduction of copper in amounts
of 0.5 to 4% in weight[65,66] leads to the decrease of the grain boundary
diffusion. Therefore, the void and hillock formation due to electromigration are also reduced. On the other hand such small addition of alloying
elements in aluminum does not change its electrical conductivity significantly. The MTF of the Al-Cu alloys depends on the method of film deposition and therefore on the grain sizes, their distribution and the orientation of the grains[67]. It has also been observed that for Al-0.5% Cu interconnections narrower than 2 μm the MTF increases with decreasing linewidth[68].
This is interpreted as due to the fact that if the linewidth dimensions get
sufficiently small, the corresponding interconnection is composed of single-
crystal regions.

In order to get a MTF as long as possible, sandwiched structures have
been tried by inserting a layer of a transition metal like Cr or Ti between
two Al-4% Cu interconnections[69]. Fine lines between 1 and 2 μm were patterned and annealed at 400^0C resulting in interconnections with MTF about two
orders of magnitude larger than the Al-Cu ones. Another approach for
increasing the MTF consists in covering the interconnections with a passivation layer, usually of silicon dioxide about 1 μm thick[70], or by anodization of the interconnect material[71].

For several purposes, among them, for electrical insulation, to fix
mobile sodium contamination and to form a smooth topography, the interconnections for MOS integrated circuits are generally deposited on a phosphosilicate glass with about 6% phosphorus composition. In these interconnections it has been observed that their electromigration resistance is
inversely proportional to the phosphorus concentration in the underlying
phosphosilicate glass. This is attributed to a corrosion effect arising
from the formation of phosphoric acid due to the reaction of moisture with
the phosphorus content in the glass[62]. To avoid the adverse effects of
moisture, a compact oxynitride film is deposited on the passivation layer.

6.2. Multilevel Interconnections

In order to achieve both higher packing densities and speed in integrated circuits for ULSI the length of the interconnect lines has necessarily
to be decreased. This is because for design features smaller than 1 μm
the delay caused by the RC time constant for 1 cm length is higher than
the MOSFET delay, even for interconnects made from aluminum as indicated
in Fig. 5. A significant reduction in interconnect length can be achieved
by a multilevel metallization. The simplest case of multilevel metallization is constituted by the two-level metallization, but in today's large
memories one can have three-level metallizations. This is not surprising
since in a relatively few number of years the chip densities in dynamic
random access memories have risen from 4K to 1 Mbit. In the two-level
interconnection scheme, the first level is used for propagation of local
signals and intergate connections and the second level for the power and
ground busses and propagation of clock signal across the whole chip. There-

fore one can use thin metallizations for the first level and consequently make easier the critical problems of step coverage in further depositions. The selection of the metal underlayer in a multilevel metallization system is very important since it sets some of the subsequent processes like, for instance, the maximum temperature of annealing.

As with the metals, it is very important to make a good selection of the intermediate dielectric between two metal layers. The dielectric should not contain pinholes to avoid shorts among the metal interconnections. It should be relatively thick but not as much as to present internal stress-cracking problems. The dielectric constant of the insulator is also very important because it affects the RC time constant of the interconnection. One very commonly used dielectric is based on silicon dioxide either undoped or doped with P, B, etc. A list of some of the most frequently used combinations of metals and dielectrics is the following[60]: $Al/SiO_2/Al$, $Ti:W/Al:Cu/SiO_2/Ti:W/Al:Cu$, $Mo/Au/Mo/SiO_2/Mo/Au$ and $Ti:W/Au/Ti:W/SiO_2/Ti:W/Au$. In the metal layers, Al and Au provide the necessary good conductivity of the interconnection. The SiO_2 can be replaced in these combinations by polyimides which as we shall see provide excellent planarization properties by spinning. Planarization is often needed because one of the most crucial problems in two-level metallizations is the proper coverage of oxide steps. At these steps the metal for the second level metallization is thinned producing quite easily electromigration failure problems. The technique most appropriate for conformal coating, i.e. without thinning in the vertical walls at the steps, is chemical vapor deposition.

The basic components of a multilevel metallization scheme are the "crossovers" between the first and second level interconnections separated by the insulator, and the "vias" which are the locations of the contacts between the metals of the two levels through a hole in the insulator. In Fig. 10 SEM photographs are presented of crossovers and vias of an VLSI bipolar circuit[60] for which the minimum feature size is 1.25 μm and in which the following metallization scheme is employed: $PtSi/Ti:W/Al:Cu/SiO_2/Ti:W/Al:Cu$.

Silicon oxide and silicon nitride are often used as interlevel dielectrics and are commonly deposited by LPCVD. If the silicon oxide is deposited by the oxidation of silane the process occurs at about 450^0C while 700^0C are needed if it is obtained by the decomposition of tetraethoxysilane (TEOS)[72]. For the silicon nitride obtained by the reaction of silane and ammonia the temperature is also between 700 and 800^0C. Therefore the low temperature (about 300^0C) plasma assisted deposition is many times preferred to minimize shorts in two-level metallization systems due to hillock growth of the first metal. The SiO_2 glass deposited over steps can show re-entrant sidewalls[67] and therefore a subsequent reflow step at temperatures above 1000^0C is needed. However if impurities such as phosphorus and boron are added, one obtains a glass called BPSG which flows at temperatures as low as 850^0C[73]. Some of these dopants (P) also act as getters for migrating ions like Na^+ and in addition tend to produce very smooth surfaces, thus facilitating the subsequent step coverage and photolithographic processes. Since about a decade, polyimides[74] have also being used as interlevel insulators which have the advantage that they are applied in liquid form and produce a planar surface once they are cured at about 300^0C. The polyimides have a dielectric constant of 3.5 which is lower than that of SiO_2 and Si_3N_4 and they are many times applied in double layers with those silicon compounds.

As the alternate layers of interconnections and insulators are deposited one over the others, the surfaces happen to become more rugged and problems like step coverage and lithography delineation become more important. Therefore, it is often necessary to apply planarization techniques

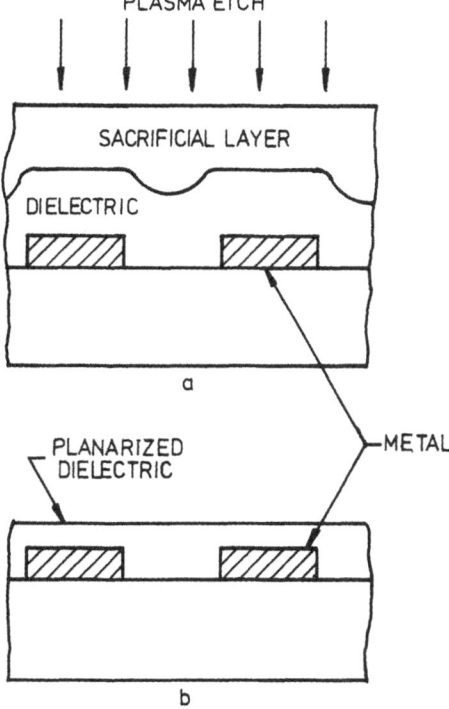

Fig. 10.   Plasma planarization. a) Planarized surface after
spinning; b) Planarized dielectric (from ref. 32).

after the deposition of dielectric layers. Some of these techniques only
achieve partial planarization, like that obtained after plasma etching.
One technique that yields a complete planarized surface is called plasma
planarization[75]. In this technique a material with low viscosity, such as
a photoresist or a polyimide is spun on the insulator material to smooth
its surface like is shown in Fig. 10 . The top layer that results after the
spinning is called a sacrificial layer and it should have an etch rate in
a plasma similar to that of the dielectric. Next, the dielectric-sacrifi-
cial layer is completely etched and the process is finished when the insulator has the desired thickness.

The contacts between two metal layers are made in the locations of
the crossovers and consist of a hole etched in the dielectric and refilled
with a metal so that results in a good contact between the two metal layers.
The resulting structure is called a via. When the area of the via lies
within the projection of the intersection of both metal layers they are
known as "nested vias" (see Fig.11a,b). On the contrary, if the vias
dimensions are larger than the width of any of the interconnnections they
are called "nonnested vias". Complete removal of the glass at the bottom
of the vias is a necessary condition so that it results in a low specific
contact resistance at the interface between the two metal layers. Sputter
cleaning of the vias before the deposition of the second metal seems to be
a good solution to avoid high resistances at the via contacts. One of the
problems with the via holes is that their walls are vertical often producing
shadowing problems and metal discontinuities at the hole walls. Therefore
several techniques had to be developed to slope the walls of the via holes.
One of such techniques that works quite well for holes larger than 1 μm
consits of a presloped photoresist which tappers or disminishes gradually

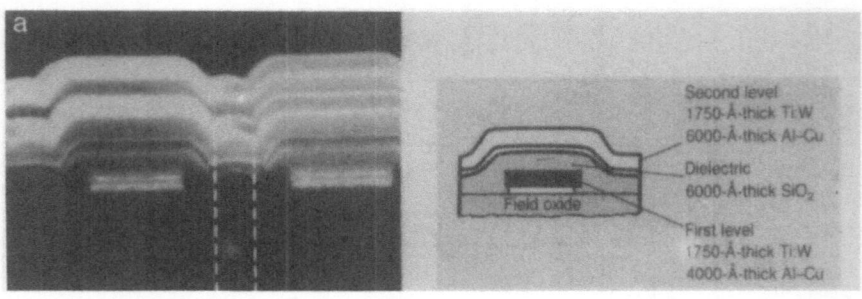

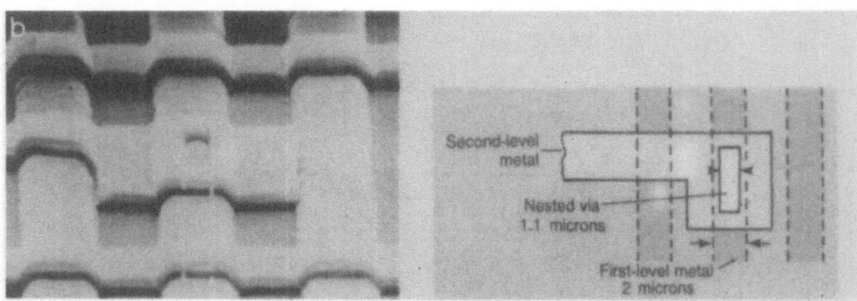

Fig. 11. SEM photographs of crossovers and bias of
a VLSI bipolar circuit[60] : a) Two-level
metallization crossovers. b) Nested vias of
two-level interconnections.

the width of the holes. In this technique the walls of the photoresist are
first sloped by baking and then the holes are tappered by means of an
anisotropic etch[32].

REFERENCES

1. D.W. Skelly, T.M. Lu and D.W. Woodruff in "VLSI Electronics Microstruc-
    ture Science. Vol. 15. VLSI Metallization". Editors N.G. Einspruch,
    S.S. Cohen and G. Sh. Gildenblat. Academic Press Inc., Orlando,(1987).
2. Y. Pauleau, Thin Solid Films, 122, 243 (1984)
3. E.K. Broadbent and C.L. Ramiller, J. Electrochem. Soc., 131, 1427 (1984).
4. E.K. Broadbent and W.T. Stacy, Solid State Technol., 28(12), 51 (1985).
5. Y. Pauleau, Solid State Technol., 30(4), 155 (1987).
6. R.S. Blewer, Solid State Technol., 29(11), 117 (1986).
7. P.B. Ghate, J.C. Blair and R.C. Fuller, Thin Solid Films, 45, 69 (1977).
8. P.S. Ho in "VLSI Science and Technology/1985", Editors W.M. Bullis and
    S. Broido, The Electrochemical Society, Pennington (1985).
9. H. Kotani, T. Tsutsumi, J. Komori and S. Nagao, Int. Electron Devices
    Meeting p. 217 (1987).
10. S.S. Cohen and G.Sh. Gildenblat, "VLSI Electronics Microstructure Scien-
    ce. Vol. 13. Metal-Semiconductor Contacts and Devices". Academic Press
    Inc.,Orlando (1986).
11. T.J. Faith, R.S. Irven, S.K. Plaute and J.J. O'Neill, J. Vac. Sci.
    Technol., A1, 443 (1983).
12. R.A. Levy, M.L. Green and P.J. Gallagher, J. Electrochem. Soc., 131,
    2175 (1984).

13. P.B. Ghate in "Thin Films and Interfaces", Editors P.S. Ho and K.N. Tu, New York, North Holland, (1982).

14. K.N. Tu, Appl. Phys. Lett., 27, 221 (1975).

15. C. Canali, C. Catellani, M. Prudenziati, W.H. Wadlin and C.A. Evans, Appl. Phys. Lett., 31, 43 (1977).

16. C.A. Crider and J.M. Poate, Appl. Phys. Lett, 36, 417 (1980).

17. A. Climent and S.J. Fonash, J. Appl. Phys., 56, 1063 (1984).

18. M.A. Nicolet and S.S. Lau, in "VLSI Electronics Microsturcture Science". Vol. 6. Editors N.G. Einspruch and G.B. Larrabee. Academic Press Inc., Orlando (1983).

19. C.Y. Ting and M. Wittmer, J. Appl. Phys., 54, 937 (1983).

20. C.Y. Ting, J. Vac. Sci. Technol., 20, 14 (1982).

21. M.E. Alperin, IEEE Trans. Electron Dev., ED-32, 141 (1985).

22. C. Arena, S. Deleonibus, G. Guegan, P. Laporte, F. Martín and J.L. Pelloie, p.577, 17th European Solid State Dev. Res. Conf., Bologna (1987).

23. Y. Taur, B. Davari, D. Moy, J.Y.C. Sun and C.Y. Ting, IBM J. Res. Develop., 31, 627 (1987).

24. F. Neppl, F. Fisher and V. Schawabe, Thin Solid Films, 120, 257 (1984).

25. K. Hieber and F. Neppl, Thin Solid Films, 140, 131 (1986).

26. M. Eizenberg in "VLSI Science and Technology/1984", Eitors K.E. Bean and G.A. Rozgonyi, The Electrochemical Society, Pennington (1984).

27. K. N. Tu, W.N. Hammer and J.O. Olowalafe, J. Appl. Phys., 51, 1663 (1980).

28. S. Swirhun, K.C. Saraswat and R.M. Swanson; IEEE Electron Dev. Lett., 5, 209 (1984).

29. M.J. Kim, R.N. Singh, D.M. Brown and D.W. Skelly in "VLSI Science and Technology/1985", Editors W.M. Bullis and S. Broydo, The Electrochemical Society, Pennington (1985).

30. A. Lindberg, M. Ostling, H. Nostróm and V. Wennstróm, p. 191, 17th European Solid State Dev. Res. Conf., Bologna (1987).

31. T. Mochizuki, Y. Tsujimaru, M. Kashiwagi and Y. Nishi, IEEE Trans Electron Devices, ED-27, 1431 (1980).

32. A.G. Sabnis in "VLSI Electronics Microstructure Science. Vol. 15, VLSI Metallization", Editors N.G. Einspruch, S.S. Cohen and G.Sh Gildenblat, Academic Press Inc., Orlando (1987).

33. A.C. Adams in "VLSI Tehcnology", Editor S.M. Sze, McGraw-Hill Book Co, New York (1983).

34. T.I. Kamins, M.M. Mandurah and K.C. Sraswat, J. Electrochem. Soc., 125, 927 (1978).

35. S.P. Murarka, "Silicides for VLSI Applications", Academic Press, New York (1983).

36. S.P. Murarka, J. Vac. Sci. Technol., 17, 775 (1980).

37. A.K. Sinha, J. Vac. Sci. Technol., 19, 778 (1981).

38. T.P. Chow and A.J. Steckl, IEEE Trans. Electron Devices, ED-30, 1480 (1983).

39. J. Denis, M. Fernández, J.P. González, J.M. Albella and J.M. Martínez-Duart, Thin Solid Films, 125, 243 (1985).

40. P.T. Schraibhand, European Patent Application 212.266.

41. K.L. Wang, T.C. Holloway, R.F. Pinizotto, Z.P. Sobczak, W.R. Hunter and A.T. Tasch, IEEE Trans. Electron Devices, ED-29, 547 (1982).

42. T.P. Chow, D.M. Brown, A.J. Steckl and M. Garfinkel, J. Appl. Phys., 51, 5981 (1980).

43. T.P. Chow and A.J. Steckl, J. Appl. Phys., 53, 5531 (1982).

44. A.G.M. Jonkers, Le Vide les Couches Minces, 42, 103 (1987).

45. M. Bartur and M-A. Nicolet, J. Electrochem. Soc., 131, 371 (1984).

46. J.M. Martínez-Duart, M. Fernández, E. Paule, A. Climent, J.M. Albella, J. Perrière and J. Siejka, Appl. Phys. Lett., 47, 579 (1985).

47. J. Perrière, J. Siejka, A. Climent, E. Navarro and J.M. Martínez-Duart, J. Appl. Phys., 61, 2656 (1987)

48. J. Crawford, Semiconductor International, p. 84, March, 1987.
49. Y. Pauleau, Solid State Technol., 30,(2), 61 (1987).
50. N. Yamamoto, S. Iwata, N. Kobayashi, K. Yagi and Wada in "VLSI Science and Technology/1984), Editors K.E. Bean and G.A. Rozgonyi, The Electrochemical Society, Pennington (1984).
51. M-A. Nicolet, Thin Solid Films, 52, 415 (1978).
52. R.W. Bower, Appl. Phys. Lett., 23, 99 (1973).
53. C.Y. Ting and M. Wittmer, Thin Solid Films, 9, 327 (1982).
54. M-A. Nicolet and M. Batur, J. Vac. Sci. Technol., 19, 786 (1981).
55. I.R. Shappirio, Solid State Technol., 28, 161 (1985).
56. M. Wittmer, J. Vac. Sci. Technol., A3, 1797 (1985).
57. J. Stimmel, J. Vac. Sci. Technol., B4, 1377 (1986).
58. R.S. Nowicki and M-A. Nicolet, Thin Solid Films, 96, 317 (1982).
59. M. Wittmer, B. Studer and H. Melchior, J. Appl. Phys., 52, 5722 (1981).
60. P.B. Ghate, Physics Today, Oct. 1986, p. 58.
61. J.A. Schwarz in "VLSI Electronics Microstructure Science". Editors N.G. Einspruch, S.S. Cohen and G. Sh. Gildenblat. Academic Press Inc., Orlando (1987).
62. M.H. Woods in "VLSI Science and Technology/1985". Editors W.M. Bullis and S. Broydo, The Electrochemical Society, Pennington (1985).
63. Y. Pauleau, Solid State Technol., 30(6), 101 (1987).
64. S. Shima, T. Moriya and M. Kashiwagi, Ext. Abst. 15th Conf. Solid State Dev. and Materials, Tokio, p. 233 (1983).
65. I. Ames, F.M. d'Heurle and R.E. Horstmann, IBM J. Res. Dev., 14, 461 (1970).
66. F.M. d'Heurle, N.G. Ainslie, A. Gaugulee and M.C. Shino, J. Vac. Sci. Technol., 9, 289 (1972).
67. A.K. Sinha, Thin Solid Films, 90, 271 (1982).
68. S. Vaidya, D.B. Fraser and A.K. Sinha, Proc. 18th Reliability Physics Symposium IEEE, p. 165, New York (1980).
69. J.K. Howard, J.F. White and P.S. Ho, J. Appl. Phys., 49, 4083 (1978).
70. J.K. Lloyd and P.M. Smith, J. Vac. Sci. Technol., 1, 2 (1983).
71. A. J. Learn, J. Appl.Phys., 3, 44 (1973).
72. A.C. Adams and C.A. Capio, J. Electrochem. Soc., 126, 1042 (1979).
73. D.S. Williams and E.A. Dein, J. Electrochem. Soc., 134, 657 (1987).
74. A.M. Wilson, Thin SOlid Films, 83, 145 (1981).
75. A.C. Adams, SOlid State Technol., 24, 178 (1981).

MULTILEVEL INTERCONNECT STRUCTURES*

Terry O. Herndon

MIT Lincoln Laboratory
Lexington, MA 02173

Introduction

My intent is to review multilevel interconnect for I.C. fabrication; to look at why multi-level is essential; to assess the current status of multilevel interconnect with regard to density, number of levels, and processing problems; to estimate future needs for multilevel interconnect and consider some possible new ways to achieve these goals. First, to establish some assumptions, Figure 1 shows an interesting relationship with regard to capacitance. CL is the line to line capacitance and CS is the line to substrate capacitance on interconnect for integrated circuits. As the metal pitch or center to center spacing goes down, the lines get narrower and therefore, the line to substrate capacitance is reduced. But at the same time, the spacing between lines is reduced and the line-to-line capacitance goes up. The result is a minimum in these two capacitances at around a 2 μm pitch. That would mean 1 μm spaces between 1 μm wide conductors which might be 1 μm thick. The other factor not shown here is that as the pitch goes down, the resistance is starting to go up due to conductor narrowing, and the RC time constant increases. One of the factors in multilevel interconnect is that usage of silicon by active devices on the substrate, ranges from 4% - 8% depending on the type of circuit and the densities. The speed of the devices are no longer the limiting factor at all. The RC time constant of the interconnect governs the total system speed. This means that the devices have to be put closer together, interconnect needs to be made denser, and at the same time reduce the RC time constant as much as possible. Therefore, I am assuming that a 2 μm metal pitch will be the reasonable expectation to be met in the future.

Figure 2 shows the effect of increasing the number of interconnect levels. As can be seen for CMOS gate arrays, a metal pitch of 10-11 microns permits fewer than 2 thousand gates on a quarter inch square chip. Three level metal would permit something like 30-35,000 gates on a $\frac{1}{4}$" square chip, assuming pitches of 2-4 microns. This is the kind of thing that is needed for some present and many future I.C. requirements. It is possible that 4 levels of interconnect are going to be needed in the near future. This would increase the number of gates to the 35-45,000 range, which would improve utilization of the silicon and maximize the system-to-single-device performance ratio. Figure 3 is an example of 2 μm metal inter-connect technology applied to an integrated circuit. These bipolar transistors have 1.5 μm contacts and are connected with conductors having a minimum width of 2 μm. It is clear

---
*This work was sponsored by the department of the Air Force. The views expressed are those of the author and do not reflect the policy or position of the U.S. Government.

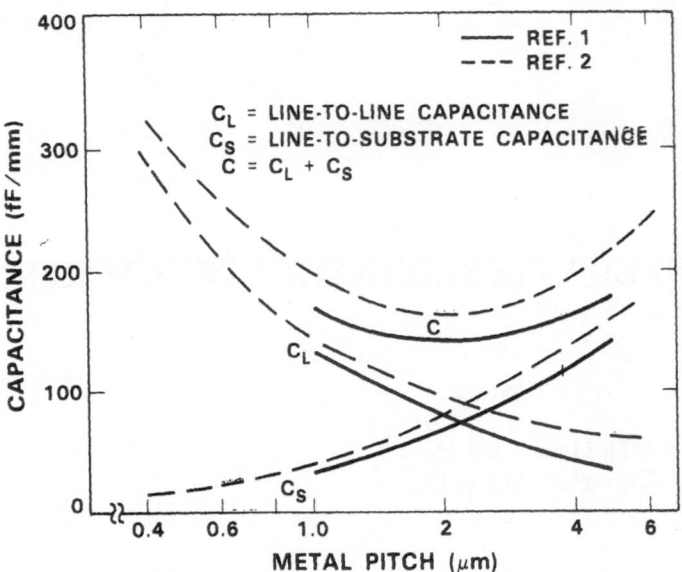

Fig. 1   Parasitic Capacitance - Pitch Relationship

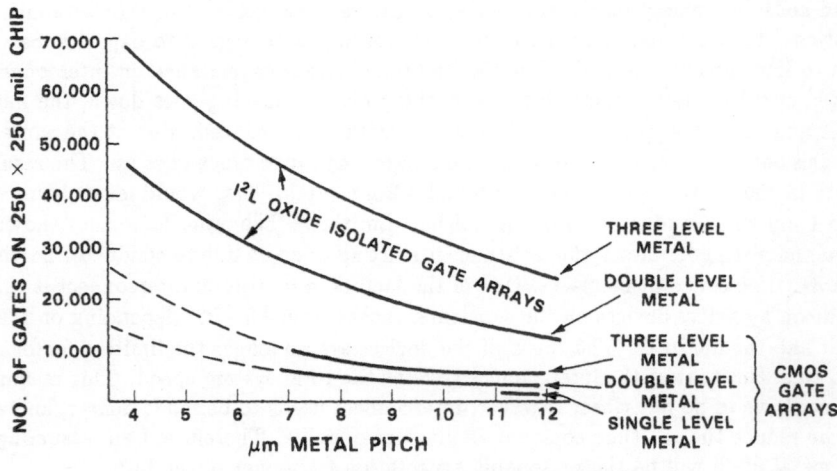

Fig. 2   Gate Density v.s. Interconnect Levels

that one level metal cannot cope with this. Transistor collectors can't be connected to anything, because they are isolated among the runs of other interconnect going from device to device. Figure 4 is an example of double level interconnect. The first level conductors are polysilicon, which are the slightly rough lines that run beneath everything and are covered by the silvery, quite rough first level metal conductors. Above the first level metal conductors is the between metal insulation, which in this case is polyimide. It is transparent and has holes etched into it called vias, to permit the smooth, silvery 2nd level metal to connect down to the rougher first level metal. Aluminum silicon copper is used for both levels, and the minimum geometries are approximately 2 μm. As one can see, the number of MOS devices

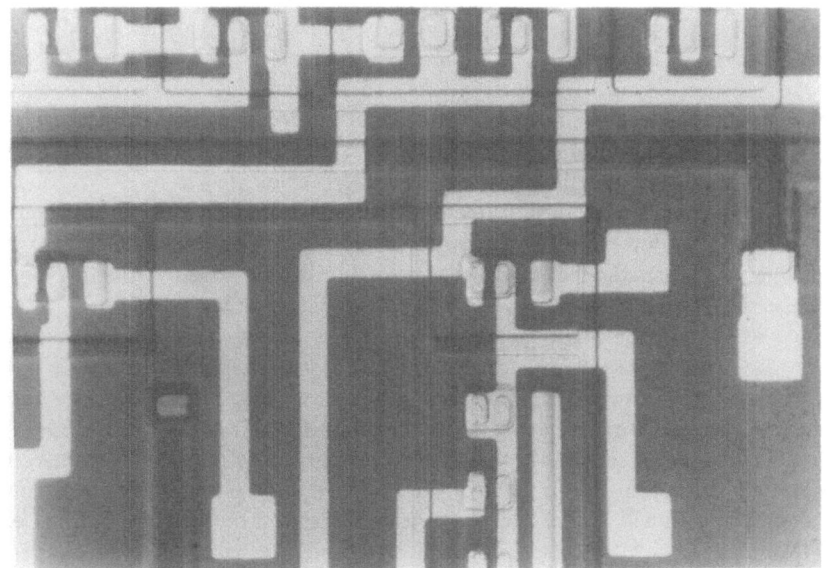

Fig. 3   Single Level Metal

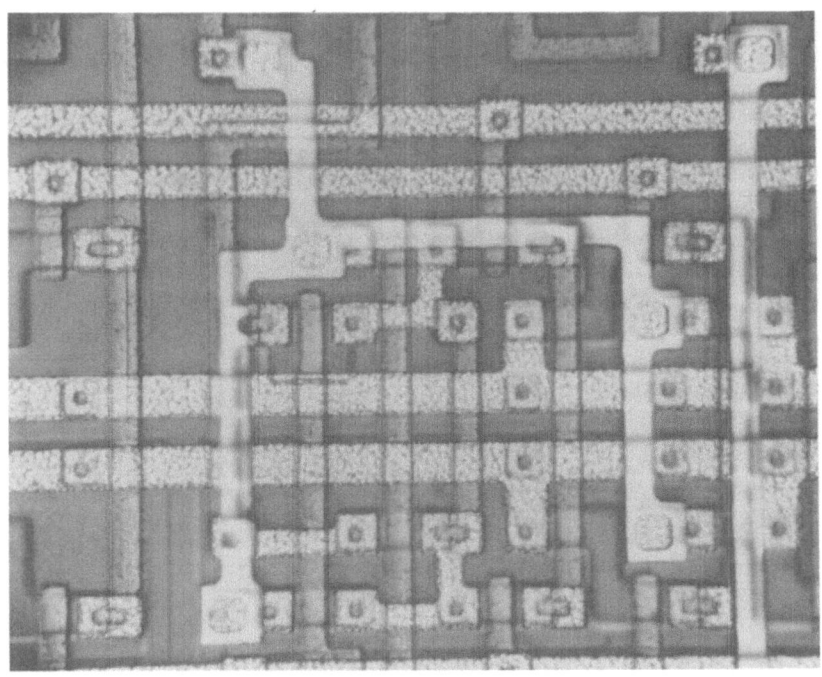

Fig. 4   Double Level Metal

has increased considerably in this view, compared to the bipolar devices seen in the previous picture. Figure 5 is an ideal four metal interconnect illustration. There are a number of important things to look at here. First, to achieve maximum density and minimum resistance, the conductors (seen in cross section) must be vertical sided. Secondly, the interconnect needs to lie on planar insulating layers, or in some way be kept completely planar in order to attain the highest resolution. This also solves a number of problems which will be discussed later, such as topography build up, lensing situations and step coverage caused by topographical considerations. Another feature in this ideal 4 metal interconnect is that the best possible situation would be for any level to be connectable to any other level. Currently, level one metal is connected to level two metal, level two metal is connected to level three metal and so forth. One does not connect level three metal directly down to level one without going through an intervening level two strap of some sort, and using up real estate. Still another ideal would be vias that are identical to the metal run widths without needing to oversize the metal around the via (nesting) while still allowing the upper metal to be misaligned slightly and yet contact the lower metal. Again, this is looking at ideal interconnect that would give the best results in terms of density and speed. Regarding RC time constants and circuit speed, if one can use four to seven levels of interconnect, one can start reducing capacitance between metal runs by separating them into the third dimension. This is an important justification for more than three levels of interconnect. Although not more than three or four levels are needed for purely interconnect purposes, one may want more than that in order to reduce capacitance. If vias can be fabricated as shown in this picture, it is clear that putting wires further away from one another into the third dimension will reduce capacitance.

Section I   METAL CONDUCTOR GENERATION

The first major component of multilevel interconnect is the metal conductor. In this section we will look at metal conductor generation which involves photoresist and both wet and dry etching techniques. Figure 6 shows photoresist lines 1 μm wide running over polysilicon which has had a 1 μm step etched in it. This illustrates a number of problems with photoresist. The first thing to note is that the photoresist just at the bottom of this 1 μm high step is approximately twice as thick as the photoresist on top of the step. This is a normal condition with regard to topographically-caused problems and it can mean that the resist exposure will never be perfect. If one wants to expose the resist properly for its thickness at the top of the step (to produce the desired line width) then insufficient light has gone through the double thick resist just over the step edge. The double thick resist in that area will be underexposed, with the result that when the resist has been developed, there will still be some etch-inhibiting resist left at the bottom of the step. Conversely, if one exposes for the maximum resist thickness (the double thick case just at the bottom of the step) too

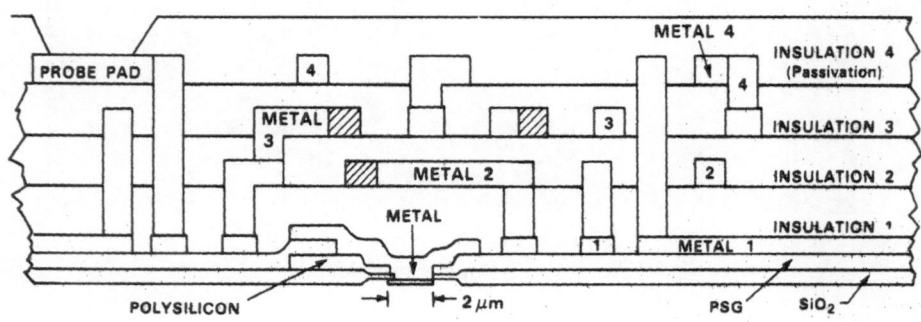

Fig.   5   Ideal Multilevel Interconnect

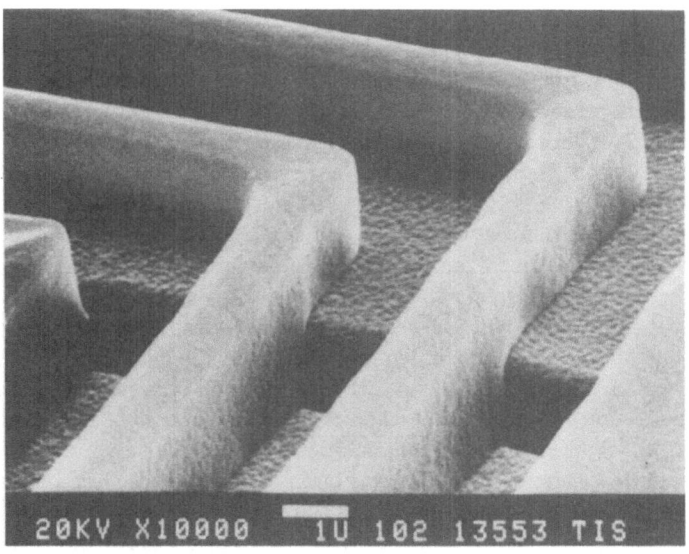

Fig. 6   Photoresist Definition/Step Coverage

much light has gone through the thinner sections of the resist which tends to undercut the resist or narrow it down. This causes a necking down of the resist where it is thinner at the top of the step. Therefore, one has resist lines that change width because the underlying topography caused the resist thickness to change. Furthermore, resist development is by wet chemistry, and the developer attacks even the unexposed resist somewhat. If the resist is developed for the proper time to remove the thinner resist, even though the thicker resist is properly exposed, there is still some residual resist left in that double thick area. So one has to develop longer than ideal for the thin sections in order to remove all of the resist in the thick layers. This again attacks the unexposed, thinner resist and causes further necking down of that resist. A number of techniques have been used to suppress these troublesome conditions, which really get to be a problem when considering resist lines that are 1 μm wide. One approach is to dye the resist somewhat so that you change its back reflectivity. Therefore, if the resist is on a shiny, smooth aluminum layer for instance, the additional reflected light from the aluminum surface will cause even further narrowing of the thinner resist layers as opposed to the thick layers. The dyed resist reduces back reflection and helps eliminate this problem. Still another approach is multilayer resist where some kind of polymer, usually non-photo sensitive, is spun on a nonplanar surface. This tends to level or smooth the topography and then a thin photoresist layer is spun on top of that. That resist layer tends to have fewer thickness variations as a result of the smoothing effect of the underlying thick polymer that was spun on. Therefore, the uniformity of line width, because of uniformity of exposure and development, is better with a double or sometimes triple layer resist of this sort. In this kind of arrangement, the resist is exposed, developed and then the underlying polymer has to be removed usually by some plasma etching process, which gives vertical sides. This is a complex technique and since the thick polymer does not fully planarize, there are still photoresist thickness variations. In general, Fig. 6 illustrates that by having a wide exposure latitude, and a proper bake/develop protocol, almost 2-1 thickness variations can be accommodated and still resolve 1 μm geometries. Proper choice of photoresist is terribly important. Another consideration is that the photoresist development is a wet process. As the resist lines get narrower the problem of photoresist adhesion to the underlying surface becomes more critical. Spray develop is essential for cleaning out small geometries, but this produces high liquid forces on the resist sidewalls which leads to lift off or tearing loose of the resist, particularly when the line width gets into the 1 μm range. This imposes a practical resolution limit on submicron production.

Figure 7 shows a standard single level interconnect and illustrates problems which arise from the etching and from resist liftoff or breakdown. In this case the metal was wet etched. The basic problem is the wet etch is isotropic, or etches in all directions. Therefore, where the metal is covered by thinner resist (the resist going over a vertical-sided step is thinner going over the top edge of the step) the etchant attack on the thin resist can break through and etch the metal across the conductor. The thin resist may also start to lift and permit undercut or "notching" in from the conductor edges. In Fig. 7 there is a clear case of undercutting going on where the lines step from a higher altitude down into a trench. Looking across the bottom of the trench and coming back up it is quite easy to see the severe undercut created by geometrical and topographical effects. If the resist is wide enough a little "notch" in the edge of the line doesn't make any difference, but as the line gets narrower and narrower this "notch" can go all the way across and cause an open. Or, the resist thinning where it goes over the edge of the step may be attacked or lifted off because of fluid forces generated by spray etching which also opens the conductor at the top of the step.

In Figure 8, the same interconnect was plasma etched. It is clear that this technique is more appropriate for etching fine geometries, since it is a gas discharge, with the constituents of the glow discharge combining with the metal interconnect, converting it into a gas which is then pumped away. Here the etching can be either completely vertical (anisotropic) or it can be relatively isotropic depending on how the flow, pressure and bias conditions are set in the plasma etching reactor. The gas viscosity is much lower than a liquid and side forces tending to lift the photoresist are reduced. Furthermore, the lower viscosity permits etching to take place in small, deep openings much more readily than with a liquid. Although the plasma process has these advantages, one problem involves across-wafer etch uniformity, where the etching may be more rapid at the edges of the wafer and slower near the center. This is the result of nonuniform gas flow coupled with etch effluent saturation locally on the wafer, depending on percentage of etched-to-unetched area. Abandonment of batch plasma etch systems in favor of single wafer etch systems may improve the control of cross wafer etching uniformity. However, since throughput must be in the range of 30-60 wafers per hour, and if one is etching 1-2 μm thick materials in a minute, or two minutes maximum, 1-2 μm/minute

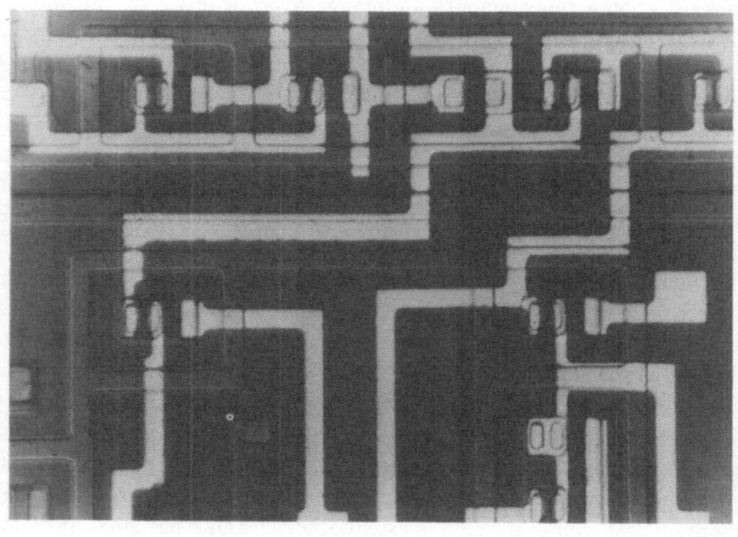

Fig. 7   Wet Etched Metal

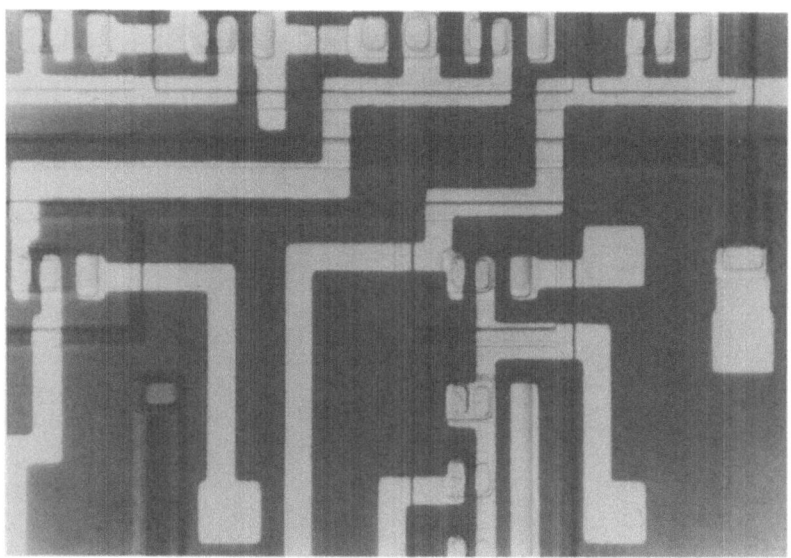

Fig. 8   Plasma Etched Metal

etch rates are required. The resulting high current densities and ion bombardment tend to attack the photoresist. This gets back to the problem of photoresist breakdown and etching where one doesn't want it to etch. Thus the photoresist must be hardened or nonerodible (Al or SiO) masking must be used. Also, at high etch rates, temperatures increase rapidly, polymerizing the photoresist which makes it very difficult to remove. This all negates some of the advantages of the 0.08-0.2 μm/min etch rates of plasma or RIE systems which do very little damage to photoresist, and don't depend strongly on resist adhesion or resist etch resistance. Thus, the high rate etching complicates the process and the benefits of the plasma or RIE etching are somewhat compromised. It would seem that the proper approach in long term is to have a more uniform small-batch etching system to allow low current densities, lower power levels and less damage to the photoresist. In this way one can maximize the benefits of the plasma process, and achieve finer geometries with wider etching tolerances. The main advantages of the plasma process are that lower viscosity gases can etch finer geometries, there are fewer forces on the photoresist and so photoresist adhesion to the substrate is not important so long as it is properly in place following development. Figure 9 illustrates some of these problems. Here, 1 μm high steps in the substrate are covered by approximately 0.7 μm thick aluminum silicon copper. The upper view shows photoresist stripes that are approximately 3 μm wide going horizontally over the surface, while the lower view shows the resist, from an angle looking directly in from the cleaved edge of the wafer. This illustrates the problem of thinning photoresist and it also shows the resist thickening problem with regard to metal step coverage. In the case of the resist one can see that it is clearly two to three times as thick right at the vertical sidewall of the step as compared to the top of the step. The top of the step is the thinnest section where the photoresist is flowing over the top corner. The photoresist normal thickness is more or less defined by the thickness on the top of the elevated step. Furthermore, looking in at the edge, one can see that the metal is thicker in the center of the slot than it is right at the corner. The metal is shadowed by the step itself. Therefore, during deposition it is thinner at the lower corner, and also thinner on the sidewalls of the vertical step. In Figure 10, the metal has been plasma etched and the upper picture shows the plasma etched metal with photoresist still in place. One can see that there has been some attack on the photoresist, but it is quite healthy and still covers the upper corner of the step. However, in the lower picture the photoresist is still on, but it is clear that the photoresist has gone away or broken down right at the upper, thin corners,

Fig. 9   Photoresist On Metal Over Steps (Unetched)

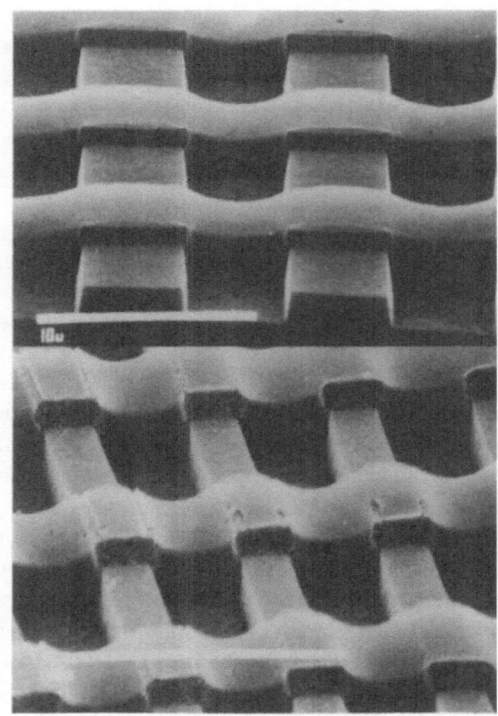

Fig. 10   Photoresist On Metal Over Steps (Plasma Etched)

going over the upper step edge in some cases. This is an example of problems with both wet and plasma etching, which are the erosion of photoresist and its thickness variations over topography. In Figure 11 the resist has been stripped and several problems arising from topographical effects are seen. The stripes of underling topography have a slot running down the middle, as seen in the upper right hand corner of Fig. 11. This slot has caused a void in the metal. One can see a sort of upside down keyhole where the slot in the substrate did not fill and self-shadowing occurred, creating a void. Also, one can see the problem with sidewall coverage. These are not vertical sided steps (they are approximately 85 degrees) but it is still clear that metal on the vertical sides is perhaps $\frac{1}{2}$ as thick as normal metal on top or down in the bottom of the slot. This step coverage problem leads to higher resistance metal and it can also produce high current density conditions causing electromigration if the metal is too thin. So, in general, the question of metal coverage of topography and photoresist coverage of the metal is a very complex issue. This is best illustrated where the narrowness of the slot, its depth, and thickness of metal coverage all affect void formation. The void in the upper right hand corner of Fig. 11 was actually formed by a shallow slot, but the width was proper to produce the void. The matter of topography and how it affects step coverage, voids, and resist thinning etc., is extremely complicated, and as far as I know no simple way of predicting topographical effects on lithography and etching exists. Vertical sidedness is a desirable feature in multilevel interconnect, but it causes a number of problems which have to be solved.

Figure 12 is an example of the resolution capability of dry etching. Here, 1 μm thick aluminum silicon copper has been etched into $1\frac{1}{2}\mu$m wide lines, with a $\frac{3}{4}\mu$m space between them. These are interdigited fingers and they are electrically isolated, demonstrating that high density metal interconnect can be achieved fairly easily. Of course, this assumes that resist breakdown and topographical problems aren't so severe that they lead to other difficulties.

Fig. 11   Etched Metal Over Steps (Etch Problems)

Fig. 12   High Density Plasma Etched Metal

There is one other thing that should be noticed in Fig. 12, which is the existence of hillocks, even though 4% copper has been added for hillock suppression. They can be seen as small bumps sitting on top of the $1\frac{1}{2}$ μm al. silicon copper lines. These suppressed hillocks are perhaps 0.02 μm high as opposed to straight aluminum or aluminum silicon where hillock growth can be as much as $\frac{1}{2}$ μm in height. This causes problems later on because the hillocks are very difficult to cover as they can be rather sharp pointed. This means that whether a CVD insulation or a spin-on polymer insulation is to be used as the intermetal dielectric, hillock suppression of some kind is necessary. Of course, as interconnect density goes up, the lower metal to upper metal shorting problems become worse because the number of metal-to-metal crossings increase. A standard technique for suppressing hillocks in aluminum is the addition of copper which can cause problems in plasma etching, since there is no known etch chemistry for etching copper and a residue is left in the plasma process. There are other techniques for suppressing hillocks, such as using titanium and aluminum in a multi-layer sandwich. This arrangement is reported not to have as many hillocks as a pure aluminum conductor. It is also been reported that the addition of silicon to polyimide intermetal insulation suppresses hillocks because the silicon tends to migrate out against the aluminum which suppresses hillocks in some stress-relief based manner.

In Figure 13, the upper image and lower image are both of aluminum silicon copper that is 3 μm thick that has been plasma etched. Normal 1 μm thick photoresist is used and as can be seen in the upper picture, the photoresist has thinned to about 0.5 μm, but remains intact. The undercut in etching 3 μm thick metal is tolerable. Looking at the lower picture one can see the classical condition with plasma etching of metal which is the spongy, very high area sidewalls. This can lead to problems in terms of corrosion because the sidewalls and the resist can absorb chlorine from the etch chemistry. When the metal comes out of the reactor the absorbed chlorine on these high area sidewalls can react with atmospheric moisture, causing localized etching. Such uncontrolled, localized etching, called corrosion, can etch all the way through the conductors and can also cause so-called rat bites or chunks to be removed from the edge of the metal. Corrosion suppression is needed in the plasma etch processes, and the standard corrosion suppression now is sulfur hexaflouride gas discharge done in the same etch reactor as the metal was patterned. This seems to work reasonably well, although there can be adhesion problems with the next overlying layer. Also, baking and stripping the resist in vacuum without exposure to atmosphere is said to eliminate corrosion. So, in summary, the metal can be etched with high resolution, good definition, vertical sides and in thick cross

Fig. 13　　Thick Plasma Etched Metal

sections. The main drawbacks have to do with step coverage over topography and with the need for high etch rates for achieving good wafer throughput, imposing severe conditions on the resist, causing resist failures.

Section II　　INTERMETAL INSULATION & STEP COVERAGE

The second major component of multilayer interconnect is the inter-metal insulation and the main characteristics to be reviewed in this section are listed in the Section Heading. The standard between metal insulator is a low temperature CVD oxide, the deposition of which is similar to snow falling on the ground on a very still day. It is highly anisotropic, coming straight down and covering horizontal surfaces very well. It covers vertical surfaces poorly. This causes problems some of which are shown in Figure 14.

Here we are looking at the cleaved edge of a wafer which has metal interconnect lines coming down from the top of the picture. The metal conductor can be seen as a narrow, horizontal shiny, rough, stripe in the black edge of the elevated section of the insulator. Since the CVD does not cover the vertical sidewalls well, one has very thin coverage at the upper corners of the metal conductor. This forces one to go to very thick insulator depositions in order to prevent shorting at the thin, upper corners between the lower metal and the upper metal. This increases the thickness by a factor of 1.5 to 2 beyond what is needed for good sidewall coverage. This leads to the effect shown in Fig. 14, which is a kind of "bread loaf", at the upper corners of the buried conductor. As can be seen, the CVD glass is reentrant where it goes vertically down into the slot between conductors. This means that step coverage for the upper layer of metal is very difficult and indeed in the example shown here, the second level conductor, that runs left to right in the top of the image is not continuous electrically. It was unable to cover the reentrant sidewall of the oxide. A number of heroic efforts are made to planarize or at least smooth this oxide. There are techniques where CVD oxide is

Fig. 14   CVD Oxide Between Metal 1 and 2

deposited for a short period of time and then etched back and deposited and etched back repeatedly, which tends to round the corners to the extent that smoother sidewall slopes are achieved. However, in the case of vertical sided metal, the problem is not easily solved and usually the process is marginally adequate at best. Furthermore, it is time consuming to do that. Currently, the use of CVD oxide generally requires as much taper on the metal sidewall as possible to get adequate step coverage which produces sidewalls that are smooth, not reentrant as in Fig. 14. Another approach is to use higher temperature metal such as tungsten for the interconnect conductor. This allows the wafers to be run up to 600-700 degrees C where the oxide will flow somewhat to give a smoother topography. There are other processes that will be discussed later which are used to flatten, smooth, or planarize the CVD oxide. In general, the use of CVD oxide involves a number of problems which are really difficult to cope with. The deposition of the CVD oxide into narrow slots is as difficult as the deposition of metal or photoresist into narrow slots because one can get into some of the same sort of problems, such as voids, that were discussed in the previous section on metal.

In Figure 15, a first metal conductor identical to that in Fig. 14 is exposed in the lower part of the picture. A second level metal conductor is visible going horizontally across the middle of the picture. A smooth surface can be see between the two second level conductors at the center and top of the image. This smooth surface is the between-metal polyimide (PI) insulation. The PI tore away from the lower metal just below and in front of the second metal conductor which runs left-right across the center of the picture. This exposed the lower metal. As can be seen, the 2nd level metal is much smoother as it goes over the 1st level metal compared to the CVD case shown in Fig. 14. Since the polyimide is spun on as a liquid is tends to fill the vertical sided topography much better than the CVD oxide. For that reason, PI has very powerful arguments in its favor as a between metal insulator. In addition, PI can be spun on as much as 10 μm or more thick. Of course it smooths, but does not planarize better than approximately 60%. Current polyimides have dielectric constants in the range of 3.5-4, as compared to CVD oxide d.c. of 4-6, and there are some newer polyimides reported having dielectric constants of 2-2.5. Whether these will work well as between metal insulators remains to be seen. However, a low dielectric constant will reduce the capacitance and therefore the RC time constants which can be very advantageous. The polyimide is easy to plasma etch, adhesion to aluminum and to SiO_2 is good, although adhesion of PI to PI can

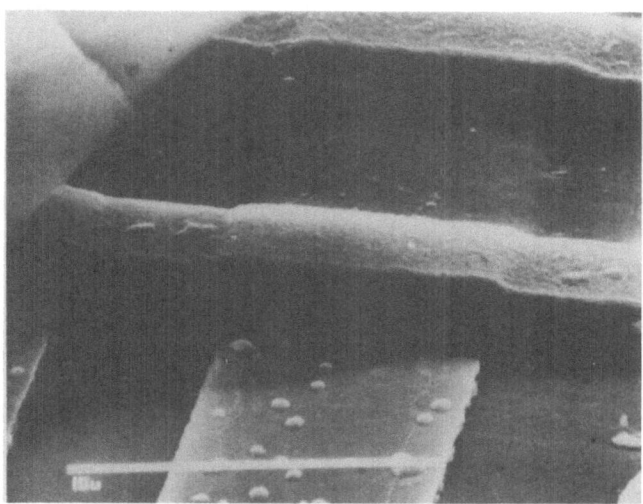

Fig. 15   Polyimide Between Metal 1 and 2

sometimes be a problem. In general, PI appears to be a very favorable material for intermetal insulation, particularly if the possible low dielectric constant can be combined with a 1-2 μm thickness, RC time constants can be minimized. However, PI does absorb water quite avidly - to the extent that it will absorb water out of silica gel. However, Senturia and others have shown that the water is absorbed only by the upper monolayers and that the water is essentially locked up in the upper 20% of the thickness. However, others have reported that the water absorption by PI can lead to corrosion of metal lying over the PI. These are all matters that have to be reviewed and further studies are needed in this area.

Figure 16 is another example of the smoothing capability of the PI. Here, two levels of plasma etched, 0.8 μm thick conductors are defined directly on top of each other, to produce a worst-case topography. PI layers, 0.8 μm thick serve as the intermetal insulation and a third level of interconnect is seen on top orthogonal to the lower levels. As can be seen, the smoothing is quite adequate for good metal definition. One other feature that can be seen here is that the top metal is wider where it goes down into the lower areas of the PI surface. This is a result of what was discussed earlier with regard to exposure of photoresist in thick sections vs. thin sections. In this case the photoresist was not quite exposed enough to remove it at the bottom of the dips in the surface, and therefore the etching of the aluminum was incomplete in the low areas of the topography which results in a wider line at that area. Of course if these lines were closely spaced then it is possible that the metal in this widened area would touch the metal in the similar widening of the adjacent line and shorting would occur.

The upper image in Figure 17 shows 3 level interconnect. The cleaved edge of the wafer can be seen at the bottom of the upper image with 4 μm wide conductors lying on the wafer surface. They are separated by 3 μm and are 0.8 μm thick. The insulating PI can be seen lying over those lines, with 2nd level metal visible coming down from the upper left toward the lower right with its ends just sticking out. These are 2 μm wide conductors with 2 μm spacing. These are covered by polyimide and the 3rd level of metal is seen coming down from the upper right to the lower left and the ends are turned up where the wafer broke away. These are 1 μm wide lines with 3 μm spacing. The lower image is a photomicrograph of the same geometry with the 4 μm 1st metal lines running left to right. The 2 μm wide 2nd metal lines are vertical and then the 1 μm 3rd metal conductors running left to right over the 4 μm conductors. It can be seen that the definition of the 1 μm wide conductor is good with little variation in line width. The PI smoothing is very successful in permitting this high density, high resolution interconnect.

Fig. 16   Polyimide Between Metal 1,2 and 3 (Smoothing)

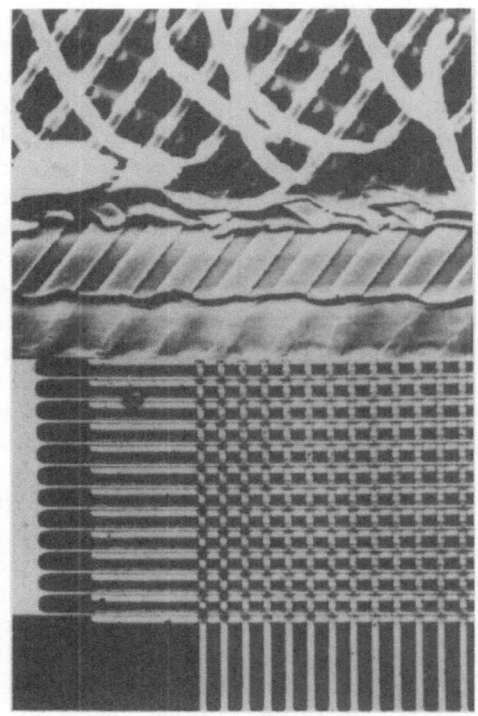

Fig. 17   Polyimide Between Metal 1,2, and 3 (1μm)

The question of making vias in the PI is addressed in two fashions. First, through wet chemical etching and secondly by plasma etching. Figure 18 shows the results of wet chemical etching where the photoresist has been exposed and developed in photoresist developer. The PI had been spun on and cured through 180 degrees C., which removed the NMP solvent. This leaves the PI as a stable, dry film that is quite soluble in the strongly basic developer. Thus, once the photoresist has been developed away from the bottom of the holes in the resist layer, the developer continues to dissolve the PI to form via holes in the PI intermetal dielectric. This is a very simple process in that the development of the photoresist also etches the via holes in the polyimide. A number of 5 μm by 25 μm vias are seen along the bottom of the picture and there are five 25 μm square vias coming down from the top of the picture. The problem is, one of those 25 μm square vias can be seen as being "blown out" or perhaps 25% too large, exposing the end of the underlying metal bar. We have found that this happens in a small percentage of the vias. It has not been explained, but is a severe enough problem one cannot afford to have even a small percentage getting too large, exposing the edges of underlying metal resulting in places where contaminates can be trapped. Therefore, the question of wet etching PI seems to have problems with control and repeatability. Another factor with any wet chemistry is if one considers tens of thousands of 1 μm square vias, 1 μm deep, the chance of getting a liquid in that many small holes consistently is small.

Therefore, one solution to this is to plasma etch the insulator as is shown in Figure 19. Here we have a 2 μm via etched in polyimide using 02 or CF402 plasma. In general, photoresists etch at the same rate or perhaps slightly faster than fully imidized polyimide. This can cause problems particularly with topography as will be shown later. This leads to the need to use a nonerodible mask, which complicates processing. In general, one approach to this is the deposition of .03 to .04 μm thick aluminum or an aluminum silicon layer on top of the PI. The photoresist is then applied, exposed, developed and the aluminum is wet etched or even developer etched. At this point the wafer goes into CF402 or straight 02 plasma, and the

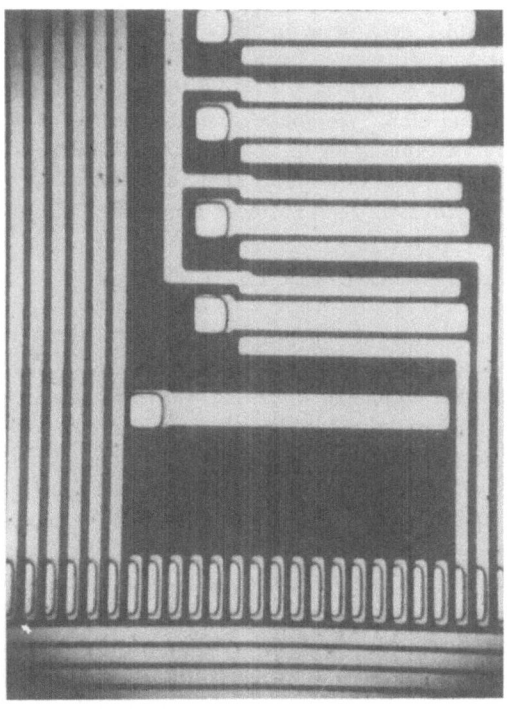

Fig. 18   Wet Etched Vias in Polyimide

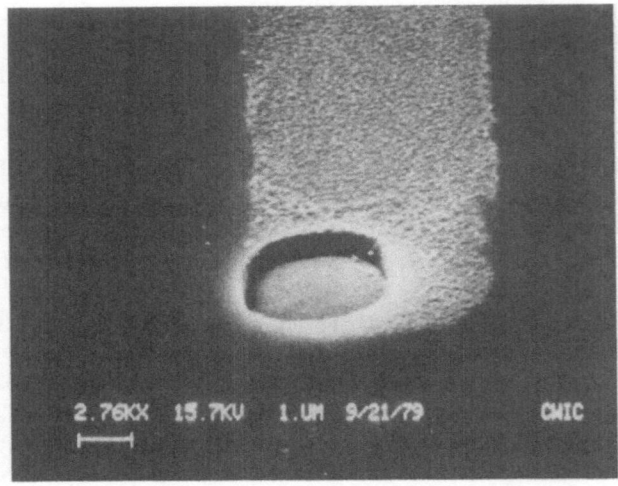

Fig. 19   Plasma Etched Vias in Polyimide

polyimide is plasma etched. If the photoresist is etched away during the polyimide etching, the etching stops on the nonerodible aluminum layer. Of course this is a very thin layer and there are possible problems with regard to particulates causing holes in this nonerodible mask, which allows holes to be etched partially into the PI. However, etching very fine geometry vias is quite easy to do in this fashion. Figure 20 shows 2nd level metal lying on PI vias going down to 1st level metal. This is a 2 μm via chain zig zag arrangement for testing via continuity. One can again see the smoothing effect of the PI, and by proper plasma chemistry one can get tapered sidewalls as shown here, or vertical sidewalls as is shown in Fig. 19.

So in general, the use of PI as a between metal insulator and the process for making vias works very well in laboratory situations. In our experience where we plasma etch metal down to PI, we have seen no cases of corrosion, if the chlorine etch chemistry has been properly suppressed, following metal etch. We don't passivate our test wafers and have had some complex wafer-scale circuits running for six years in room ambient without any failures. This is certainly not life testing, but it does imply that the question of water absorption-caused corrosion may be less of a problem than has been supposed.

Section III   TOPOGRAPHICAL EFFECTS

In this Section, we will look at topographical effects such as step coverage, thickness doubling over steps, topographical-caused lenses and topographical growth. Figure 21 shows 2nd level metal coming over polyimide which is covering 1st level metal at the upper right of the photograph. Second level metal is coming down into a large, common via step in the PI and this illustrates several problems. First, the smoothing of the polyimide over the 1st metal feature is quite evident in terms of 2nd level metal having very good step coverage going over the buried 1st level metal in the upper right hand corner of the picture. However, step coverage problems and thickness doubling is evident where the metal comes over the nearly vertical side of the PI cut. As was discussed earlier the metal is at least twice as thick right at the step, when looking at it from a vertical, or anisotropic perspective. This causes the photoresist exposure problem as well as problems with plasma etching of the metal at the step. Remember, the photoresist, as exposed, will leave some residue at this step where the photoresist would have been thicker. Secondly, the metal is much thicker vertically, in terms of anisotropic or vertical etching process. Therefore, one would normally be etching through 0.75 μm of metal on the horizontal surfaces, but at the vertical step the metal is

Fig. 20   Metal 2 Via Chain on Polyimide

Fig. 21   Unetched Metal 2 on Polyimide Sidewall

close to 1.4 μm thick, which requires overetching. Overetching, unless very anisotropic, means undercut of the metal and narrowing of the lines. In this picture one can see horizontal tabs of metal between the 2nd level conductors on the nearly vertical sidewalls of the PI. This is unetched 2nd level metal and at least a 20% overetch is required to remove these islands of metal which can short the conductors together.

In Figure 22 we see 2nd level metal going over a 85-87 degree step in polyimide. We see the thinning of the 2nd metal as it goes over the vertical sidewall of the PI. The metal here is at least twice as thick on the horizontal than where it goes over the vertical step. Remember, the 2nd metal thickness - in terms of vertical thickness as seen by anisotropic etch - is a function of the step height. A 1 μm high, vertical sided step covered by 1 μm thick metal produces at least 2 μm thick metal right at the step sidewall. At the same time, the thickness through the metal, horizontally, might be only 0.3 μm. This strange condition makes the metal harder to etch while also reducing its current carrying capacity.

Figure 23, summarizes a number of problems caused by topography. In the upper picture we see topographical growth resulting from stacking metal on top of metal with intermetal insulation between the layers. This growth is occurring at approximately one and a half times as fast as the thickness increase. Also illustrated is a very important phenomenon which I call lensing. This can create some interesting surprises, and is not easy to predict until the circuit is completed and the etching is finished all the way through 2nd level metal. As is shown in the upper image, 2nd metal is going over 1st metal, and a line is being imaged on top of the 2nd level metal and intermetal insulation mountain. An image can be positioned on a concavity-in this case metal 3 is to be etched into a line just at the concave top. If this image is the proper width, right at the edge of that concave slot, one will get light undercut, as is shown by the incoming rays being reflected off the sides of the concavity which will expose the resist between the two incoming rays causing that feature to narrow. We have seen cases which produce a very heavy notch or even an open where the metal line is going over stacked via sets like this, and has been exposed by such lensing conditions. Increasing the number of metal levels increases the possibility of this happening, with even more pronounced results, unless the design can avoid stacking either in the via areas or at close spaced crossings. This effect results in a repeated, single open or notch in a conductor which is not caused by imaging processes or particulate defects.

The middle picture in Fig. 23 shows a via problem where, as the via gets smaller and deeper, the step coverage becomes worse, i.e., the metal is thinner and filling the via with metal becomes more problematic. Another factor, which has to always be considered is

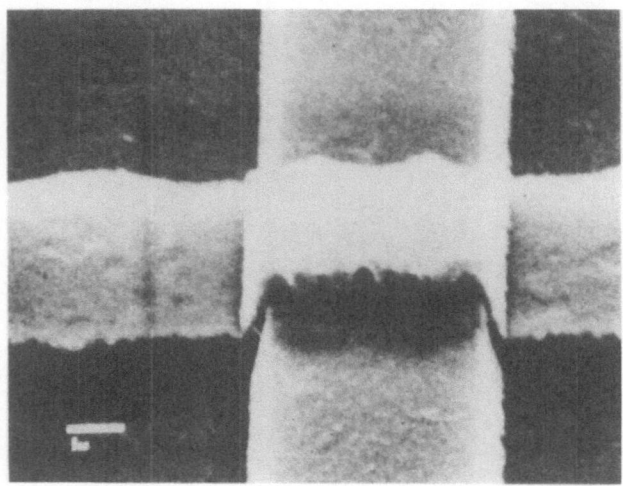

Fig. 22   Metal Step Coverage Over Polyimide

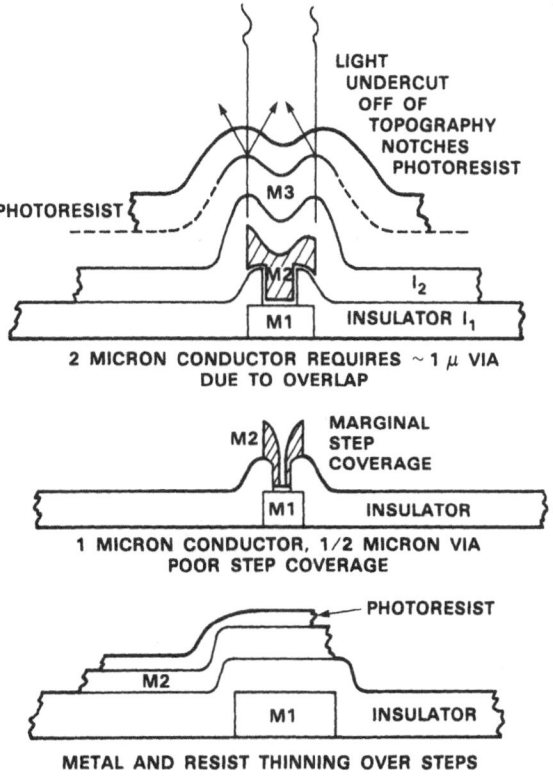

2 MICRON CONDUCTOR REQUIRES ~ 1 μ VIA
DUE TO OVERLAP

1 MICRON CONDUCTOR, 1/2 MICRON VIA
POOR STEP COVERAGE

METAL AND RESIST THINNING OVER STEPS

Fig. 23   Lensing - Via Step Coverage - Photoresist Thinning

the cleaning of metal 1 at the bottom of the via so that metal 2 will make good electrical connection. With a deep, small via, the more difficult it is to assure that all of the intermetal insulation has been removed and that 1st metal is satisfactorily cleaned up, so that 2nd metal will make good electrical contact, or act as a proper starting surface for CVD tungsten. These are very serious problems when one contemplates having circuits with 10,000-50,000 vias which all have to work. Increasing intermetal interconnect densities with smaller vias will certainly lead to an increase in via failures because of these characteristics. The bottom image is an illustration of metal 1 with conformal insulation going over the vertical part of the metal 1 sidewall getting thicker by a factor of 2 on top of the metal vis-a-vis at the upper corner of metal 1. Then the photoresist going on top of metal 2 thinning the most right over the upper corner of the metal 2 step. All of these condition are worse case, and simply cause problems with each step.

Figure 24 is an attempt to illustrate a real life situation, where vias are to be opened in polyimide intermetal insulation going over 1st metal that has PSG insulation below it and polysilicon conductors below that. The upper picture shows the cross hatched 1st level metal with very vertical sides, having been plasma etched, and the 1st metal is sitting on top of the stack of polysilicon, and PSG towards the left side. One can see that the polyimide has thinned from about 5,000 angstroms nominal thickness to about 2,000 angstroms on the upper corner of this stacked, highest topographical mesa. Also, on top of this metal the PI is thinned to 4,000 angstroms. This clearly shows that the PI smooths and does not planarize. True planarization is very difficult to achieve. Also, the resist, which is nominally 20,000 angstroms thick, has thinned to approx. 16,000 angstroms on top of this mesa or protrusion in the circuit. To the right we are going to open a via hole, shown here as 4 μm x 4 μm. The photoresist has been exposed, but hasn't cleaned out fully, or the exposure wasn't quite right. If the via to be opened were 1 μm square, this would be a more serious problem since the

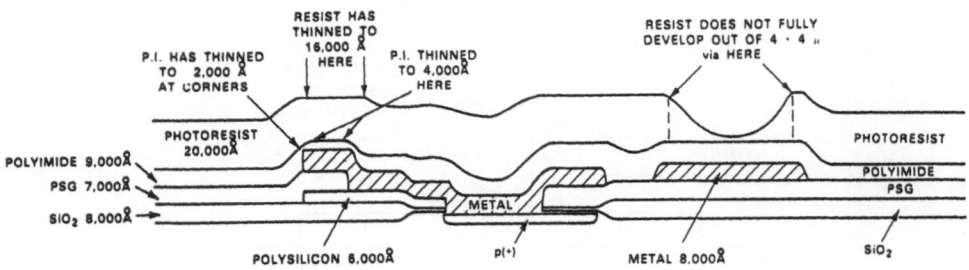

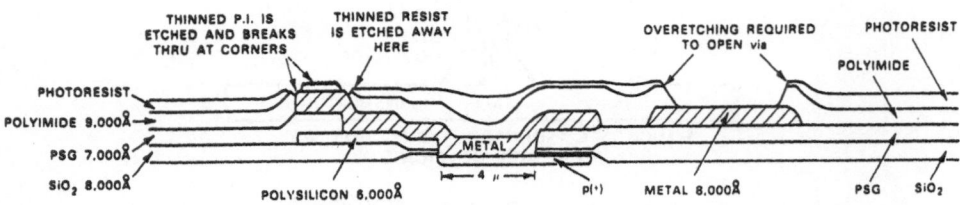

Fig. 24   Photoresist Breakdown On Polyimide Over Topography

development, even with proper exposure, is problematic because of the difficulty of getting a liquid to the bottom of very tiny vias. If you have bubble formation or poor fluid flow you will not develop out the small opening in the photoresist properly. Furthermore, this happens to be a high altitude location and it is conceivable that somewhere else on the circuit there could be a via sitting over a lower area, surrounded on each side by higher topography. This would mean the resist would be thicker over the lower area, which would require that exposure be lengthened in order to properly expose the thicker resist in that topographically-caused hole. Thus, the high altitude vias would get larger in dimension because they would be overexposed. This is a problem that gets worse as the topography differences increase and the via sizes get smaller. In the lower image, we have plasma etched the via in the polyimide, and had to overetch some to fully open that via and clean out all of the PI which was left due to the improperly exposed or improperly developed photoresist which defined the hole originally. Looking to the left, on top of the highest mountain, we see what happened because of the via needing an overetch. Here we have broken through the thinnest areas in the photoresist and polyimide. This exposes the 1st level metal which means that shorting can take place between 1st and 2nd level metal in this area. It also means that etching of 1st level metal can take place during the overetch of the 2nd level metal. Photoresist and PI etch at similar rates, with photoresist often etching at a slightly higher rate than polyimide with certain photoresists. It is therefore essential to use a non-erodible mask on the PI beneath the photoresist to prevent these problems from happening. This gives plenty of margin for overetch of PI and prevents any possible problems with topographically - caused thinning of PI and photoresist.

An example of these problems is shown in Figure 25. The upper SEM is of 2nd level metal lying on PI, with 2 vias visible, lower left and center. One can barely see the 1st level metal going horizontally beneath the upper via, and there is a polysilicon stripe going vertically through the center of the picture. The 1st metal stripe crosses the polysilicon stripe, just to the right of the upper via, and a rough square is visible at the corner of the 2nd level metal as it goes towards the via. There is a rough square on the lower right hand corner of the picture. These areas are over the highest features in this particular circuit. The photoresist was etched through and the PI was then etched as seen by the extreme roughening. This caused electrical shorting between 1st and 2nd level metal. The lower picture is simply a

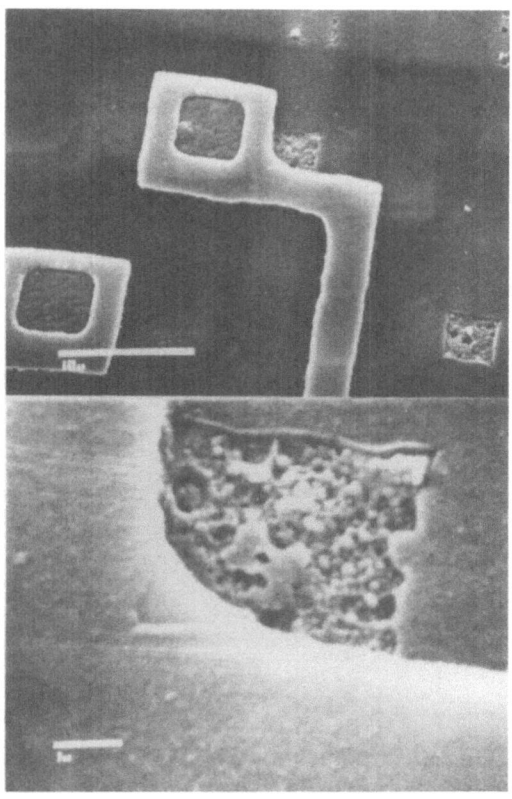

Fig. 25   Polyimide Etch Attack From Topo.- Caused PR Thinning

close up of this rough PI which has nearly been etched away over the buried mesa top that caused the thinning of the PI and photoresist flowing over it. These thinner polymer sections were etched away during the overetch for opening vias in the polyimide. This is a very severe problem, and it was interesting that one couldn't see these rough areas in the microscope - they only showed up in the SEM. It illustrates the topographically caused problems in etching and the need for non-erodible mask to prevent this kind of thing from happening. As the number of layers of interconnect build up, these conditions become much worse, reinforcing the need for good smoothing and particularly for good planarization if possible.

In summarizing current multilevel interconnect status, it is my opinion that metal definition, etching and conductivity is satisfactory for some time to come, on planar surfaces. Metal via fill is poor, and some solution is needed which will be discussed later. The insulators insulate well, but are stuck with having to do the planarization also and are thus in trouble, because they won't smooth or planarize readily. The vertical interconnect is essential but vias have limits on size, depth, ease of clean-out, but again this is an insulator related problem.

Section IV POSSIBLE SOLUTIONS TO MULTILEVEL INTERCONNECT PROBLEMS

In this Section, possible solutions to the multilevel interconnect problems will be discussed, emphasizing planarization, smoothing and via fill. This will essentially review current techniques being used to solve the topographical problems and make it possible to do high density 2,3,4 level metal interconnect. Figure 26, our ideal 4 level interconnect is shown simply to remind us of the goal, which is very small, vertical sided vias, complete planarity and the ability to put metal 4 down to metal 1 directly without going through intervening levels, with vertical sided metal to provide highest density and current carrying capacity. At this point, the effects of planarization should be discussed. In Figure 27, the upper drawing illustrates conformal between-metal insulation where the insulation is equal thickness over bumps and sidewalls with no smoothing, or planarization. The middle image depicts smoothed between metal insulation, which is basically what everyone talks about. The word planarized between metal insulation is used a great deal, but it is inaccurate, since no intermetal insulation currently in use planarizes perfectly. It smooths, which is very important and very useful, because smoothing allows the next level of metal to cover vertical sided steps with fewer difficulties. This middle picture shows PI between-metal insulation, but re-flow techniques used on inorganic insulators can achieve similar kinds of smoothing. The bottom image shows

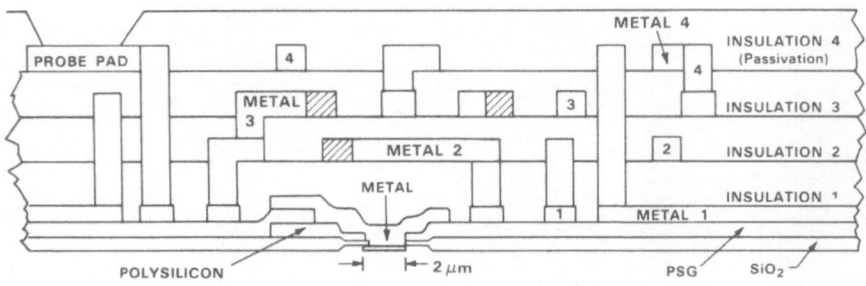

Fig. 26   Ideal Four Level Interconnect

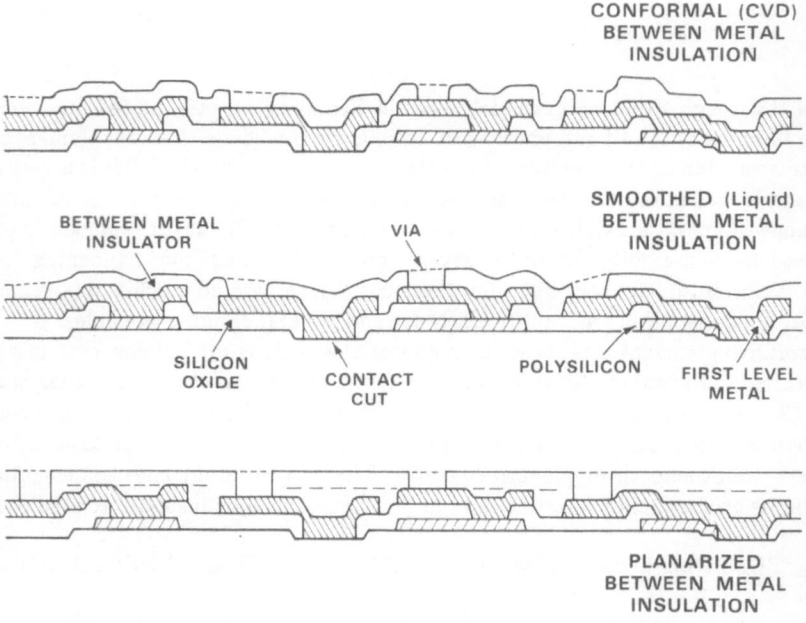

Fig. 27   Conformal/Smoothed/Planarized Topography

planarized between metal insulation. This is truly planar. Whatever is going on below is not reflected on the surface above. Some researchers have been able to achieve this with very considerable effort, and very complex processing techniques. However, it carries with it some disadvantages along with its planar advantage. First, as you can see, the vias will be different depths. This is a very difficult problem which generally has not been addressed. Secondly, the insulation thickness varies, thin over high altitude features, thick over features in valleys.

Figure 28 illustrates one very simple approach to smoothing. Here, the the CVD insulation over vertical sided metal is shown, having the normal reentrant or breadloafed profile. If one simply plasma etches this, the sharper corners etch fastest and smooth to some extent as shown in the lower figure. However, thickness is being lost and this process is not smoothing very well, as can be seen to the left and right where the breadloaf has become nearly vertical sided rather than reentrant. Such problems mean that this approach doesn't work very well and is not really used much. Figure 29 shows the problem with trying to planarize using a spin on material. The three sets of images show constant 1.4 μm thick features either as white side views or as cross hatched end views. These various width features are then coated

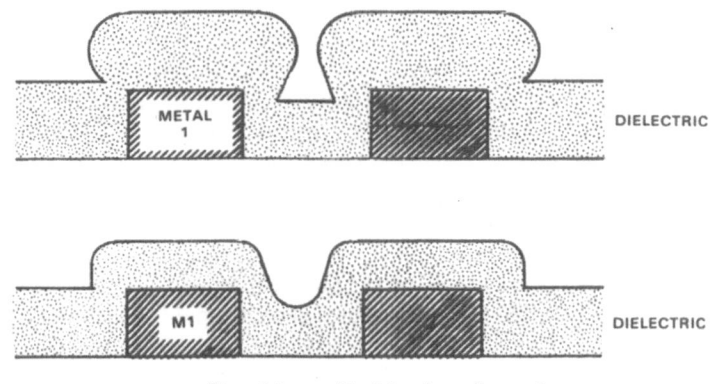

Fig. 28   Etchback to Smooth

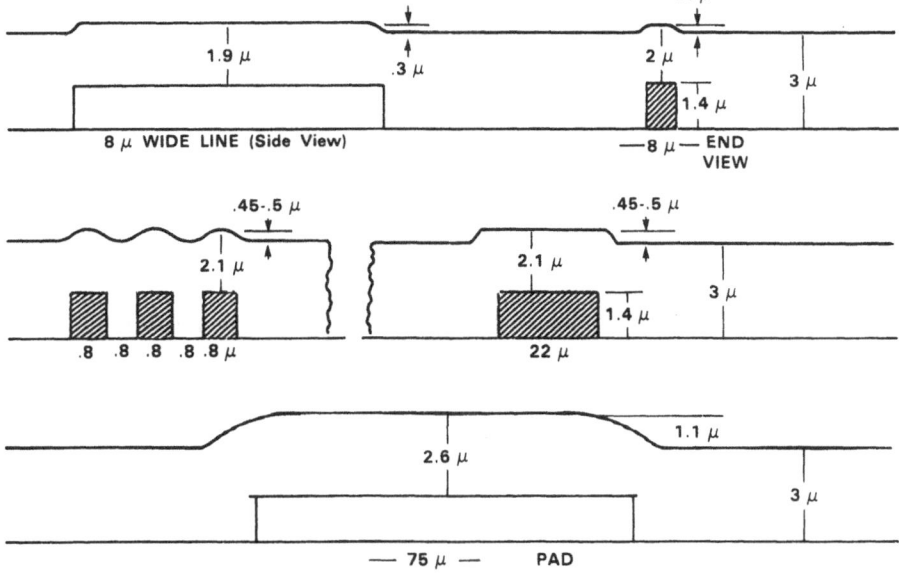

Fig. 29   Smoothing Characteristics of Spin-On Polymer

with 2 μm thick resist over 1 μm thick PI, for a total 3 microns of spin-on material over these features. Actual measurements show 2 μm of insulator above the 1.4 μm high conductor. Looking at the middle picture, you see that the 8 μm wide conductors have 8 μm spaces between them, and the insulation thickness increases by 1,000 angstroms to 2.1 μm. This is also true on the 22 μm wide conductor. In the bottom figure, however, the 75 μm wide pad starts looking like an infinitely wide surface to the PI, and the insulator thickness increases to 2.6 μm. The viscosity, solids content, drying and spinning characteristics of this PI are such that a 150 μm wide feature would produce 3 μm of insulator on top of it, so there would be no planarization. Once the dimension of the lower topography becomes wide enough to look infinite to the spin on the material, that material will build as thickly on top of that topography as it would on a planar, featureless surface. This is a very serious problem, in that it means one is having to etch different thicknesses of spin on material. Non-conformal, or planar insulation gives a better surface on which to print fine geometries above the lower features, but the non-conformal condition brings with it the disadvantage of having different thicknesses of insulator to etch holes through for vias to make connection between lower metal and upper metal.

Figure 30 shows a widely reported technique which is possibly used in production. Here the insulator has quite a bit of topography as it lies over the lower metal interconnect as outlined in the upper image. A very thick organic layer is spun on top of it. This tends to bury and smooth to some extent. One would have to spin many microns of organic on top of 1 μm steps to fully planarize, which is not practical. Normally a few microns of an organic layer are spun over 1 μm steps, and this will smooth or 'planarize' to a maximum of 60%. Plasma etching of the organic layer down to the insulator results in the organic layer ideally filling the spaces between the CVD insulation topography. Now, if the etch rate of the planarizing layer and the etch rate of the insulator can be closely matched, one continues etching the insulator and the organic, which is protecting the trench bottoms, to produce the idealized result in the bottom picture.

Since the oxide insulation is conformal, or of uniform thickness on top of all topography, the question is, are all topographic features at the same altitude? In the unlikely case that this is true, then an organic layer which fully planarizes is to be desired. Thus, the plasma etchback exposes all topographic peaks simultaneously and the inorganic is etched off the

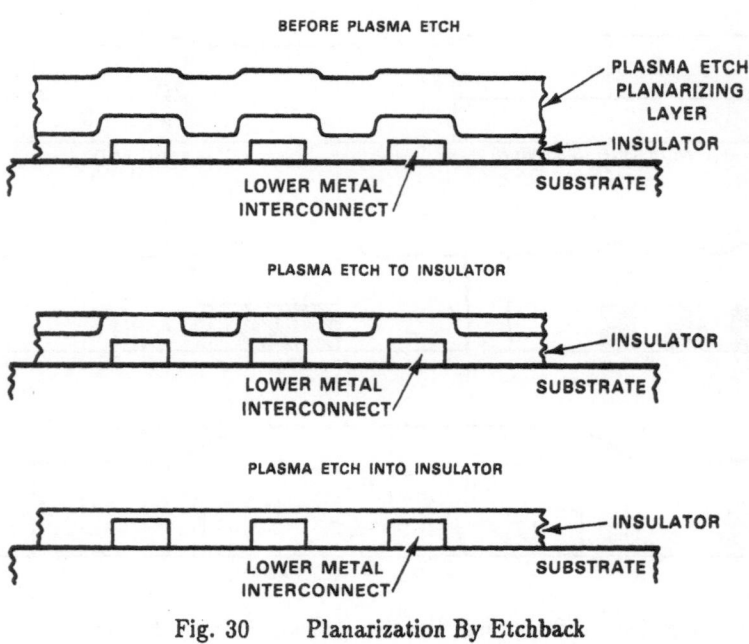

Fig. 30   Planarization By Etchback

tops of all underlying topography uniformly. Ideal levelling is accomplished if the insulation thickness is identical to the height of the topographical features. However, in real life the conformal oxide insulation covers topographic features which are at different elevations. Now, if the organic layer planarizes perfectly, the inorganic on the highest topographic elevations will be exposed to the plasma first during etchback, and possibly etched away before the inorganic on the lower topographic elevations is even exposed to the plasma. This is because a planar layer over topographic peaks that are not planar differs in thickness, being thinner over high altitude peaks, and thicker over lower peaks. In this case, the organic insulator can be removed from the highest topographic peaks only, as there is no point in removing it from the lower elevations unless repeated depositions and etchbacks are used. This means that the underlying topography is planarized only by an amount equal to the thickness of the inorganic oxide removed from the top of the highest elevations. Also, after the next CVD deposition, the oxide will be thicker on the lower, unetched topography by the amount of the new oxide thickness. This means nonuniform via depths and the problem of overetching shallow vias while waiting for the deep vias to clear, getting good step coverage in deep vias, etc.

If it is desired to obtain uniform removal of an inorganic oxide on topographic peaks having differing altitudes, a conformal organic is preferred, as it will be of uniform thickness over all topography and permit uniform inorganic removal everywhere. However, this will not planarize the underlying topography at all. It will only maintain uniform insulation thickness over topographical peaks, regardless of their altitude differences. In general, it is impossible to planarize topography with a sacrificial etch-back technique, except for the ideal case outlined in Fig. 30.

The big problem here is that complete conformality is not often achieved with a spin-on material, as has been noted previously. Also, finding a planarizing spin-on organic which will etch at precisely the same rate as the underlying insulator is very difficult. There are many cases where a slight difference in etch rate of the planarizing layer vis-a-vis the underlying layer, leads to very strange effects in the final etched result where fences are formed along the former vertical sides of the underlying features. Figure 31 shows the sacrificial etch-back layer over the dielectric insulation and one can see what often happens. The sacrificial layer is non planar and as it etches down, it etches more or less rapidly than the dielectric. In this case it etched faster than the dielectric and at the end of the etch, when the sacrificial layer is gone, the dielectric features are diminished, but still remain. Most importantly, the reentrant characteristic of the dielectric has not been removed. So there is a lower profile, but the reentrant corners of the dielectric on the left and right in the lower image would still be difficult to cover with the next metal layer. Figure 32 is an SEM view of the CVD oxide in the left hand picture before planarization. The right hand image shows the amount of planarization that has been achieved after three coat/etchback sequences using a sacrificial

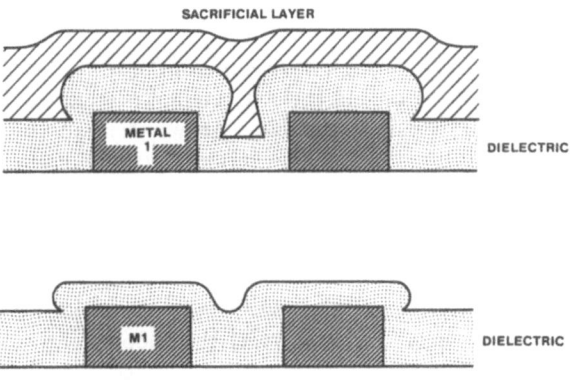

Fig. 31   Sacrificial Layer Etchback

319

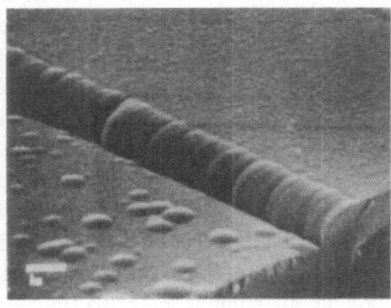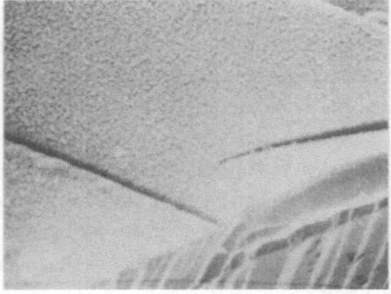

Fig. 32   Planarization of Intermetal Dielectric

layer. Here the sacrificial layer etched more rapidly than the CVD oxide, so it was etched all the way off, then coated again, then etched again, each time going further into the CVD oxide. After a third coat and etchback, considerable planarization was achieved.

Figure 33 shows the end view of the same topography before planarization on the left and after planarization on the right. In this case multiple sacrificial etchback worked reasonably well. However if one were to look at wide conductors or wide topography elsewhere on the circuit one would see that those features would not be as well planarized as in this case. Furthermore, this required repeated coatings and etch-backs of the sacrificial layer, which is a relatively complex process. In Figure 34, we see one of the problems that takes place with the sacrificial spin-on layer and what can happen. In the upper image the topography is clearly buried by the smoothing sacrificial layer. In the middle photograph the same topography is shown after the sacrificial layer has been etched off. If one looks closely at the interior corners of the topography little dark lines are visible, which are shown closeup in the lower photograph. This is a case where the sacrificial layer was etching faster than the vertical sided, underlying topography. The sacrificial layer was etched off the top of the topography, but there is still some sacrificial material on the lower level, and one can see where it has shrunk back from the inside corners of the topography. In the lower photograph the sacrificial layer has been etched away completely and it is clear where the pull back of the sacrificial layer allowed the etching of the underlying surface at the inside corner of the topographic feature which was being planarized. This problem depends on the amount of shrink that takes place during the etching of the sacrificial layer, which in turn depends on the temperature reached during the plasma etch and the thermal stability of the sacrificial polymer. A high speed process can become quite hot and considerable shrink of spin-on polymer can occur.

In Figure 35 the upper picture illustrates an example of the vertical sided insulator before etching. A sacrificial layer was spun on and the etch rate of the sacrificial layer was matched to the rate of the CVD oxide as closely as possible. Part way through the process, the lower photograph in Fig. 35 shows the oxide mesa having been etched, with the sacrificial layer clearly lying to the left and right of it, almost planar with the oxide top. This shows that the two materials are etching at almost the same rate. In Figure 36 the CVD oxide can be seen to have been etching a bit faster than the sacrificial layer as the rough topography of the oxide is slightly below the surface of the sacrificial layer. The CVD oxide is essentially planarized and the sacrificial layer is selectively stripped to show the result in the bottom picture. Here the flattened CVD oxide topography looks rough with the smooth substrate to the left which had been protected by the sacrificial material. However, a "fence" remains, which appears as a white line around the edge of the features which were once the sidewall of the topography.

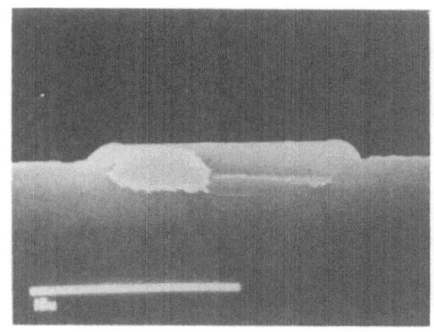 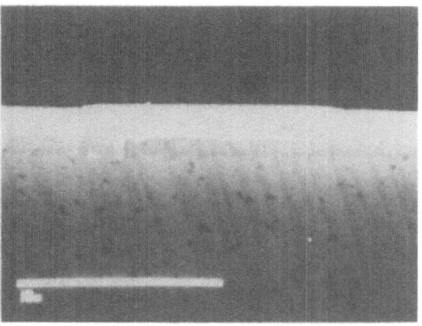

BEFORE PLANARIZATION   **AFTER PLANARIZATION**

Fig. 33   Planarization of Intermetal Dielectric (End View)

Fig. 34   Sacrificial Layer Shrink After Etch (Gap)

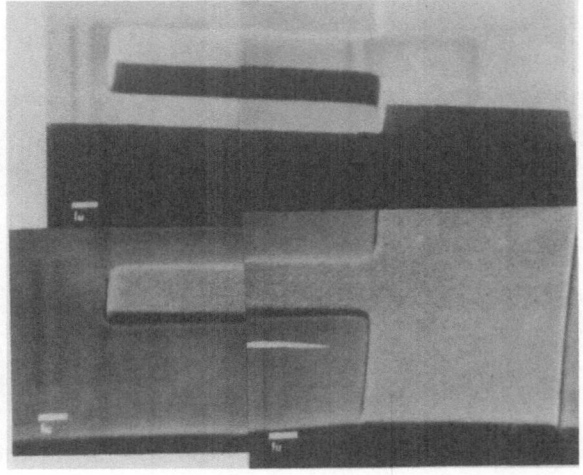

Fig. 35   Sacrificial Layer Fence Problem

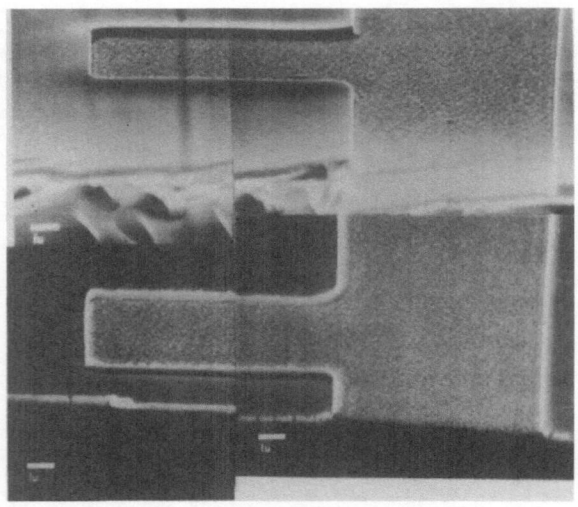

Fig. 36   Sacrificial Layer Fence Problem II

This "fence" is approximately 4000 angstroms high, perhaps a 1000 angstroms wide and is of unknown composition, other than it seems to result from a reaction involving the sacrificial material and the etching of the oxide. It is essentially impossible to remove with the various wet and plasma processes tried. Thus, even slightly imbalanced etch rates between the oxide and sacrificial layer can leave a high, vertical "fence" or wall around the former edge of the planarized topography.

Figure 37 is a drawing showing Spin-On Glass (SOG) being used to help smooth CVD oxide. CVD oxide can be put over metal runs, and then Spin-On Glass applied. SOG is a very low viscosity material which on the plane surface, may build only 2000 angstroms, but it has the nice feature that it piles up in the vertical sides and the reentrant corners and therefore tends to smooth the topography. It is a material which has to be baked through a series of carefully controlled steps up through 450 degrees C. to prevent its cracking in these thick cross sections which build at the sides of vertical steps. If cracking can be minimized, then another layer of CVD oxide can be put down which is much smoother. The concern with SOG cracking is not a mechanical problem since such cracks nominally are 100 - 1,000 angstroms wide, and the SOG would support the overlying CVD. However, the cracks may act as trapping areas for etching products, which could later cause problems in terms of circuit reliability. Figure 38 is an end view SEM photograph of this kind of situation, where the metal is seen with the first level of CVD oxide covering it. Then the black SOG fills the space between the two conductors, and it can also be seen as a very thin black layer on top of the CVD oxide. On top of this another layer of CVD oxide has been put down, which is seen to be much smoother than the CVD oxide beneath the SOG, but it is not planarized - merely much smoother and tends to have the look of a spin-on material, This means that the next level of metal can achieve good step coverage. Figure 39 illustrates another complex smoothing process involving lift-off techniques. Here an organic photoresist layer is defined on top of the metal run, and CVD dielectric is deposited. A wet etch undercuts the dielectric as shown in the second image from the top, exposing the edge of the photoresist. The photoresist can then be dissolved, taking the breadloafed CVD top off the metal run, which

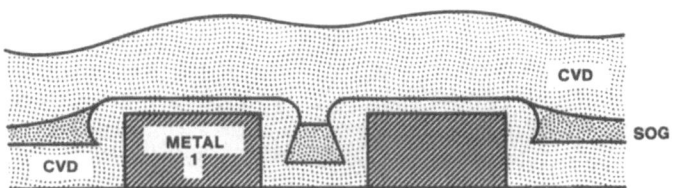

Fig. 37   CVD/SOG/CVD Planarization

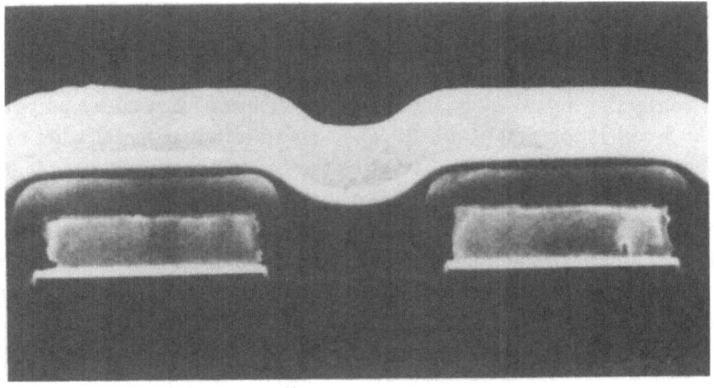

Fig. 38   Spin-On Glass End View

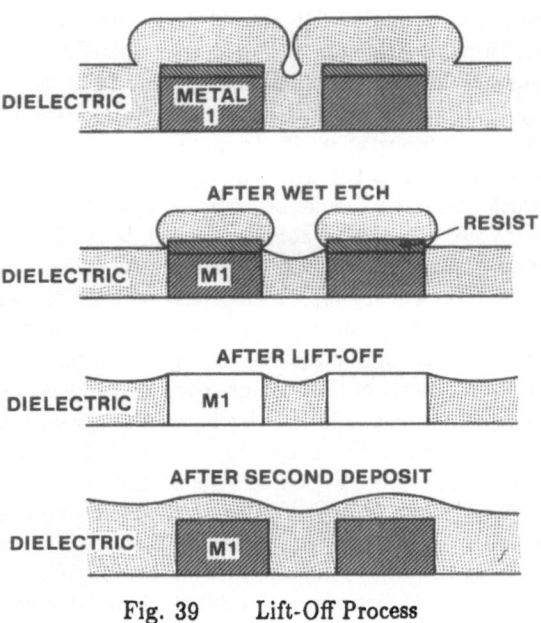

DIELECTRIC METAL 1

AFTER WET ETCH

DIELECTRIC M1 — RESIST

AFTER LIFT-OFF

DIELECTRIC M1

AFTER SECOND DEPOSIT

DIELECTRIC M1

Fig. 39   Lift-Off Process

gives a nearly planar surface. Then a second deposit of CVD oxide is put on, as shown in the bottom image. This again is a very complex process which requires careful control of etching, resist thickness and dimensions, alignment and CVD thickness. In any of the planarization processes involving the etchback of the sacrificial layer, another factor has to be kept in mind which is illustrated in Figure 40, showing polyimide shrinkage. The spun on polyimide in the left hand image planarizes fairly well. However, once the solvents are baked out and the material is fully cured it shrinks more (20 - 25%) in the thicker sections beside the metal conductor than it does in the thinner sections on top of the metal conductor. This of course serves to accentuate topographical features instead of planarizing them. Quite frequently in etch back processes involving sacrificial layers, the organic sacrificial layer must be baked at relatively high temperatures in order to endure the temperatures reached during a production plasma process. So this is a condition that has to be considered in any sacrificial etch back technique.

We now look at what is happening in real life situations by reviewing some processes being employed by manufacturing organizations. Figure 41 illustrates a three level interconnect arrangement used by IBM. The drawing shows the lower level of metal being covered by a nitride layer, which is done to isolate the polyimide from the surface of the wafer and from the contact cuts. Vias are etched in the nitride layer, PI is spun on, vias again opened, metal 2 is generated contacting metal 1 through the vias. This is followed by polyimide 2, then vias, and metal 3 which contacts metal 2. Then on to terminal metal where solder bump technology is employed to connect the circuit to the backplane or package. It is interesting in this situation that the polyimide is apparently stable enough to accommodate the forces and temperatures involved with the solderball reflow process that IBM uses.

The next figures illustrate a double layer refractory metal process which was developed and used in production by Hewlett Packard. Figure 42 summarizes the process. I am assuming that CVD blanket tungsten is used for the interconnect, with metal 2 being 1.8 μm thick. Figure 43 shows the overview of the die, and Figure 44 is a magnified view of the outlined area of Fig. 43, with the inter-metal dielectric removed. As one can see, the one and half micron first metal lines with one micron spaces between them are quite tightly packed, and

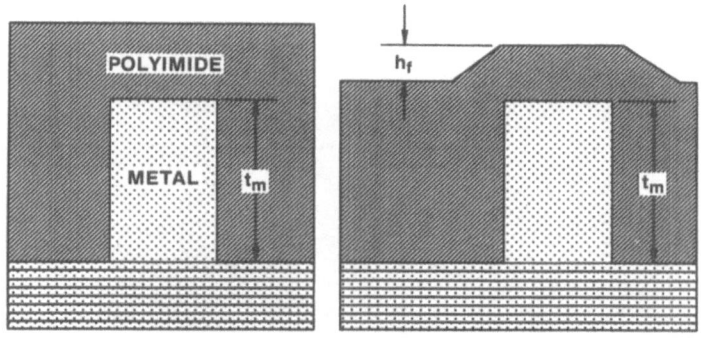

Fig. 40   Polyimide Shrinkage

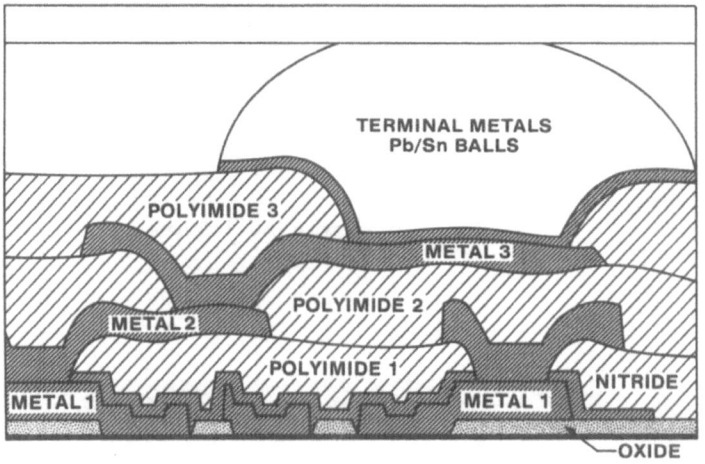

Fig. 41   IBM TLM Process With Polyimide

METAL 1

TUNGSTEN	4000 Å
LINE/SPACE	1.5 µm/1.0 µm
ETCH UNDERCUT	.1 µm/EDGE

METAL 2

TUNGSTEN	18,000 Å CVD
LINE/SPACE	5 µm/3 µm
Rs	.04Ω/□
ETCH UNDERCUT	.5 µm/EDGE

Fig. 42   HP Double Layer Refractive Metal

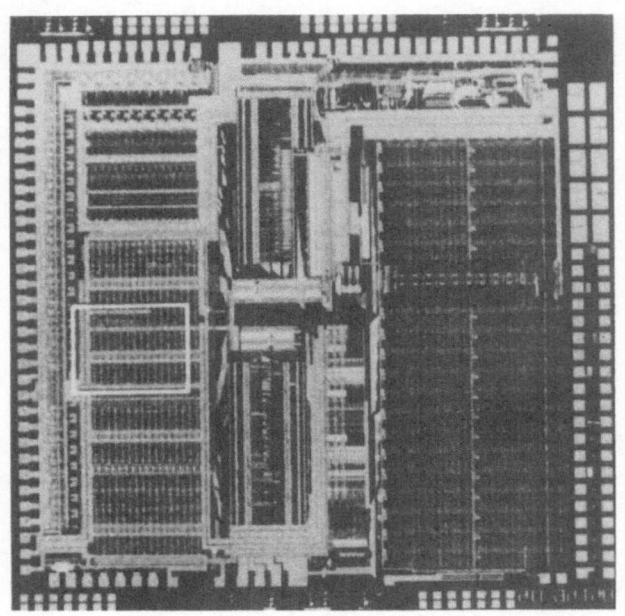

Fig. 43　　HP Device Die Overview

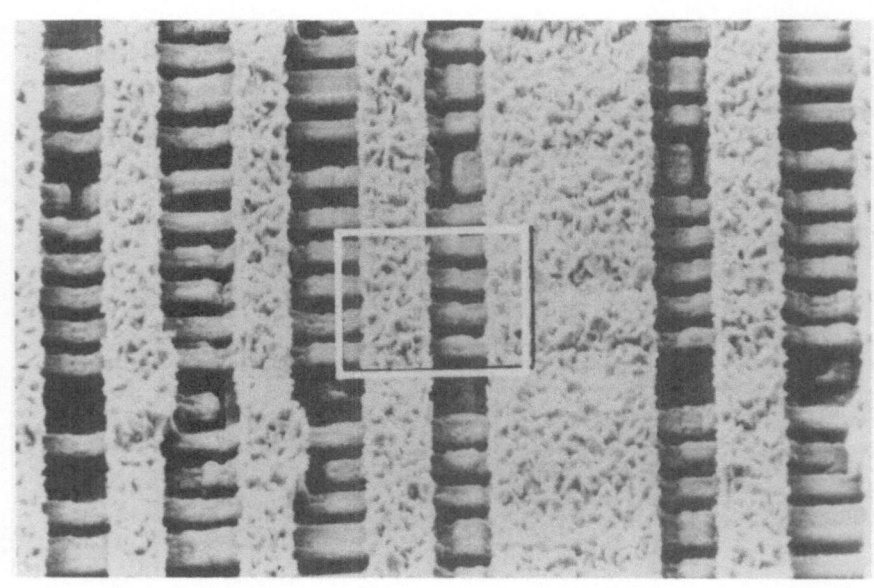

Fig. 44　　HP Closeup of 2-Level Tungsten

one suspects that this process is being used to its maximum capability. Figure 45 is the area enclosed by the white box in Fig. 44, and the first level tungsten is rather smooth, while the 1.8 μm thick second level tungsten is very rough. The thick second level tungsten is probably needed to cover steps produced by the topography of the first level metal and also to provide a low enough series resistance for the needs of the circuit design. Figure 46 shows the first level metal making contact to the silicon below. Figure 47 shows the second metal to first metal via connections, and Figure 48 is a cross-section view with first metal conductors seen as small rectangular whitish areas in the dark CVD oxide insulating layer, and the second metal resembling cauliflower on top of this. This process was in production for several years

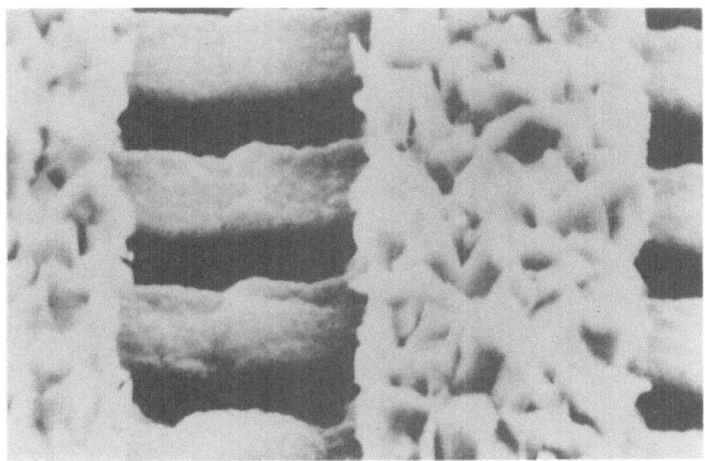

Fig. 45   HP Closeup of 2-level Tungsten II

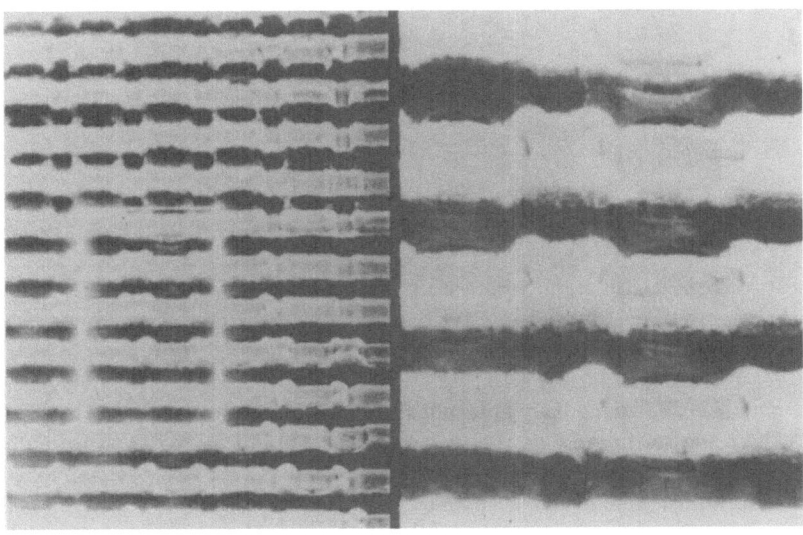

Fig. 46   HP Contact Closeup

Fig. 47   HP Via Closeup

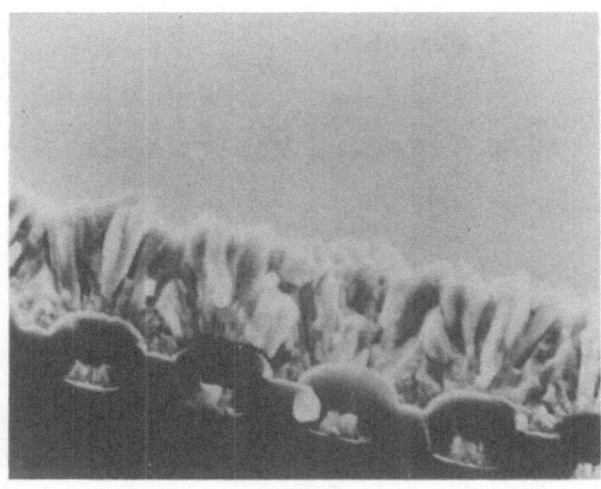

Fig. 48   HP 2 Level Tungsten Cross Section

and illustrates the considerable effort which commercial manufacturers expend to produce high density, multi-level interconnect. It appears that all of this metal has been plasma etched from the anisotropic nature of the metal sidewalls.

Figure 49 is an example of selective CVD tungsten deposition technology. This shows contact cuts in oxide which have been filled with CVD tungsten where the selective process permits the tungsten to grow on the underlying silicon and not on the silicon dioxide insulator. Thus, this offers considerable advantage in filling vias and overcoming some of the via problems which were described earlier. Figure 50 is looking at selective CVD tungsten which has been deposited in trenches in the silicon dioxide insulating layer with the lower image showing the end view, illustrating how well the 1 μm thick CVD selective tungsten

Fig. 49   Selective Tungsten in Holes

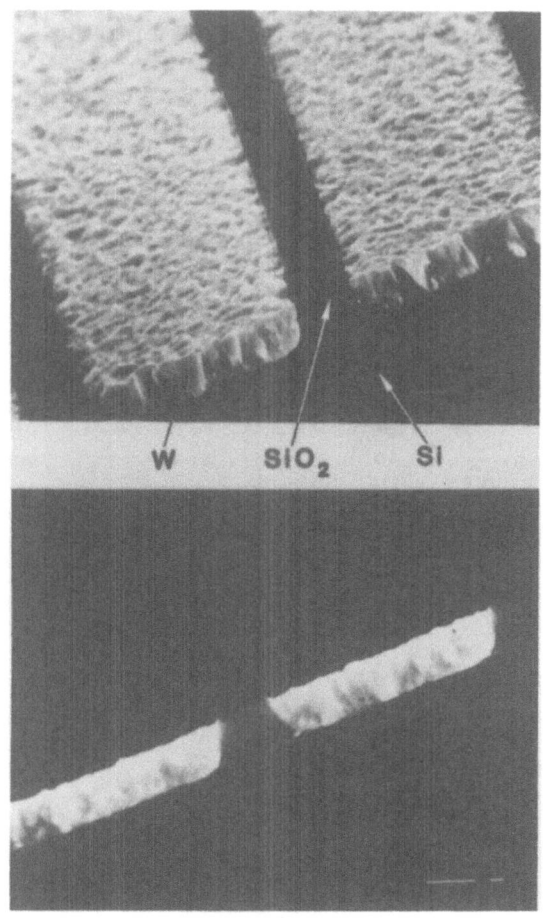

1 μm Thick

Fig. 50   Selective Tungsten in Trenches

planarizes the trenches. Figure 51 is an example of blanket CVD tungsten via fill in the left hand image and selective CVD tungsten via fill in the right hand image. As can be seen in the left hand image, blanket CVD tungsten deposits, on the bottom and sides of the narrow slot filling it partially, but leaving a void in the middle. However, if the CVD tungsten is selective and grows from the bottom of the slots, as is shown in the right hand photograph, the via is uniformly filled. Once the slots or vias are planar, a blanket CVD tungsten layer can be deposited on top to give a planar profile. The selective CVD tungsten process offers considerable promise particularly when coupled with blanket CVD tungsten or even when paired with aluminum interconnect. In the hybrid case of aluminum and tungsten the CVD tungsten can be used as a vertical via fill and the aluminum can be used for the horizontal interconnect. The basic problem with using tungsten interconnect is the high resistance of the tungsten compared to aluminum. This means the need to use thicker metal or somehow be able to cope with a higher RC time constant. If selective CVD tungsten is used only to fill vias, with aluminum for horizontal runs, then the series resistance of the short, tungsten via fill is a small addition to the resistance of the circuit and good planarity is achieved. However, a considerable amount of work is needed to fully prove these techniques for production. The questions which have to be answered are- how small a via hole can be filled with the selective CVD process? How repeatable is the process? Also, there is the very difficult problem of the CVD tungsten selective process requiring good, clean silicon at the bottom to begin with. Can tens of thousands of via holes be made with uniform bottoms for starting a CVD tungsten selective process? One of the recent advances in the CVD tungsten process has been the use of silane rather than hydrogen in the selective CVD tungsten reaction. It has been reported that the selectively is better, deposition rates are as much as 1 μm/min and problems experienced with the selective CVD process eating into the underlying silicon are eliminated. To date the selective CVD process has not been reliably selective-it can spontaneously become blanket tungsten on areas of the silicon dioxide insulator. It also has a deposition rate on the order of a few hundred Angstroms/min and attacks the underlying silicon which can cause problems with the contact.

The next five figures show a process developed by IBM to achieve complete planarization. It was found to have problems in prototype tests which prevented its use in production, but is an instructive example of ideal planarization, as well as the kinds of unexpected results which

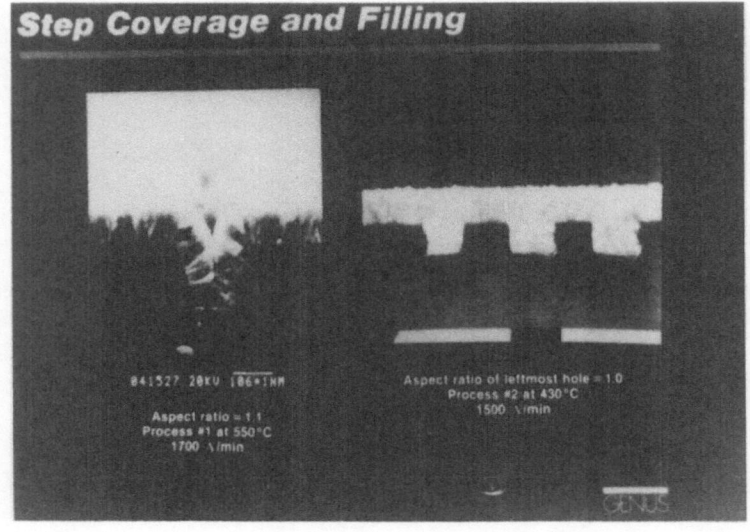

Fig. 51   Blanket/Selective Tungsten Via Fill

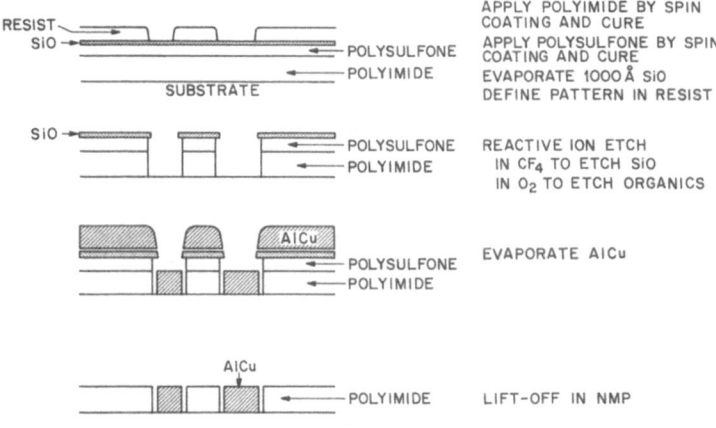

RESIST
SiO
POLYSULFONE
POLYIMIDE
SUBSTRATE

APPLY POLYIMIDE BY SPIN
COATING AND CURE
APPLY POLYSULFONE BY SPIN
COATING AND CURE
EVAPORATE 1000 Å SiO
DEFINE PATTERN IN RESIST

SiO
POLYSULFONE
POLYIMIDE

REACTIVE ION ETCH
IN CF$_4$ TO ETCH SiO
IN O$_2$ TO ETCH ORGANICS

AlCu
POLYSULFONE
POLYIMIDE

EVAPORATE AlCu

AlCu
POLYIMIDE

LIFT-OFF IN NMP

Fig. 52   IBM Multilevel Concept I

arise in dealing with complex systems. The top image in Figure 52 shows the substrate coated first with Polyimide, then with Polysulfone, then with SiO_2 and then finally with defined photoresist. In the next image the SiO_2 has been etched in CF_4, and the Polysulfone - Polyimide layers etched in oxygen. You will notice that the SiO_2 is undercut and overhangs the organic Polysulfone and PI slightly. In the next image AlCu has been deposited and since the SiO_2 overhang causes a break in the deposition, the AlCu deposited on the bottom of the openings is not connected to that lying on the upper surfaces. In the bottom image, the Polysulfone is lifted off in NMP, taking with it the SiO_2 shadow mask layer and the overlying AlCu. This leaves the AlCu at the same height as the PI thickness which gives a completely planar surface. However, note that the AlCu does not fill the slots in the PI completely and there are very small gaps between the AlCu first level metal interconnect and the sides of the slots in the PI.

Figure 53 shows a continuation of the process where PI is again spun on, Polysulfone is spun on top of that, then another SiO_2 layer and finally resist is defined. The middle image of Fig. 53 shows the same etch of the organics and the SiO_2 to form a shadow mask overhang with the SiO_2. Again AlCu is deposited in this hole and the Polysulfone lift off material is removed in NMP taking the SiO_2 and AlCu with it. This forms what would be a via fill, which is planar with the PI upper surface, and which does not fill the via hole out to its sides. The process is then repeated in Figure 54, and a second level of metal interconnect is put down going through the same deposition, etching, metal deposition and lift off steps. This again makes a very planar surface up through second level metal. At the bottom of Fig. 54, all levels are shown, with complete planarity being maintained at the upper surface of the PI insulator.

Figure 55 is a SEM view of a cross section showing four levels of metal put down using this technique. At the bottom of the figure on can see five small rectangular metal 1 lines, with second level metal going on top of them horizontally in the center of the figure. Then, with wider third level metal runs seen as three larger rectangles and finally, fourth level metal on top of that. Some via fill metal shows as three dark, rectangular, vertical columns. Looking at the surfaces of second, third, and fourth level metal going over the underlying topography it is quite clear that this is a very flat arrangement. This is a tour de force demonstration of the kind of planarity desired.

Figure 56 is an oblique view of this circuit and again one can see the high resolution, dense lower metal with the intermediate and widest top level metal clearly shown. This is a perfect example of what is desired, and represents the goal of multi-level interconnect.

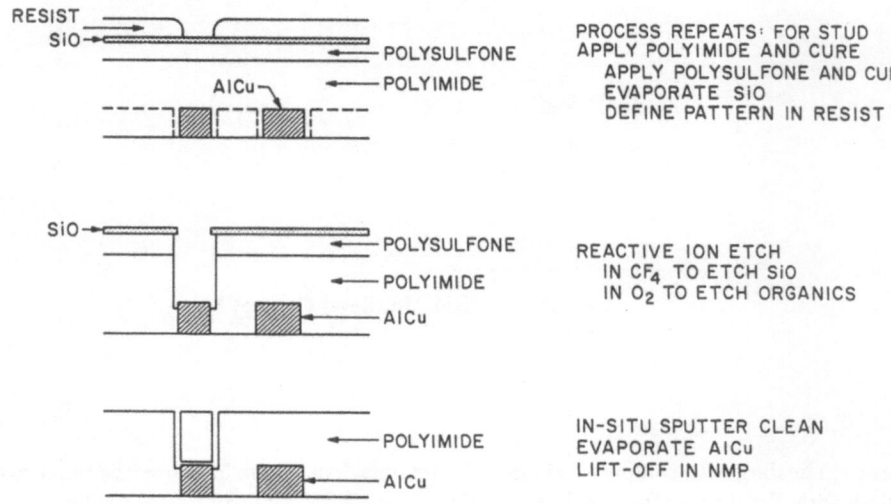

RESIST
SiO
POLYSULFONE
POLYIMIDE
AlCu
AlCu

PROCESS REPEATS: FOR STUD
APPLY POLYIMIDE AND CURE
    APPLY POLYSULFONE AND CURE
    EVAPORATE SiO
    DEFINE PATTERN IN RESIST

SiO
POLYSULFONE
POLYIMIDE
AlCu

REACTIVE ION ETCH
    IN CF$_4$ TO ETCH SiO
    IN O$_2$ TO ETCH ORGANICS

POLYIMIDE
AlCu

IN-SITU SPUTTER CLEAN
EVAPORATE AlCu
LIFT-OFF IN NMP

Fig. 53   IBM Multilevel Concept II

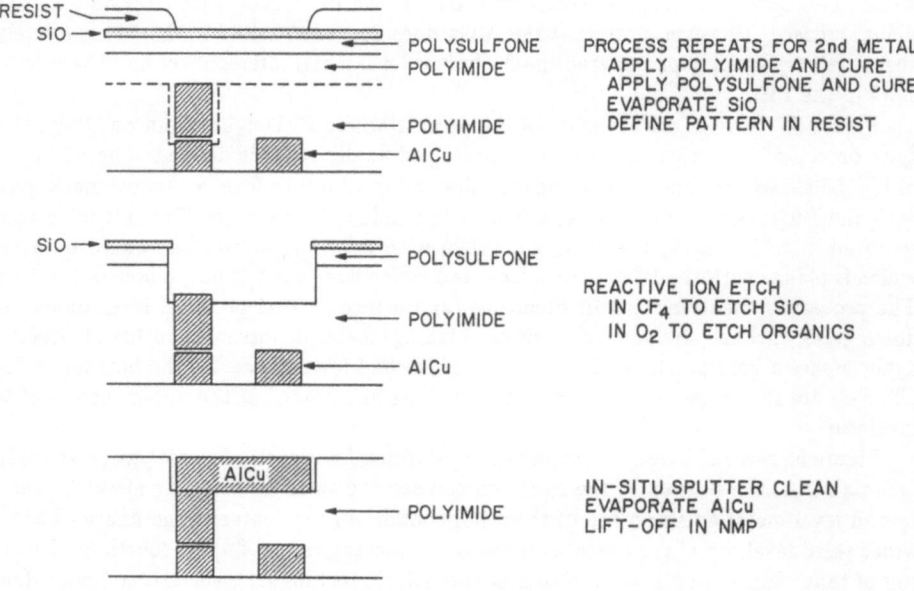

RESIST
SiO
POLYSULFONE
POLYIMIDE
POLYIMIDE
AlCu

PROCESS REPEATS FOR 2nd METAL
    APPLY POLYIMIDE AND CURE
    APPLY POLYSULFONE AND CURE
    EVAPORATE SiO
    DEFINE PATTERN IN RESIST

SiO
POLYSULFONE
POLYIMIDE
AlCu

REACTIVE ION ETCH
    IN CF$_4$ TO ETCH SiO
    IN O$_2$ TO ETCH ORGANICS

AlCu
POLYIMIDE

IN-SITU SPUTTER CLEAN
EVAPORATE AlCu
LIFT-OFF IN NMP

Fig. 54   IBM Multilevel Concept III

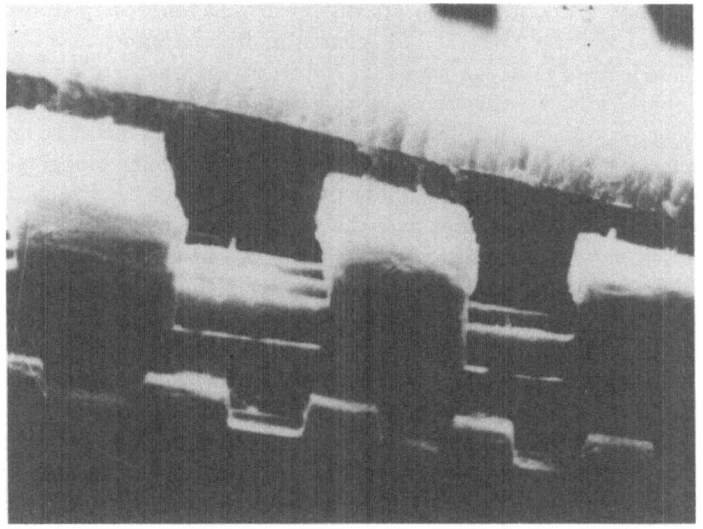

Fig. 55   IBM Multilevel Concept IV

Fig. 56   IBM Multilevel Concept V

It is very interesting that the work done on this process looked very hopeful through the prototype stage. However, it seems there were two problems which kept it from going into production. It was previously noted that the metal did not tightly fill to the sides of the holes. Apparently there were problems with the PI layer that went on over this filling those very small interstices, which meant leaving voids at the sides of the metal. This could result in trapping of chemicals which might later lead to reliability problems and also cause potential step coverage problems for the next metal interconnect layer in the case of the via studs. Another very interesting, unpredicted difficulty was the poor adhesion of PI to itself in the upper levels of this interconnect. It seems that the polyimide did not adhere to the underlying, cured polyimide adequately to keep the circuits from delaminating. This problem was not easily solved, certainly had not been foreseen, and kept this process from going into production. Of course the final comment on this very interesting demonstration is that it is a many, many, many step process involving a very complex technology. This simply underscores the fact that need for planar, multi-level interconnect is so great that manufactures are looking at any process that will work, almost regardless of its complexity.

A technique which is currently being investigated to help with planarization is shown in Figure 57, and one equipment maker has a machine available to do this. The upper SEM cross section shows metal going over the five steps forming the underlying topography and one can see the metal almost bread loafing going over it, creating severe topographical dislocations. This process scans an Excimer laser over the wafer surface melting the Al, causing it to become liquid and flow into the topography as is shown in the bottom cross section. The topography of the metal has been planarized almost completely, vis a vis the topography over which it is lying. This is an interesting process and does a good job of planarization in this illustration. The Excimer laser spot size is much smaller than the wafer, so must be scanned back and forth and is pulsed, which means there have to be overlaps of the illumination spots in the scan direction and sideways to achieve full melting.

Several observations have to be made about this process. First, it may be rather time consuming to planarize metal on a wafer by this technique since it is essentially a serial process. Secondly, what happens to the grain size, the oxide formation and the conductivity of the metal as a result of this reflow? Thirdly, what happens with regard to the overlap of the XMER laser traces in terms of the grain size and the electrical characteristics of the double melted areas that inevitably occur with the overlap. Fourth, the reflectivity of the aluminum changes as a result of oxide thickness and surface roughness, so there may be areas that reach too high a temperature in the overlap regions. Does the liquid aluminum tend to spike through thin gate oxides? It also has the effect of making for interesting etching problems in

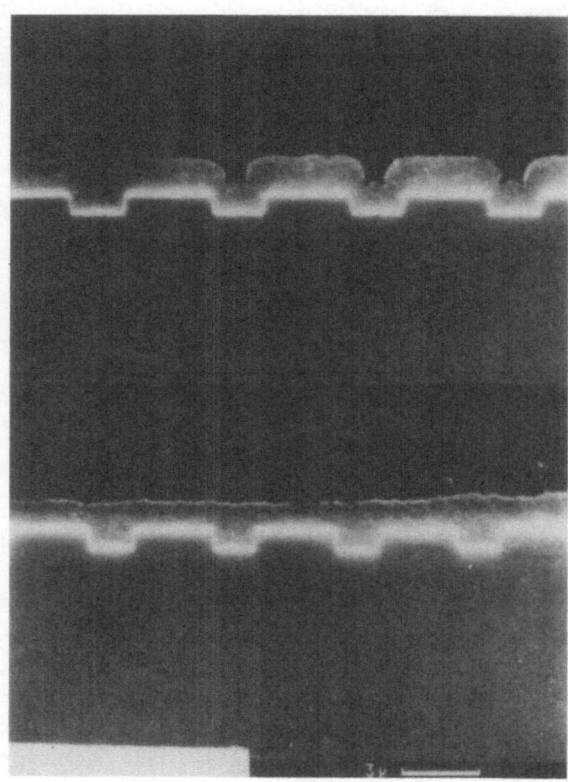

Fig. 57   Aluminum Reflow by Laser Melting

that you have to etch metal that is of differing thickness, and furthermore the conductivity of the metal changes over the topography, being thinner on top of the steps and thicker in the channels between the steps resulting in loss of some current carrying capacity on top of the steps. Again, this is a difficult process which may help in planarization but it would be nice to find another way to achieve the same results with less complexity.

Section V   SPECULATION ON FUTURE MULTILEVEL INTERCONNECT

It is now appropriate to look at some new techniques to increase density, or which may be applicable to planarization, as well as some concepts which would be interesting to test out. The basic fact is that on an I.C. chip, perhaps 5-10% of the silicon is occupied by gates. Once the chip is packaged and put into a system, the gates occupy less than 1% of the system area. Perfect multilevel interconnect might permit 50% of the silicon to be used by gates at the chip level. This increases the die silicon usage by a factor of 10-20, but the gate occupancy of the system real estate would still be less than 10% because of the immense area used up by wire bonds, packages, plugs and interdevice wiring. Not only must on-die densities be improved, but system densities must be greatly increased to fully utilize the results of higher density at the die level.

One approach to improving this situation is outlined by the following six figures. In Figure 58 we see what is called a laser link cross section. This shows first metal covered by a material called link insulator which may be amorphous silicon or silicon nitride. Then a layer of PI covers this link insulator which surrounds the first metal, a via is etched in the PI intermetal insulator, along with normal M1 and M2 vias and the second metal layer is put on top of that. If the second metal is hit with a laser beam and melted with the melt hole going through the second metal and the link insulator and into the first metal, an electrical connection is formed between the first metal and the second metal. This connection is believed to be due to the alloying of Al through the link insulator material down to the first metal. It is not due to simple melting of second metal with the melt running through the link insulator to touch the first metal. Figure 59 shows the application of this. First level metal is shown running horizontally separated by PI intermetal insulation from the second level metal which runs vertically. There are three openings in each cross-section, one opening is through the PI to expose the first level metal. The second opening is through the PI under the second

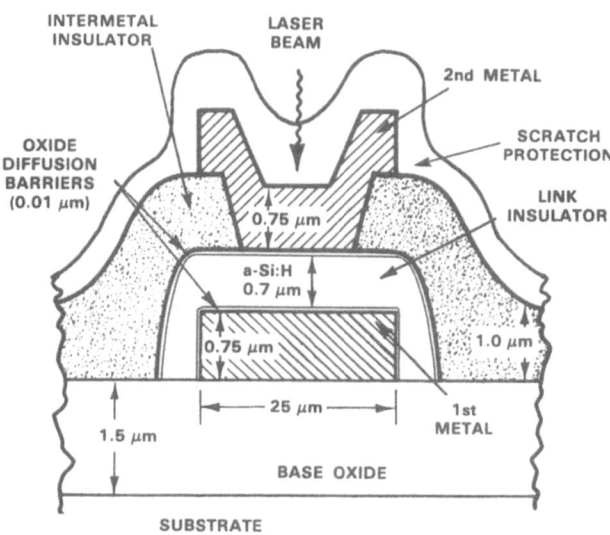

Fig. 58   Laser Link Cross Section

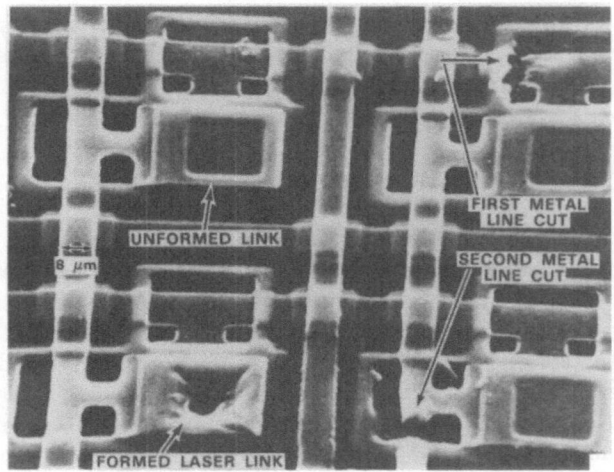

Fig. 59   Laser Links - Cut - Linked -Normal

level metal, allowing the second level metal to go down to the underlying oxide. The third opening is the via under the second level metal, which is lying on the link insulation, and is seen as a square. This is the link area, where the second metal is separated from the first metal only by the link insulator, as can be seen in the upper left, titled "unformed link".

In the upper right hand link one can see the laser has melted and opened the exposed, first metal conductor. In the lower left hand link the formed laser link has been made where the laser pulse hit the link metal, melted a hole in it, melted a hole in the link insulator, and melted into the first layer metal pad beneath these. This makes a connection between the first metal and the second metal. Then in the lower right hand one can see another area where the second metal line has been cut. This capability allows the wiring of integrated circuits directly on the die at integrated circuit scale, since you can connect first level metal to second level metal and disconnect second level metal and disconnect first level metal. This technique is used on wiring up circuits on wafer scale integrated circuit dice.

Figure 60 shows an example of this where the wafer scale die, approximately 1.8 inches by 1.7 inches is seen in a gold package. The logic circuits are in the squares all over the die with spaces between them in which metalization runs across the entire die horizontally and vertically. This across die metalization allows each logic square to be connected to another logic square with the laser linking process to structure a complete system from these many logic circuits. This wafer-scale die is cut from a standard 3 inch diameter wafer as shown in Figure 61. The die is in the center, with test devices surrounding it.

Figure 62 is a close-up of four logic circuits with the cross-die interconnect lines running horizontally and vertically between the logic circuits. Around each logic cell one can see probe pads. The logic circuits are each individually probe tested for functionality. The working logic circuits are mapped into a computer which contains an algorithm which determines where to make connections between metal and where to cut metal to hook up, for instance, the upper right-hand logic circuit to the lower left-hand logic circuit. This is done automatically on a precision computer-driven XY table. Thus the die is put on the table, aligned to the laser beam which is focused through a microscope, and a shutter is used to generate laser pulses to make or break connections in the horizontal or vertical runs of interconnect to wire one logic circuit to another.

Figure 63 is a more enlarged close-up which shows the laser link-interconnect field on the right beside a bit of the logic circuit on the left with the probe pads to the right of center and large in/out driver transistors near the center of the picture. This laser linking process allows a large number of logic circuits to be inter-connected at high density on a given wafer-scale

Fig. 60   Wafer Scale Die In Package

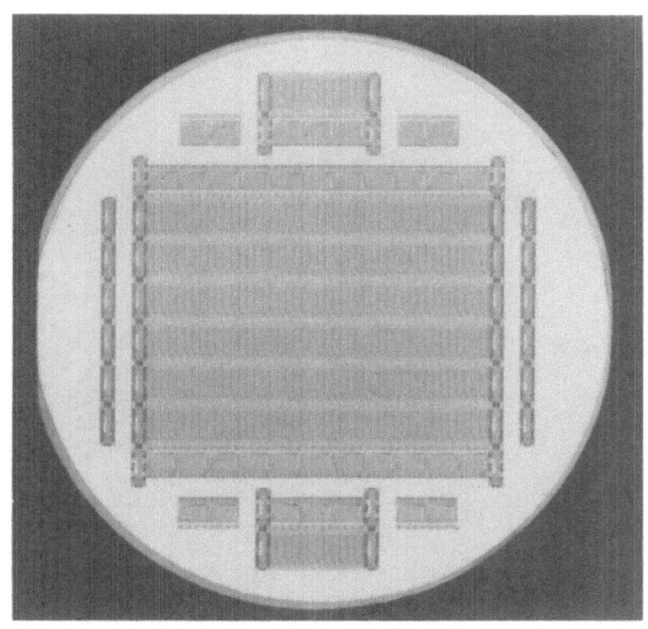

Fig. 61   Wafer Scale Wafer

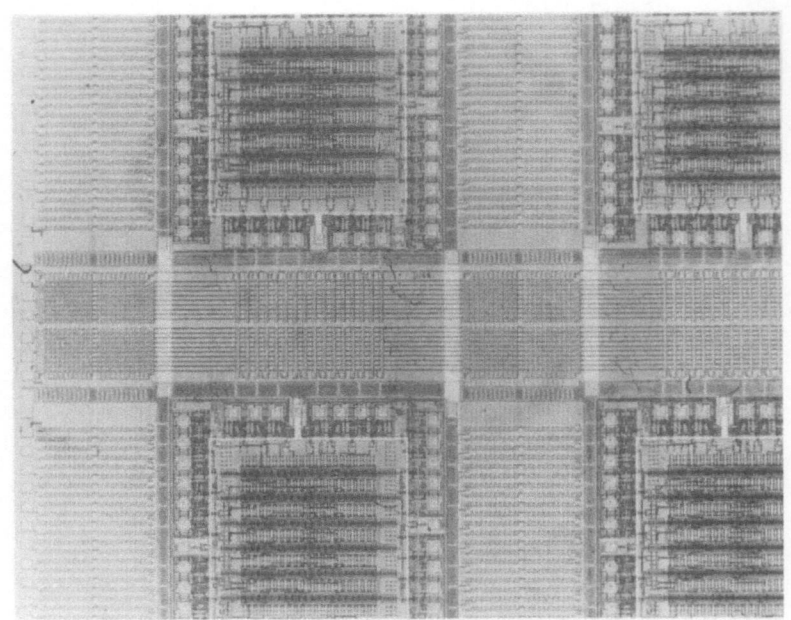

Fig. 62   Wafer Scale Link Overview

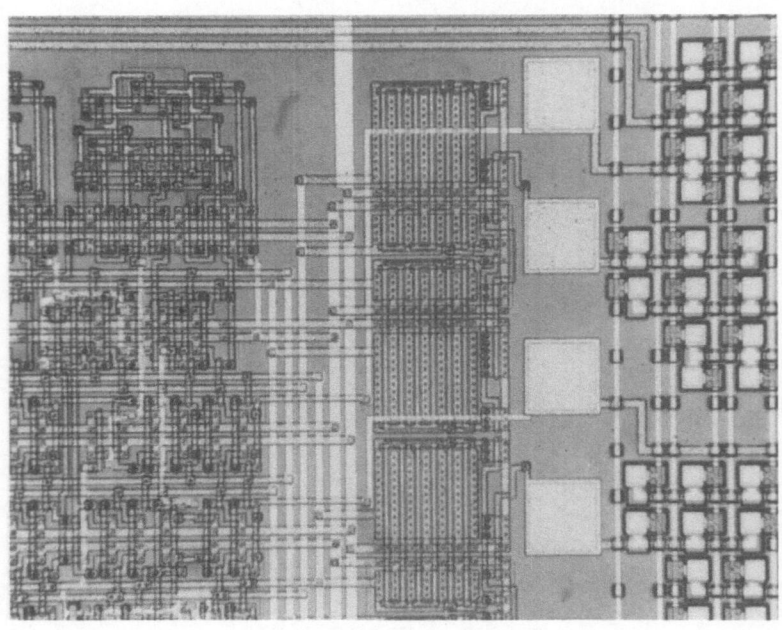

Fig. 63   Wafer Scale Link Closeup

die using two levels of interconnect. This is not a planarization process but it is a technique permitting high density on a large die, promising higher speed and system density.

It is also conceivable to achieve better packing density than currently possible when putting individual small die into hybrid packages and wire bonding them together. This might be another approach to increasing the density of gates in a system by not only increasing the density on the die, but also increasing the density from die to die. As shown in Figure 64, instead of very large wafer scale die, one considers the more normal die size ranging from $\frac{1}{8}$" to $\frac{1}{2}$" square. In the right hand drawing on Fig. 64, the die has micro-beam leads protruding beyond its edge. Before etchback, the die carries its micro-beam leads nearly to the die edge, at the same dimensions and line-to-line spacing as used elsewhere on the integrated circuit. Then, the die is etched back allowing the micro-beams to protrude over the edge. Here it is put face up into a silicon PC receiver, which has recesses etched in it and interconnect which is at the same density as is found on the die. This permits the dice to be placed very close to each other, and the silicon backplane could have double level metal if necessary, to permit impedance-matched high frequency line over ground plane interconnect. The left hand image in Figure 65 shows the micro-beams coming to within a few microns of the edge of the IC die. The right-hand figure shows the micro-beam protruding some 60 μm out from the edge of the IC die. This is accomplished by masking the interconnect on the IC die and placing it face down in a plasma etch system in which 60 μm is etched off of its back and its sides to allow the microbeams to hang out over the etched-back edge. In the right-hand photo one can just see a horizontal dark line approximately half way up the picture which is the edge of the IC die. The micro-beam protrusion beyond the edge of the die is dependent on the metal thickness and micro-beam width, both of which affect how stiff it is. These micro-beams are 2 μm wide with 5 μm spaces between them and if one etches the die much more than 60 μm back from their ends, they become flimsy, and are difficult to work with during alignment.

During this etch-back, the micro-beam ends bend up from the surface of the die. In the right-hand picture all micro-beams bend toward the viewer by 10 to 15 μm, probably due to stress in the metal. Thus, if the die were put on the Si PC receiver face down the micro-beams ends would contact the Si PC receiver leads and pressing the die against the Si PC receiver would deflect and load the micro-beams ends. This is needed during the laser weld process to keep the micro-beam molten end pressed against the Si PC receiver molten line at

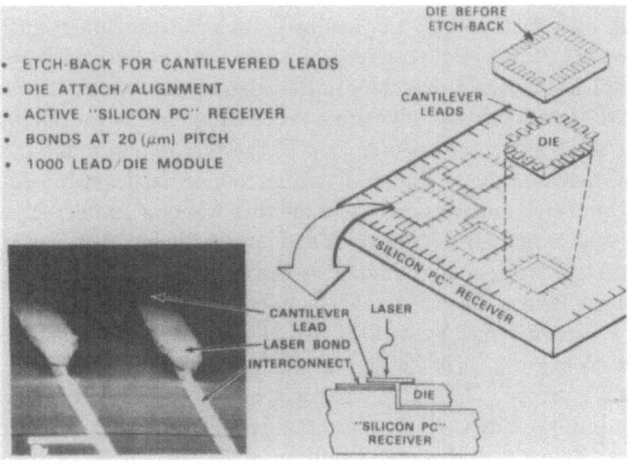

Fig. 64   Laser Bonded Hybrid Overview

Fig. 65   Laser Bonded Hybrid - Microbeam Etchback

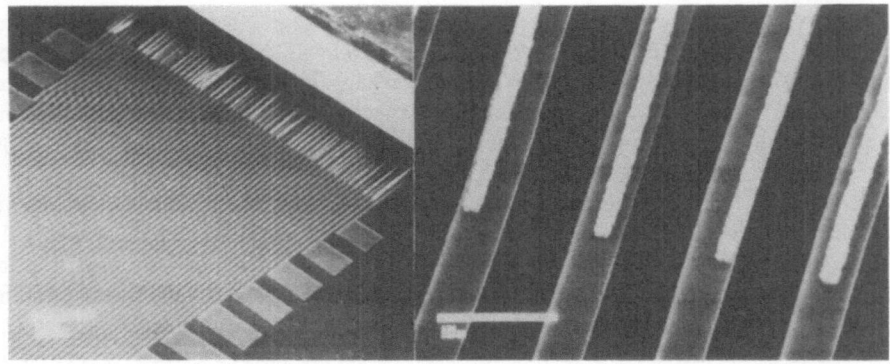

Fig. 66   Laser Bonded Hybrid - Alignment

the moment both lines melt to help the two melt zones flow together. In Figure 66, the left hand picture shows the test Si PC receiver which has straight lines with probe pads on them. At the upper right hand corner one can see the edge of the test IC die and protruding out from under the white band of that edge are a set of micro-beams touching the Si PC receiver conductors. The right picture is a closer view of the thin, white micro-beams coming down from the top and touching the Si PC receiver interconnect leads. All metal is Al, 1% Si, 2% Cu, and the 1.2 μm thick micro-beams are 2 microns wide on 10 micron centers. The Si PC receiver metal is approximately 0.8 μm thick and 5 μm wide on 10 μm centers. It is clear that if the microbeams are stiff enough, alignment can be done with very fine wires protruding from the edge of the IC die. Basically this alignment is done with an XYZ system where the die is held with a vacuum chuck by its back and viewed by CCTV connected to a microscope angled at about 45 degrees. The IC die is manipulated with the XY system while being observed on the TV screen. Once alignment is achieved the die is moved in the Z direction toward the surface of the Si PC receiver until one observes the micro beams touching the Si PC receiver lines. A low viscosity glue has been placed on the Si PC receiver prior to this alignment. The bent down micro-beams press against the Si PC receiver leads and the die is held in proper location until the adhesive has dried. The welding is done in an argon or argon-helium atmosphere using an argon laser, the beam of which is shuttered and focussed through a microscope. In the left picture of Figure 67, the laser melted the micro-beam on either side, causing a weld between the micro-beam and the receiver. In the right image, the 2 μm diameter argon laser spot melted the receiver lead and the micro-beam just at the end of the micro-beam.

The problem with this technique can be illustrated by the right-hand picture. The Si PC receiver line is well heat sinked to the substrate and takes considerably more energy to melt than does the micro-beam which is not heat sinked at all. The ratio of laser power is about 10 to 1, making this a very difficult problem to cope with since the micro-beam leads will melt back very easily and essentially vaporize before the receiver lead is melted at all. This meant that one had to be very careful in positioning the laser relative to the microbeam so that the main energy of the laser beam would be concentrated on the receiver lead and only a small amount would go into the end of the micro-beam. This is a very position-sensitive way of doing it and not satisfactory. Another approach was to pre-melt the end of the micro-beam to form a larger, more massive ball of Al and then weld that larger thermal mass to the receiver lead. This worked, but it had the problem that, with the close center-to-center spacing of these leads, the light hitting the spherical end of the premelted micro-beam could very easily reflect on to an adjacent beam and evaporate it.. This happened frequently, where one would try to weld a premelted spherical-ended micro-beam to the receiver lead, hit it with a laser pulse and see the adjacent lead simply disappear, because enough of the light was reflected sideways to vaporize the adjacent lead. Another approach which proved to be more satisfactory is shown in Figure 68. Here the micro-beams are approximately 8 μm wide on 20 μm centers and the receiver leads are approximately 3 μm wide. Now the receiver lead has

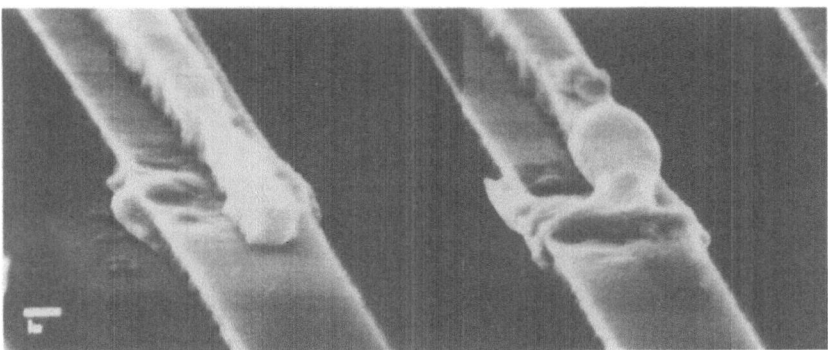

Fig. 67   **Laser Bonded Hybrid - Weld Closeup**

Fig. 68   **Laser Bonded Hybrid - Wide/Narrow Microbeams**

a higher thermal impedance to the substrate and the wider micro-beam has higher thermal mass for more heat absorption during the melt period. Furthermore, the wider micro-beam is stiffer in the left-right direction which makes it sturdier for the alignment procedure. Also, the more massive micro-beam can supply metal to the weld during the laser hit. The upper picture of Figure 69 shows 2 micro-beams coming down over the narrower PC receiver leads with a close-up of each of these shown right below. The PC receiver beam narrows slightly during the melt process and the liquid Al rises up in the third dimension, while the micro-beam supplies some Al to the melt site on the receiver lead. The narrowest obtainable laser pulsewidths (80 us) seemed to give more repeatable welds, which were made with one pulse. Attempts were made to premelt the receiver lead by hitting it with laser pulses, starting away from the end of the micro-beam and moving the laser into the end of the micro-beam, essentially causing a pre-melted liquid to move a long the receiver lead up to the micro-beam end and then the micro-beam would be melted with the metals flowing together. This didn't seem to provide much advantage, was hard to repeat, and requires high accuracy motion. Furthermore, any premelting changed the surface profile and oxide condition, randomizing the melting parameters. It is probable that the most interesting approach would be to use a solder plate on the aluminum alloy, reducing the need for aluminum welding in exchange for a solder flow approach.

These experiments demonstrate the possibility of using a very high density between-die interconnect, along with the possibility of a hybrid system with the Si PC receiver used for a mix of Si or Gallium Arsenide devices where it could incorporate its own active devices for repeater purposes. If a particular die did not function, it could conceivably be removed by cutting the micro-beams with a laser, then aligning and welding another die to the Si PC receiver leads. This is simply a micro-wire bonding approach which would give an order of magnitude increase in system density.

Figure 70 shows another approach to a similar concept. In this case the upper right hand photo shows two chips with lines running between standard bonding pads on those IC chips. The technique here is to wire the chips together using a laser beam. The chips are placed into

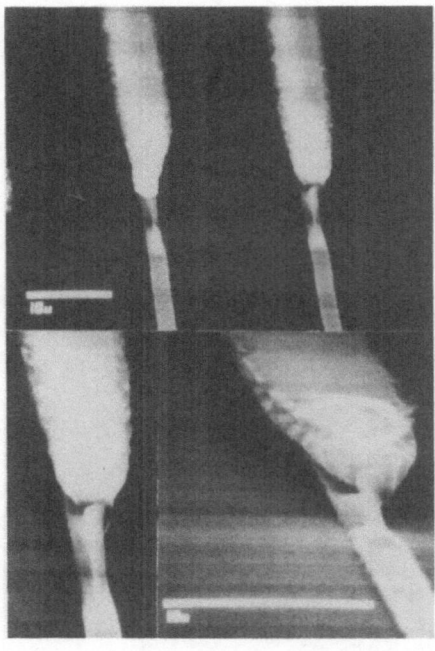

Fig. 69   Laser Bonded Wide Microbeams/Narrow Welds

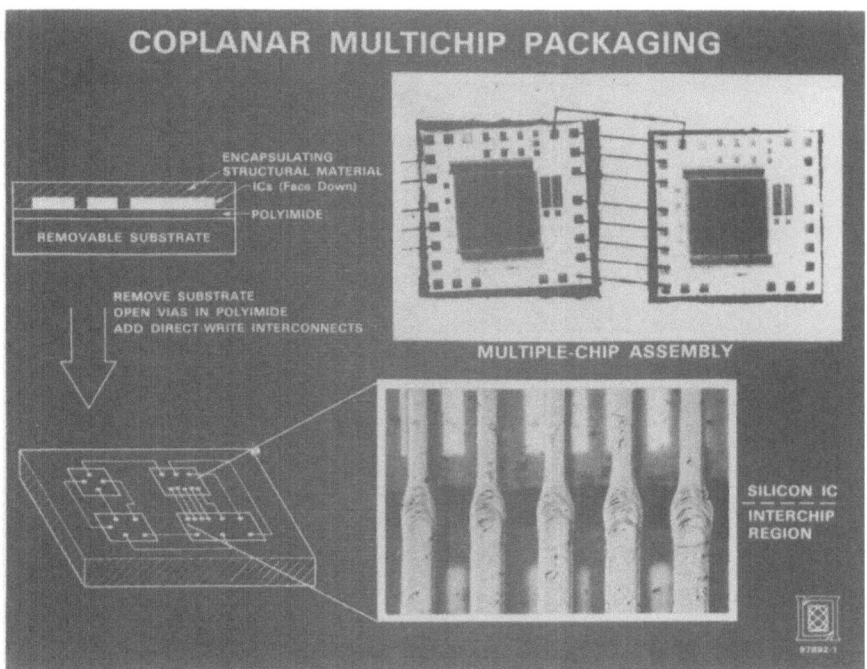

Fig. 70   Laser Deposited Tungsten Lines (Interconnect)

a common surface as is shown in the drawings in the upper and lower left images and then the metalization which connects the pads from chip to chip is drawn on using a laser. This is done in a reaction chamber containing a tungsten hexaflouride-silane atmosphere. A laser is focused on the surface of the package carrying the dice,the laser is turned on and moved along the surface, heating the surface of the package to the point of causing a CVD tungsten layer to deposit on the pad and across the surface of the package to the other pad. Therefore, in the case of the upper right hand photograph, the laser would be positioned on the lower right pad of the left hand die and moved horizontally from the left hand die to the lower left hand bottom pad on the right hand die, depositing tungsten from the thermal reaction of the atmosphere and the heated die/package surfaces. The process could be repeated, forming the nearly horizontal stripes seen in the photograph between the two die. This requires that the die have good planarity on the surface in terms of focusing the laser and that the surface be open in the pad areas for making electrical connections. It also means that the reactive atmosphere in which this is done is carefully controlled, and it is a serial process. Again it is an attempt to increase the density of between-die interconnect.

Figure 71 is an example of the same technique of laser heating causing a CVD like reaction used to customize ICs. An IC may need to have some connections made or removed as was cited previously in the laser linking example. Here, the laser can open lines and it can also deposit interconnect lines between existing metal sites. This allows micro-wiring to take place on the IC surface and offers the advantage of higher density interconnect structuring than was possible with the laser link approach discussed earlier.

Moving from increasing interdie density back to the principle subject of on-die density improvement, Figure 72 illustrates the problem with lensing and tographic build-up in the left hand sketch and below that on the left hand side, the difficulty of getting metal into very small via openings. The right hand drawing shows a concept of, "Wouldn't It Be Nice Not To Open A Hole In The Intermetal Insulation In Order To Make A Connection Between Layers Of Metal?" This would reduce lensing problems, help make things planar and potentially permit much smaller vertical interconnect. Also, it might permit reduction of metal real

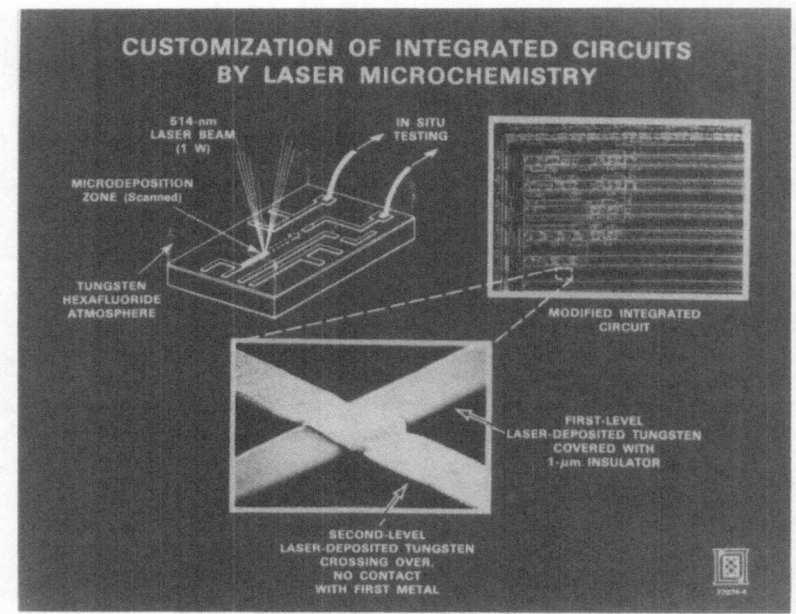

Fig. 71   Laser Deposited Tungsten Lines (Customizing)

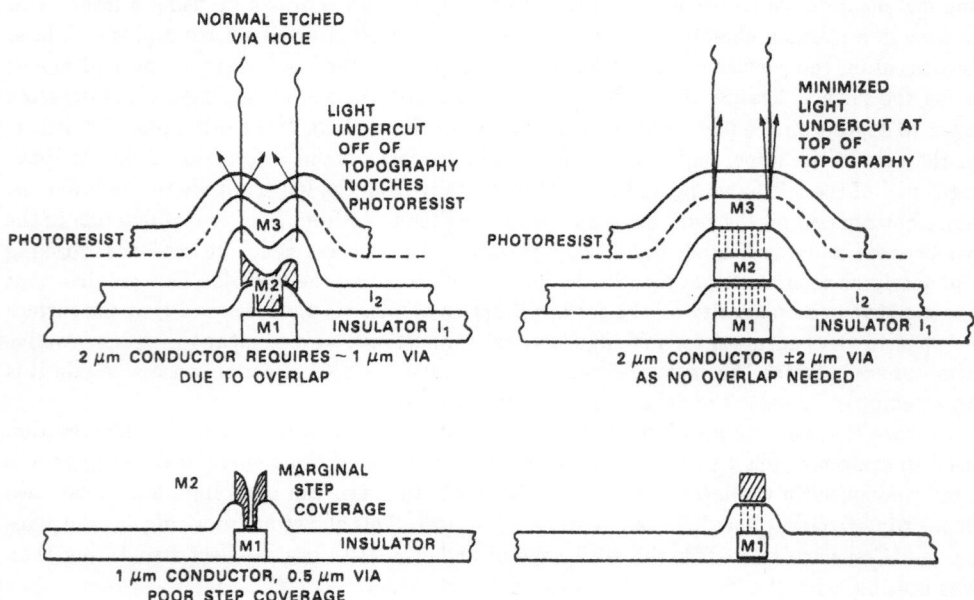

Fig. 72   Implanted Planar Vias

estate usage, by eliminating "nesting" of vias within a wide metal pad. One technique for possibly accomplishing this would be to convert the insulator locally into a conductor where the via would normally exist. The insulator might be modified in some fashion to make it conductive, and even if the conductivity of this specially treated via area were not as high as the metal interconnect, it would be short and the contribution of its increased resistance to the overall interconnect resistance would be minimal.

Figure 73 outlines an implantation approach to accomplish this conversion. Here the silicon nitride insulator has been covered with an aluminum ion implant mask having openings etched in it where vertical connections between the first metal and the second metal are to be made. These openings are where via holes would normally be etched in the intermetal insulator. The silicon nitride is implanted with silicon, using doses of 1E17 to 1E18 atoms /cc., which makes the silicon nitride very silicon rich.

The implants are at 200 kv to implant deeply, and 90 kv to implant the intermediate layer and at 25 kv to fill the upper layer of the silicon nitride. The implant mask is etched away and second level metal is deposited, with AlSiCu being used for both first and second level metal. Thus, second level metal lies over the small areas of silicon implanted intermetal silicon nitride insulation. The circuit is then sintered at 425 degrees C to alloy the Al into the Si and the Si into the Al from the metal on both sides of the silicon nitride. This forms an ohmic connection between first and second metal in a perfectly planar fashion. Implants of one micron diameter have resulted in resistance between first and second metal of one ohm or less. Figure 74 shows what is normally a via chain, where one can see the wide rough second metal lying over rectangles of narrower first metal with continuous, transparent silicon nitride between. In the center upper part of the picture, one can see two second metal rectangles lying over the first metal where no implantation was done. In the lower half of the picture, second metal can be seen lying over first metal and four small, round dots are visible near the ends of the second metal bars. These are the sites of 1 μm implant areas in the SiN lying beneath the second metal. Figure 75 is a close up of two implanted planar via sites at either end of the second metal rectangle lying over the ends of two first metal bars. The lower picture is a more oblique view which clearly shows hillock bumps in the top metal. The implanted area is a very smooth, circular island about 2 microns in diameter in the center of the picture. It is clear that this implanted area in the silicon nitride is quite planar. The underlying implanted region in the SiN somehow suppressed the hillock growth in the second

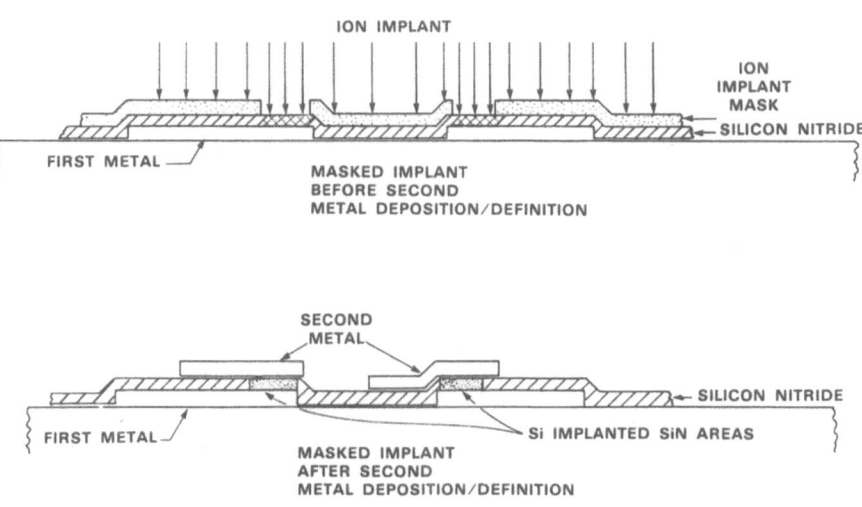

Fig. 73   Implanted Planar Vias - Cross Section

Fig. 74   Implanted Planar Vias - 1μm Chain

Fig. 75   Implanted Planar Via - Closeup

level metal so that the second level metal clearly shows where the implanted region is. Auger analysis shows that after sinter, the implanted region in the silicon nitride contains Al and nitrogen. It is presumed the that Al alloyed into the implanted silicon nitride and silicon migrated into the Al, which suppressed the hillocks.

Figure 76 is a picture of a focused ion beam implanted silicon area in silicon nitride covered with L- shaped second metal. This is a 0.75 μm square via and first to second level metal via resistance, checked with a four point probe, was 0.75 ohms. Figure 77 summarizes the results of implanting Si into various materials. It defines the lower conductor, the resistivity of that material, the insulator material, thickness and refractive index, which in all cases shows the CVD insulators were deposited silicon rich. Also, the implanted ion dose, the upper conductor composition, the sinter temp. and time along with the via size and resistance are outlined. It is interesting to note that this not only works with AlSiCu on both sides of the insulator, it also works with metal on the upper surface of the insulator and polysilicon or N+ silicon beneath, as in a device contact. However, it does not seem to work with P+ silicon beneath the insulator. Furthermore, as seen in the right hand column, this technique works with SiO_2 intermetal insulation. However, it must be noted that the implantation dose had to be an order of magnitude greater than with SiN. It is assumed that the oxygen content of the SiO_2 is responsible for forming an aluminum oxide during the sinter and that one must implant even more silicon in order to overcome this problem. It should also be noted that the insulator thickness is between 0.25 and 0.3 μm in order to achieve Si implantation clear through the insulator at 200 KV. This is an inadequate intermetal insulator thickness and techniques would have to be evolved to implant through 0.75 - 1 μm thick intermetal insulation. One possibility is implanting Be, which also produces a conductive planar vias, although the mechanism is not understood. Be, being much lighter, will penetrate 1 μm thick SiN or SiO_2 at 200 KV. This simply another example of a somewhat futuristic, far-out technique, which might be used to increase planarity and density on interconnect for multilevel systems.

In Figure 78 a curiosity is shown were a laser was used to define or write a PI line. The PI precursor, polyamic acid, is spun on the wafer and baked at about 150 degrees C, which is the point that the solvents are driven off. At this stage the material is still polyamic acid, and as may be recalled, it would be possible to remove this material with a positive photo resist developer. In this case, an argon laser beam was scanned across the polyamic acid layer, raising its temperature, driving off the OH groups and imidizing it to create PI. The test chip was put into a strong base solution which dissolved away the polyamic acid layer

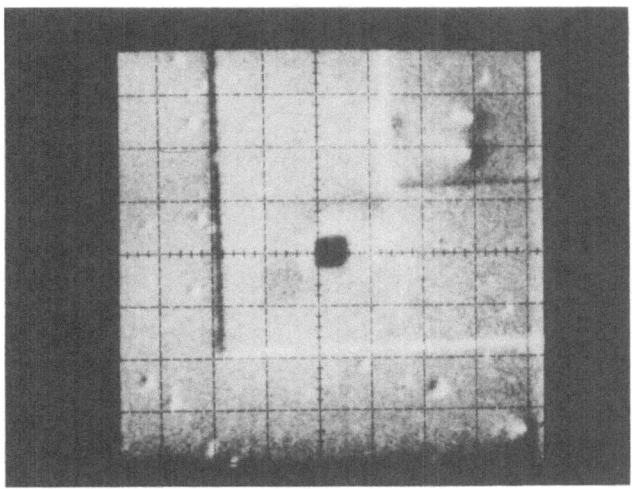

Fig. 76   Implanted Planar Via - Focused Ion Beam Implant

SUMMARY - PLANAR VIA IMPLANT/MATERIALS

	Al-1%Si-2%Cu	N+ Silicon	P+ Silicon	PolySilicon	Al-1%Si-2%Cu
LOWER CONDUCTOR	Al-1%Si-2%Cu	N+ Silicon	P+ Silicon	PolySilicon	Al-1%Si-2%Cu
RESISTIVITY	0.4Ω/sq	45Ω/sq.	80Ω/sq	20Ω/sq	0.4Ω/sq
INSULATOR	PECVD Si_3N_4	PECVD Si_3N_4	PECVD Si_3N_4	PECVD Si_3N_4	PECVD SiO_2
INSULATOR THICK.	0.25 Micron	0.27 Micron	0.28 Micron	0.3 Micron	0.25 Micron
INS. REF. INDEX	2.10-2.25	2.20-2.25	2.20-2.25	2.20-2.25	1.51-1.53
IMPLANT ATOM	Silicon	Silicon	Silicon	Silicon	Silicon
IMPLANT DOSE	6×10^{16}/25KeV 1×10^{17}/90KeV 2×10^{17}/200KeV	6×10^{16}/25KeV 1×10^{17}/90KeV 2×10^{17}/200KeV	6×10^{16}/25KeV 1×10^{17}/90KeV 2×10^{17}/200KeV	6×10^{16}/25KeV 1×10^{17}/90KeV 2×10^{17}/200KeV	5×10^{17}/25KeV 1×10^{18}/80KeV 2×10^{18}/180KeV
UPPER CONDUCTOR	Al - 1%Si	Al - 1%Si	Al - 1%Si	Al - 1%Si	Al - 1%Si
SINTER TEMP.	425°C	450°C	450°C	450°C	450°C
SINTER TIME	0.5 Hour	0.5 Hour	0.5 Hour	0.5 Hour	0.5 Hour
VIA SIZE	1 Micron	1 Micron	-	1 Micron	3 Micron
VIA RESISTANCE	1 Ohm	1 Ohm Approx.	Same as no implant	1 Ohm, Approx.	1.5 Ohm

Fig. 77   Implanted Planar Via - Summary

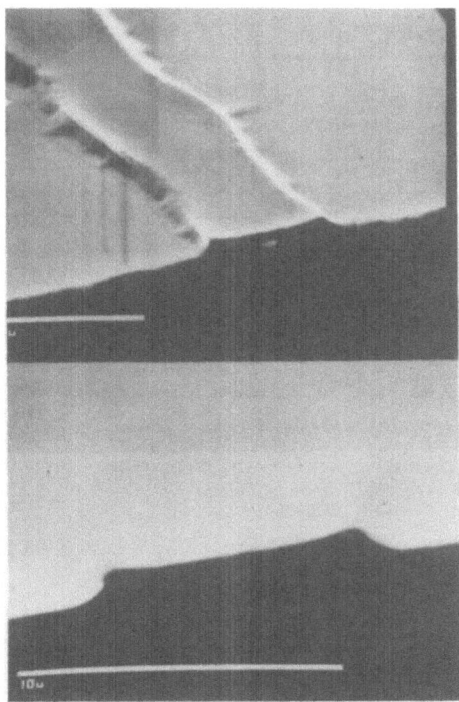

Fig. 78   Thermally Defined Polyimide Line

where the solvent had been evaporated off, leaving the base-resistant PI stripe. The lower picture shows a cross section of this PI line. Essentially this is a thermally written PI line and it is clear that the striations coming out of the edge of this line are a problem but the sidewalls are fairly steep. This might be a technique which could be used for some kind of side to side metal planarazation. Its conceivable that one could spin on PI and expose it to a pattern that was the reverse of a normal metal pattern. This would require some sort of thermal exposure which would allow the PI to be removed from the tops of the interconnect and leave it in the areas between lines of the metal interconnect. This would planarize the metal interconnect if the PI lying between the metal was the same thickness as the metal. It is simply a concept and might be useful in some situations.

Moving to even more esoteric and untried areas in Figure 79, we see an ideal topography to be planarized in the upper line drawing. In the the lower drawing it is clear that if the metal conductors had dummy metal left between them a nearly planar topography could be achieved with a spin-on polymer. This assumes that the slots between adjacent metal areas would be of uniform width. If the slots between the dummy metal and metal conductors were 2 or 3 μm in width the spin-on insulator would be almost planar. However, as can be seen in the summary of pros and cons, this would not work well with CVD insulators unless the slots were very narrow and it would produce high inter metal capacitance which would be fatal.

Furthermore, it would be useful primarily on first level metal because with upper level metal it might restrict placement. It involves more complex mask design, could lead to shorting and is based on the assumption that all metal is at the same altitude, which of course isn't true, because there is as much as 0.5 μm of topography beneath first level metal.

In Figure 80 the upper drawing shows the same conductor topography to be planarized, and in this case photo resist is defined in the spaces between the metal. The CVD oxide is etched off of the metal tops, which leaves the CVD oxide as shown with X's across it in

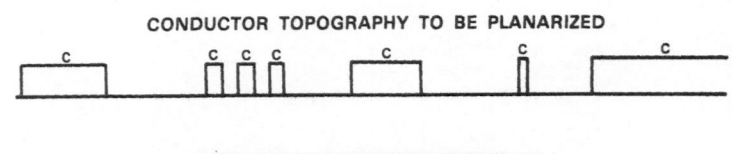

CONDUCTOR TOPOGRAPHY TO BE PLANARIZED

UNIFORM TOPOGRAPHY APPROACH I

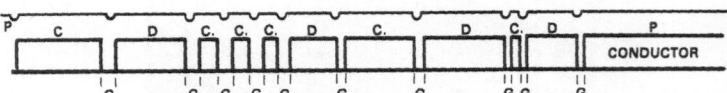

LEAVING DUMMY METAL (D) BETWEEN ACTIVE CONDUCTOR METAL (C) PRODUCES
CONSTANT GAP WIDTHS (G) WHICH IS THEN SMOOTHED/PLANARIZED BY SPIN-ON
INSULATOR (P)

PRO   a. NEARLY PERFECT PLANARIZATION BY SPIN-ON INSULATOR
         b. WOULD IMPROVE METAL ETCH DUE TO UNIFORM ETCH LOADING

CON   a. WOULDN'T WORK WITH CVD INSULATORS
         b. WOULD PRODUCE HIGH INTERMETAL CAPACITANCE
         c. WOULD GET IN THE WAY OF VIAS AT UPPER LEVELS
         d. PROBABLY TEND TO CAUSE SHORTING
         e. MORE COMPLEX MASK DESIGN

Fig. 79   Dummy Metal - Planarizing Concept

CONDUCTOR TOPOGRAPHY TO BE PLANARIZED

UNIFORM TOPOGRAPHY APPROACH II

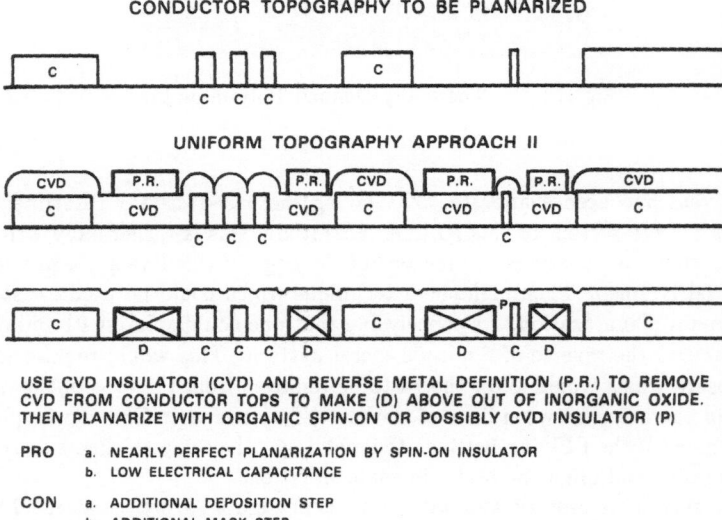

USE CVD INSULATOR (CVD) AND REVERSE METAL DEFINITION (P.R.) TO REMOVE
CVD FROM CONDUCTOR TOPS TO MAKE (D) ABOVE OUT OF INORGANIC OXIDE.
THEN PLANARIZE WITH ORGANIC SPIN-ON OR POSSIBLY CVD INSULATOR (P)

PRO   a. NEARLY PERFECT PLANARIZATION BY SPIN-ON INSULATOR
         b. LOW ELECTRICAL CAPACITANCE

CON   a. ADDITIONAL DEPOSITION STEP
         b. ADDITIONAL MASK STEP
         c. ADDITIONAL ETCH STEP

Fig. 80   Reverse Metal Definition CVD Etch - Planar. Concept

the bottom drawing. The CVD oxide has been etched so that there are gaps between it
and the metal sidewalls. This assures that all of the topography is removed from above the
metal corners and in the vicinity of the metal corners. This leaves narrow gaps in the CVD
oxide, which is similar to the dummy metal case shown in Fig. 79. A spin-on organic can
planarize and smooth this condition very well, whereas putting another layer of CVD oxide
would not work as well. The reason for etching slots in the CVD oxide is to get rid of fences
which form because of differences in etch rate between CVD and photo resist, and also to
take care of misalignment problems between the image which will create the photoresist and
the underlying topography. This wide etch would also take care of the variable breadloafing
that would occur with the CVD oxide over the underlying, vertical sided metal. This has the

disadvantage of requiring additional masking and etching steps, and an additional deposition step, but it would give fairly planar results on this kind of topography.

Now some way out concepts, all unproven, but presented in an attempt to encourage thought as to how one might go about this kind of thing. Figure 81 shows Al anodization by ion implant. Implant resist is defined on the Al where metal conductors are to be. Then oxygen might be implanted in the exposed metal to form an Al oxide-like material which would insulate the unimplanted Al conductor stripes from one another. This would be completely planar conceptually, and would not require any metal etching. How much oxygen would be needed to do this? Could a satisfactory implant resist be found? What would the dielectric characteristic of the Al oxide and the capacitance/leakage be, vertically and horizontally? Furthermore, if this material is fully oxidized, it is likely that the volume taken up by the Al oxide would be greater than the initial pure Al volume and one would have either highly stressed material or swelling in these areas, which would form topography.

While daydreaming, look at Figure 82. Again the top drawing represents the features to be planarized. The question here is, would there be any way to produce a selective CVD insulator, that grows on SiO_2, but not on metal? After all, there is selective tungsten that grows on silicon but not on SiO_2. As depicted in the lower drawing, if we could just grow something up from the bottom between each conductor we could fill the ares with Selective CVD insulating material. This would give a perfectly planar surface to begin with and a spin-on organic or CVD inorganic material could be deposited on top. The problem is that no one knows how to do this, but it would be a ducky way to solve the planarization situation.

In Figure 83 a hybrid approach is illustrated which is similar to the previous proposal, only selective tungsten is used to fill slots in previously etched insulator. In the upper picture one would have the devices with selective tungsten deposited in contact holes through the SiO_2 down to the transistors. Then a very thin layer of blanket CVD silicon, perhaps 500 to 1000 angstroms thick, could be put down. The next step would be to etch that silicon into an interconnect-like pattern with the silicon remaining where the metal interconnect is to be. This silicon would be a bit wider than the interconnect metal, and it would also overlap but not cover the selective tungsten via fill plugs.

Then a CVD SiO_2 or PI layer would be deposited on top and trenches etched into the insulator using the interconnect pattern. In other words, instead of metal being etched away,

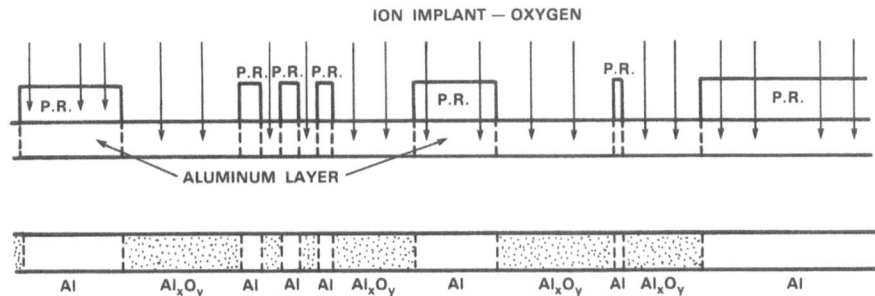

ION IMPLANT — OXYGEN

USING A PHOTOSENSITIVE ION RESIST (P.R.) IMPLANT POSSIBLY O OR O_2 TO FORM AN Al_2O_3-LIKE INSULATOR (Al_xO_y) BETWEEN CONDUCTORS (Al)

PRO a. TOTALLY PLANAR REGARDLESS OF INSULATOR
   b. NO METAL ETCHING

CON a. AMOUNT OF OXYGEN NEEDED (Implant Time?)
   b. RESIST THICKNESS AND DURABILITY?
   c. DIELECTRIC CHARACTERISTICS OF Al_xO_y?

Fig. 81 Aluminum Anodization by Implant - Planar. Concept

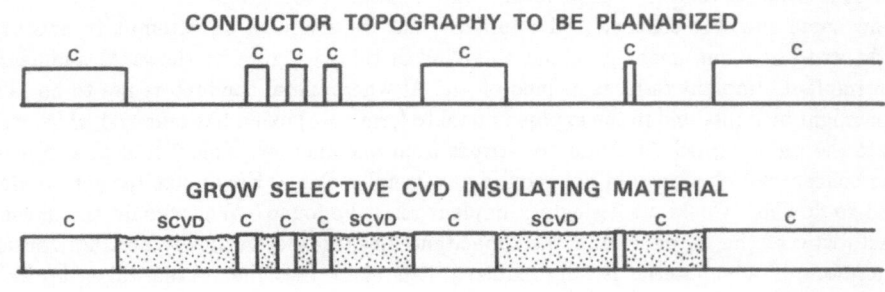

CONDUCTOR TOPOGRAPHY TO BE PLANARIZED

GROW SELECTIVE CVD INSULATING MATERIAL

IF A SELECTIVE CVD (SCVD) PROCESS COULD BE DEVISED WHICH WOULD PERMIT AN INSULATOR TO BE GROWN ON Si OR SiO$_2$ AND NOT ON METAL, THEN THE CONDUCTOR LEVELS (C) WOULD BE PLANARIZED

PRO   a. PERFECT PLANARIZATION POTENTIAL
         b. LOW ELECTRICAL CAPACITANCE

CON   a. NO PROCESS IS KNOWN TO EXIST
         b. ADDITIONAL DEPOSITION STEP

Fig. 82   Selective CVD - Planarizing Concept

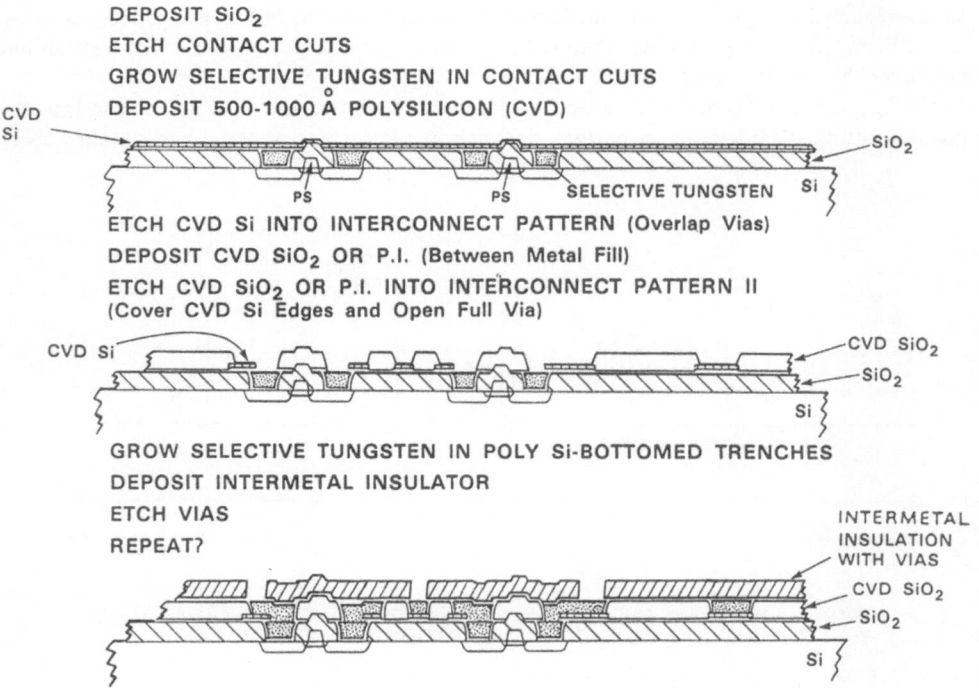

DEPOSIT SiO$_2$
ETCH CONTACT CUTS
GROW SELECTIVE TUNGSTEN IN CONTACT CUTS
DEPOSIT 500-1000 Å POLYSILICON (CVD)

ETCH CVD Si INTO INTERCONNECT PATTERN (Overlap Vias)
DEPOSIT CVD SiO$_2$ OR P.I. (Between Metal Fill)
ETCH CVD SiO$_2$ OR P.I. INTO INTERCONNECT PATTERN II
(Cover CVD Si Edges and Open Full Via)

GROW SELECTIVE TUNGSTEN IN POLY Si-BOTTOMED TRENCHES
DEPOSIT INTERMETAL INSULATOR
ETCH VIAS
REPEAT?

Fig. 83   Selective Tungsten Trench Deposition - Planar. Concept

slots would be opened in the insulator down to the silicon/tungsten plug level. These slots would be identical to normal the metal interconnect. Now, selective CVD tungsten could be grown in these trenches up to the top of the insulator, giving a nearly planar surface. Intermetal insulation can be put down, openings etched again and the steps repeated if one wanted an all tungsten interconnect. However, the proposal here would be to use polysilicon as the first level of interconnect, selective CVD tungsten as the second level of interconnect for short runs, and then aluminum could be used on the third and fourth levels of interconnect. This would give high resolution capability for the third level of metal, with medium resolution at the fourth level and could increase density considerably.

In conclusion, I hope to have given an accurate view of the needs for multilevel I.C. interconnect, the goals it must meet in the future, and current state of technology and problems. I sincerely hope that some of the newer, untried approaches reviewed will trigger thinking leading to some much-needed breakthroughs in multilevel I.C. interconnect.

ACKNOWLEDGEMENTS

I would like to thank Laura Rothman of IBM. Douglas Crook and Eugene Zeller of HP, Robert Blewer of Sandia Nat. Labs., Jack Raffel and Daniel Ehrlich of Lincoln Labs. and Steven Senturia of MIT for their comments and permission to use certain graphic material in this talk. Also, to the many unacknowledged scientists whose efforts have created the material reviewed here, we all owe our thanks, as their work forms the basis of the IC revolution. Many thanks to Linda Bohunicky for her dedicated dictation-taking, and Julius Zolotarevsky who read all of this and helped planarize my convoluted writing. Finally, orchids to Roland Levy of Bell Labs, for his excellent organization, thorough patience and cheerfulness in dealing with all aspects of the 1988 NATO Advanced Study Institute.

INTERLEVEL DIELECTRICS FOR REDUCED THERMAL PROCESSING

F. S. Becker

Siemens AG Corporate Project MEGA

Otto-Hahn-Ring 6, D-8000 Munich 83, Germany

I. INTRODUCTION

The invention of the planar integrated circuit 29 years ago has triggered a development race which resulted in chips with millions of active components /1/. In spite of the dramatically increased packing density made possible by modern technology, chips have constantly grown in size, as illustrated in Fig.1 for five generations of dynamic random access memories (DRAM's) /2/. In the course of development, the purely planar approach has already been replaced by designs comprising several conductive layers. The advent of Megabit-DRAM technology, however, marks a decisive turning point, as the third dimension is now actively used by extending the storage capacitor into the substrate /3/. Figure 2 shows a cross-section through the cell array of the Siemens 4Mbit-DRAM. These tendencies in chip development, which are now a common feature of DRAM-cell concepts /4/, have several technological consequences of conflicting nature:

1.) Larger chips require more attention to the problem of fast, low-loss signal transport. This means additional levels of metallization as well as reduced step angles of the topography to be covered. This is because steep slopes can lead to enhanced resistivity of polycide lines or can even become a reliability problem due to the bad step coverage of sputtered aluminum.

2.) For multilevel metallization systems, the temperature budget is restricted to temperatures below 450° if aluminum is used.

3.) The patterning of 0.5 - 0.8 µm structures by means of optical lithography is only possible if the underlayer is planarized as far as possible.

4.) As the shrinking of minimal features applies mainly to the lateral, not to the vertical dimensions of the structures, the topography variations are increased. This problem can be aggravated by future more elaborate three dimensional cell concepts.

5.) The temperatures allowed for the planarizing flow steps of interlayer dielectrics prior to aluminum deposition are reduced to prevent outdiffusion of shallow junctions as well as warpage of the larger wafers ($\geqq$ 150 mm diameter).

As a consequence, it can be stated that future highly integrated devices will contain more dielectric layers planarized, however, with a lower temperature budget. Possible solutions to this problem fall mainly into two categories:

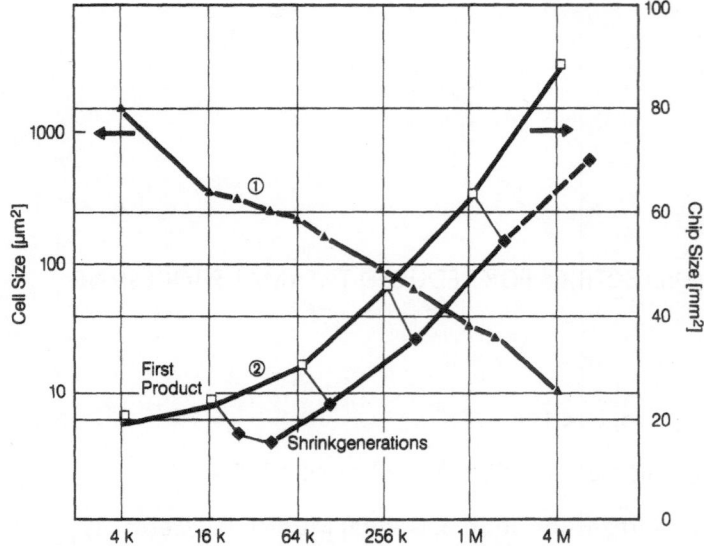

Fig. 1 Development of cell size (1) and chip size (2) for different DRAM-generations. In spite of the increased packing density, (as reflected by the reduced cell size), the chips are getting larger with each generation. (Ref. 2)

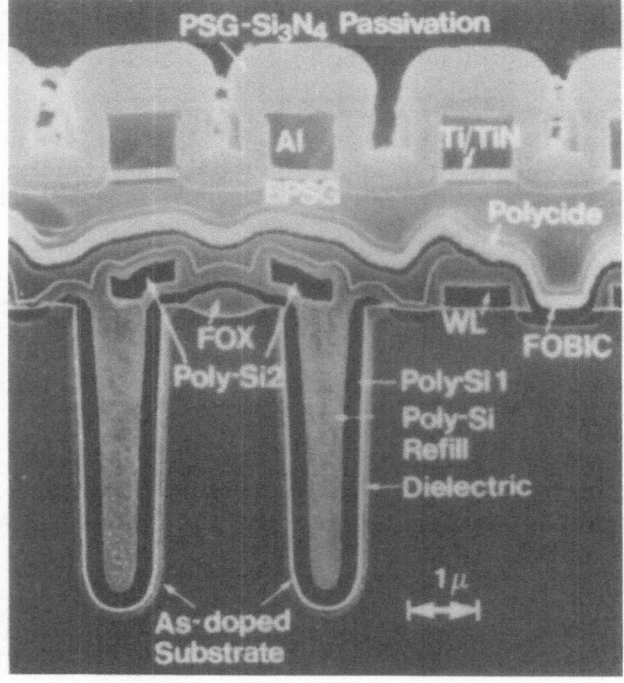

Fig. 2 Cross section through the cell array of the Siemens 4M-DRAM.

1.) High-temperature ($\geq$ 800 °C) heat-treatments of an oxide, mixed with dopants to lower the viscosity, for planarizations prior to aluminum deposition.

2.) Low-temperature ($\leq$ 450 °C) planarization by a combination of chemical vapor deposition (CVD), and application of liquids like photoresist, polyimide or spin-on glass and subsequent etch-back. Bias sputtering of SiO_2 or CVD deposition in combination with bias etch-back are other principles that can be realized when aluminum is already present.

The advantages and limits of these different approaches are being evaluated in the following sections.

Table 1. Criteria for Dielectrics

To achieve: Films that feature	To avoid: Depositions that feature
● Low dielectric constant	● Complicated setups
● Low particle counts	● Low through—put
● Low compressive stress	● High investment and
● Uniform thickness	maintenance costs
● Good etchability	● Poor repeatability
● Good flow properties	● Narrow process window
● Good step coverage	
● High dopant levels possible	
● Homogeneous dopant distribution	
● Easy dopant analysis	

II. PRE-METAL PLANARIZATION PROCESS-FLOWS

When analyzing the complex problem of the optimization of the planarization process sequence, it is helpful to concentrate on three major questions:

- What do we need ?
- How do we get it ?
- How do we know we got it ?

Focusing on these three questions, we start with the requirements for the dielectric, which is mainly borophosphosilicate glass (BPSG) in this section. Next, the deposition processes and the possibilities of optimizing the effectiveness of the flow-step are investigated. The last aspect to be considered is the analytical methods required to enable a meaningful characterization of the films before and after the flow-step.

II. 1 Requirements

The desired properties of an interlayer dielectric as well as the drawbacks to be avoided are listed in Table 1. Whereas the positive requirements refer to the films, the problems to be avoided are mainly related to the deposition systems. As an in-depth discussion of the different reactor types is beyond the scope of this article, the reader interested in technical details is referred to the literature quoted in Table II.

II. 2 Deposition process

The state-of-the-art deposition principle for interlayer dielectrics is CVD. In the last twenty years, the standard material used has been silicon dioxide, doped with different elements. The dopant, which has generally been phosphorus, /5-9/ acts in a way to

* reduce the film stress /5, 10/
* modify the wet etch rate /5/
* provide gettering ability for sodium /6/
* soften the glass during a heat-treatment ("flow-step") /7-9/.

As the phosphorus content, which determines the viscosity reduction of the phosphosilicate glass (PSG), can not be raised above 8 weight% without causing stability problems, the drive towards lower process temperatures has led to a fast acceptance of borophosphosilicate glass (BPSG) /10-68, 97-99/.
Originally developed as a passivation layer /10/, BPSG has enabled a flow temperature reduction of about 200°C. Table II summarizes the BPSG literature which, in spite of being extensive, is probably not exhaustive. Apart from the original approach based on atmospheric pressure (AP) CVD /10-28/, other methods have been developed which work at low pressure (LP) /29-38,/ or are plasma-enhanced (PE) /39-46, 66, 97/, respectively. To avoid silane and other dangerous hydrides, organic molecules like tetraethylorthosilicate (TEOS) /47-58, 67/, diacetoxyditertiarybutoxysilane (DADBS) /66/ or tetramethylcyclo-tetrasiloxane (TMCTS) /68/ have been successfully employed as well. Totally different approaches like the use of spin-on glass /60/ or a boron drive-in /61/ have not met with general acceptance.

Table II has already provided a short evaluation of the different processes. In the following, a more detailed characterization will be undertaken using the criteria listed in Table I.

a) Low dielectric constant. As integrated circuits consist of several alternate layers of conductive and insulating materials, undesired parasitic capacitance can occur. This leads to signal delay /100/ and/or increased ambiguity of data in the case of DRAMs. To limit this effect, interlevel insulators should have a dielectric constant k as low as possible. In this respect, silicon dioxide (k = 3.5-4) compares favorably with e.g. silicon nitride (k = 7.5) or alumina (k = 10.2) /100/. In addition to the earlier-mentioned possibility of tailoring the oxide properties by adding dopants, the low dielectric constant makes SiO_2 a nearly ideal material. Therefore, this section deals only with interlayer dielectrics based on silicon dioxide with practically identical k values.

b) Low particle counts. Particles usually stem from different sources, e.g.:

* Wafer handling
* Flaking of deposits from the reactor system
* Gas-phase nucleation etc.

Table 2A. BPSG Deposition Processes

Source Materials		Process Type	Temperature [°C]	Evaluation	Refs.
Silicon	Dopants				
SiH_4/N_2+O_2 (Silane)	B_2H_6/N_2 PH_3/N_2	AP	350–430	+:State-of-the-art process; good thickness & dopant uniformity –:Large deposition system; complicated wafer-handling; particle problem, bad film step coverage	10–28
SiH_4+O_2	B_2H_6/SiH_4 PH_3/SiH_4	LP	400	+:Improved step coverage, lower particle counts –:Expensive quartz-ware and setup or single-wafer system, resp., lower upper dopant level. B_2H_6 decomposition & handling	29–33
SiH_4+O_2	BCl_3 PH_3	LP	425	+:BCl_3 more stable than B_2H_6; use of standard LTO system possible –:Complicated setup (injector, caged boats), Cl in films and deposition system	34–38 42
$SiH_4/Ar+ N_2O$ (+NO)	B_2H_6/Ar PH_3/Ar	PE	300–450	+:No complicated quartz-ware –:Process control difficult (many parameters), Film quality ?	39–46, 66, 97
spin-on BPSG		spin-on	1.) 190 2.) 630	+:Simple deposition process, high temperature only for reflow –:Glass shrink, BPSG quality	60
PSG-doping in B-atmosphere		AP + drive-in	750	+:Deposition = flow step, simple equipment ––:Effectivity depends on film thickness	61

Table 2B. BPSG Deposition Processes

Source Materials Silicon	Dopants	Process Type	Temperature [°C]	Evaluation	Refs.
$Si(OC_2H_5)_4$ (TEOS)	$B(OCH_3)_3$ $PH_3/Ar+O_2$	LP	620–700	+:Good step coverage, stable films with high dopant level –:Control of vapors not standard, caged boats and injector still required	47–54
$Si(OC_2H_5)_4$	$B(OCH_3)_3$ $PO(OCH_3)_3$	LP	650–700	+:No injectors or caged boats required, no dangerous media –:Regulation of 3 vapors; low P content, low depos.rate	48–50
$Si(OC_2H_5)_4$	$B(OCH_3)_3$ $P(OCH_3)_3$	LP	650–700	+:No injectors or caged boats required; no dangerous media; good step coverage –:Regulation of 3 vapors; strong film inhomogeneities;	48–50 53–58
$Si(OC_2H_5)_4$	$B(OC_2H_5)_3$ $PO(OCH_3)_3$	LP	640–700	+:No injectors or caged boats required, no dangerous media –:(Not enough information available)	67
$SiO_6C_{12}H_{24}$ (DADBS)	$B(OCH_3)_3$ $P(OCH_3)_3$	LP	450–550	+:No dangerous media, lower depos. temperature –: Lower film density, DADBS not stable uponheating	59
$Si_2O_4C_4H_{16}$ (TMCTS)	$B(OCH_3)_3$ $P(OCH_3)_3$	LP	550–650	+:No dangerous media, vapor pressure of TMCTS > TEOS –:(Not enough information available)	68

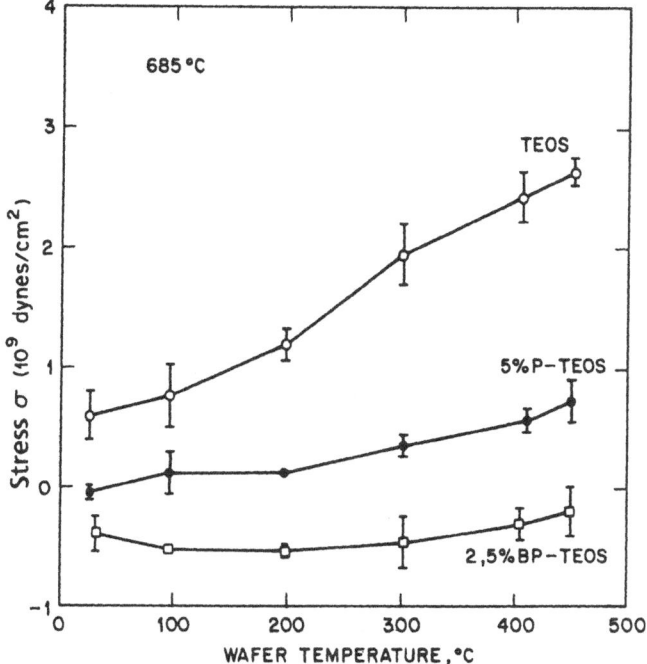

Fig. 3   Stress reduction in LP TEOS-Oxides
(deposited at 685 °C) by increased dopant
content. TEOS-BPSG exhibits a low and
slightly compressive stress (Ref. 47).

Whereas the first two factors depend strongly on the system configuration
and produce comparatively large particles, the last one is process-dependent
and contributes mainly particles below 1 μm^2. In traditional APCVD, particles
originate from the layers that build up on the wafer tray and from the gas
nozzle /19-21/. More refined constructions feature automatic cleaning of the
conveyor belt, as well as improved nozzle designs. Particles, however, remain a
matter of constant concern, because the principle of wafer pick-up does not
lend itself easily to automation.
Good results have been obtained with low-pressure (LP) systems /29/, which are
less prone to gas-phase reaction. Flaking of the loose deposit from the caged
boat, however, requires frequent cleanings and the wafer handling is critical as
well. The information available on LPCVD systems working without boats is of
too preliminary a nature /31, 32/ to enable a conclusive statement. When using
TEOS as silicon source instead of silane, deposits with very good adhesion and
low particle values are obtained /48-50, 52/. Care has to be taken, however, that
the vigorous PH_3 oxidation , which catalyzes the TEOS-decomposition, does not
cause gase phase reaction.

c) Low compressive stress. Film stress originates from the different thermal
expansion coefficients of the silicon wafer and the oxide layer. It can bend the
wafer and lead to cracks or even detachment of the films. Therefore it is
desirable that the stress be low and slightly compressive.

The degree of stress is strongly dependent on the deposition principle, the
process conditions, and the amount of dopant present in the film.
Consequently, the stress values measured after deposition range from tensile
/11-13, 19, 24, 30-33/ to compressive /24, 29, 39, 47, 68/. The incorporation of
dopants reduces the stress /12, 29, 47/, as illustrated in Fig. 3 for TEOS-BPSG. If
the temperatures used during a subsequent densification or flow step are high
enough, stress can be greatly relaxed by the onset of viscous flow. Figure 4
demonstrates this effect for APCVD /24/.

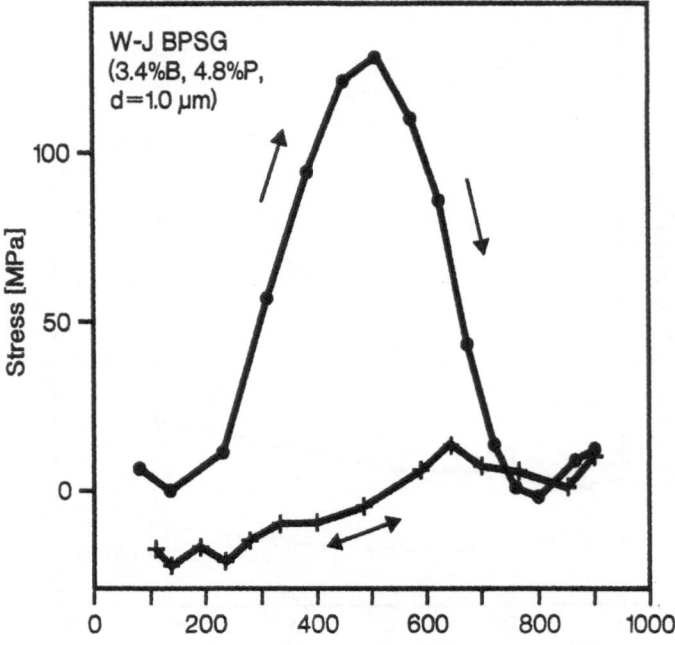

Fig. 4   Stress evolution of AP silane-BPSG. The decrease in tensile stress above 500 °C indicates the onset of plastic flow (Ref. 24b).

    d) Uniform layer thickness. This is generally one of the first problems to be addressed in CVD, as thickness inhomogeneities cause problems for the photolithography. The reason is that the film acts as an antireflective coating and a thickness variation changes the intensity of the light reflected from the wafer surface. In addition, the requirements regarding anisotropy and selectivity of the following etch-step are more severe if the layer to be structured varies in thickness over the wafer. In the case of a more or less planarized BPSG, a special situation prevails, as the glass flow inherently results in considerable local thickness variations, which are in fact a serious challenge for the dry etching process.

As APCVD has been studied for over twenty years /10/, optimization of the systems has enabled good uniformities (< ±2% for 100 mm wafers /19/) at high deposition rates.

The many parameters influencing PECVD make it difficult to reach a generally valid conclusion for this type of deposition. It can be assumed, however, that the difficulties connected with the control of so many variables pose serious problems as regards the long-term repeatability.

LPCVD of doped oxides normally requires caged boats to obtain good thickness uniformities /29, 49, 50, 52/. Especially elaborate constructions are used in the Anicon reactor /29/. This process yields good uniformities, but is like all silane LPCVD extremely sensitive to even minor variations of the quartz-ware. This problem can be greatly alleviated by using organic reagents /48-50, 52/, but the general trade-off between improved uniformities and reduced deposition rates remain. Systems working without boats are a promising aproach /31, 32/, but according to our experience, many problems have still to be solved before the system really lives up to the promises of its vendors. In the future, however, the implementation of wafer diameters beyond 150 mm will aggravate the situation for batch systems working with fragile and expensive quartz-ware, which will render single-wafer processing increasingly attractive. The optimum solution could be the combination of advanced single-wafer technology with the advantages of depositions based on the pyrolysis of organic molecules /69, 70/.

e) Good etchability. Usually, the patterning of BPSG is performed by anisotropic dry-etching, but isotropic wet-etching is used for some applications as well.Whereas the addition of dopants tends to increase the dry-etch rate /50, 97/, the effect is more complex for the case of HF containing aqueous solutions /11, 12, 21, 29, 30, 34, 39, 40, 47-50, 52, 59/. Here, the etch rate increases with the phosphorus content and is reduced by the addition of boron /11, 29, 39-40, 47-50, 52/. This effect can be used to establish a desired etch rate independently of the glass composition /98/, but it prevents the simple dopant determination feasible in the case of binary glasses /5, 71-73/. It is possible, however, to check for deviations in the dopant content, as it is highly improbable that both the B and P concentration change in a way to neutralize the mutual effects on the etch rate. In addition, studies of BPSG wet-etch rates can shed insight into the complex process of glass formation, because the preparation methods strongly influence the result.

Samples prepared by low temperature SiH_4 CVD, for example, show a pronounced reduction of the etch-rate with increased temperature of densification /12, 21, 30, 39/, while TEOS-BPSG remains practically unaffected by a heat-treatment /52/. This hints at a higher glass stability of as-deposited TEOS-BPSG, a finding in accordance with other observations /50/.

f) Good flow properties. To facilitate the understanding of the many parameters bearing on the flow results, these are listed in Table III broken into three categories.

Before discussing the relative importance of the factors under consideration, it is necessary to define the two potential applications of flow possible in integrated circuits:

* Flow: Smoothes or even planarizes the device topography prior to the BPSG patterning.
* Reflow: Performed after the contact-hole etching through the BPSG, it serves to round-off the upper part of the glass to improve the step-coverage of sputtered aluminum, as shown in Fig. 5.

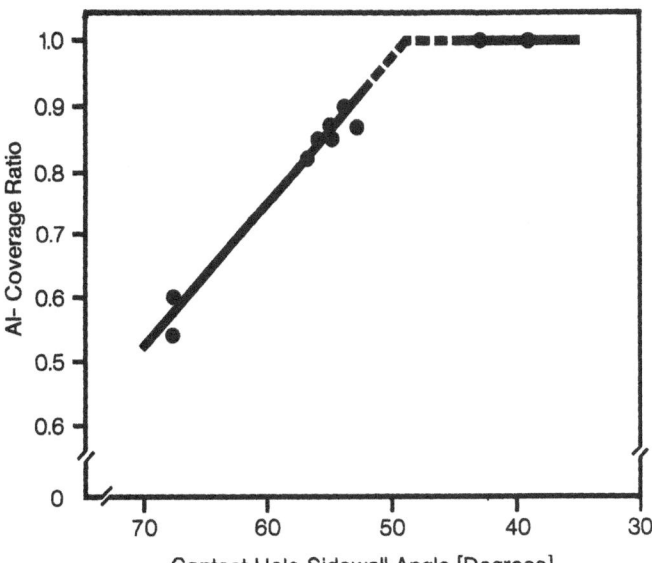

Fig. 5   Improved aluminum step coverage (defined as the ratio of the thickness on the side-walls to the thickness on the horizontal surface) due to increased tapering of the contact hole wall. (Ref. 29)

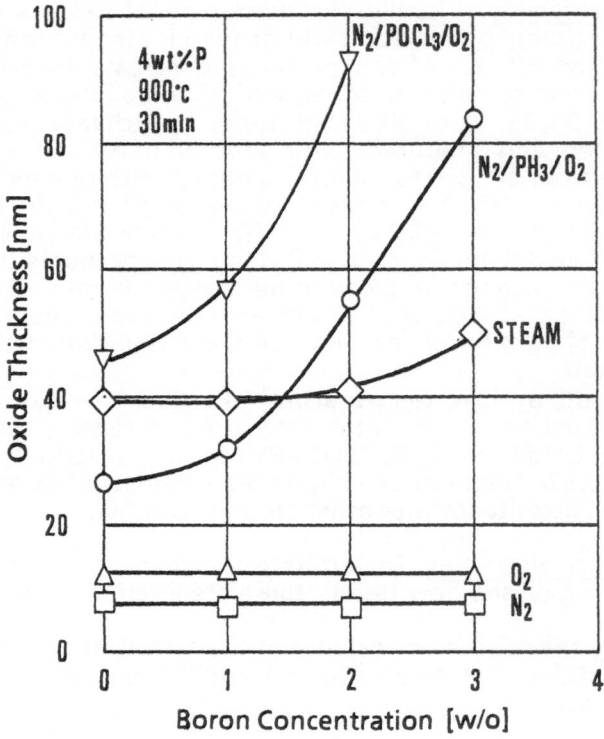

Fig. 6   Thickness of the silicon oxide grown in
the center of a 100 x 100 μm^2 contact
window as a function of the reflow
atmosphere and the boron content of
the BPSG. Increased oxide thickness can
translate into higher contact resistance
because the boron doped oxide is
difficult to remove completely (Ref.26).

According to these applications, two different methods of flow evaluation
have been proposed, both using scanning electron microscopy (SEM). The first is
based on depositing the film to be tested over a defined structure and
measuring the step angle reduction caused by the glass flow /12, 13, 18, 21, 25,
30, 33, 34, 36, 38, 39, 41, 48-50, 52-55, 59-64/. In the second type of experiment,
the rounding of the upper part of a hole etched into BPSG is taken as a measure
to define the flow /11, 12, 14, 26, 29, 36-40, 43, 45, 51, 53, 54, 60, 64/.
This second test is simpler, but only of limited value, because in many devices the
consequences of a reflow step for the contact hole properties cannot be
tolerated. Figure 6 demonstrates the influence of the boron content of BPSG on
the contact resistance /26/. In addition, the special mechanisms governing the
flow (c.f. Sect. II - 3) make the results of reflow experiments extremely sensitive
to geometrical parameters like depth, width and spacing of the contact holes
/43, 45, 51, 52/. This effect, as illustrated in Fig. 7, can even produce spherical
overhangs instead of tapered slopes if the viscosity reduction is carried too far.
But even when using the first method to evaluate glass flow properties, all the
parameters listed in Table III have to be controlled to get comparable results.
For glasses with high dopant levels, the total thermal budget has to be taken
into account, because the ramp-phase alone can be sufficient to induce flow, as
shown in Fig. 8. Figure 9 demonstrates the influence of the structure pitch on
the flow result. It can be seen for example that it is much easier to fully planarize
a closely spaced structure than an isolated polysilicon stripe.

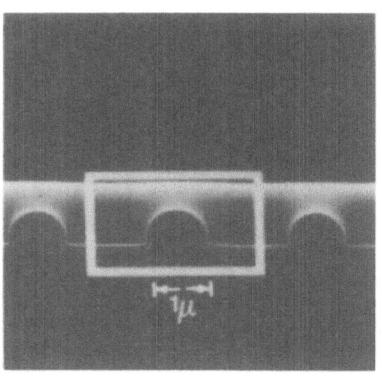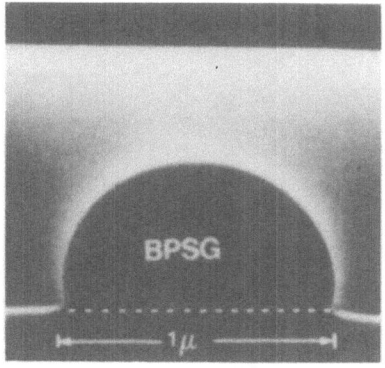

Fig. 7   Shaping of a contact hole area of a 4M DRAM after a 1100 °C, 2 s pulse in N$_2$ (Eaton ROA system). The BPSG (5.5% B, 3.5% P) is drawn out of the narrow regions between the contacts and the remaining material shows a tendency to form spherical overhangs.

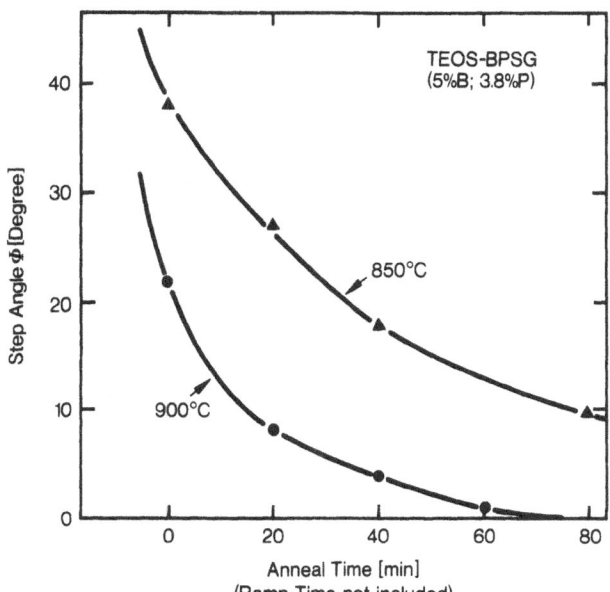

Fig. 8   Tapering of TEOS-BPSG layers deposited over a polysilicon test structure /49/ by the combined effect of the furnace temperature ramp and the anneal step. The ramp conditions were as follows: 10min. move-in at 800°C, ramp-up at 10°C/min., ramp-down at 3.3°C/min. This ramp-cycle alone is enough to reduce the step angle from 80° after deposition /49, 50/ to less than 40°.

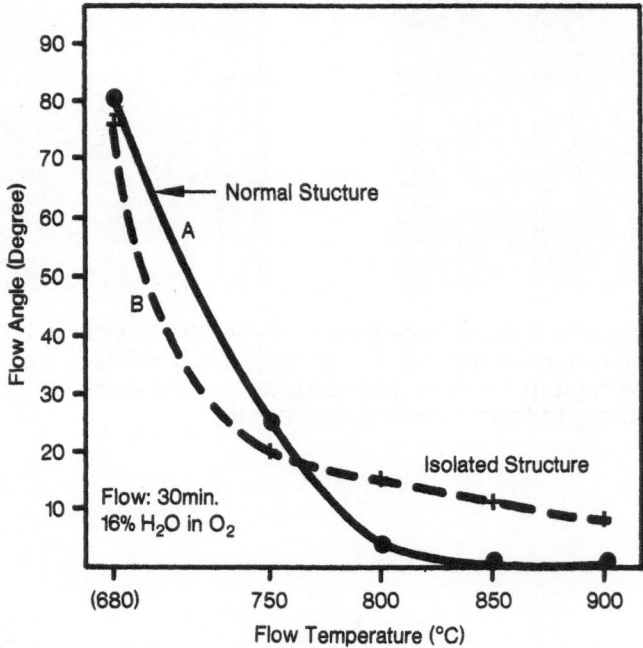

Fig. 9　Influence of the test structure pitch on the flow result for TEOS-BPSG (6.6% B, 3.5% P) in wet atmosphere. An isolated polysilicon stripe is more difficult to planarize (Ref. 50).

Table 3. Factors Influencing the Flow Result

A) Influence of the Topography	➡ Structure pitch
	➡ Structure hight
	➡ Step angle
B) Film Properties	➡ Dopant content
	➡ Step coverage
	➡ Film thickness
	➡ Glass structure
C) Parameters of the Flow Step	➡ Temperature
	➡ Atmosphere
	➡ Pressure
	➡ Time
	➡ Method (Rapid anneal/furnace)

Another important point to be considered is the glass structure. Due to the fact that the parameters of Table III differ widely in the published literature, it is nearly impossible to compare the flow properties of the described glasses. Experiments with different BPSG types conducted by the same authors show, however, that considerable differences in the flow properties can occur even if the glass compositions are nominally similar /34, 46/. Especially critical is PE-BPSG, where the flow has been found to depend on the reactor type /46/, or not to occur at all without a special pre-annealing step /43/.

Two film properties of special importance, i.e. the step coverage and the maximum dopant content achievable, will be treated separately in the following.

g) Good step coverage. Step coverage is defined as the ratio of film thicknesses on the vertical and the horizontal part of the topography. It is governed by pressure, temperature and chemistry of the deposition process /50, 74/.

Step coverage can best be studied by using deep and narrow trenches as test structures /31, 32, 48-50, 52, 57/. Comparisons have shown that depositions based on the pyrolysis of organic molecules like TEOS result in a superior step coverage of 60 - 80 % /48-50, 52, 57, 74, 75/, whereas the values for silane-CVD are typically below 30 % /31, 32, 49, 50, 52, 57, 74/. Care has to be taken, however, that the delineation etch required to enhance the SEM contrast of the cleaved samples does not reduce the thickness of the film to be studied. Especially silane-BPSG with its higher etch-rate can thus be stripped completely in the lower part of the trench, resulting in misleading step-coverage values of practically zero (c.f. Ref. 57, Fig. 11a).

In normal topographies with structure height-to-width ratios ≤ 1, the most critical point is the reentrant angle, which is formed by the vertical and lower horizontal surfaces of the film. If this angle is greater than 90°, stronger flow is required to achieve a tapered surface. On a topography with closely spaced protruding parts, touching of the upper region of the layer can even result in voids which are a potential hazard during subsequent etching.

Under this aspect too, TEOS compares favorably with silane as a silicon-source for CVD. If a TEOS-BPSG film with a certain thickness is deposited over a vertical step of the same height, the reentrant angle is typically about 80° /49/ and can

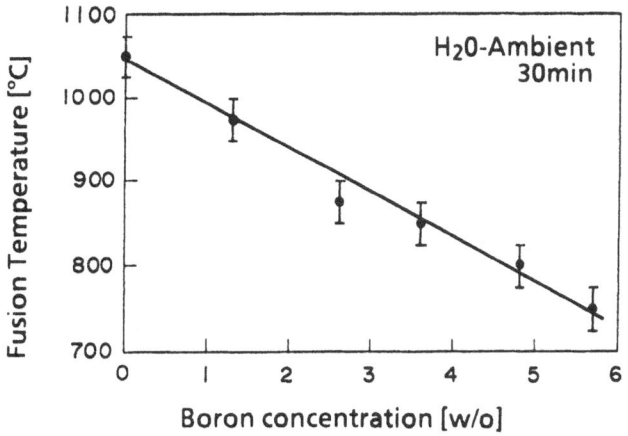

Fig. 10 Fusion temperature dependence of BPSG containing 4% P on the boron content. Fusion was defined as the temperature required to obtain a step angle of about 20° (Ref. 18).

367

be further reduced by raising the film thickness /50, 57/ without risking film cusping. In contrast, reentrant angles $\geq 90°$ are typical for all kinds of low temperature silane-CVD. The situation is worst for APCVD, which constitutes a serious disadvantage of this process with regard to future applications in ULSI devices.

h) High dopant level. Changing the dopant content constitutes the most powerful means to tailor the BPSG flow properties. Measured in weight % (w/o), boron is about twice as effective in reducing the glass viscosity as phosphorus /40/. (Sometimes, the dopant concentrations are given in mole % (m/o). As a rule of thumb it can be said that 1 w/o P equals about 1 m/o P and 1 w/o B corresponds to about 3 m/o B /28a, 33, 65/)

Figure 10 shows the step angle reduction possible by increased boron content /18/. As PSG had been the state-of-the-art material before, quite often BPSG with higher concentrations of phosphorus than boron is still used. To optimize the flow, this ratio should be reversed. The glass transition curves depicted in Fig. 11 demonstrate, that for glasses with a boron content above 4 w/o, a reduction of the phosphorus level is not detrimental to the flow. This is important, because the addition of either dopant impairs the film stability by increasing the susceptibility of the material to attack by moisture. The result can be the formation of metaboric acid crystals of considerable size on the BPSG surface /12, 19, 21, 29/. These particles are dissolved by immersion of the wafer in methanol /19/ or ethanol /21/, but the procedure depletes the surface of boron and is surely not of the kind to be recommended for ULSI applications.

Thus the possibility of glass viscosity reduction by increased dopant content is limited, but the upper stability level is strongly dependent on the glass deposition principle. When comparing stabilities, it is very important to clearly define the boundary conditions like atmosphere humidity, storage time,

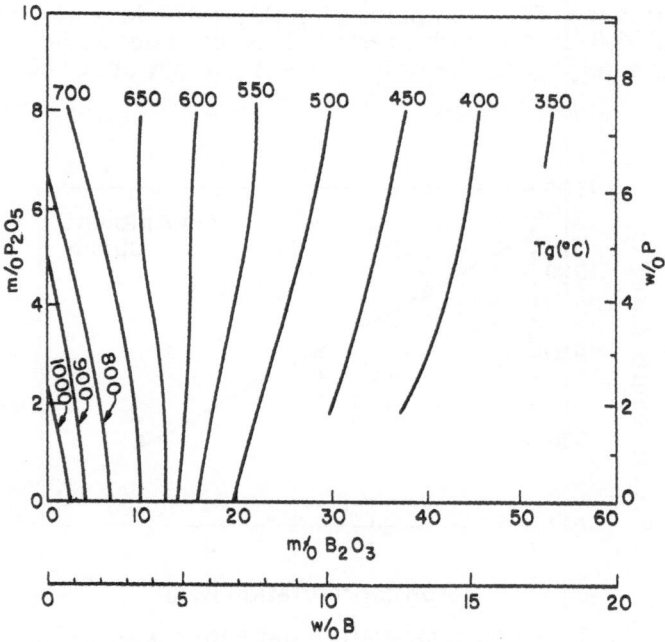

Fig. 11 Glass transition isotherms for bulk borophosphosilicate glass systems. To obtain the flow temperature, a constant 288 ($\pm$ 16)°C must be added (Ref. 65).

concentrations of both phosphorus and boron and the thermal history (anneal ?). As 3 w/o phosphorus is assumed to be sufficient to getter sodium /53, 65/, the comparison of such samples with different amounts of boron is a meaningful tool to characterize BPSG. If we define as the stability requirement for undensified films a minimum time of 24h, the following rules of thumb can be given:

* Silane-LPCVD   /29/:   Stable up to ~ 4 w/o B
* Silane-APCVD /11-28/:   Stable up to ~ 5 w/o B
* TEOS-LPCVD /48-50,52/:   Stable up to ~ 6 w/o B

Higher limits have been claimed for APCVD /28b/, but have not been confirmed by our own observations. The superior stability of TEOS-BPSG /50/ can be a attributed to the higher deposition temperature, which leads to the formation of a glass whose structure is much closer to the final stage.
This result can be achieved in part by heat treating silane-BPSG immediately after deposition in an atmosphere containing a certain amount of oxygen. Such a requirement for a strict time coupling is feasible, but of course undesirable for practical applications. A second problem, which has been encountered when trying to produce samples with high B content in an Anicon reactor /29/, is the difficulty to attain B concentrations above 4 w/o. This could be due to the fact that the high diborane concentrations required in the process gas (30-60 % B_2H_6 in SiH_4) are unstable and decompose in the course of time, depending on the storage temperature /42/. This problem is circumvented by using BCl_3 instead /34-38/, but the chlorine is detrimental to the LPCVD system and is incorporated into the BPSG, where it can lead to stability problems /34/.
A possible solution could be the substitution of diborane by trimethylborate (TMB), which has been reported to improve the step-coverage as well /76/, but the information existing is to scarce too enable a final statement.
To summarize these findings, it can be concluded that the choice of a BPSG deposition principle has important consequences for the resulting flow.
Therefore this aspect, as well as technical issues, should be considered when deciding on a certain BPSG process.

i) Homogeneous dopant distribution. The amount of boron and phosphorus can vary over the wafer or across the layer thickness. If the variations are too strong, differences in the flow properties and the etch rates can be the consequence. Silane-BPSG /e.g. 12,19,29/ usually exhibits better homogeneity across the layer than TEOS-BPSG /49,50, 52/, because the precise regulation of liquid sources with low vapor pressure /50/ is not yet state-of-the-art. Inhomogeneities over the wafer can occur in APCVD if the uniformity of the gas feed through the nozzle system is impeded by e.g. clogging. In LPCVD using caged boats /e.g. 29, 34, 50/, depletion of the process gases can result in dopant variations between the center and the edge of the wafer. Usually this problem can be overcome by adjusting the process parameters, but it will call for more careful consideration with larger wafers ($\geqq$ 150 mm). For the time being, however, all processes listed in Table II should be able to meet the requirements.

k) Easy dopant analysis. As this can be a time - determining factor in development as well as production control, great lengths have been taken to find a fast, accurate, and if possible, non-destructive analysis technique. The advantages and drawbacks of the approaches investigated will be discussed in section II-4.

II.3. Optimization of the flow-step

For a better understanding of glass flow, it is helpful to start with a short discussion of the relevant mechanisms /24, 33, 45, 50-52, 55, 77-79/.

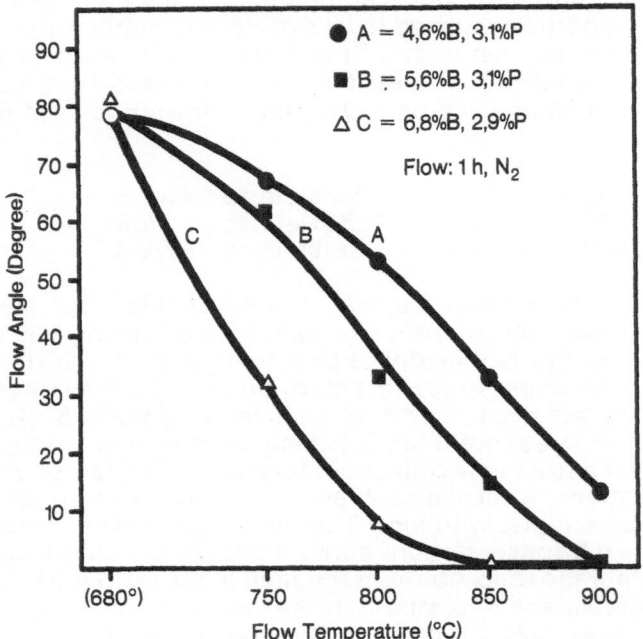

Fig. 12   Flow inprovement by higher temperature
for TEOS-BPSG deposited at 680 °C with
different boron contents.

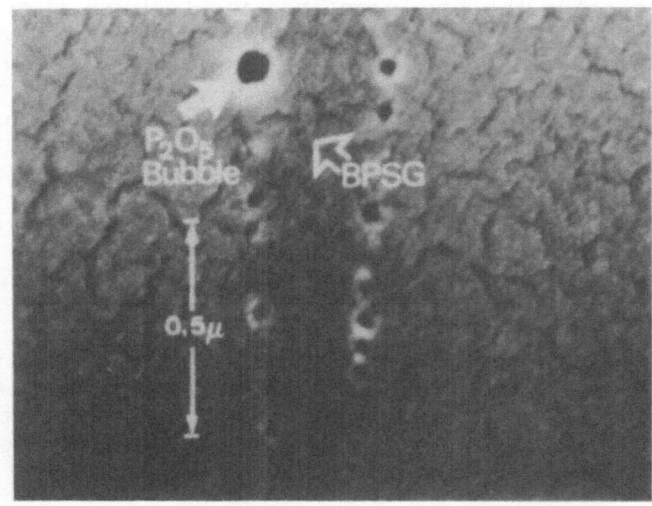

Fig. 13   Formation of P_2O_5 bubbles in TEOS-BPSG
(5.3% B, 3.5% P) deposited into deep
trenches and heat treated at 950 °C, 1h, N_2.
No delineation etch was performed on the
cleaved samples.

The smoothing of the glass surface happens, because the system wants to satisfy the minimum surface-free-energy condition /77/. The dominant mechanisms have been found to be surface diffusion and viscous flow /77/. For lower viscosities, the surface tension tries to reduce the surface area by concentrating the material in spheres, hence the spherical overhangs observed in reflow experiments /51, 52/. As the wafers stand upright in the furnace, gravitation is normally not a driving force. An exception could be rapid anneal, where low viscosities are realized for very short times with the wafers in a horizontal position /52/. Dopants reduce the glass viscosity, because their oxides distort the SiO_2 structure /53, 54, 65/.

The effectiveness of the parameters listed in Table III -C to promote glass flow varies as well as their negative impact on the device. Therefore, a discussion of the possible flow enhancement factors has to include the consideration of unwanted side-effects as well.

a) Temperature. As glass viscosity fits an Arrhenius-type equation, a temperature rise is very efficient in breaking the cation-oxygen bonds /78/. Figure 12 shows the increased planarization effected by higher anneal temperatures for three glasses with different boron content /50/. For practical applications, however, an upper temperature limit of about 900-950° has been found because bubbles appear due to the increased fugacity of the P_2O_5 /51-53, 55, 57, 59/ or the release of N_2 in the case of PE-BPSG /39/. This effect, which is cosmetically undesired if not also electrically harmful, is illustrated in Fig. 13. Deterioration of the glass quality by phase separation is another possible matter of concern /53, 55/.
Even more stringent are the limitations derived from the requirements to preserve shallow junctions in the devices and to minimize wafer warpage. Therefore, the 850° - 900° temperature range can be assumed as the upper limit tolerable for future ULSI planarization processes.

b) Atmosphere. To aid in breaking the cation-oxygen bond and to soften the glass, two principles have been found effective. The first uses additional P_2O_5 deposited on the BPSG by either reacting PH_3 /25, 26 /, $POCl_3$ / 26 / or PBr_3 / 62 / with oxygen. The upper BPSG layer is thus enriched in phosphorus and the glass is flown at the same time. Afterwards, the P_2O_5 is selectively etched off, leaving a smoothed surface. The drawbacks of this method, however, are increased complexity and the formation of comfit-like particles as shown in Fig. 14 / 25 /.

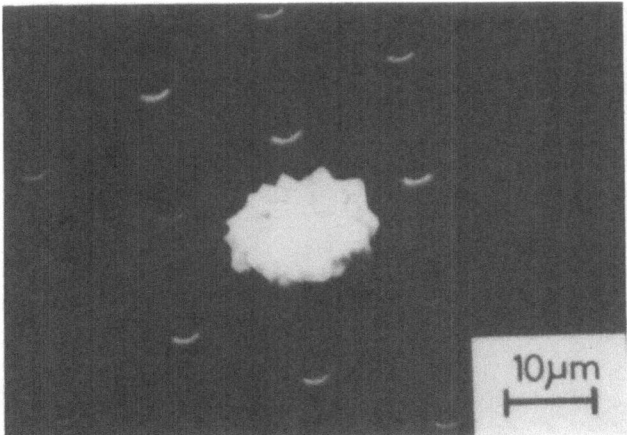

Fig. 14 BPO_4 particle formed on BPSG (3.6% B, 4.0% P) after flow in PH_3/O_2 ambient (Ref. 25).

They consist of BPO$_4$ and are insoluble in H$_2$0/HF. The particles can be avoided by reducing the boron content of the BPSG, / 62 / or by covering it with a thin layer of either PSG or SiO$_2$, but these solutions both obviate at least partially the initial advantage.

The second approach makes use of the fact that the cation-oxygen bond cleavage can be promoted by the presence of water vapor /80/. Various authors have demonstrated the superior flow possible in H$_2$0 atmosphere / 12, 13, 18, 21, 28-30, 33, 34, 38, 39, 49, 50 /. It becomes evident by comparing Fig. 9, curve A with Fig. 12, curve C. The wet atmosphere improves the chemical stability of the films as well, an important aspect especially for silane-BPSG.
The disadvantage, however, is that underlying silicon structures like transistor gates can become oxidized to an undesirable extent. This can be minimized by working with a reduced H$_2$0 concentration / 49, 50 / or even eliminated by placing a thin nitride layer below the BPSG, which can serve as a barrier against unwanted out- diffusion of dopant as well / 38 /.
If such a nitride layer can be tolerated in the device, the H$_2$0 atmosphere is surely the best means to achieve excellent flow.

c) Pressure. It has been shown for PSG / 81, 82 / and BPSG / 63, 64 / that further considerable flow enhancement is obtained by combining steam atmosphere with high pressure. Figure 15 demonstrates the tapering possible at a temperature of only 700° if the process is carried out at 10 atmospheres   /63, 64/. The necessity to protect the silicon is of course even more urgent under these conditions.
As fab lines tend to be conservative by nature, further objections· are to be expected from that side against high pressure technology, which probably will be accepted only if inevitable.

d) Time. Especially at higher viscosities, longer time helps to planarize the glass by enabeling better diffusion / 77 /. Unfortunately, however, the same holds for the out-diffusion of dopants in the device. Therefore, it is advisable to keep the process time short, the ultimate possibility being rapid anneal.

e) Rapid anneal. A period of very low viscosity, even if only short, exerts a decisive influence on the flow result. Therefore, several authors have successfully applied rapid anneal to shape the BPSG surface / 13, 14, 19, 21, 30-33, 36, 43, 45, 46, 49, 50, 52 / while preserving shallow junctions / 14 /. Figure 7 has shown that even 2 sec can be sufficient to flow a low viscosity glass. If longer

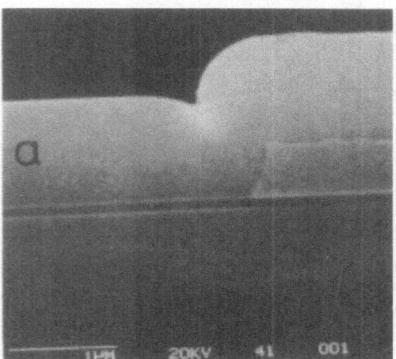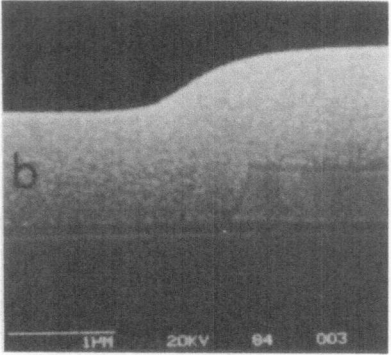

Fig. 15   Influence of pressure on the flow in wet ambient for BPSG (5% B, 3% P).
a) 1 atm, 700 °C, 38 min   b) 10 atm, 700 °C, 30 min.
Whereas the flow is negligible at 1atm, good tapering is achieved at 10 atm (Ref. 63).

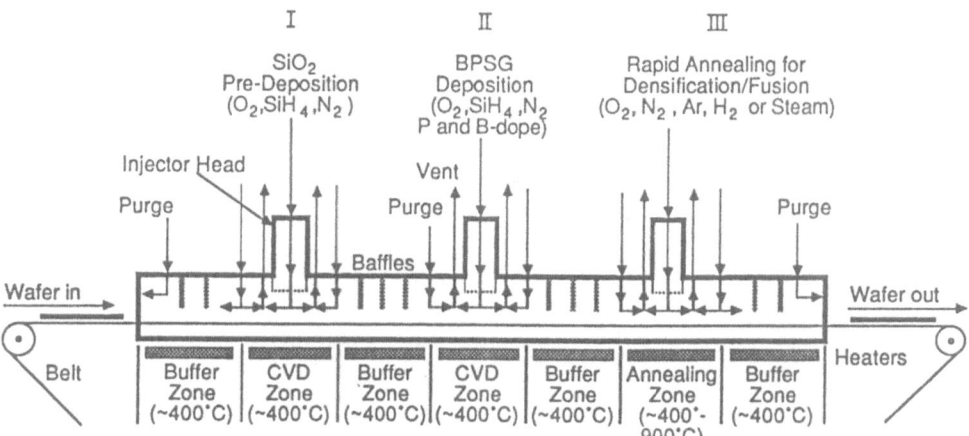

I   II   III

SiO$_2$ Pre-Deposition (O$_2$,SiH$_4$,N$_2$)   BPSG Deposition (O$_2$,SiH$_4$,N$_2$ P and B-dope)   Rapid Annealing for Densification/Fusion (O$_2$, N$_2$, Ar, H$_2$ or Steam)

Fig. 16   Possible BPSG deposition system with integrated flow-step by rapid anneal for the production of densified and fused films (Ref. 33).

times (>20 sec) are used, a slight flow improvement has been observed in wet atmosphere / 30, 33 /.

As flow or reflow takes place at a late state of wafer processing, an estimation of all the consequences of a short heat pulse on the device performance is difficult and beyond the scope of this article. Further investigations destined to elucidate this point will help to facilitate the acceptance of rapid anneal in a production environment. Another favorable factor will be the trend towards single wafer processing caused by the projected further increase in wafer size beyond 150 mm. A combination of deposition and in-situ anneal, as already proposed for APCVD / 33 / (Fig. 16), would present a unique possibility for advantageous rapid anneal application.

II. 4. Film analysis

Standard measurements of film thickness or particle density can be performed with the same equipment used for the characterization of undoped SiO$_2$ / 75 /. Special methods have to be developed for the evaluation of glass flow and dopant concentration.

The state-of-the-art tool to determine the flow has been SEM investigation of cleaved samples (c.f. section II. 2.f). As this is a cumbersome and destructive technique, profilometry has been tried as an alternative /55/. It seems to work for regular and symmetric test patterns, but is probably not sensitive enough for complex topographies.

The situation is even worse as regards the problem of dopant analysis. Great lengths have been taken to establish a fast and non-destructive method, but these efforts have met with only limited success.

As both B$_2$O$_3$ and P$_2$O$_5$ show distinct infrared (IR) absorptions /83, 84/, this technique has been frequently tried / 12, 13, 19-21, 24, 27, 28, 30, 38-40, 48-50, 52, 57, 59, 60 /. Unfortunately, however, the main phosphorus peak at 1320 cm^{-1} is comparatively weak and overlaps with the much stronger boron peak at 1400 cm^{-1} / e.g. 27, 50, 52 /. In addition, considerable departure from the peak-form of binary glasses is often encountered when another dopant is present. The addition of phosphorus, for example, distorts the boron peak as demonstrated in Fig. 17 for TEOS-BPSG / 50 /. The same problem has been observed for AP-silane-BPSG /85/, but not for LP-silane-BPSG /50/. Silane films, on the other hand, cannot be measured at all without a prior densification step

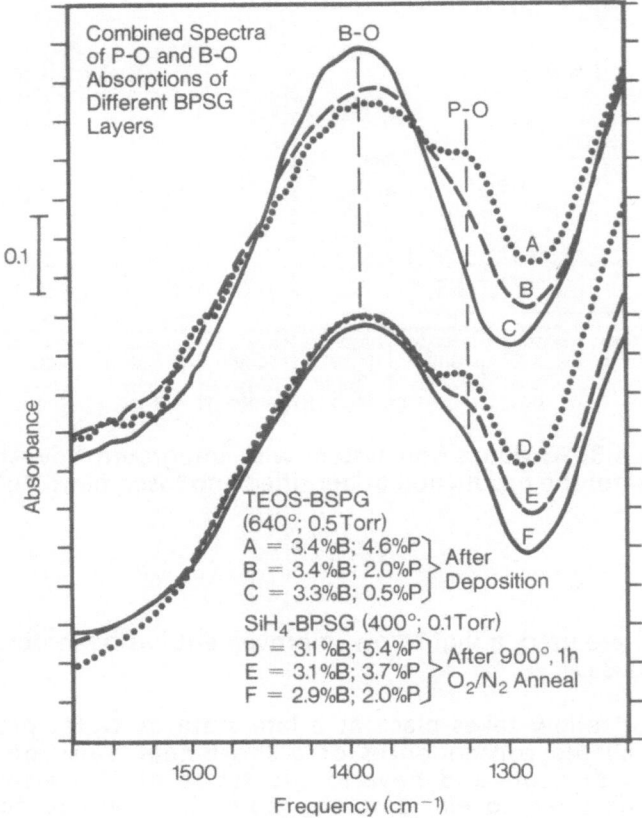

Fig. 17 Influence of a variation of the phosphorus
content on spectra of BPSG films with
constant boron content produced by TEOS
decomposition (A-C) and SiH₄-oxidation
(D-F). The intensity of the B-0 band at
1400cm⁻¹ remains unchanged for different
phosphorus values, whereas for the TEOS-
BPSG films, a decrease in the B-0 absorption
with increasing phosphorus content is
observed. (Ref. 50)

/ 12, 19,21-24, 39 / to resolve the absorption bands. The results published so far
permit the conclusion, that IR technique can be used to measure the boron
content / 21, 39, 40 / if the phosphorus concentration remains more or less
constant. To determine the phosphorus, it is possible to use the refractive index
/ 40, 44, 55 /, but the accuracy is not very good.

If destruction of the samples can be tolerated, a wide variety of analysis
techniques is at the disposal of the process engineer. The compositional
uniformity of BPSG in terms of boron and phosphorus content has been
investigated by secondary mass spectroscopy (SIMS) / 12, 13, 15, 17, 19, 21, 29,
31, 32, 34, 48-50, 52, 59, 62 /, Rutherford backscattering (RBS) / 17 / or Auger
spectroscopy /38, 41/. Other methods employed to determine the dopant
concentrations are ion chromatography (IC) / 40, 48-50, 52, 56 /, wet-chemical
analysis /11, 12, 17, 21, 24, 26-32, 34, 38, 42 /, electron-microprobe / 11, 12, 15,
17, 40, 41, 43, 44, 55-58/, spectrophotometry /13, 18, 56/, plasma-spectroscopy

/ 38, 46, 58, 59 / or x-ray fluorescence / 12, 20, 21, 38, 40, 42, 79 /. The last method has mostly been used for the determination of phosphorus, but the feasability has been demonstrated for boron as well / 86 /.

The agreement between all these methods has often been unsatisfactory and it is impossible to say with assurance which one should be recommended. The most straightforward approach is probably to "calibrate" the existing in-house analysis method by establishing the required flow and to take published concentration data just as a guide-line.

II. 5. Other methods involving oxides

This section shortly reviews approaches not based on BPSG. Unfortunately, however, these concepts either do not meet all the requirements or suffer from problems of integration into existing process flows.

The properties of arsenosilicate glass (AsSG) have been investigated / 87,88 / and AsSG has even been employed in devices /89/. The advantage of this alternative, however, is not evident because arsine is a very dangerous process gas and comparatively high As concentrations are required to achieve glass flow /88/. Figure 18 illustrates this lack of flow in comparison to BPSG for a glass of normal As-concentration deposited with a much safer, TEOS-based chemistry /90/.

Another element studied is germanium, which has the advantage of not acting as a dopant source / 54, 65, 91-93 /. Its oxide readily substitutes for silicon in the silicon-oxygen network /54/, which reduces the danger of phase separation. Due to the same property, however, high Ge concentrations are required to cause good glass flow and such glasses are highly hygroscopic /54, 65/. It should be mentioned here that the high reactivity of Ge has been used successfully to flow a germanosilicate glass by reducing the GeO_2 in H_2 atmosphere /91/, but this technique is difficult to integrate into existing process flows.

Another approach to improve flow has been the implantation of As into PSG /94/ but this involves the use of high-cost equipment and the implantation efficiency at higher steps is doubtful.

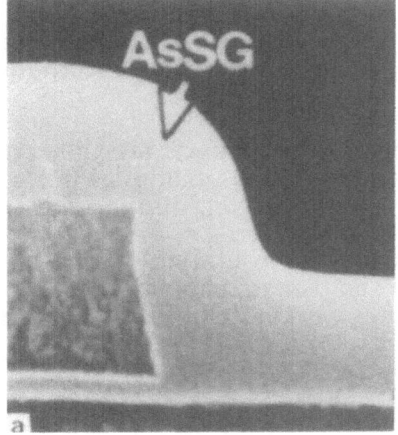 

Fig. 18   Comparison of the tapering achievable with a) TEOS AsSG (12 w/o As), flown 1 h, 1000°, N_2 and b) TEOS-BPSG (5% B, 3.8% P), subjected to only the temperature ramp up to 900°, N_2 (see Fig.8). In spite of the much higher thermal budget, the AsSG flow is only negligible. Poly-Si thickness is 0.5 µm.

If at an earlier stage of processing the planarization of a comparatively regular structure is desired, a combination of CVD-deposition and photoresist etchback has been employed / 95 /.

II. 6. Conclusion

In the last years, BPSG has clearly emerged as the mainstream-technology and will maintain this dominant position. Developments will focus on improved, but probably more complicated reactor designs (low pressure, single wafer). With the advent of technical solutions for the regulation of organic liquids, increased use will be made of the superior film quality possible with these processes. The investigation of silicon compounds containing already one or more dopants /96/ could be a further important step in that direction.
The flow of these high-quality glasses will be increasingly performed by rapid anneal. If devices are designed in a way to permit flow in wet atmosphere, temperatures could be lowered to about 800° using conventional furnaces.
If necessary, a further reduction by about 100° could be obtained with high-pressure technology. This will narrow, but not eliminate the gap between high-temperature processing and the temperature range compatible with Al or W metallization. Therefore, the development of low-temperature planarization processing will remain of highly topical interest.

III. MULTILEVEL - METAL (MLM) CONCEPTS

In this part of the article, the planarization methods applicable to structures incorporating aluminum will be reviewed. A generally valid judgement, however, is nearly impossible due to the complexity of the matter. In contrast to the clearly defined BPSG technology discussed in section II, no dominant mainstream principle has evolved in MLM. The main reasons are twofold:

1. As memory designers up to now have tended to avoid MLM, this technology lacks the high degree of standardization imposed by high-volume production of devices with similar structures. Therefore, many of the mechanical or electrical MLM characterizations reported are pertinent only to a specific device.

2. To achieve good planarization without high temperatures, most techniques involve a combination of CVD deposition, spin-on of a liquid and subsequent etch-back . As a wide variety of processes is available for each of these steps, many of the experiments published differ vastly concerning the technologies employed.

Exhaustive discussions covering all the relevant aspects are the topic of specific conferences (e.g. V-MIC) and are beyond the possibilities of this paper. What shall be attempted is to give an overview of the relevant principles and to provide some insight into the main problems to be solved.
First, the questions related to CVD depositions of insulating layers shall be treated, followed by a discussion of the spin-on planarization methods. A third part is dedicated to technological approaches working with a combination of deposition and in-situ etch-back.

III.1. Deposited layers for MLM

Two contradictory requirements determine the choice of insulating layers for MLM:

* On the one hand, quality demands on the as-deposited films are even higher than in the case of pre-metal insulators, because no high-temperature anneal to improve film quality is permitted.

Table 4. Low Temperature CVD Deposition Processes

Method	Process Gases	Film	Evaluation	Refs.
PE—CVD	$TEOS+O_2$	SiO_2	+: Good step coverage, high deposition rate —: radiation damage possible	69,70,101 154
	$Silane+N_2O$ or $O_2 (+PH_3)$	SiO_2 (PSG)	+: Versatile state—of—the—art process, low stress —: Step coverage, radiation damage possible	102—107 123,127 135,136,148
	$Silane+NH_3$	Si_3N_4	+: Nitride good barrier: state—of—the—art process —: Stress, radiation damage possible, high H_2 content	102,121
LP—CVD	$TEOS+Ozone$	SiO_2	+: Good step coverage —: Films are soft and hygroscopic	69,108
	$Silane+O_2$ $(+PH_3,B_2H_6)$	SiO_2 (PSG,BSG,BPSG)	+: Improved step coverage, compressive stress —: Low depos rate with caged boats	29,109— 112,122,126
	$DADBS (+TMP)$	SiO_2 (PSG)	+: Good step coverage —: Depos.Rate low at $\leq 450^0$,film quality poor	113—115
AP—CVD	$Silane+O_2$ $(+PH_3,B_2H_6)$	SiO_2 (PSG,BSG,BPSG)	+: Established Process, high deposition rate —: Poor step coverage, tensile stress, particles	e.g.23,28 111
	$Silane+O_2+AsH_3$	As SG	+: ? ; —: Dangerous process gas	87
	$Disilane+O_2$	SiO_2	+: Improved step coverage —: Expensive and dangerous process gas	116,133
Photo-CVD	$Silane+O_2$ or N_2O	SiO_2	+: Good step coverage: —:systems not suitable for production	104,117,118
Ion Beam Depos.	$SiO+O$	SiO_2	+: No problems with step coverage; —: Exotic technique	119

377

* On the other hand, the properties of films deposited below 450 °C are normally inferior in comparison to those of high-temperature CVD. This holds especially for critical parameters like step-coverage, stress, chemical stability and electrical quality.

Table IV compares different deposition principles. The layers normally consist of doped or undoped oxides and/or silicon nitride, which exhibits excellent passivation properties against humidity and sodium. The short evaluation given in Table IV shall be detailed in the following discussion concentrating on some salient criteria.

a) Step coverage. Step coverage is especially critical in MLM, because the combination of narrow metal patterns and thick deposited dielectric can lead to film cusping, a potential hazard during subsequent etch-back. As mentioned before (see II.2.g), step coverage is strongly dependent on the deposition principle and the process gases used. Worst is APCVD, while DADBS-LPCVD or TEOS-Ozone LPCVD are best, but produce films of otherwise unsatisfactory quality at temperatures below 450 °C /69, 108, 113-115/. Disilane APCVD /116/ is promising, but the high cost and the dangers connected with this compound have impeded the general acceptance.

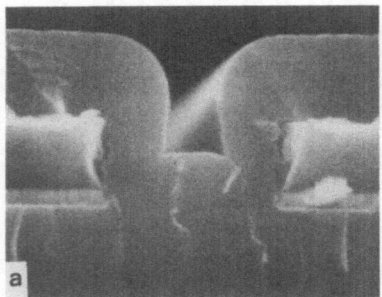 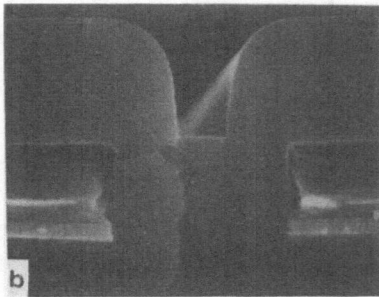

Fig. 19   Comparison of the step coverages of PECVD-SiO$_2$ films deposited using as silicon source silane (a) and TEOS (b). (Ref. 101)

Therefore, the efforts have been concentrated on PE CVD, because this method is equally applicable to the deposition of oxides and nitrides or a mixture of both /102/. PE-Nitride usually has a better step coverage than oxide, but the considerable content of H$_2$ /102/, the stress and the higher dielectric constant /101, 100/ compromise this advantage. Recent investigations /69, 70, 101, 154/ have shown that a distinct improvement in step coverage can be brought about by replacing silane by TEOS as the silicon source for the deposition of plasma oxide, as shown in Fig. 19. Note the lack of oxide overhangs in the case of TEOS. Some concern exists as regards the purity of the process /133/ so further studies will be required.
Another approach to avoid the negative reentrant angle is the optimization of the silane-PE process parameters /104/. Even more effective is the use of photo-CVD /104/, as demonstrated by Fig. 20. Photo CVD is, however, still in its technical infancy and the problems connected with the uniform irradiation of many large-size wafers preclude any rapid break-through in the near future. It goes without saying that the same holds for futuristic concepts like ion beam deposition /119/.

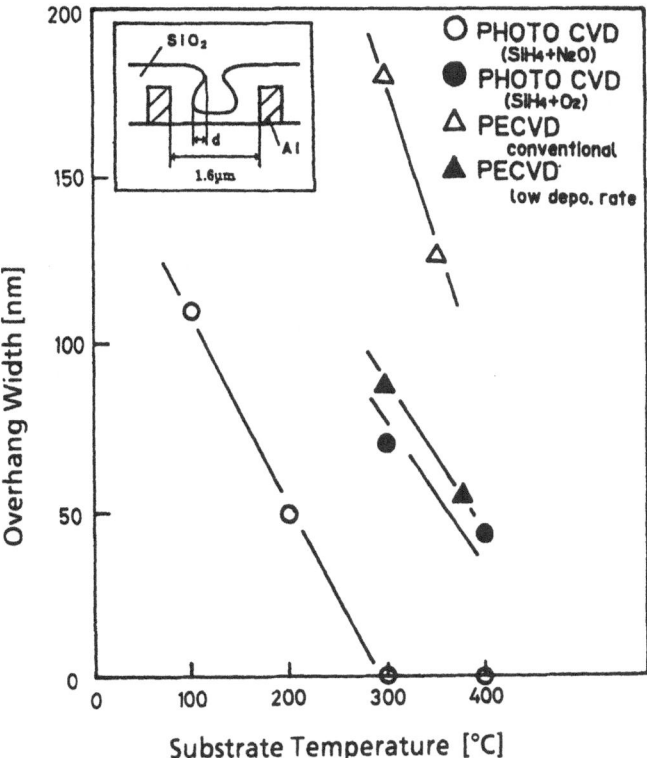

Fig. 20 Step coverage characteristics (as defined by the amount of overhang d) for SiO₂ films deposited by PECVD and Photo-CVD under different conditions. (Ref. 104)

b) Stress. This factor plays an important part in MLM because aluminum is less stable than for example poly-silicon and stress can create reliability problems like hillock growth or reduced electromigration resistance /102, 142 /. In the case of oxides, phosphorus /109, 111/, sometimes together with boron /12/ is used to reduce stress, with phosphorus having the additional benefit of immobilizing sodium ions. In contrast to LPCVD /109, 111/ APCVD produces films with undesired tensile stress /111/. As regards PECVD, stress is low and compressive for oxide and can vary from tensile to compressive for nitride /102/. It can be influenced by substrate temperature, gas composition, pressure, power and frequency /102/. Care has to be taken, however, not to degrade other desired film properties (step coverage, uniformity, deposition rate).

c) High deposition rate. Under special conditions , 300 °C can be already sufficient to promote aluminum hillock growth /135/. Therefore, the temperature budget allotted to the dielectric depositions should be kept as low as possible. The total dielectric thickness deposited, however, has to be comparatively high, because part of it is consumed during the planarizing etch-back. This leads to the requirement of high deposition rates, an important factor for a good through-put as well. Normal silane-LPCVD working with caged boats /29/ is unable to meet this demand, but other constructions offer better prospects /31, 32/. For PECVD as well as APCVD, sufficient deposition rates have not been found to be a problem.

d) Electrical quality. Ideal insulators should have a low dielectric constant as well as high electrical resistance and breakdown values. As the first property is already determined by the choice of the material, only the latter ones can be influenced by the process. The breakdown values reported for low-temperature CVD, in spite of showing a wide spread, are generally inferior to those of high temperature films /50, 75, 100, 106/, let alone thermal oxide. Due to the high film thickness, however, no problems should arise if care is taken to avoid excessive local film thinning in the course of the etch-back procedure.

III.2. Planarization concepts

The processes discussed in this section are usually applied in conjunction with CVD deposited dielectrics. A liquid is spun on which fills gaps in the CVD-layer and planarizes the surface. A major advantage is the simplicity of the procedure in comparison to CVD, and the ease of integration into existing process flows. Thickness and uniformity of the layers can be adjusted by modifying the viscosity of the liquid or by changing the rotational speed of the spin-on equipment. Careful optimization is required for the associated processes like curing steps or the etch-back. For the latter process, end-point detection by means of emission-spectroscopy is essential /121, 123/, because usually no defined etch-stop exists. The liquids used have normally been either photoresist, spin-on glass, or polyimide. The following discussion will therefore use this classification to evaluate the different approaches under consideration.

a) Planarization with photoresist (PR). /120-124, 128, 129, 140/ This is a well established process /120/. Because the PR shrinkage is only minimal, good planarizations can be achieved on regular patterns /95/. The situation becomes more difficult, however, if structures on a severe topography have to be smoothed, due to the high planarization effectiveness of this principle. For example, conductor lines on protruding parts of the device can be attacked by the etching long before the process reaches the lower parts still submerged in resist. Consequently, variations of the photoresist thickness translate directly into variations of the residual dielectric thickness.

In the course of processing, the relation of the exposed surfaces of resist and dielectric varies, which can provoke unwanted changes in the etch-rate /120, 121, 133/. In addition, the resist has to be eliminated completely even in small gaps to avoid reliability problems and the formation of polymers requires frequent cleanings of the etching system. Therefore, some authors have omitted the resist and limited their efforts to improving the step coverage of the dielectric by deposition and etch-back of a thicker film /125-127, 148/. If this principle is carried further, only spacers are left which also present a means to taper the sidewalls /148/. Both of these concepts, however, do not really planarize the structure and are not applicable to closely spaced lines /126, 148/.

The potential of an optimized process-flow using photoresist planarization is demonstrated in Fig. 21 for a three layer metallization /123/.

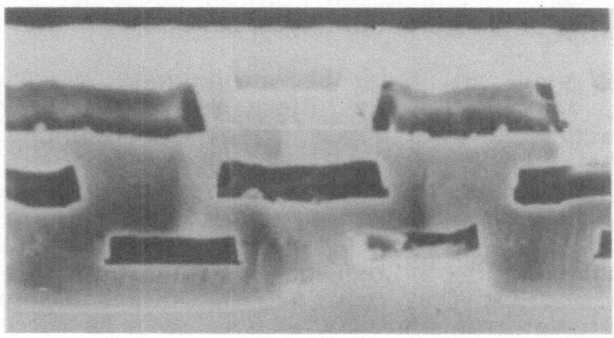

Fig. 21   Three layer metallization realized by a combination of PECVD-SiO$_2$ deposition and photoresist planarization. (Ref. 123).

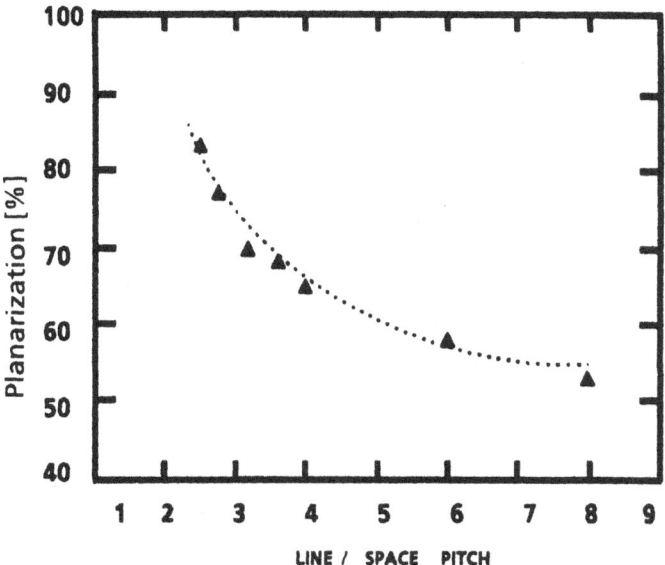

Fig. 22 SOG planarization as a function of metal
pitch for 0.5 μm CVD oxide. (Ref. 148)

b) Planarization with spin-on glass (SOG). /104, 128-139, 148/ Spin-on glasses
are liquid mixtures of a solvent and compounds containing mostly silicon,
carbon and oxygen. Upon curing, the solvent evaporates and a polymerization
takes place which produces a solid film. An oxygen plasma treatment can help
to accelerate this process. The common SOG types consist of either silicates or
polysiloxanes /148/ . Because the films are brittle and shrink during the curing
cycle, high tensile stress and cracks are a big problem. To combat this effect,
polysiloxanes with various attached organic groups have been developed /139,
148/. The resultant films are more elastic but have a higher carbon content after
curing /133, 139/, which can lead to increased leakage currents /135/.

Temperatures of 900 °C transform SOG into a real glass /139/, but this procedure
is of course not compatible with aluminum.

Due to the film shrinkage, SOG is mostly applied in thin layers /130, 133, 135,
138, 148/ ; if a thicker film is needed, multiple coatings with intermediate curing
steps have to be performed /135/. The planarization effectiveness of SOG is
dependent on the structure pitch /135, 148/, as illustrated in Fig. 22. In contrast
to photoresist, SOG planarizes only partially /135/, with the best results obtained
for closely packed structures. Problems with changing etch-rates or
contamination of the etch reactor are less urgent if SOG is used instead of resist.
Another major advantage of SOG is the fact that remaining residues can be
tolerated on the device, which permits the filling of small gaps between
neighboring oxide structures. As curing below 450°C is not sufficient for a
complete rearrangement of the SOG molecule, outgassing is a potential hazard
during subsequent Al sputtering under vacuum /133, 139/. To prevent this effect
as well as water absorption or adhesion failures, SOG is often used in a sandwich
layer between two CVD-films /104, 130, 135, 136, 138/. SOG can be doped /60,
135, 137/ to tailor the etch rate or to provide gettering ability.
Problems to be watched are particles originating from the dispensing system
and blistering of the film upon curing. Optimization of these process steps as
well as the choice of the right SOG material are key factors in determining the
success. Details regarding such sensitive know-how, however, are normally not
disclosed in the literature.

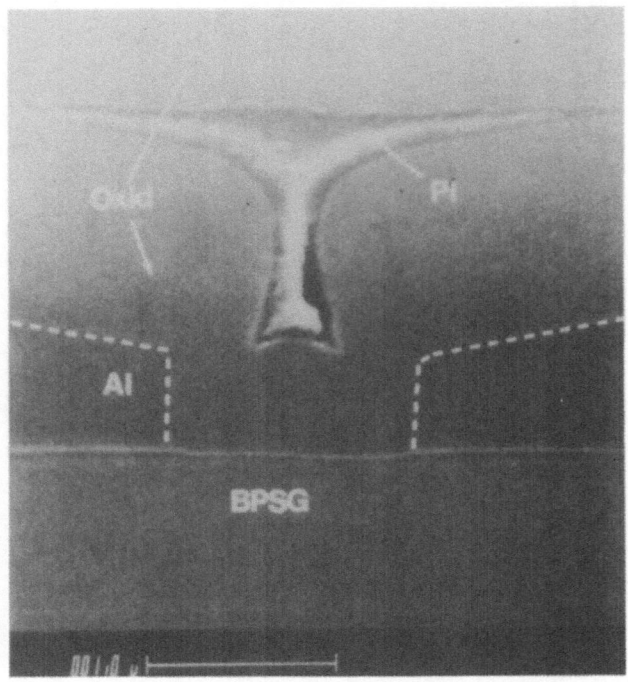

Fig. 23   Demonstration of the refill capability of PI
for narrow structures. Top layer is LPCVD
oxide. (Ref. 147)

c) Planarization with polyimide (PI) /137, 140-147, 153,/. PI and SOG
technologies share some common features like the use of low-cost equipment
and the strong film shrinkage during the curing step. The main difference lies in
the nature of the film, because PI is an organic material which decomposes if
heated above 500 °C, but unlike SOG can never yield a true silicate. Due to this
fact, PI remains elastic upon curing, the films exhibit low stress and cracking is
not such a problem /132/. Therefore, PI can be applied in thicker layers than
SOG. It has excellent fill properties for narrow voids as illustrated in Fig. 23. Up
to now, however, semiconductor manufactures have been reluctant to accept
such a material incorporated permanently into a device, and military customers
used to refuse it altogether for reliability reasons /144/. This is because water
released from the highly hygroscopic PI can increase corrosion sensitivity, /142/
metal adhesion is often poor without a special promoter, and large shifts in flat-
band voltage have been observed /137/. If properly encapsulated with nitride,
most of these problems can be overcome /145/. Figure 24 shows a three-level
metallization realized in polyimide technology.
Another argument in favor of PI is its versatility. It can be screen-printed for
passivation purposes /144/ and photosensitive PI can be directly patterned
without additional etching steps /133, 144, 153/ , but these special materials are
still difficult to handle due to their limited shelf-life /153/. As advances are
taking place rapidly in this field, it can be expected that polyimides are
developed that do not exhibit the problems outlined before, while featuring
better resistivity and breakdown values as well as low dielectric constants.

From the process simplicity point-of-view, the combination of deposition and planarization in the same system is very attractive, but the final worth of such a principle is measured by its impact on device yield and through-put.
Bias-sputtering is an approach which has been investigated by several authors /148-150/. Its effectiveness depends on balancing the deposition-to-etch ratio in such a way, that high points of the topography are levelled out while grooves are filled. This is possible because the etch-rate is angle dependent and vertical slopes are attacked preferentially /149/. Nevertheless, the process is painfully slow and yields high particulate counts /148/, because the reaction products are not volatile. The final oxide thickness is strongly dependent on the metal pitch and radiation damage adversely affects the device characteristics /148, 150/.

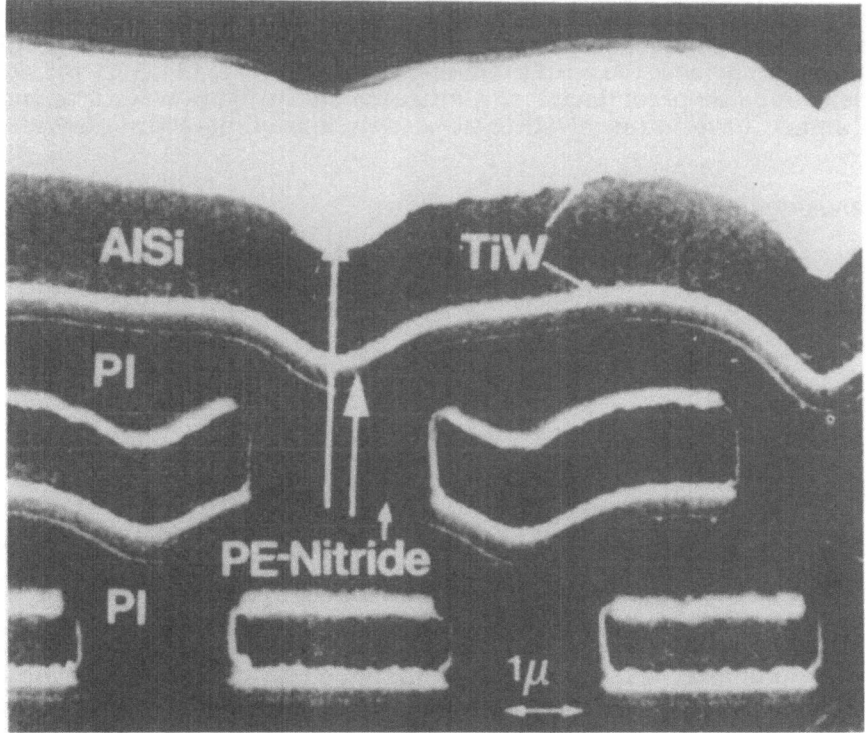

Fig. 24   Three layer metallization in polyimide technology. For process details see Ref. 145 (Photo courtesy of H. Eggers).

Electron cyclotron resonance (ECR) has been proposed as an alternative /133, 149, 151/, but the system seems too complicated for production equipment. Another approach using a sequential deposition and etch-back has been realized in a plasma reactor /152/. The advantage is that deposition and etching parameters can be optimized separately, but informations concerning electrical results were not given.

III. 4. Conclusion

The 4Mbit generation has marked the beginning of mulilevel metallization acceptance for DRAM technology /122/. The unabated drive towards higher density and speed will make several layers of metal an essential part of future DRAM designs. As a consequence, the variety of approaches existing today will be reduced in favor of a few standard process flows, which helps to settle the remaining reliability issues as well.

Manufacturability aspects will be a decisive factor in deciding on the technology to be used. This favors the combination of plasma deposition and planarization by either polyimide (PI) or spin-on glass (SOG), because the equipment is state-of-the-art. Development of more refined spin-on materials will reduce the difference between PI and SOG by increasing for example, the organic groups attached to the SOG molecule.

Especially for development applications, systems containing different chambers for sequential deposition and etching /e.g. 69, 70,/ are a very promising approach.

Intrinsic to any multilevel metallization, however, will be a suitable conductor, which can fill deep and narrow holes with vertical walls. The consensus of opinions does not favor aluminum CVD /155/ due to many unsolved problems. Therefore, tungsten is the most promising candidate /132, 134, 142, 145/. As the problems regarding conductor and insulator are in many ways intertwined, both aspects have to be solved to achieve multilevel metallization with high yield.

Acknowledgement

I am indebted to G. Higelin, U. Fritsch, D. Allred, and V. Grewal for helpful discussions.

References

1   J.A. Meindl: Opportunities for Gigascale Integration;
     Sol. State Technol. 30 (12), 85 (1987)
2   W. Beinvogl and E. Hopf: The Development of a 4Mbit-
     DRAM; To be published in Advances in Solid State
     physics
3   H. Sunami, T. Kure, N. Hashimoto, K. Itoh, T. Toyabe,
     and S. Asai: A Corrugated Capacitor Cell (CCC); IEEE
     Trans. Electron Devices 31, 746 (1984)
4   P. Chatterjee: Trench and Compact Structures for
     DRAMs; IEDM Technol. Dig. 1986, 128 (1986)
5   W. Kern and R.C. Heim: Chemical Vapor Deposition of
     Silicate Glasses for Use with Silicon Devices, II.
     Film Properties; J. Electrochem. Soc. 117, 568 (1970)
6   P. Balk and J.M. Eldridge: Phosphosilicate Glass
     Stabilization of FET Devices; Proc. IEEE 57 (9), 1558
     (1969)
7   W.E. Armstrong and D.L. Tolliver: A Scanning Electron
     Microscope Investigation of Glass Flow in MOS
     Integrated Circuit Fabrication; J. Electrochem.
     Soc. 121, 307 (1974)
8   A. Naumann and J.T. Boyd: Phosphosilicate Glass Flow
     for Integrated Optics; J. Vac. Sci. Technol., 17, 529
     (1980)

9   R.A. Bowling and G.B. Larrabee: Deposition and Reflow
    of Phosphosilicate Glass; J. Electrochem. Soc. 132,
    141 (1985)

10 -W. Kern and B. Mead: Silicate Glass Coating of Semi-
    conductor Devices; United States Patent, No. 3481 781
    (1969)

11 C. Ramiller and L. Yau: Borophosphosilicate Glass
    for Low Temperature Reflow; Semicon West Techn. Proc.
    Vol. 5, p. 29, (1982)

12 W. Kern and G.L. Schnable: Chemically Vapor-Deposited
    Borophosphosilicate Glasses for Silicon Device Appli-
    cations; RCA Review 43, 423, (1982)

13 H. Matsui and M. Yoshimaru: Low Temperature Glass
    Flow; Proc. SEMI Technol. Symp. Tokyo, Dec. 84;
    p 1-37, (1984)

14 H. Nishimura, Y. Suizo, and T. Tsujimaru: A Rapid
    Isothermal Annealing for a VLSI Interconnection
    Technology; ECS Ext. Abstr. 84-2, 760, (1984)

15 P.K. Chu: Quantitative Depth Profiling of Boron and
    Phosphorus in Borophosphosilicate Glass; Chem. Phys.
    (Springer Ser.), Vol. 36, p 332, (1984)

16 Y. Chasset, P. Launay, and J.L. Liotard: Statistical
    Analysis of Planarized Double Metal Structures;
    Proceedings 3rd. Int. VLSI Multilevel Interconnection
    Conf., IEEE Cat. No. CH 2337-4/86, p.114, (1986)

17 P.K. Chu and S.L. Grube: Quantitative Determination
    of Boron and Phosphorus in Borophosphosilicate Glass
    by Secondary Ion Mass Spectrometry; Anal. Chem.,
    Vol. 57-6, 1071, (1985)

18 M. Susa, Y. Hiroshima, K. Senda, T. Kuriyama,
    S. Matsumoto, S. Terakawa, and T. Takamura: Boro-
    phosphosilicate Glass Flow for Solid State Imager
    Application; J. Appl. Phys. 58 (10), 3880, (1985)

19 W. Kern and R.K. Smeltzer: Borophosphosilicate
    Glasses for Integrated Circuits; Solid State
    Technol.28 (6), 171,(1985)

20 C. Dornfest: The Effect of Reducing Deposition
    Temperature in an Atmospheric Pressure BPSG Process;
    ECS Ext. Abstr. 85-1, 347, (1985)

21 W. Kern, W.A. Kurylo, and C.J. Tino: Optimized
    Chemical Vapor Deposition of Borophosphosilicate
    Glass Films; RCA Review 46-2, 117, (1985)

22 W. Kern: Deposited Dielectrics for VLSI; Semicond.
    Int.Vol. 8 (7), 121, (1985)

23 V. Teal, C. Ha, and A. Chowaniec: BPSG as Interlevel
    Dielectric for Double-Level Metal Process; Proc.
    Second. Int. IEEE VLSI Multilevel Interconnection
    Conf., IEEE Cat. No. 85 CH 2197-2, p 273, (1985)

24 a) P.H. Townsend and R.A. Huggins: Stresses in BPSG-
    Films During Thermal Cycling; ECS Ext. Abstr. 86-2,
    520, (1986)
    b) P.H. Townsend, personal communication

25 M. Susa, Y. Hiroshima, K. Senda, and T. Takamura:
    Borophosphosilicate Glass Flow in a PH_3-O_2 Ambient,
    J. Electrochem. Soc. 133, 1517, (1986)

26 H. Ozaki, S. Mayumi, S. Ueda, and M. Inoue: Contact
    Resistance Behavior in Borophosphosilicate Glass;
    Proc. 4th Int. VLSI-Multilevel Interconnection Conf.,
    IEEE Cat. No. CH-2488-5/87, p 323, (1987)

385

27   K.H. Hurley: Process Conditions Affecting Boron and
      Phosphorus in Borophosphosilicate Glass Films as
      Measured by FTIR Spectroscopy; Sol. State Technol. 30
      (3), 103, (1987)

28   a) K.H. Hurley, L.D. Bartholomew, and D.T. Bordonaro:
         BPSG Films Deposited by APCVD; Semicond. Int. 10
         (10),91, (1987)
      b) Watkins-Johnson Progress Report BPSG, (1987)

29   A.J. Learn, and B.Baerg: Growth of Borosilicate and
      Borophosphosilicate Films at Low Pressure and
      Temperature; Thin Sol. Films 130, 103, (1985)

30   J.S. Mercier and N. Shah: Rapid Isothermal Fusion of
      BPSG Films in a Steam Ambient; ECS Ext. Abstr. 86-2,
      846, (1986)

31   J.C. Mitchener, and I. Mahawili: Low Pressure CVD-
      High Quality SiO_2 Dielectric Films at High Deposition
      Rates; Proc. 3rd Int. VLSI-Multilevel Interconnection
      Conference, IEEE Cat. No. CH 2337-4/86, p.298, (1986)

32   J.C. Mitchener: LPCVD Forced-Convection Flow and the
      Deposition of SiO_2 Dielectric Films; Proc. Techn.
      Symposium Semicon Osaka, Japan, p 222,(1987)

33   J.M. Mercier: Rapid Flow of Doped Glasses for VLSI-
      Fabrication; Sol. State Technol. 30 (7), 85, (1987)

34   T.C. Foster, G.C. Hoeye, and J. Goldman: A Low
      Pressure BPSG Deposition Process; J. Electrochem.
      Soc. 132, 505, (1985)

35   T.C. Foster, J.C. Goldman, and G.W. Hoye: Process for
      Deposition of Borophosphosilicate Glass; U.S. Patent
      4.557.950, (1985)

36   R.L. Baker and S.R. Jennings: The Transient Reflow
      of Phospho- and Borophosphosilicate Glasses; Proc.
      3rd Int. VLSI-Multilevel Interconnection Conf. IEEE
      Cat. No. 86-CH-2337-4, 484, (1986)

37   C.C. Wong and W.A. Brown: Substrate Effects on Flow
      Characteristics of BPSG Deposited by LPCVD; ECS Ext.
      Abstr. 86-1, 318, (1986)

38   P.B. Johnson and P. Sethna: Using BPSG as an
      Interlayer Dielectric; Semicond. Int. 10 (10), 80,
      (1987)

39   I. Avigal: Inter-Metal Dielectric and Passivation
      Related Properties of Plasma BPSG; Solid State
      Technol. 26 (10), 217, (1983)

40   J.E. Tong, K. Schertenleib, and R.A. Carpio: Process
      and Film Characterization of PECVD Borophosphos-
      silicate Films for VLSI Applications; Solid State
      Technol. 27 (1), 161, (1984)

41   S. Shanfield and S. Bay: Process Characterization of
      PSG and BPSG Plasma Deposition; J. Electrochem. Soc.
      131, 2202, (1984)

42   K.C. Ray Chiu, S.V. Dunton, and W.R. Snow: Modeling
      of BPSG Film Deposition; In "Reduced Temp. Processing
      for VLSI", R. Reif, G.R. Srinivasan, Editors. ECS PV
      86-5, 175, (1986)

43   J.R. Gigante, J.M. Geneczko, and R.N. Ghoshtagore:
      Rapid Reflow of Borophosphosilicate Glass; in Reduced
      Temp. Processing for VLSI, R. Reif, G.R. Srinivasan,
      Editors. ECS PV 86-5, 160, (1986)

44   W.D. Partlow and B.C. Samuels: Dependence of Plasma
      Assisted BPSG Deposition on Substrate Properties;
      J. Electrochem. Soc. 134, 1740, (1987)

45   A. Tissier, A. Poncet, and J.F. Teissier: Glass
     Reflow Modeling for Process Optimization; Proc. 17th
     ESSDERC Bologna, p.453, (1987)

46   M. Brillouet, C. Masurel, P. Normandon, Y. Pauleau,
     and J.M. Temerson: Study of W-AL Interconnection
     Scheme for VLSI Multilevel Metallization; Proc. 4th
     Int. VLSI Multilevel Interconnection Conf. IEEE Cat.
     No. CH 2488 -5/87, 13, (1987)

47   G. Smolinsky and T.P.H.F. Wendling: Measurement of
     Temperature Dependent Stress of Silicon Oxide Films
     Prepared by a Variety of CVD Methods, J. Electrochem.
     Soc. 132, 950, (1985)

48   F.S. Becker, D. Pawlik: A New LPCVD Borophosphos-
     silicate Glass Process Based on the Doped Deposition
     of TEOS Oxide; ECS Ext. Abstr. 85-2, 380, (1985)

49   F.S. Becker, D. Pawlik: A New LPCVD Process Based on
     the Doped Deposition of TEOS Oxide, in "Reduced Temp.
     Processing for VLSI", R. Reif, G.R. Srinivasan,
     Editors. ECS PV 86-5, 148, (1986)

50   F.S. Becker, D. Pawlik, H. Schäfer, and G. Staudigl:
     Process and Film Characterization of Low Pressure
     Tetraethylorthosilicate Borophosphosilicate Glass;
     J. Vac. Sci. Technol. B 4 (3), 732, (1986)

51   R.A. Levy, and K. Nassau: Reflow Mechanisms of
     Contact Vias in VLSI Processing; J. Electrochem. Soc.
     133, 1417,(1986)

52   F.S.Becker, and S. Röhl: Low Pressure Deposition of
     Doped SiO$_2$ by Pyrolysis of Tetraethylorthosilicate
     (TEOS), Part I: Boron and Phosphorus Doped Films;
     J. Electrochem. Soc. 134, 2923, (1987)

53   R.A. Levy, and K. Nassau: Viscous Behavior of Phos-
     phosilicate and Borophosphosilicate Glasses in VLSI
     Processing; Sol. State Technol. 29 (10), 123, (1986)

54   R.A. Levy, and K. Nassau: Viscous Behavior of Phos-
     phosilicate, Borophosphosilicate and Germanophospho-
     silicate Glasses in VLSI Processing; Reduced Temp.
     Proc. for VLSI, R. Reif, G.R. Srinivasan Editors, ECS
     PV 86-5, 132,(1986)

55   D.S. Williams, and E.A. Dein: LPCVD of Borophospho-
     silicate Glasses from Organic Reactants;
     J. Electrochem. Soc. 134, 657, (1987)

56   M.C. Hughes, and D.R. Wonsidler: Chemical and
     Electron Microprobe Analysis of Borophosphosilicate
     and Phosphosilicate Glasses; J. Electrochem. Soc.
     134, 1488, (1987)

57   R.A. Levy, P.K. Gallagher, and F. Schrey: A New LPCVD
     Technique of Producing Borophosphosilicate Glass
     Films by Injection of Miscible Liquid Precursors;
     J. Electrochem. Soc. 134, 430, (1987)

58   R.A. Levy, and T.Y. Kometani: Analysis of Borophos-
     phosilicate Glass Films by Inductively Coupled Plasma
     Atomic Emission Spectroscopy; J. Electrochem. Soc.
     134, 1565, (1987)

59   R.A. Levy, P.K. Gallagher, and F. Schrey: Low
     Pressure Chemical Vapor Deposition of Borophospho-
     silicate Glass Films Produced by Injection of
     Miscible DADBS-TMB-TMP Liquid Sources;
     J. Electrochem. Soc. 134, 1744 (1987)

60   S.L. Chang, K.Y. Tsao, M.A. Meneshian, and
     H.A. Waggener: Spin-on BPSG and its Application to
     VLSI, Proc. 3rd. Int. Symp. on VLSI Science Technol.

ECS Proc., Vol. 85-5, 231, (1985)

61 C.Y. Fu: A Novel Borophosphosilicate Glass Process;
    IEDM Techn. Dig., 602, (1985)

62 V.V.S. Rana, A.S. Harrus, F.A. Stevie, A.S. Manocha,
    and A.K. Sinha: The Flow of Borophosphosilicate Glass
    in PBr$_3$ Ambient; ECS Ext. Abstr. 87-2, 1530, (1987)

63 S.P. Tay, and J.P. Ellul: High Pressure Technology
    for Silicon IC Fabrication; Semicond. Int. 9 (5),
    122, (1986)

64 S.P. Tay, and J.P. Ellul, M.I.H. King: Application of
    High-Pressure Technology to VLSI Fabrication; Mat.
    Res. Soc. Symp. Proc. Vol. 71, 467, (1986)

65 K. Nassau, R.A. Levy, and D.L. Chadwick: Modified
    Phosphosilicate Glasses for VLSI Applications;
    J. Electrochem.Soc. 132, 409, (1985)

66 R.U. Martinelli and R.E. Enstrom: Reliability of
    Planar In GaAs/InP Photodiodes Passivated with
    Borophosphosilicate Glass; J. Appl. Phys. 63, 250
    (1988)

67 S. Ohashi, Y. Shioya, S. Nokai, Y. Matsuda, and
    K. Yonagida: Low Pressure BPSG Deposition Using TEOS,
    TMP, and TEB; ECS Ext. Abstr. 88-1, Abstr. 162 (1988)

68 J.C. Schumacher Co, preliminary data sheet (11.1987)

69 D.N.K. Wang, S. Somekh, and D. Maydan: Advanced CVD
    Technology; In "Proc. 1st Int. Symp. on Ultra Large
    Scale Integration; S. Broydo and C.M. Osburn,
    Editors, ECS PV 87-11, Pennington, N.J. p. 712 (1987)

70 K. Law, J. Wong, and D.N.K. Wang: A Single System
    Approach to Intermetal Dielectrics; Semi Technol.
    Symposium Proc. Vol., p. 154, Semicon Tokyo, Japan
    12.1987

71 A.S. Tenney and M. Ghezzo: Etch Rates of Doped Oxides
    in Solutions of Buffered HF; J. Electrochem. Soc.
    120, 1091 (1973)

72 W. Kern: Wet-Chemical Etching of SiO$_2$ and PSG Films,
    and an Etching-Induced Defect in Glass-Passivated
    Integrated Circuits; RCA Review 47, 186 (1986)

73 L.R. Plauger: Etching Studies of Diffusion Source
    Boron Glass; J. Electrochem. Soc. 120, 1428 (1973)

74 R.M. Levin and K. Evans-Lutterodt: The Step Coverage
    of Undoped and Phosphorus-Doped SiO$_2$ Glass Films;
    J. Vac. Sci. Technol. B1, 54, (1983)

75 F.S. Becker, D. Pawlik, H. Anzinger, and A. Spitzer:
    Low-Pressure Deposition of High-Quality SiO$_2$ Films by
    Pyrolysis of Tetraethylorthosilicate; J. Vac. Sci.
    Technol. B5 (6), 1555 (1987)

76 C. Arena, personal communication

77 N. Hashimoto, Y. Yatsuda, and S. Mutoh: Glass Flow
    Mechanism of Phosphosilicate Glass and its
    Application in MOS Devices; Jpn. J. Appl. Phys. 16,
    Suppl. 16-1, 73 (1977)

78 C.R. Hammond: Fusion Temperatures of SiO$_2$-P$_2$O$_5$ Binary
    Glasses; Physics and Chemistry of Glasses, Vol. 19,
    p. 41 (1978)

79 R.A. Levy, S.M. Vincent and T.E.Mc Gahan: Evaluation
    of the Phosphorus Concentration and its Effect on
    Viscous Flow and Reflow in Phosphosilicate Glass;
    J. Electrochem. Soc. 132, 472 (1985)

80 R.F. Bartholomew: Water in Glass; in "Treatise on
    Materials Science and Technol. Vol. 22, p. 75 (1982)

388

81 R.R. Razouk, L.N. Lie: Pressure Induced Phospho-
    silicate Glass Flow; ECS Ext. Abstr. 82-1, 138 (1982)
82 N. Kajiwara and K. Tanigawa: Reflow of Phospho-
    silicate Glass Employing High Pressure Oxidation; JST
    News Vol. 2 (3), 16 (1983)
83 A.S. Tenney and M.Ghezzo: Composition of Phospho-
    silicate Glass by Infrared Absorption;
    J. Electrochem. Soc. 120, 1276 (1973)
84 E.A. Taft: Infrared Absorption of Chemical Vapor
    Deposited Borosilicate Glass Films; J. Electrochem.
    Soc. 118, 1985 (1971)
85 K. Krishnan, personal communication
86 M. Schuster, L. Müller, K.E. Mauser, and R. Straub:
    Quantitative X-Ray Fluorescence Analysis of Boron in
    Thin Films of Borophosphosilicate Glasses; Thin Sol.
    Films 157, 325 (1988)
87 G.W.B. Ashwell and S.J. Wright: Arsenosilicate Glass
    as an Interlayer Dielectric; Semicond. Int. 8 (1),
    132 (1985)
88 G.W.B. Ashwell and S.J. Wright: The Reflow of
    Arsenosilicate Glass; in "Proc. 2nd Int. VLSI
    Multilevel Interconnection Conf., IEEE Cat. No. 85 CH
    2197-2, p. 285 (1985)
89 N. Hoshi, S. Kayama, T. Nishihara, J. Aoyama,
    T. Komatsu, and T. Shimada: 1.0 /um CMOS Process for
    Highly Stable Tera-Ohm Polysilicon Load 1Mb SRAM;
    IEDM Technol. Dig. 1986, 300 (1986)
90 F.S. Becker and H. Treichel: LPCVD of Arsenic and
    Boron Doped SiO_2 Films as Diffusion Sources Using
    TEOS and Organic Compounds; In "Proc. 6th European
    Conf. on CVD" Jerusalem, R. Porat, Editor, p.207 (1987)
91 T. Ogino and Y. Amemiya: A Planarization Technique
    Utilizing Oxide Flow During H_2 Treatment of a SiO_2-
    GeO_2 Film; Jpn. J. Appl. Phys. 25, 1115 (1986)
92 A. Iqbal, W.I. Lehrer, and J.M. Pierce: Phospho-
    germanosilicate Glass Films for VLSI Devices; ECS
    Ext. Abstr. 83-2, 359 (1983)
93 F.C. Chien, R.L. Brown, G.N. Burton and M.B.Vora:
    A Two Micron Metal Interconnect Process over Severe
    Topography; Semicond. Int. 8 (3), 78 (1985)
94 M. Furukawa and T. Hara: Rapid Heating Reflow of
    Phosphosilicate Glass Enhanced by As Ion Implanta-
    tion; Jpn. J. Appl. Phys. 25, L 795 (1986)
95 A.T. Mitchell, C. Huffman, and A.L. Esquivel: A New
    Self-Aligned Planar Array Cell for Ultra High Density
    Eproms; IEDM Technol. Dig., CH 2515-5/87, p. 548 (1987)
96 H. Treichel, F.S. Becker, D. Fuchs, and Th. Kruck:
    A Novel Borosilicate Glass (SiOB-BSG) by Low Pressure
    Decomposition of a Monomolecular Liquid Precursor;
    Journal de Physique, Colloque C4 Supplement au n° 9,
    Tome 49, p.541 (1988)
97 F. Gualandria, G.U. Pignatel, S. Rojas, and
    J. Scannell: Borophosphosilicate Glass Etching in a
    Fluorinated Plasma; Proc. 7th Int. Symp. Plasma
    Chemistry, Vol. 3, p. 1048 (1985)
98 C.L. Hooker, D.W. Tomes: Silicon Dioxide Etch Rate
    Control by Controlled Additions of P_2O_5 and B_2O_3; US.
    Patent No. 3913126 (1975)
99 E.P. van de Ven: High Throughput PECVD Using Continous
    Processing; to be presented at Semicon Osaka (Japan),
    July 1988

100 A.C. Adams, R.S. Benton, W.J. Bertram, H.J. Levin-
 stein, W.Q. Mc Knight, J.J. Rubin and B.A.ter Haar:
 High Density Interconnect for Advanced VLSI
 Packaging; in Ref. 69, p.485

101 B.L. Chin and E.P. van de Ven: Plasma TEOS Process
 for Interlayer Dielectric Applications; Sol. State
 Technol. 31(4), 119 (1988)

102 A.C. Adams: Plasma-Assisted Deposition of Dielectric
 Films; in "Reduced Temperature Processing for VLSI,
 R. Reif and G.R. Srinivasan, Eds.,ECS PV 86-5,p.111(1986)

103 S.J.H. Brader and S.C. Quinlan: Scaled PECVD Oxide as
 an Interlayer Dielectric for CMOS DLM Processing; in
 Ref. 16, p.58

104 T. Fujita, K. Yano, S. Tanimura and T. Ueda: Photo CVD
 Technology for Interlevel Dielectrics in Submicron
 VLSI's; in Ref. 26, p.285

105 J.M. Blum: Chemical Vapor Deposition of Dielectrics:
 A Review; in "Proc. 10th Int. Conf. CVD, G.W. Culler
 and J.M. Blocher, Eds. ECS Proc.Vol.PV 87-8,p.476(1987)

106 J. Batey, E. Tierney and T.N. Nguyen: Electrical
 Characteristics of Very Thin SiO_2 Deposited at Low
 Substrate Temperature; IEEE Electron Dev. Letters EDL
 8(4), p.148 (1987)

107 J.H. Houskova, K.K.N. Ho and M.K. Balazs:
 Characterizing Plasma Phosphorus Doped Oxides;
 Semicond. Int. 8(5), p.236 (1985)

108 K. Maeda and J. Sato: Very Low Temperature Chemical
 Vapor Deposition of Silicon Dioxide Films Using Ozone
 and Organosilane; Denki Kagaku 45(10), p.654 (1977)

109 A.J. Learn: Phosphorus Incorporation Effects in
 Silicon Dioxide Grown at Low Pressure and
 Temperature; J. Elechochem. Soc. 132; p.405 (1985)

110 B. Gorowitz, R.H. Wilson and T.B. Gorczyca: Recent
 Trends in LPCVD and PECVD; Sol. State Technol.
 30(10), p.97 (1987)

111 M. Shimbo and T. Matsuo: Thermal Stress in CVD PSG
 and SiO_2 Films on Silicon Substrates; J. Electrochem.
 Soc. 130, p.135 (1983)

112 B.R. Bennett, J.P. Lorenzo, K. Vaccaro and A. Davis:
 Chemical Vapor Deposition of High-Quality Silicon
 Dioxide at 100-300°C; ECS Ext. Abstr. 87-1, p.370 (1987)

113 J. Andrews and G. Smolinsky: Dielectric Breakdown
 Strength of LPCVD SiO_2 Deposited at 500°C; in "Proc.
 10th Int. Conf. CVD", G.W. Cullen and J.M. Blocher,
 Eds., ECS Proc. Vol. PV 87-8, p.497 (1987)

114 G. Smolinsky: The Low Pressure Chemical Vapor
 Deposition of Silicon Oxide Films in the Temperature
 Range 450 to 600°C from a New Source:
 Diacetoxyditertiarybutoxysilane; Symp. VLSI
 Technol., IEEE CAT.No. CH 2318-4 (1986)

115 G. Smolinsky: LPCVD of SiO_2 Films Using the New
 Source Material DADBS; in "Proc. 10th Int. Conf.
 CVD", G.W. Cullen and J.M. Blocher, Eds.; ECS Proc.
 Vol. PV 87-8, p.490 (1987)

116 Y. Mishima, M. Hirose and Y. Osaka: Direct
 Photochemical Deposition of SiO_2 from the $Si_2H_6+O_2$
 System; J. Appl. Phys. 55(4), p.1234 (1984)

117 J. Marks and R.E. Robertson: Silicon Dioxide
 Deposition at 100°C Using Vacuum Ultraviolet Light;
 Appl. Phys. Lett. 52(10), p.810 (1988)

118 K.J. Scoles, A.H. Kim and M.H. Jiang: Deposition and Characterization of Silicon Dioxide Thin Films Deposited by Mercury-Arc-Source Driven Photon-Activated Chemical-Vapor Deposition, J. Vac. Sci. Technol. B 6(1), p.470 (1988)

119 Y. Minowa and H. Ito: SiO2 Films Deposited on Si by Dual Ion Beams; J. Vac. Sci. Technol. B 6(1),p.473(1988)

120 A.C. Adams and C.D. Capio: Planarization of Phosphorus-Doped Silicon Dioxide; J. Electrochem. Soc. 128, 423 (1981)

121 S. Mayumi, K. Fujiwara, S. Nishida, S. Ueda, and M. Inoue: Etch-Back Planrization Technique for Multilevel Metallization; Jpn. J. Appl. Phys. 27(2), 280 (1988)

122 T.D. Bonfield, R.J. Gale, B.W. Shen, G.C. Smith, and C.H. Huffman: A One Micron Design Rule Double Level Metallization Process; in Ref. 16, p.71

123 H. Fritzsche, V. Grewal, and W. Henkel: An Improved Etch-Back Process for Multilevel Metallization and its Reliability Results for CMOS-Devices;in Ref.16,p.45

124 W. Geiger and A. Sharma: An Optimized Planarization Process for a Multi Layer Interconnect,in Ref.16,p.128

125 T. Abraham: Reactive Facet Tapering of Plasma Oxide for Multilevel Interconnect Applications;in Ref.26,p.115

126 J.S. Mercier, H.M. Naguib, V.Q. Ho, and H. Nentwich: Dry Etch-Back of Overthick PSG Films for Step-Coverage Improvement; J. Electrochem.Soc.132,1219 (1985)

127 B. Lee, A. Pierfederici, and E.C. Douglas: Dielectric Planarization Techniques for Narrow Pitch Multilevel Metallization, in Ref. 26, p.85

128 D. Flowers: Processing Requirements for Multilevel Interconnect Fabrication; in Ref. 16, p.78

129 W. Kern: Pertinent Inorganic and Organic Dielectric Systems for Multilevel Interconnection; VLSI Multilevel Interconnection State-of-the-Art Seminar, 6. 1986, Visuals Booklet, p.119 (1986)

130 J. Kiefer Elliot: Current Trends in VLSI Materials. Part.2: Dielectrics; Semicond. Int. 31(4), 150 (1988)

131 A.N. Saxena and D. Pramanik: Planarization Techniques for Multilevel Metallization; Sol. State Technol. 29(10), 95 (1986)

132 A.N. Saxena and D. Pramanik: Manufacturing Issues and Energing Trends in VLSI Multilevel Metallizations; in Ref. 16, p.9

133 K. Skidmore: Techniques for Planarizing Device Topography; Semicond. Int. 11(4), 114 (1988)

134 U. Fritsch and G. Higelin: Sub Micron CMOS Two Level Metal Process with Planarization Techniques; Submitted to VLSI Multilevel Interconnection Conf. 1988

135 A. Rey, D. Iafond, J.M. Mirabel, M.C. Tacussel, and M.F. Coster: A Double Level Aluminum Interconnection Technology with Spin-on Glass Based Insulator; in Ref. 16, p.491

136 P. Elkins, K. Reinhardt, and R. Tang: A Planarization Process for Double Metal CMOS Using Spin-on-Glass as a Sacrificial Layer; in Ref. 16, p.100

137 G.E. Whitwell and T.E. Wade: The Performance and Processing of a New Spin-on Polysiloxane Interlevel Dielectric Material; in Ref. 16, p.292

138 J.K. Chu, J.S. Multani, S.K. Mittal, J.T. Orton, and
    R. Jecmen: Spin-on-Glass Dielectric Planarization for
    Double Metal CMOS Technology; in Ref. 16, p.474
139 C.H. Ting, H.Y. Lin, P.L. Pai, and W.G. Oldham:
    Planarization Process Using Spin-on-Glass; in Ref.
    26, p.61
140 L.K. White: Planarization Properties of Resist and
    Polyimide Coatings; J. Electrochem. Soc. 130, 1543 (1983)
141 K. Mitsuhashi, K. Shiozaki, K. Ohtake, M. Koba, and
    K. Awane: Etch-Back Planarization Technology for
    Interconnection of Stacked Structure; in Ref. 69, p.557
142 G.C. Schwartz: Interconnection Metallization for
    Advanced Bipolar Devices; in Ref. 69, p.493
143 H. Umezaki, N. Koyama, R. Suzuki, and H. Matsuyama:
    Planar Process for 16 Mb Bubble Memory Devices Using
    Thermal Reflow Type Polyimide; ECS Ext. Abstr. 87-1,
    338 (1987)
144 P. Burggraaf: Polyimides in Microelectronics;
    Semicond. Int. 31(3), 58 (1988)
145 H. Eggers and K. Hieber: Recent Development in
    Multilevel Interconnect Technology; IEDM Technol.
    Dig. 1987, 200 (1987)
146 D.C. Hofer: Organic Dielectrics for Multilevel
    Interconnection; VLSI Multilevel Interconnection
    State-of-the Art Seminar June 17, 1987 Santa Clara,
    Visuals Booklet p.391
147 V. Grewal, A. Gschwandtner, and G. Higelin: A Novel
    Multilevel Metallization Technique for Advanced CMOS
    and Bipolar Integrated Circuits; in Ref. 16, p.107
148 H.M. Naguib, C. Jang, T.F. Klemme, K. Wong,
    A. Rangappan, W.W. Yao, and R.T. Fulks: The
    Evaluation of Planarization Techniques for Double-
    level Metallization in 1.2 Micron CMOS
    Technology, in Ref. 26, p.93
149 H. Kitahara: Planarization of SiO_2 Insulating Inter
    layers; Semi Technol. Symp. Proc., Dec.1987 Tokyo,p.162
150 Y. Hazuki and T. Moriya: A Damage Free Perfect
    Planarization Method Using Bias Sputtered SiO_2; in
    Ref. 16, p.121
151 M. Doki, K. Takasaki, K. Fujino, and Y. Ban: ECR
    Plasma CVD of Insulator Films-Low Temperature Process
    for ULSI; ECS Ext. Abstr. 87-1, 366 (1987)
152 E.J. McInerney: An In-Situ Planarized PECVD Silicon
    Dioxide Interlayer Dielectric; in Ref. 16, p.467
153 F.De Geyter, G. Brasseur, and F. Coopmans: Comparison
    of Physical Characteristics of Five Different
    Polyimide Films; in Ref. 16, p.319
154 H.J. Thoma, W.T. Cochran, A.S. Harrus, H.P.W. Hey,
    G.W. Hills, C.W. Lawrence, and J.L. Yeh: A 1.0 /u
    CMOS Two-Level Metal Technology Incorporating Plasma
    Enhanced TEOS; in Ref. 26, p.20
155 R.A. Levy, M.L. Green, and P.K. Gallagher:
    Characterization of LPCVD Aluminum for VLSI
    Processing; J. Electrochem. Soc. 131, 2175 (1984)

LOW TEMPERATURE SILICON EPITAXY FOR NOVEL DEVICE STRUCTURES

John Ogawa Borland

Epi Applications Lab
Applied Materials, Inc.
3050 Bowers Ave
Santa Clara, CA   95054

INTRODUCTION

With the continued scaling of devices down to the submicron level, many new device processing issues have arisen that can inhibit device speed and degrade device performance. For bipolar these processing issues encompass; 1) new deep isolation structures (>1.0um), 2) sidewall base contact structures, 3) thin base regions (<0.1um), 4) dual n-type and p-type buried layer autodoping and 5) shallow emitter junction formation. For CMOS the processing issues center around; 1) latch-up immunity, 2) new advanced DRAM and SRAM cell structure designs, 3) shallow source/drain junction formation with optional LDD structures and 4) self-aligned contact hole refill and planarization techniques for back-end metallization interconnect technology. For BiCMOS you have the combination of all of the above effects in terms of processing issues with very complex processing steps in order to optimize both bipolar and CMOS device performance. Therefore, this paper will present some very attractive new advanced device structures that are possible through the use of low temperature/low pressure (LT/LP) production worthy epitaxial growth techniques that address these ULSI submicron device processing issues.

LOW TEMPERATURE EPITAXY

Recently, much attention has been focused on low temperature (<900°C) single crystal silicon epitaxial growth with very abrupt epi/substrate transition width in order to achieve very thin epitaxial layers. For advanced bipolar device fabrication transition widths <0.1um with total epilayer thickness of <1.0um is desirable. For submicron CMOS technology transition widths <0.2um and total epilayer thickness of 2um will be necessary since both the well depth and device processing temperatures are

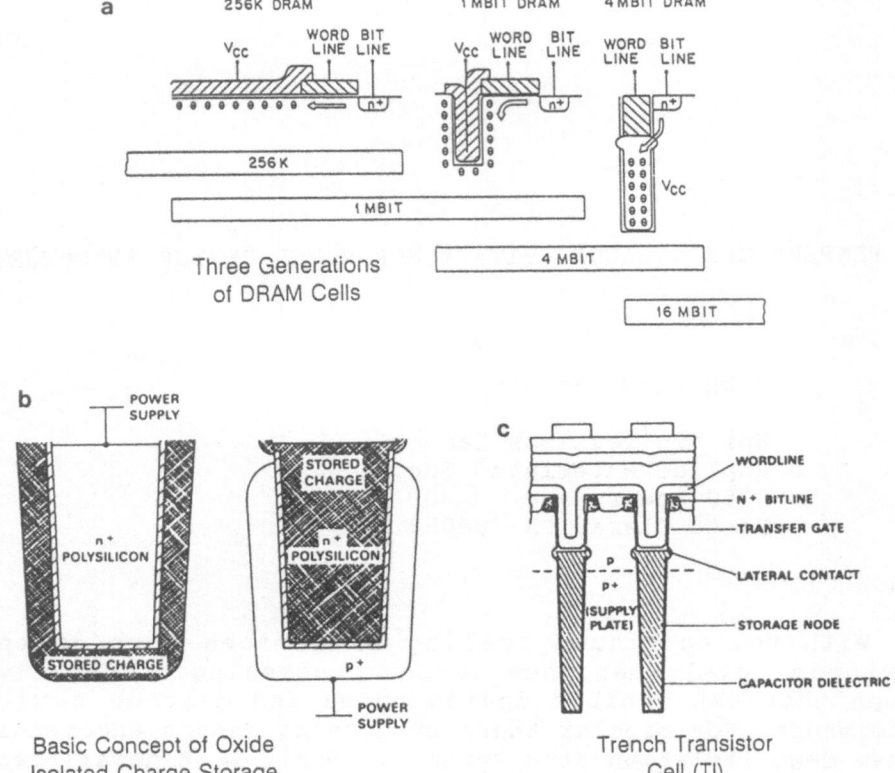

Figure 1. New DRAM cell designs, (a) three generation of DRAM inovation using the third dimension, (b) basic concept of oxide isolated charge storage and (c) trench transistor cell[1].

reduced causing minimal substrate dopant out-diffusion (low thermal budget). An example of thin epi need for CMOS is shown in Fig.1 (4Mega Bit DRAM cell structure) which uses thin p/p+ epitaxial structures for the DRAM trench storage capacitor cell structure[1]. Two factors that influence the epi/substrate transition width and thereby limit the epilayer thickness are: 1) autodoping from the buried layer structure (bipolar and BiCMOS devices) or heavily doped substrate (CMOS devices) and the re-incorporation of the autodoped dopant and 2) dopant out-diffusion from the substrate during epitaxial processing at elevated temperatures (>1000°C).

Historically, low temperature silicon epitaxial growth at temperatures below 1000°C have been achieved by using SiH_4 in conventional chemical vapor deposition (CVD) epi reactors. Problems associated with SiH_4 epi deposition are safety, particle related defects and bell jar coating at temperatures above 1000°C. New epitaxial growth techniques have recently reported single crystal silicon epitaxial growth at temperatures below 850°C[2-5]. These new techniques include Molecular Beam Epitaxy (MBE), Plasma Enhanced CVD (PE-CVD), Photo and/or Laser Assisted CVD (P/LA-CVD), Ion Cluster Beam CVD (ICB-CVD) and Ultra Low Pressure CVD (ULP-CVD). With the exception of MBE systems, all of these other epi reactor systems are one of a kind, user built epi reactors and are being used in R&D labs only. Although these new epitaxial growth techniques results in low temperature single crystal silicon epitaxial growth with very abrupt epi/substrate transition width, the epilayer quality is moderate to poor by ULSI standards. Typically high density of dislocations >10^3 defects/cm^2 are observed in the epilayer along with high levels of impurities and deep level traps making them impractical for ULSI technology. However, in the last two years, significant breakthroughs in conventional CVD epitaxial processing techniques have been achieved through the realization of high quality low temperature (<900°C) low pressure (<40torr) SiH_2Cl_2 epitaxial growth in commercially available reduced pressure barrel epi reactors[6-10].

Growth Kinetics

A plot of growth rate versus temperature for SiH_2Cl_2 in an AMC-7810 at various pressures (760 torr, 200 torr, 80 torr and 27 torr) is shown in Fig.2[11]. Note that at a temperature of ~950°C, the epitaxial growth mechanism changes from gas phase diffusion (>950°C) to surface reaction limited growth (<950°C) independent of pressure[12]. Above 950°C the growth rate is limited by the gas phase

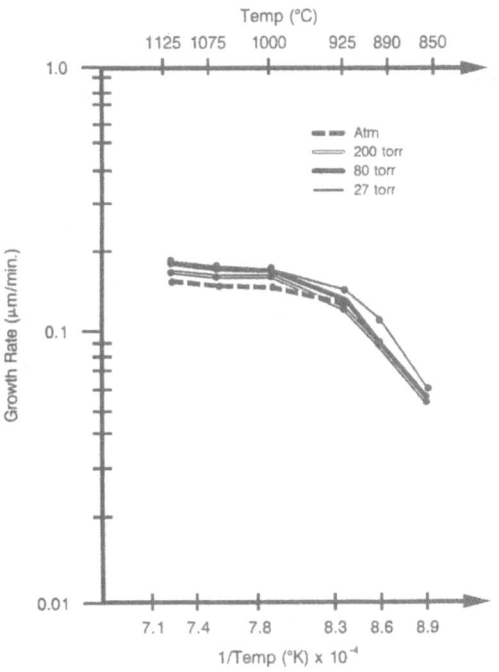

Figure 2. Growth rate versus temperature at various pressures[11].

diffusion of the silicon species to the substrate surface, therefore, growth rate is fairly independent of temperature. Below 950°C the growth rate is strongly dependent on temperature and limited by the surface reaction kinetics of the silicon species at the substrate surface.

Native Oxide Removal

Two key factors in achieving high quality low temperature silicon epitaxial growth are; 1) low pressure epitaxial growth techniques and 2) maintaining clean substrate surfaces free of native oxide. In 1978, Duchimen[13] reported the effects of pressure on the single/poly deposition transition temperature (see Table 1) using a user built epi reactor. Pressures 30 torr and below were required for <900°C single crystal epi growth. Then in May of 1985, Nagao et al.[14] using an AMC-7800RPX epi reactor reported on the importance of reduced pressure (40 torr) pre-epi hydrogen (H_2) bake on the quality of 930°C epilayers. They also showed low pressure deposition improved the epilayer quality (Fig. 3). The in situ pre-clean in the epi reactor to remove native oxide prior to epi growth is critical, and low pressures improve (enhance) H_2 (hydrogen) removal/reduction of native surface oxide. This was reported in August of 1985 by Borland and Drowley[15]. Using an AMC-7800RPX they reported on lowering the in situ pre-epi H_2 baking from 1150°C down to 950°C at 25 torr prior to epi growth at 826°C and still maintained effective H_2 removal of the native surface oxide resulting in excellent low temperature epitaxial structures. This observed native oxide reduction at 950°C, 25 torr with just a 5 minute H_2 bake is not by the traditional etching mechanism of SiO_2 by H_2 annealing. Kugimiya and Hirofuji[16] reported that at 950°C the etching rate of SiO_2 by H_2 annealing is only 0.3A/min (Fig.4). A model describing this H_2 reduction (erosion) of the SiO_2/Si (native oxide/silicon) interface using an AMC-7800-RPX and AMC-7810 epi reactor was reported by Liu, Chan and Borland[17] covering the temperature range from 1150°C to 900°C and pressures from atmosphere (760 torr) to 25 torr. The H_2

Table 1. Effects Of Pressure On The Single/Poly Transtion Temperature[13]

Hydrogen Pressure (Torr)	1050°C	1000°C	950°C	900°C	850°C	800°C
760	Mono	Mono	Poly	Poly	Poly	Poly
250	Mono	Mono	Poly	Poly	Poly	Poly
70	Mono	Mono	Mono	Poly	Poly	Poly
30	Mono	Mono	Mono	Mono	Mono	Poly
10	Mono	Mono	Mono	Mono	Mono	Mono

Process Depo Pressure (Torr) Temp. (°C)	conventional	with prebaking 40 Torr	reduced pressure deposition		
	760	760	200	80	40
1080	◯	◯	◯	◯	◯
1030	◯				◯
980	✕				◯
930	✕	△	△	◯	◯

✕ ; Poly-Si   △ ; S-Si   ◯ ; excellent S-Si

Comparison of crystalline film quality. ◯ excellent single-crystalline silicon. △ single-crystalline silicon. ✕ polycrystalline silicon.

S. Nagao, K. Higashitani, Y. Akasaka and H. Nakata,
Journal of Applied Physics, Vol. 57, No. 10, p. 4589, May 15, 1985

Figure 3. Crystalline film quality in reduced pressure silicon epitaxy at low temperature[14].

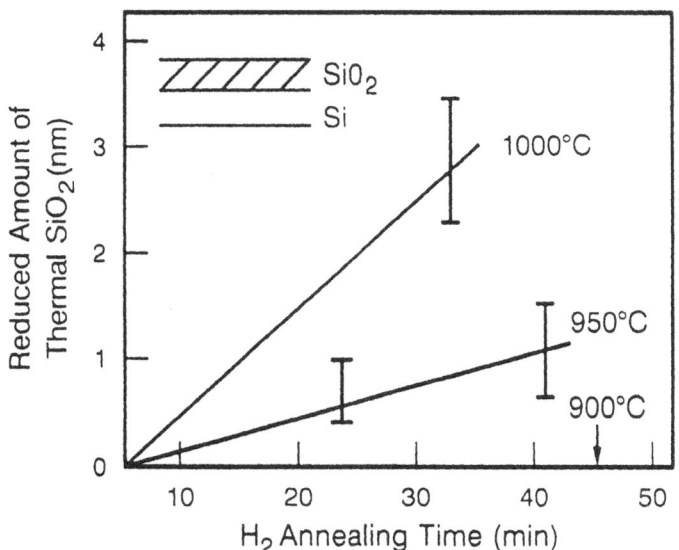

K. Kugimiya and Y. Hirofuji, Japanese Journal of Applied Physics
Vol. 24, No. 5, May, 1985 pp 518-523

Figure 4. Etching rate of SiO₂ by H₂ annealing[16].

erosion model of the SiO_2/Si interface is based on the oxide undercutting and lifting effect observed with high temperature SEG processing[6,15] (Fig.5). SIMS analysis reported by Goulding and Borland[18] showed effective native oxide removal with no evidence of oxygen nor carbon at the interface when using a 5 min. 950°C/30torr H_2 bake (Fig.6). In May of 1988, Borland, Murali and Wei[19] showed effective native oxide removal for SEG processing using a 5 min. H_2 bake operating an AMC-7810 at 850°C/10torr.

Epilayer Quality

The crystalline quality of the low temperature/low pressure (LT/LP) epilayers have been examined by various techniques to determine both structure perfection and

Figure 5. Oxide undercutting and lifting effect[17].

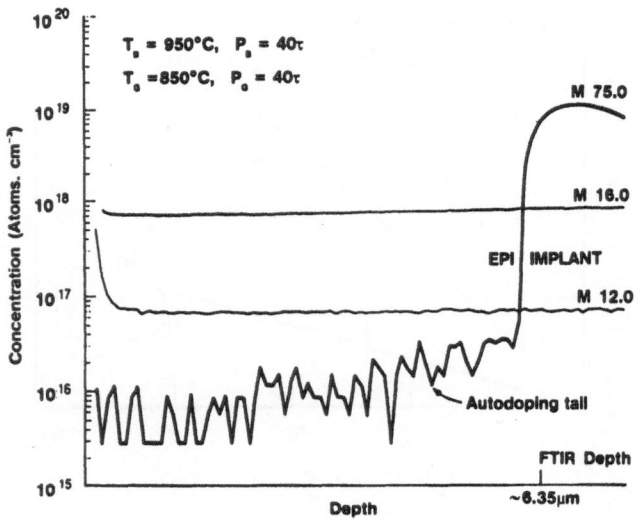

Figure 6. SIMS analysis for oxygen and carbon at the epi/substrate interface[18].

electrical integrity. Borland et al.[12,15] used electron channeling pattern analysis to determine the crystallinity of their 2.5um thick epilayers. A 950°C atmospheric pressure SiH_2Cl_2 epilayer was single crystal, however, optical microscopy shows it to be very defective (Fig.7a). An 833°C atmospheric pressure epilayer was poly-crystalline with no defined surface orientation (Fig.7b). And, an 826°C, 25torr pressure epilayer was single crystal with a (100) epilayer surface orientation (Fig.7c). Nagao et al.[14] used planar Transmission Electron Microscopy (TEM) and

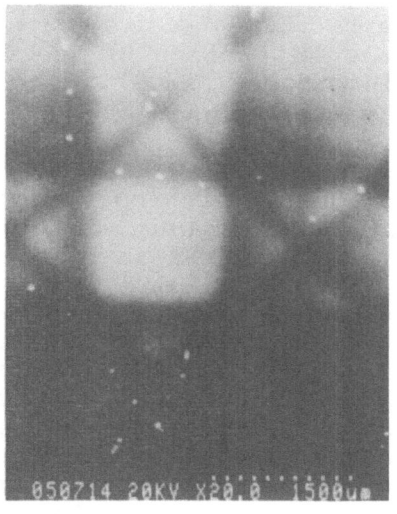

a

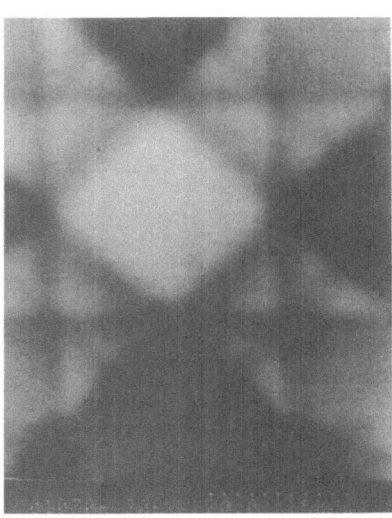

b   c

Figure 7. Electron channeling pattern analysis of epilayer
crystallinity, (a) 950°C/760torr,
(b) 833°C/760torr and (c) 826°C/25torr[12].

optical microscopy on Sirtl defect etched samples (Fig.8) to analyze the low temperature (930°C) epilayer quality described in Fig.3. At atmospheric pressure they observed poly-deposition. At 200 torr pressure they observed single crystal deposition with a high density of stacking faults in the epilayer. Depositions at pressures of 80torr and 40torr were required for high quality defect free single crystal epilayers. They also reported a 40 torr rather than atmospheric pressure pre-epi bake at 930°C temperature followed by an atmospheric deposition at 930°C did result in single crystal but defective epilayer. Similarly, operating at 850°C, Goulding and Borland[18] showed that at pressures >80torr a degradation in epilayer quality occurs and resulted in the broading of the p/p+ transition width due to enhanced boron diffusion by the defects induced at the epi/substrate interface (Fig.9).

Epitaxial Intrinsic Gettering

Wright chemical defect etching reported by Borland and Schmidt[9] showed that the lowest defect densities were observed in epilayers grown on intrinsically gettered epi substrate wafers. 826°C/25torr, 100mm,p/p+ epitaxial structures with post epitaxial intrinsic gettering had 4 defects per wafer while 100mm n/n+ epitaxial structures with pre-epitaxial intrinsic gettering had 4 to 6 defects/cm². Various electrical measurements were also conducted to determine the epilayer quality. MOS capacitors were fabricated using 0.0750um thermal oxides and both minority carrier generation lifetimes and oxide leakage yield measurements were made on the n/n+ and p/p+ epi

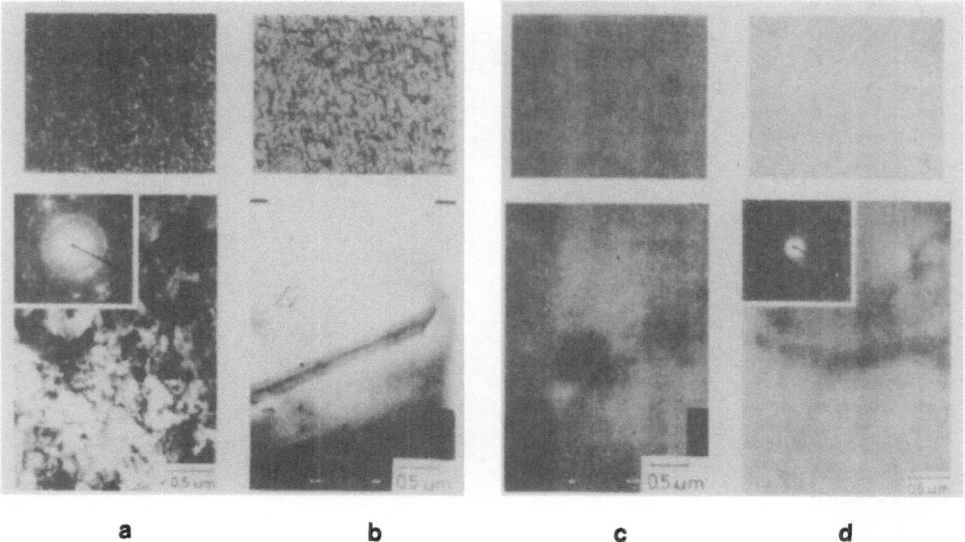

a   b   c   d

Figure 8. TEM and optical micrographs of epi grown at 930°C, (a) 760torr, (b) 200torr, (c) 80torr and (d) 40torr [14].

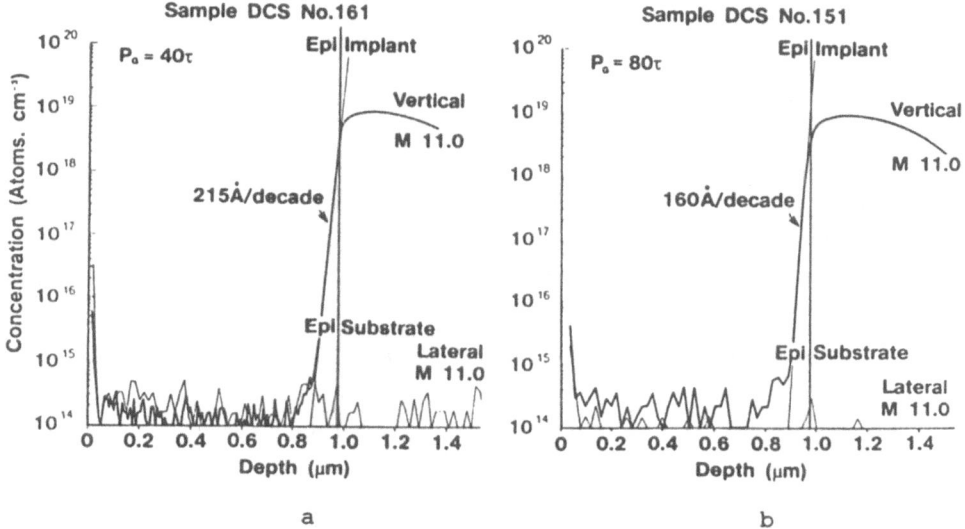

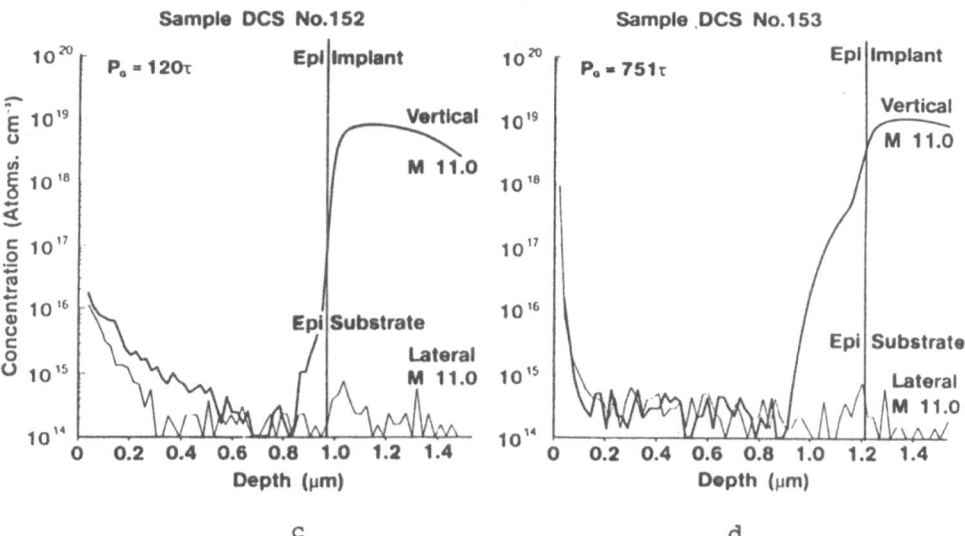

Figure 9. SIMS analysis showing boron enhanced diffusion at the epi/substrate interface due to defects at 850°C and pressures of, (a) 40torr, (b) 80torr, (c) 120torr and (d) atmospheric pressure[18].

wafers. Lifetimes for pre-epi intrinsic gettered n/n+ epilayers varied between 55usec to 791usec with the average being 363usec. This was higher than those typically observed on standard atmospheric high temperature non-intrinsic gettered n/n+ epilayers. Lifetimes for post-epi intrinsic gettered p/p+ epilayers averaged 37usec. Oxide leakage measurements on standard p/p+ and n/n+ epi-wafers purchased from several silicon epi wafer suppliers typically yield 40% to 80% while the 826°C, 25torr non-intrinsic gettered p/p+ epi wafers yielded 69%, post-epi intrinsic gettered p/p+ epi wafers yielded 81% and pre-epi intrinsic gettered n/n+ epi wafers yielded 87%.

Similar low defects levels have also been reported by Borland et al.[20]. Using an AMC-7810 at 875°C/34 torr pressure, they grew single crystal epilayers on 125mm (5") n^+ (100) substrates and compared the effects of either an 1150°C/34 torr, 5min. H_2 pre-epi bake versus an 950°C/34 torr, 5min. H_2 pre-epi bake for effective native oxide removal. Table 2 shows the extremely low defect levels observed on the epi structures as revealed by chemical defect preferential crystallographic etching. Junction leakage current results are shown in Table 3 for the average of 54 point measurements per 5" epi-wafer. Note again the excellent results on the samples that received only the 950°C pre-clean treatment. MOS generation lifetime results are also shown in Table 3.

Table 2. Epilayer Crystal Quality[20]

		Stacking Faults	Shallow Etch Pits	Oxidation Induced Stacking Faults (1000°C 3 Hour Wet Oxidation)	Haze
1150°C H_2 Bake 875°C Epi	Lot-1	<1	None	<10	None
	Lot-2	<1	None	<10	None
950°C H_2 Bake 875°C Epi	Lot-3	<1	None	<10	None
	Lot-4	<1	None	<10	None
Bulk Wafers	Lot-5	<1	None	<10	None
	Lot-6	<1	None	<10	None

Table 3. 875°C Epi Electrical Results[20]

		Junction Leakage Current at 5.5 Volts Peripheral Length 31-1cm Junction Area 7.87x10⁻³cm²		MOS Generation C-t Lifetime
		Leakage Current	Percent (<1nA)	Lifetime
1150°C H_2 Bake 875°C Epi	Lot-1	242 pA	66.7%	299 μsec
	Lot-2	355 pA	74.1%	261 μsec
950°C H_2 Bake 875°C Epi	Lot-3	259 pA	75.9%	395 μsec
	Lot-4	242 pA	79.6%	339 μsec
Bulk Wafers	Lot-5	199 pA	90.7%	365 μsec
	Lot-6	246 pA	85.2%	734 μsec

AUTODOPING, TRANSITION WIDTH AND DOPANT INCORPORATION/ REINCORPORATION

There are numerous epitaxial processing factors that affect autodoping, dopant re-incorporation and dopant diffusion thereby limiting the epi/substrate transition width. Sirnivasan[21], Chang[22], Cullen et al.[23] and Graef et al.[24] have all recently reviewed autodoping and showed that autodoping is not only dependent on the epitaxial deposition conditions (temperature and pressure) but also can be dependent on the in situ pre-deposition cleaning conditions (temperature, pressure, time, H_2 bake or HCl etch). These studies were done at conventional CVD deposition temperatures above 1000°C and pressures above 80 torr.

Studies down to the 850°C temperature range and pressures as low as 25 torr were recently reported by Borland et al.[11,20] and Goulding and Borland[18]. Their analysis were achieved by 4-point probe measurements, spreading resistance (SRP) concentration depth profiles and secondary ion mass spectroscopy (SIMS) dopant depth profiling. They observed that reducing the boron buried layer or p+ substrate pre-epi baking temperature from 1150°C to 950°C reduces boron autodoping (Fig.10) while for arsenic buried layers or n+ arsenic substrates, high temperature/low pressure depositions reduces arsenic autodoping. Also, low pressure deposition enhances boron incorporation rate (Fig.11) so the combination of low temperature/low pressure bake (<950°C/<30torr) with low temperature/low pressure deposition (<950°C/<40 torr) are required to totally eliminate boron autodoping and solid phase diffusion. Note the sharp transition width and elimination of boron lateral autodoping in Figs.12&13.For thicker layers, an LT/LP (850°C/40torr) epi cap followed by a high temperature/low pressure (HT/LP) >1050°C/<80torr

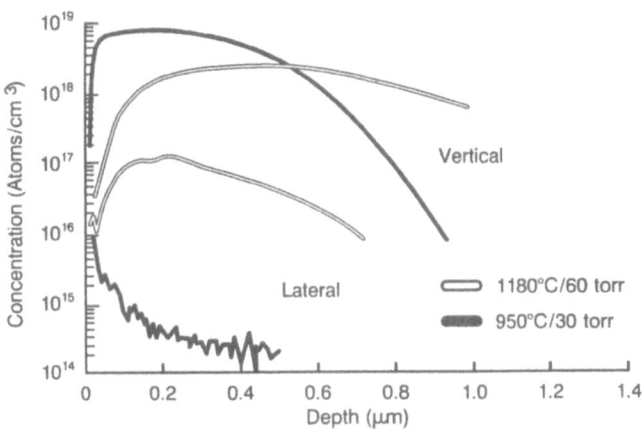

Figure 10. Boron vertical and lateral SIMS profile of dopant outdiffusion and re-indiffusion relating to a 5 minute H_2 bake with no epi growth[18].

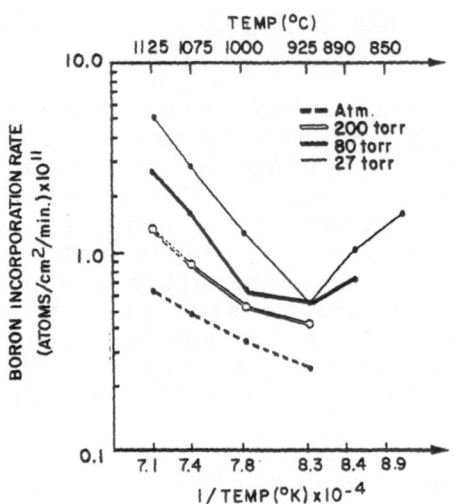

Figure 11. **Boron dopant incorporation rate versus temperature at different pressures[11].**

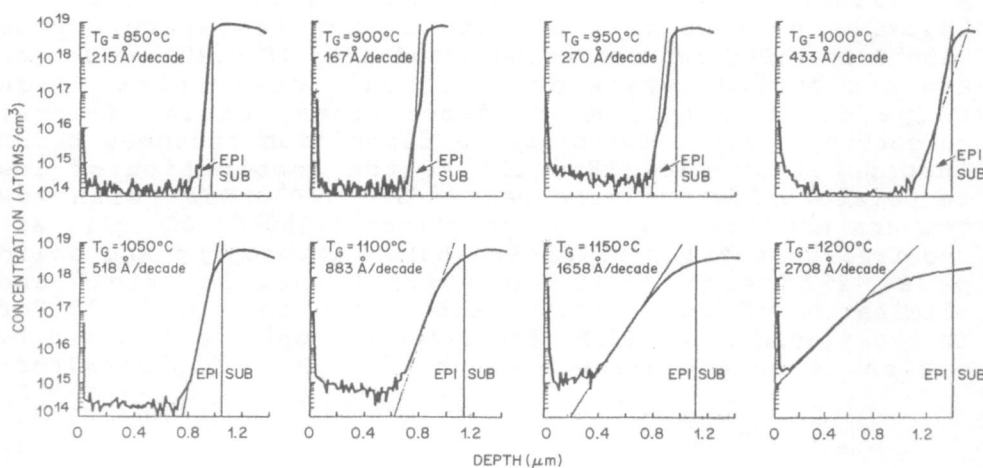

H₂ Prebake, 950°C, 30 torr; Epitaxial Deposition Pressure, P_G = 40 torr;
Total Carrier Gas Flow Rate, 77 SLM; DCS Flow Rate, 700 sccm

Figure 12. **Boron vertical SIMS profile showing boron buried layer updiffusion for epi growth pressure of 40torr and temperatures from 1200°C down to 850°C[18].**

primary epi growth also eliminates boron autodoping (Fig.14). Some recent data on phosphorus and arsenic dopant incorporation rate as reported by Borland et al.[11] are shown in Figs.15&16.

Arsenic (As) dopant incorporation rate versus 1/Temp is shown in Fig.16. At temperatures below 1000°C, As dopant incorporation mechanism seems to be independent of both pressure and temperature, however, at higher temperatures (>1050°C), the incorporation mechanism is

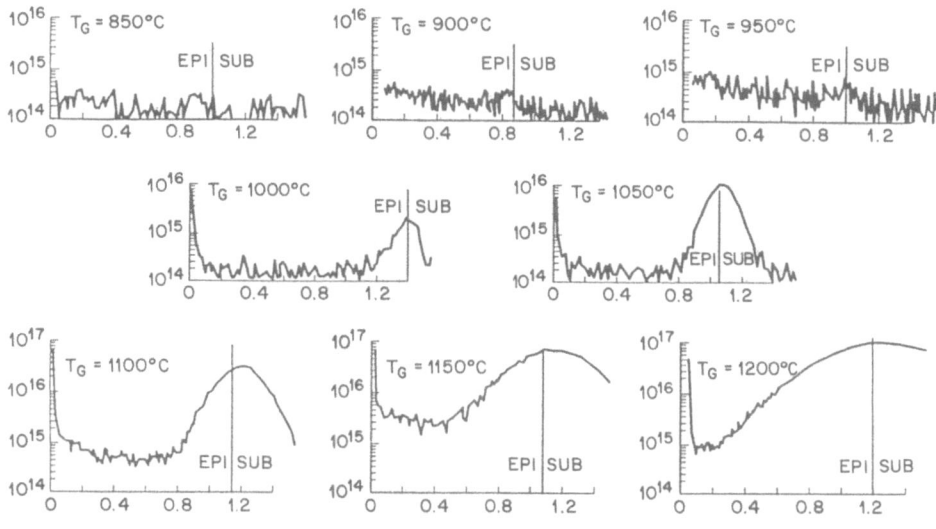

H₂ PREBAKE, 950°C, 30 torr; EPITAXIAL DEPOSITION PRESSURE, P_G = 40 torr;
TOTAL CARRIER GAS FLOW RATE, 77 SLM; DCS FLOW RATE, 700 sccm

Figure 13. Boron lateral SIMS profile showing boron buried layer lateral autodoping for epi growth pressure of 40torr and temperatures from 1200°C down to 850°C[18].

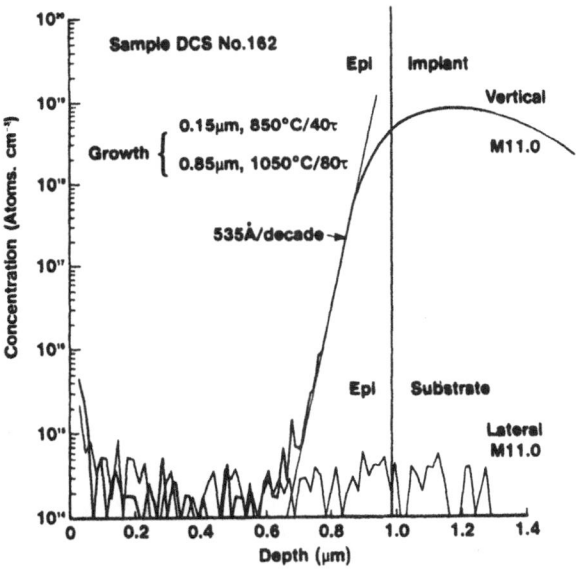

Figure 14. Boron vertical and lateral SIMS profile for a two-step low temperature cap, high temperature growth epi process[18].

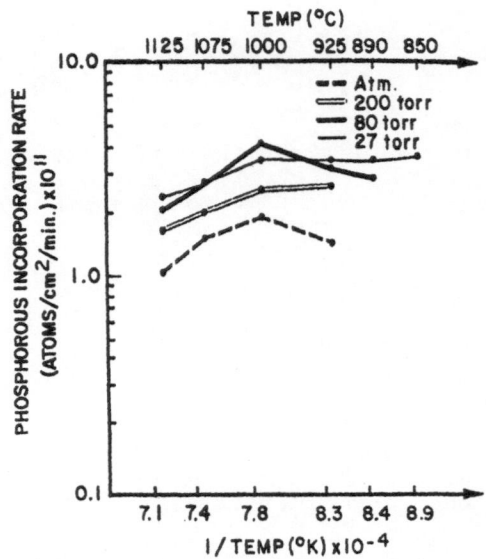

Figure 15. Phosphorus dopant incorporation rate versus
temperature at different pressures[11].

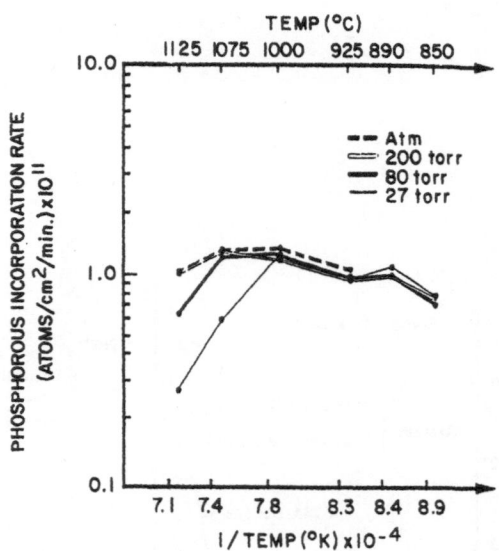

Figure 16. Arsenic dopant incorporation rate versus
temperature at different pressures[11].

strongly dependent on both pressure and temperature. For
example, at 1075°C, lowering the deposition pressure from
80 torr to 27 torr had a pronounced effect of retarding the
As dopant incorporation rate by a factor of 3. This
observed retardation of As incorporation during high
temperature/low pressure (HT/LP) epi processing can have a
significant impact on advanced bipolar and BiCMOS
processing technology that use heavily doped As n⁺ buried
layer structures. The vertical and lateral As autodoping
profiles are shown in Fig.17a&b. The lateral As autodoping

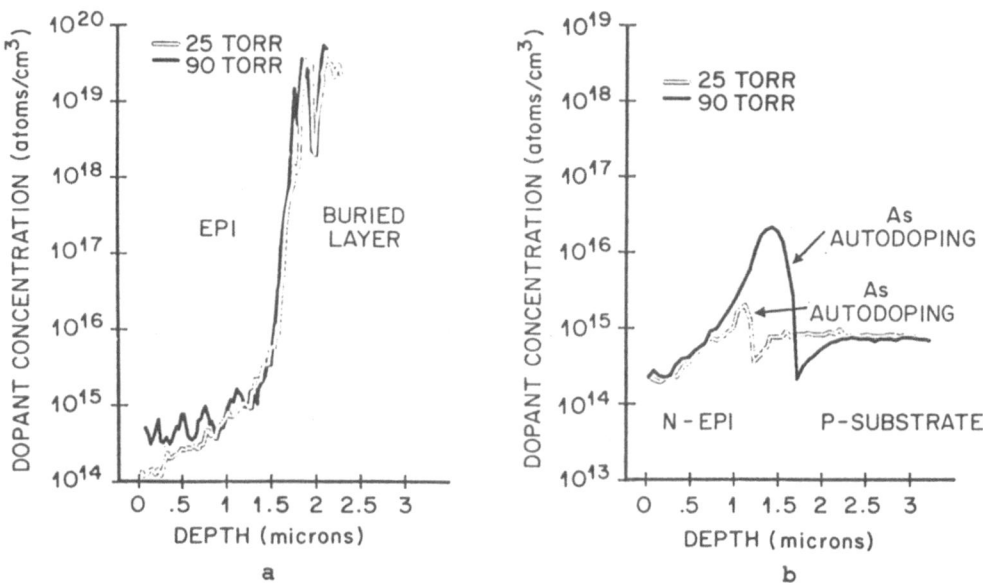

Figure 17. SRP analysis of arsenic buried layer with epi grown at 1080°C and either 90torr or 25torr pressure, (a) vertical autodoping profile and (b) lateral autodoping profile[20].

peak off the buried layer for a typical 1080°C/80torr epi deposition is $2 \times 10^{16}/cm^3$. Reducing the deposition pressure from 80 torr to 25 torr and keeping the temperature at 1080°C resulted in the reduction the lateral As autodoping peak by over an order of magnitude down to $1.8 \times 10^{15}/cm^3$ (Fig.17b).

A comparison between antimony (Sb) and arsenic (As) buried layer transition width and diffusion at different temperatures and pressures is shown in Table 4 as reported by Borland et al.[20]. Note the very abrupt transition width achievable by LT/LP epi growth techniques. Also, a plot of measured As buried layer updiffusion versus temperature is shown in Fig.18. The most significant reduction in updiffusion occurs by lowering the deposition temperature from 1120°C to 1000°C. Only a small reduction in updiffusion occurs below this temperature when maintaining an 1180°C pre-epi H_2 bake. It was shown earlier to minimize As autodoping, HT/LP epi processing is required so lateral autodoping as a function of deposit seal/cap temperature and primary deposit temperature was studied. The results are given in Table 5. From these results it is evident that even with reduced pressure processing, the deposit seal and flush steps are essential to minimize autodoping. The most significant result, however, is that the primary deposition temperature can be lowered by 100°C without increasing autodoping if a high temperature deposit seal/cap is used.

Table 4. Low Temperature/Low Pressure (LT/LP) Epi/Substrate Transition Width[20]

Antimony (Sb) Buried Layer		Measurement Method	Transition Region Width	
			10^{17} to 10^{19}/cm^3	Buried Layer Peak to Epi Flat
Process 1:	1150°C H$_2$ Bake	SRP	0.30μm	0.60μm
	1080°C Atm Epi	SIMS	0.115μm	0.335μm
Process 2:	1150°C H$_2$ Bake			
	1080°C 25 torr Epi	SRP	0.20μm	0.50μm
Process 3:	1150°C H$_2$ Bake			
	900°C 25 torr Epi	SIMS	0.025μm	0.12μm
Process 4:	950°C H$_2$ Bake			
	900°C 25 torr Epi	SIMS	0.0175μm	0.0925μm
Arsenic (As) Buried Layer				
Process 1:	1150°C H$_2$ Bake			
	1080°C Atm Epi	SRP	0.75μm	1.25μm
Process 2:	1180°C H$_2$ Bake			
	1080°C 80 torr Epi	SRP	0.15 μm	0.45 μm
Process 3:	1100°C H$_2$ Bake	SRP	0.075μm	0.20μm
	900°C 25 torr Epi	SIMS	0.040μm	0.15μm

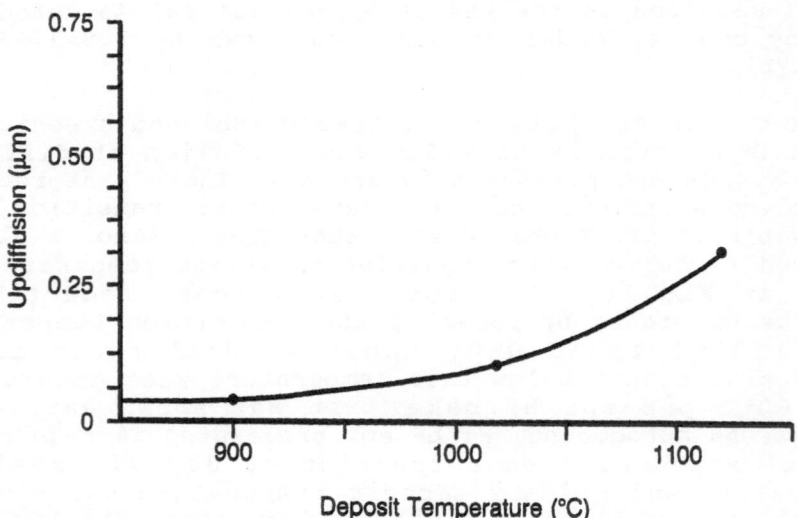

Figure 18. Arsenic buried layer updiffusion versus temperature[20].

Table 5. Effects Of Deposit Seal (Cap) On Arsenic Buried
Layer Lateral Autodoping Peak[20]

Deposit Seal Temperature (°C)	Primary Deposit Temperature (°C)	Lateral Autodoping Peak (atoms/cm^3)
1120	1120	1.8E15
NA	1120	3.4E15
1120	1020	1.8E15
1020	1020	6.0E15
NA	1020	1.2E16

BURIED LAYER PATTERN TRANSMITTANCE

Buried layer pattern transfer for a 3.5 um thick (100) epilayer was studied at different deposition pressures and temperatures[9]. The control non-epi buried layer test structure is shown in Fig.19a. An atmospheric 1080°C epilayer (0.12 um/min growth rate) is shown in Fig.19b where pattern washout occurred. Reducing the pressure to 9 torr resulted in the 1080°C epilayer (0.15 um/min growth rate) shown in Fig.19c with good pattern transfer. Lowering the deposition temperature to 900°C (0.12 um/min growth rate) at 9 torr resulted in pattern distortion (see Fig.19d). When the temperature was further lowered to 826°C (0.06 um/min growth rate) at 10 torr an improvement in pattern transfer due to the decrease in growth rate and surface reaction rate limited growth regime (see Fig.19e).

SELECTIVE EPITAXIAL GROWTH

Several new attractive applications of low temperature/low pressure (LT/LP) silicon epitaxial growth techniques are in the areas of advanced CMOS, bipolar and BiCMOS processing technologies where the realization of new advanced (novel) device structures are required to reduce traditional processing steps, complexity and costs. This has been achieved through the use of selective epitaxial growth (SEG) techniques which requires LT/LP epitaxial growth techniques[25,26].

In the last few years, great interest has been generated in the selective growth of single crystal silicon in seed windows of an SiO_2 mask. The primary aim has been to develop an advanced dielectric isolation structure to replace the conventional LOCOS approach due to scalability limits. This increased interest in selective epitaxial

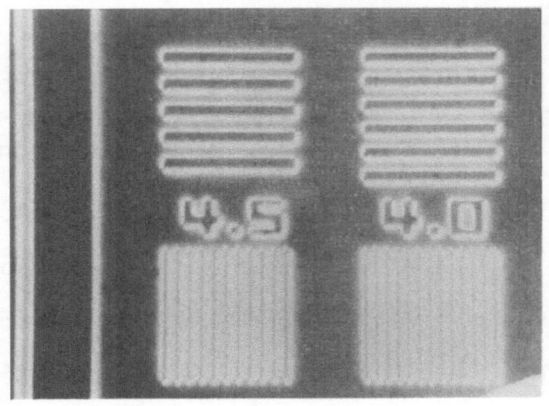

a

b

c

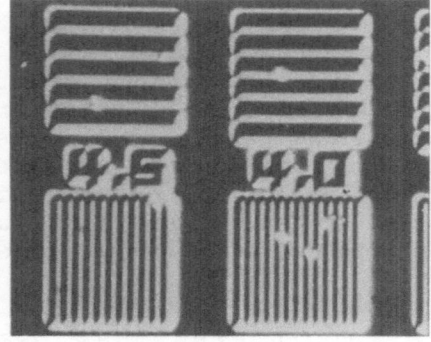

d

e

Figure 19. Effects of pressure, temperature and growth rate
    on buried layer pattern transmittance,
        (a) control (no epi), (b) 1080°C/760torr,
        (c) 1080°C/9torr, (d) 900°C/9torr and
        (e) 826°C/10torr.

·growth (SEG) and its derivatives (simultaneous single/poly deposition (SSPD) and epitaxial lateral overgrowth (ELO)) have led to several key processing breakthroughs that are currently changing the future direction and usage of silicon epitaxy.

SEG Historical Review

Historically, local oxidation of silicon (LOCOS) has been the technique used to laterally isolate n-channel and p-channel devices. One of the main limitations to LOCOS is its scalability due to lateral oxidation under the nitride mask forming the birds beak encroachment structure(Fig.20). Chen et al.[27] reported this scalability limit to be ~2.5um for LOCOS with conventional CMOS Epi technology. Besides scalability limitations, LOCOS also has step coverage/non-planarity problems from the birds beak structure and shallow isolation depth (~0.25um) latch-up immunity problems. A solution to scalability down to submicron device separation and deep isolation for latch-up immunity was the combined use of silicon trench isolation with an epitaxial structure as reported by Yamaguchi et al.[28] (Fig.21). The use of a 6um deep trench was not sufficient to decouple the lateral parasitic bipolar transistor through lateral beta reduction (B_L) and the use of a p/p+ epitaxial structure was not sufficient to reduce substrate resistance (Rs) and prevent forward biasing for latch-up turn on. However, the combined reduction in Rs by the epitaxial structure and enhanced B_L reduction by the trench penetrating down into the substrate low lifetime and low resistance region sufficiently decoupled the lateral parasitic bipolar transistor and prevented latch-up down to a 0.5um design rule n-well CMOS technology. This stressed the importance of a deep isolation structure with an epitaxial structure. However, there are three main limitations to trench isolation: 1) step coverage/non-planarity for narrow and wide trenches, 2) sidewall inversion and 3) process complexity and void formation.

A new isolation technique that combines a deep isolation with an epitaxial structure all in one step and eliminates the problem of step coverage/non-planarity and

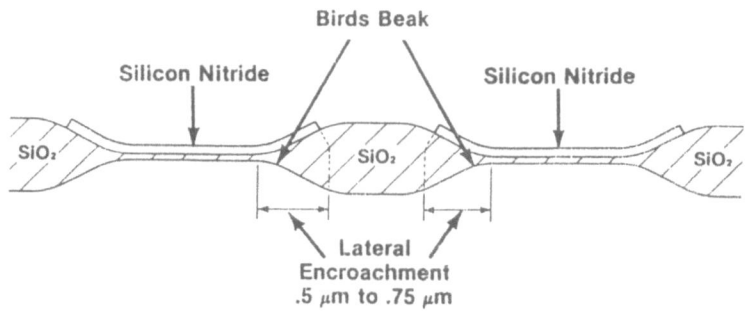

Figure 20. Conventional LOCOS isolation structure.

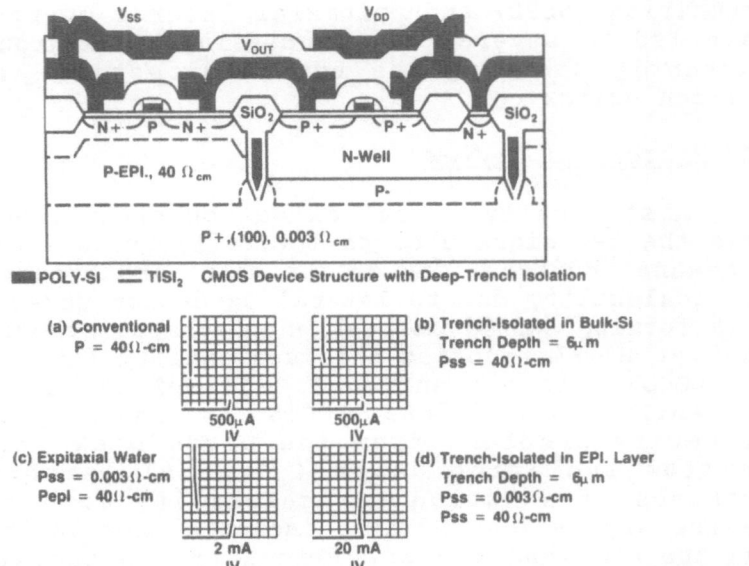

POLY-SI ═ TISI$_2$ CMOS Device Structure with Deep-Trench Isolation

Parasitic SCR Latchup Characteristics for Various Structures.   T. Yamaguchi et. al., IEEE INT. Electron Device Meet., p.522, 1983

Figure 21. Latch-up prevention by the combined use of a p/p+ epi structure with deep trench isolation[28].

process complexity associated with trench isolation is
Selective Epitaxial Growth (SEG) (Fig.22). The basic
concepts of SEG have been around for 27 years since Joyce
and Baldrey[29] first reported on it in 1962. However, its
full potential was not observed until reduced pressure epi
reactors were developed in the late 70's and reduced
pressure SEG was investigated in the early 80's. Tanno et
al.[6] first reported on reduced pressure SEG using SiH$_2$Cl$_2$
with HCl additive in an AMC-7800-RPX epi reactor down to
900°C, 20 torr pressure for (100) and (111) silicon
substrates using a poly-silicon or nitride sidewall coating
to prevent H$_2$ and HCl attacking the oxide sidewall
resulting in oxide undercutting and lifting. The need for a
side wall coating to prevent undercutting and lifting could
be eliminated by going to a 950°C, 20torr H$_2$ bake as
reported by Borland and Drowley[15]. Also, the local loading
effects, growth rate variation across a wafer as a function
of oxide window size and silicon to oxide surface area can
be eliminated by optimizing the amount of HCl additive as
reported by Ishitani et al.[30] and Pagliaro et al.[8]
(Figs.23&24). Stivers et al.[31] and Manoliu and Borland[32]
reported SEG epilayer thickness uniformities of +/-2% and
+/-5% respectively on 100mm SEG epi-wafers.

<u>Selectivity</u>

The effects of HCl additive on non-selective versus
selective deposition was first reported by Jackson[33] in
1965. Borland and Drowley[15] reported on the effects of
pressure, temperature and HCl showing that selectivity

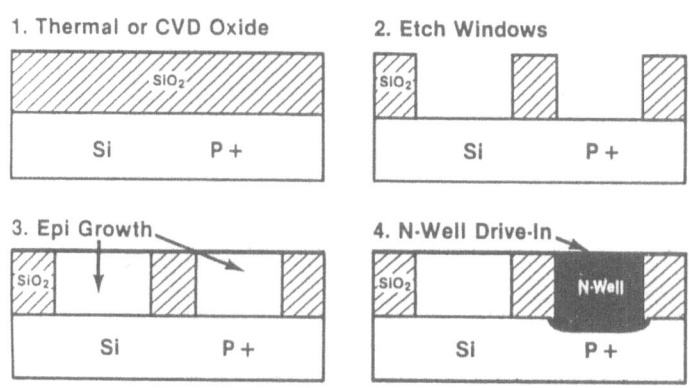

Figure 22. Basic selective epi growth process for an n-well CMOS structure.

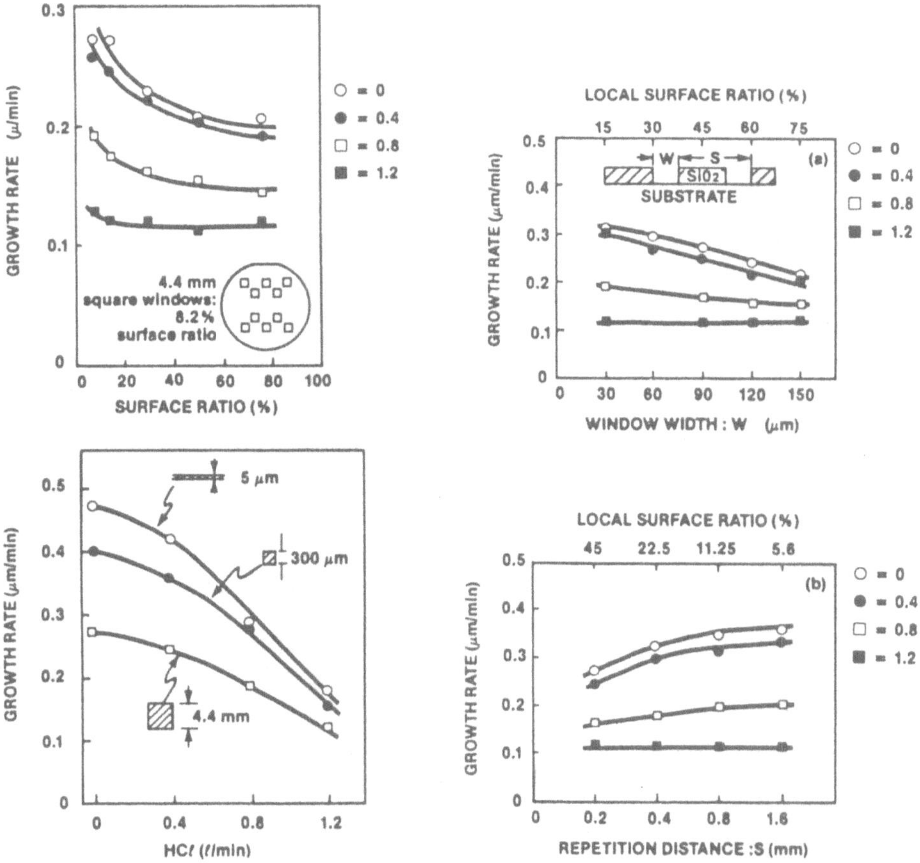

Figure 23. Local loading effects in SEG processing[30].

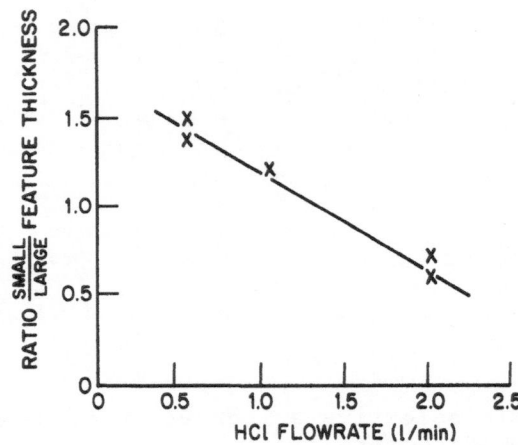

Figure 24. Effects of HCl flow on epi thickness variation
ratio for small (5um) and large (50um)
windows[31].

improved with decreasing temperature and pressure requiring
less HCl additive (Fig.25). This was also reported by
Pagliaro et al.[8] and Matsumoto[34]. Decreasing SEG deposition
temperature also improved the surface planarity as reported
by Voss et al.[35]. Ting et al.[36] observed this at 875°C and
Pagliaro et al.[8] observed this at 850°C.

Si/SiO₂ Sidewall Interface

The effects of the oxide sidewall orientation on SEG
faceting has been reported by Borland and Drowley[15] and
Endo et al.[37]. For a (110) sidewall orientation on (100)
substrates a (311) facet plane is formed for deposition
temperatures >930°C and a (111) facet plane for
temperatures <930°C[36]. For silicon islands surrounded by
oxide, sidewall facets form for <110> sidewall orientation
while for <100> sidewall orientation there are no sidewall
facets, only corner facets due to the <110> direction
component at the corners of intersecting (100) sidewalls
(Fig.26). For oxide islands surrounded by silicon, gradual
sidewall facets form for (110) sidewall orientation and no
sidewall facets nor corner facets form for (100) sidewall
orientation. Borland and Drowley[15] also observed sidewall
induced defects for (110) orientation and none for (100)
orientation as revealed by cross-sectional TEM. This has
also been reported by Endo et al.[37] along with the
associated sidewall leakage currents (Figs.27&28). Nagao et
al.[38] reported that (110) orientation sidewall induced
defects can be eliminated by low temperature (<950°C) SEG
conditions. Using an AMC-7900, Jastrzebski et al.[39]
reported on the effects of temperature, pressure, HCl
additive and oxide sidewall orientation on the leakage
current of edge MOS transistors (Fig.29). They also showed
the effects of using a sacrifical oxide and POCl₃ gettering
(Fig.30).

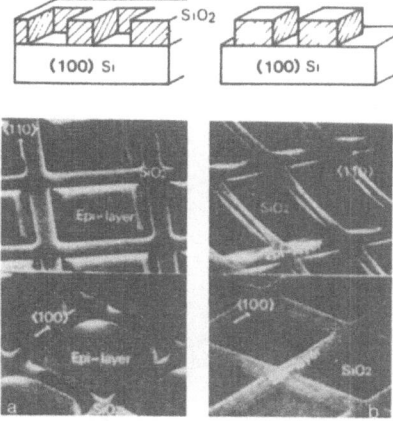

Figure 25. SEG selectivity optimization at 25torr as a function of HCl flow and temperature[15].

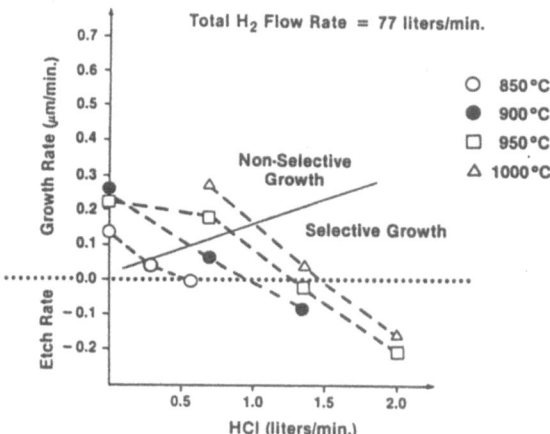

Figure 26. Effects of sidewall direction and pattern on faceting, (a) silicon square structures and (b) oxide square structures[37].

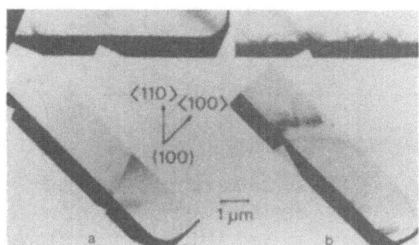

Figure 27. Planar TEM showing sidewall defects for (110) versus (100) sidewall orientation, (a) Si₃N₄ sidewall and (b) SiO₂ sidewall[37].

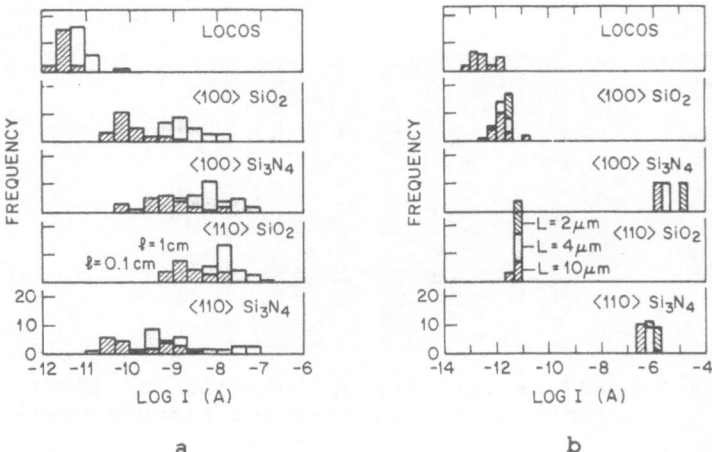

a

b

Figure 28. Leqakage current histograms for n-channel
MOSFET's, (a) junction leakage and
(b) subthreshold leakage[37].

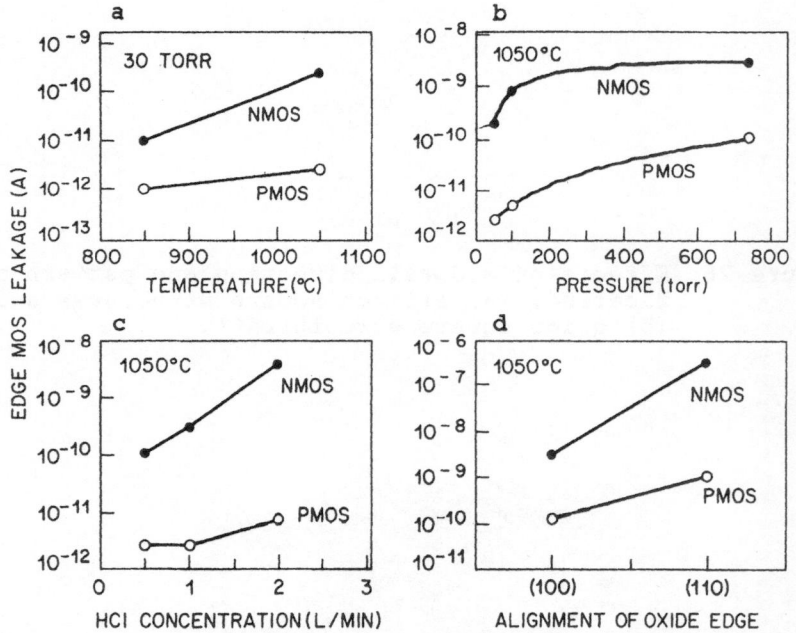

Figure 29. Typical leakage current of edge MOS transistors
made in SEG as a function of (a) temperature,
(b) pressure, (c) HCl concentration and
(d) oxide sidewall orientation[39].

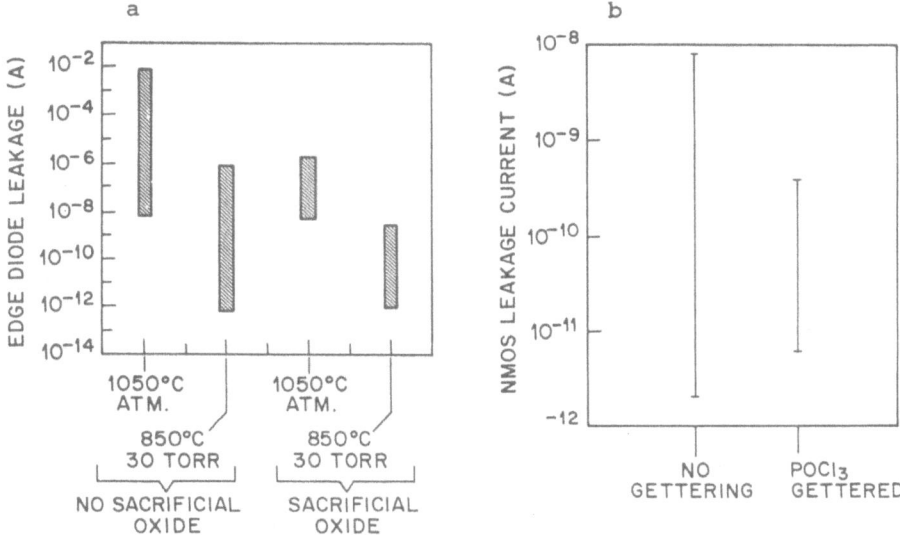

Figure 30. Leakage current of, (a) edge diodes in SEG at 1050°C versus 850°C with or without sacrificial oxide, and (b) edge MOS transistors in SEG with or without phosphorus gettering[39].

Current Limitations and Solutions

Although there have been significant breakthroughs in SEG process development over the last 5 years, there are still some observed problems associated with SEG processing that must be resolved before it is widely accepted in production. These deal with the Si/SiO_2 side-wall interface and relate to enhanced side-wall junction leakage[20,31,32,34,38] and side-wall inversion causing a "kink" effect in the n-channel device subthreshold characteristics[31,32,36,37,38,40] (Fig.31). An improvement in side-wall leakage has also been observed by Stivers et al.[31] by going to lower SEG deposition temperatures (875°C, at 25 torr pressure). This has also been reported by Matsumoto et al.[34,40] and they correlated this effect to a significant reduction in the Si/SiO_2 side-wall interface defect by lowering the SEG deposition temperature from 1000°C to 900°C (Fig.32)[34]. Elimination of the Si/SiO_2 side-wall inversion and n-channel device "kink" effect is possible through proper SEG side-wall doping control ($>10^{17}/cm^3$) as shown by Manoliu and Borland[32].

Another concern associated with SEG is the corner faceting along the <110> direction. This faceting effect can be minimized and eliminated by optimizing the SEG growth conditions and device layout design. This has been shown by Stivers et al.[31] (Fig.33), Fong and Borland[41] (Fig.34), and Borland et al.[19] (Fig.35).

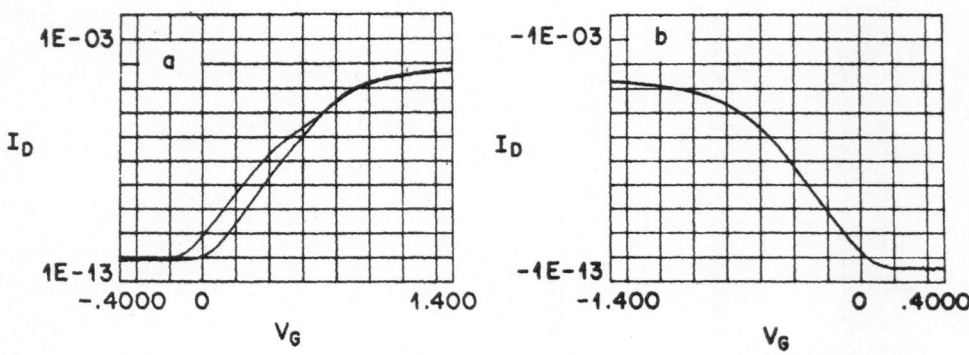

Figure 31. "Kink" effect in n-channel device subthreshold
characteristics, (a) n-channel MOSFET and
(b) p-channel MOSFET[14].

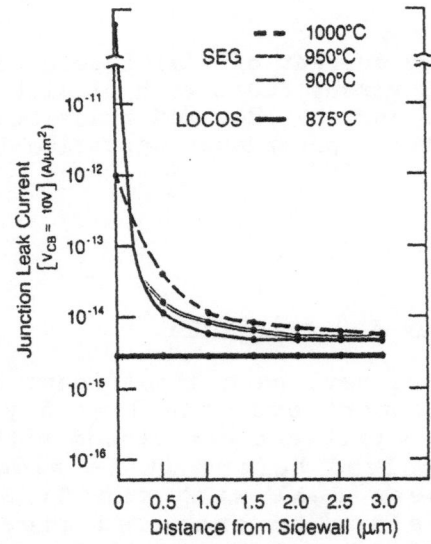

Figure 32. Effects of SEG temperature on junction leakage
current[34].

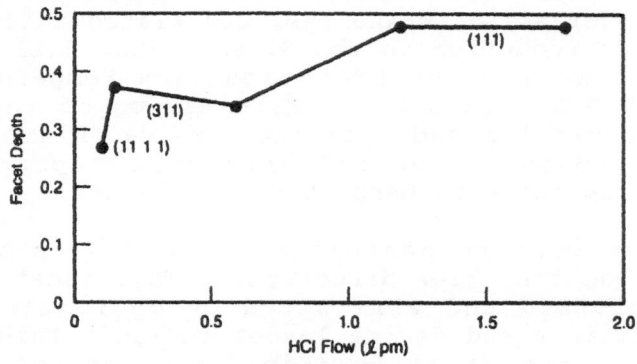

Growth Condition is 950°C and 25 Torr

Figure 33. Effects of HCl flow on facet depth[31].

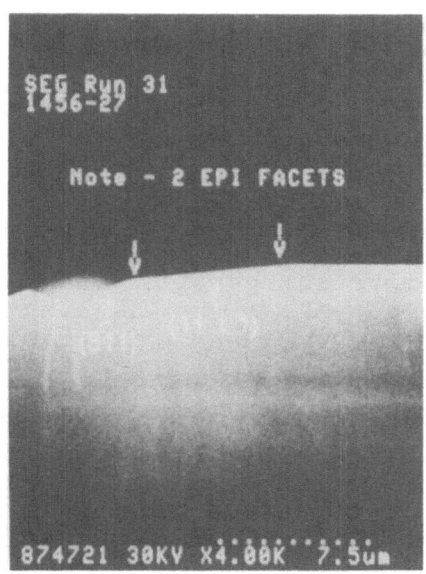

Figure 34. Formation of both the (311), 25° facet plane and the (11 1 1), 7° facet plane[41].

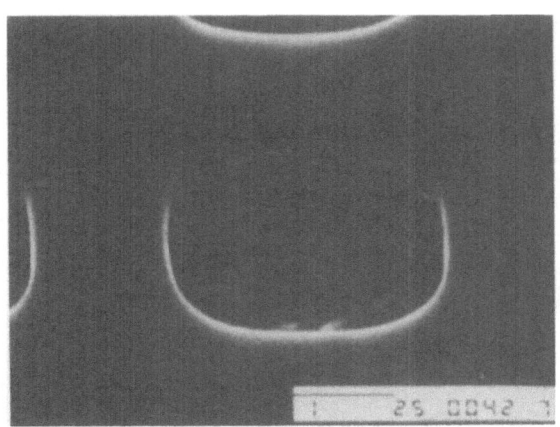

Figure 35. Facet free SEG grown at 850°C/10torr pressure[19].

Applications For Novel (Advanced) Device Structures

In spite of these limitations, novel CMOS, bipolar and BiCMOS device structures have been achieved through the use of SEG processing. There are basically two types of SEG material requirements. One type requires SEG of excellent crystal quality since the selectively grown silicon is used in the active device area (examples include lateral and vertical isolation and certain forms of selective doping).

The other type is insensitive to SEG material/crystal quality since the selectively grown silicon is used in the non-active device area (examples include trench refill and selective doping for contact refill). Further description of the various areas in which SEG processing is being applied to achieve novel device structures are described below.

Lateral Isolation. One attractive area for SEG application is lateral CMOS, bipolar and BiCMOS device isolation. An important benefit of SEG isolation is its scalability which is well below the line width resolution of current production lithographic equipment. Kasai et al.[42] showed a CMOS structure where the n-channel and p-channel devices are isolated by an SEG structure 0.25um wide, 4.0um deep as shown in Fig.36. SEG is also being used to refill deep trench isolation structures for high speed bipolar devices[43,44] (Fig.37). This SEG refilling technique is very attractive because it is insensitive to the trench profile resulting in planar refill, free of void formation and can refill varying trench widths simultaneously as reported by Silvestri[43].

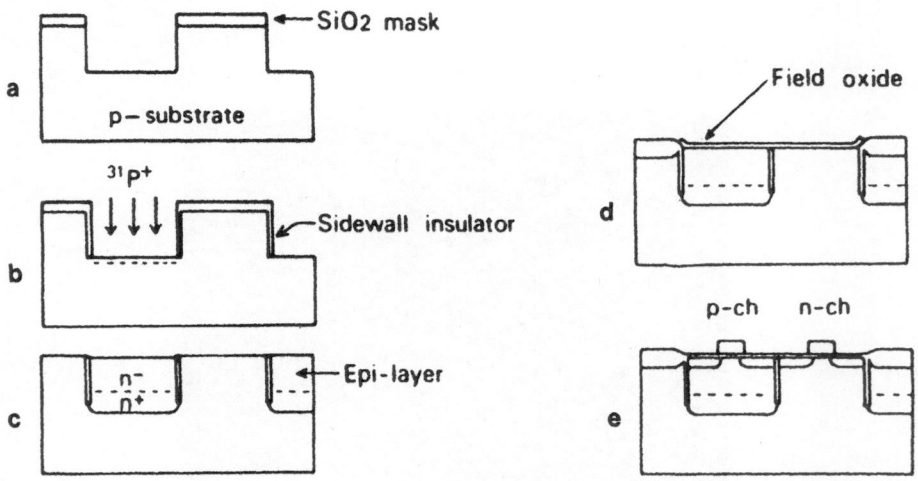

Figure 36. SEG CMOS isolation structure 1/4um wide, 4um deep[42].

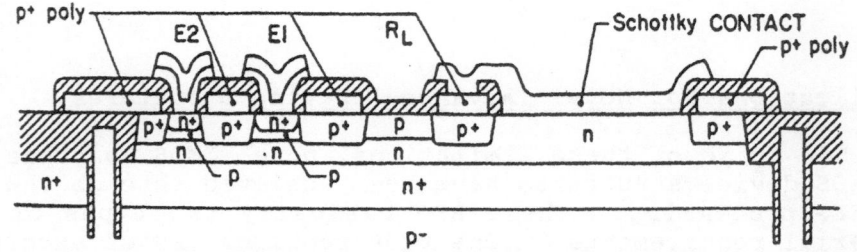

C. T. Chuang, D. D. Tang, G. P. Li, E. Hackbarth, R. R. Boedeker, SSDM 1986, p. 267, Aug. 1986

Figure 37. SEG refilled trench isolation structure for high speed bipolar RAM cell[44].

SEG used for trench isolation refill also provides narrow non-encroaching isolation without the etchback difficulties associated with the more traditional polysilicon refill approach. Unlike the traditional conformal polysilicon refill approach, using SEG allows trenches of greatly varying widths to be refilled simultaneously. Trenches of varying widths, but constant depth, can be etched in silicon using a hardmask typically oxide, nitride or some combination. Results from Fong and Borland[41] using an AMC-7810 for SEG trench refill is shown in Fig.38. They used an in situ 1000°C 5 min H_2 bake at 25

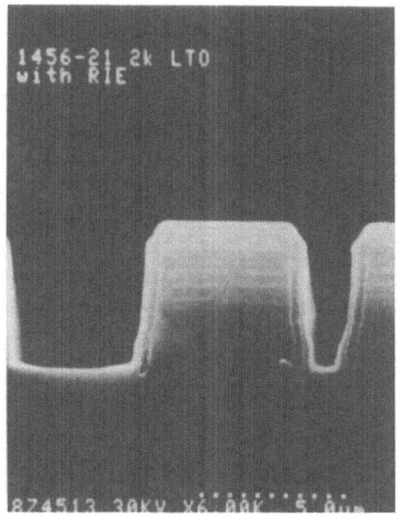

a

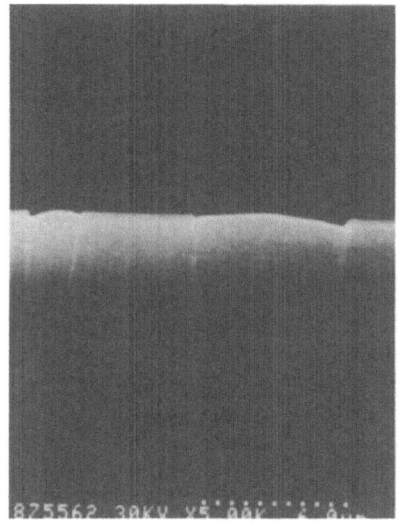

b

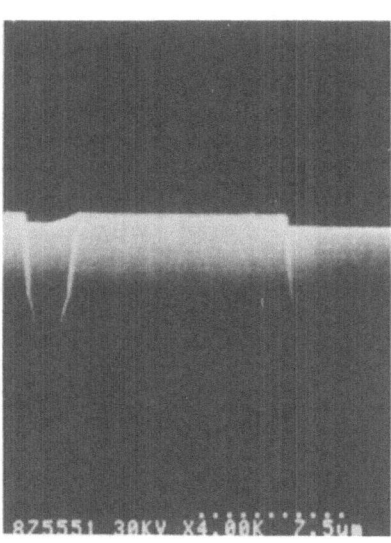

c

Figure 38. SEG trench isolation refill, (a) trench etched in silicon 5um deep, (b) 950°C/25torr SEG refill and (c) 875°C/25torr SEG refill[41].

torr followed by SEG growth at either 950°C or 875°C at 25 torr. The trench structures are 5um deep with widths varying from 1.5um to >150um. The fabrication steps to form the trench structures are as follows. Fig.38a shows trenches etched in silicon to a depth of 5um after thermal oxidation resulting in an oxide liner along the sidewalls 0.1um to 0.2um thick after anisotropic etching to remove the oxide from the bottom of the trench providing a site for silicon epi nucleation. Then, epi is selectively grown to refill the trenches terminating the growth when planarity is achieved. Note the uniform trench refill independent of trench width. The uniformities were within +/-5% and growth rates ~0.23um/min. Results from a 950°C, 25 torr SEG refill is shown in Fig.38b and a 875°C, 25 torr SEG refill is shown in Fig.38c.

Selective Doping. Another attractive application of SEG is in the formation of selectively doped structures. Independent n-well and p-well CMOS structures with retrograde wells can be formed as shown in Fig.39 without the use of ion implantation and high temperature thermal heat treatments through graded epi techniques or buried layer epi techniques[15]. Also, independent bipolar npn and pnp transistors can be formed by SEG for complementry bipolar as described by Matsumoto[45], and for BiCMOS applications, independent bipolar and CMOS device doping levels are possible as described by Favreau[46] and O et al.[47]. SEG has also been used to form thin bipolar base structures and sidewall base contact bipolar transistors[48,49,50] (Fig.40). Binder et al.[51] reported on using SEG to heavily dope the side-wall of a 4 Mega Bit DRAM trench capacitor cell structure.

Selective doping is also very attractive for SEG usage in back-end device processing for shallow junction

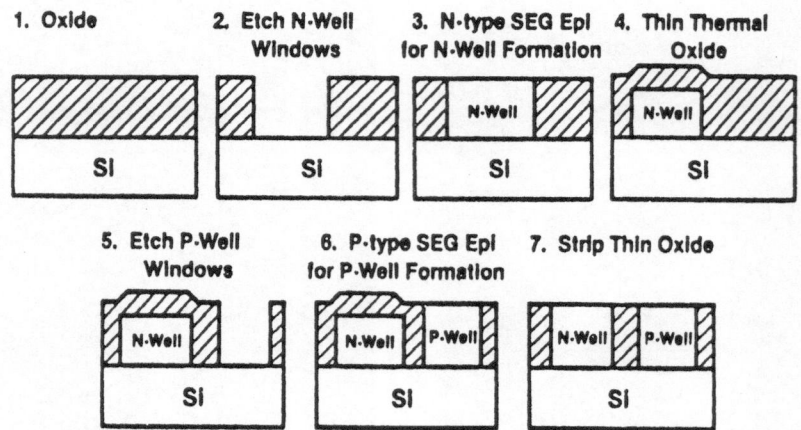

*Retrograde Wells are possible by Buried Layer Epitaxy or Graded Epitaxy Techniques

Figure 39. Two-step SEG for independent n-well and p-well CMOS formation[15].

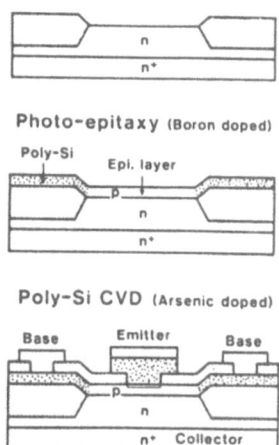

Photo-epitaxy (Boron doped)

Poly-Si CVD (Arsenic doped)

T. Sugii, T. Yamazaki, T. Fukano and T. Ito, 1987 Symposium on VLSI Technology, IEEE Cat. No. 87 TH 0189-1, p. 35, May 1987

Figure 40. Simultaneous formation of thin base region with sidewall base contact bipolar structure by SSPD (simultaneous single/poly deposition)[48].

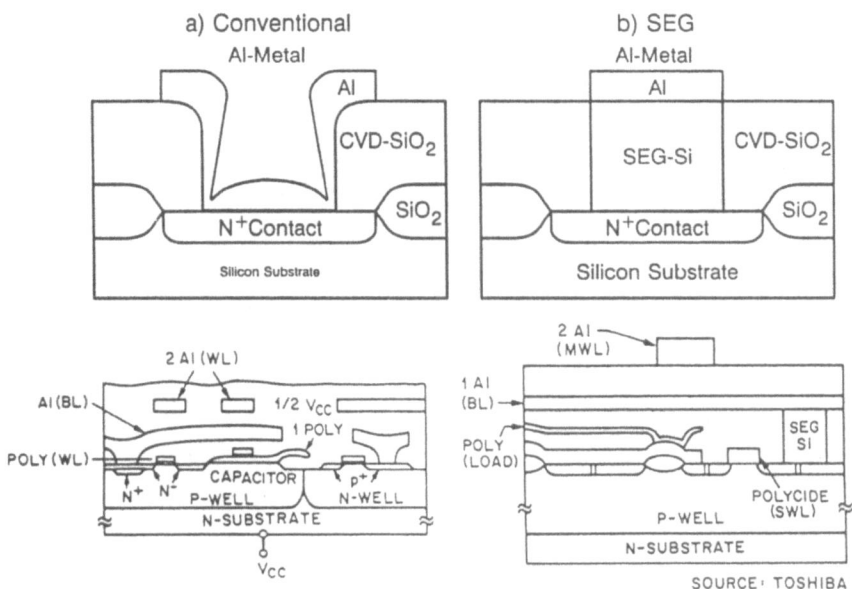

Figure 41. SEG self-aligned contact refill and planarization for 1 Mega Bit SRAM technology, (a) conventional and (b) SEG[53&54].

formation (CMOS source/drain and bipolar emitter) and self-aligned contact refill and planarization for interconnect. Fig.41 shows a selective heavily phosphorus-doped SEG structure for self-aligned contact refill and planarization for a 1 Mega Bit SRAM device[52,53,54,55]. This approach gives excellent contact yields and is not troubled by problems seen with selective tungsten such as contact-substrate shorts (worm tunneling defect) and metal surface morphology. NMOS and PMOS source/drain contact refill can be accomplished by using the independent selective (n-type/p-type) doping described in Fig.39, or by the use of metal silicidation techniques. Bipolar emitter contacts are also possible replacing the poly emitter formation process.

<u>Silicon On Insulator</u>. The third area for application of SEG and its derivatives is in the formation of future 3-D silicon on insulator (SOI) structures. Two proposed 3-D SOI DRAM cell designs using SEG techniques are shown in Figs.42&43[56,57]. Also, elevated source/drain SOI structures have been achieved by simultaneous single/poly deposition (SSPD) as shown in Fig.44 where poly silicon is selectively deposited over an oxide and single crystal silicon is selectively deposited over the exposed silicon regions. A totally isolated SOI structure is shown in Figs.45&46 using Epitaxial Lateral Overgrowth (ELO)[17,40].

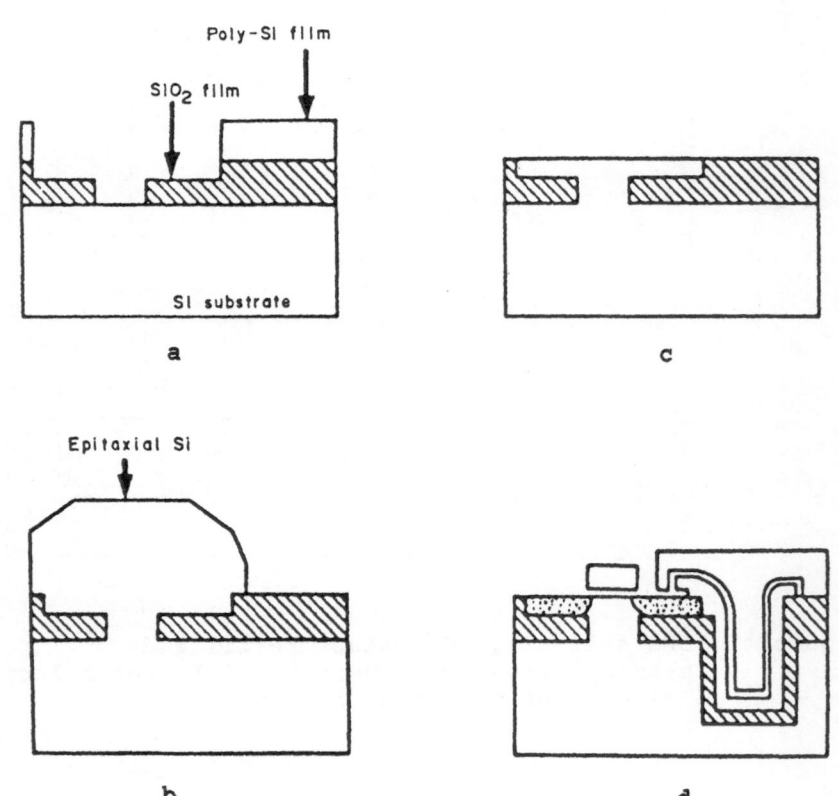

Figure 42. SEG TOLE (transistor on lateral epi) cell structure for 16 Mega Bit DRAM technology, (a) device and seed area formation, (b) epi lateral overgrowth, (c) preferential polishing and (d) capacitor formation[56].

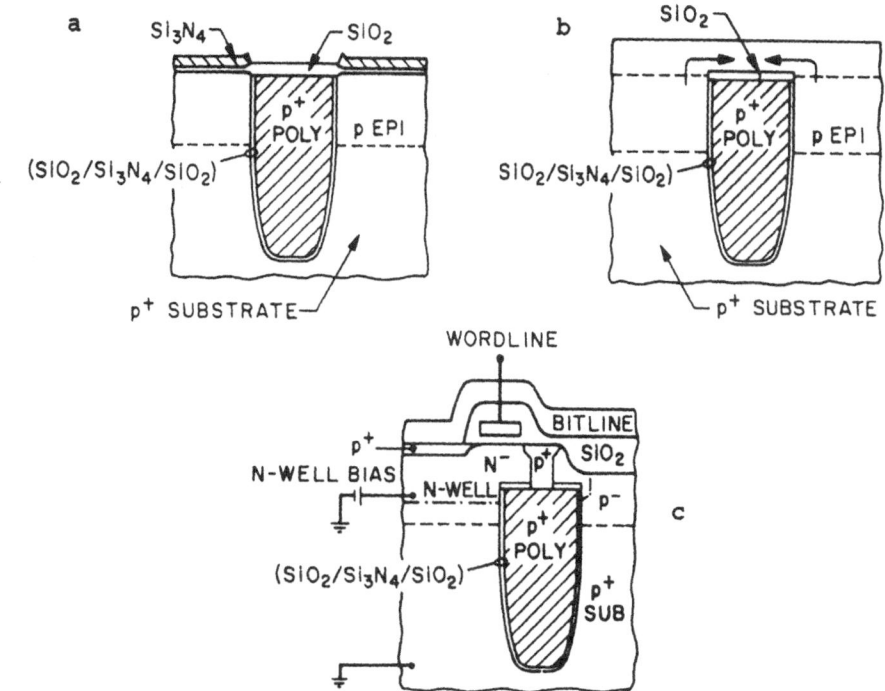

Figure 43. SEG EOT (epi over trench) cell structure for 64
Mega Bit DRAM technology, (a) after self-aligned
cap oxidation, (b) after selective epitaxial
overgrowth and (c) DRAM cell formation[57].

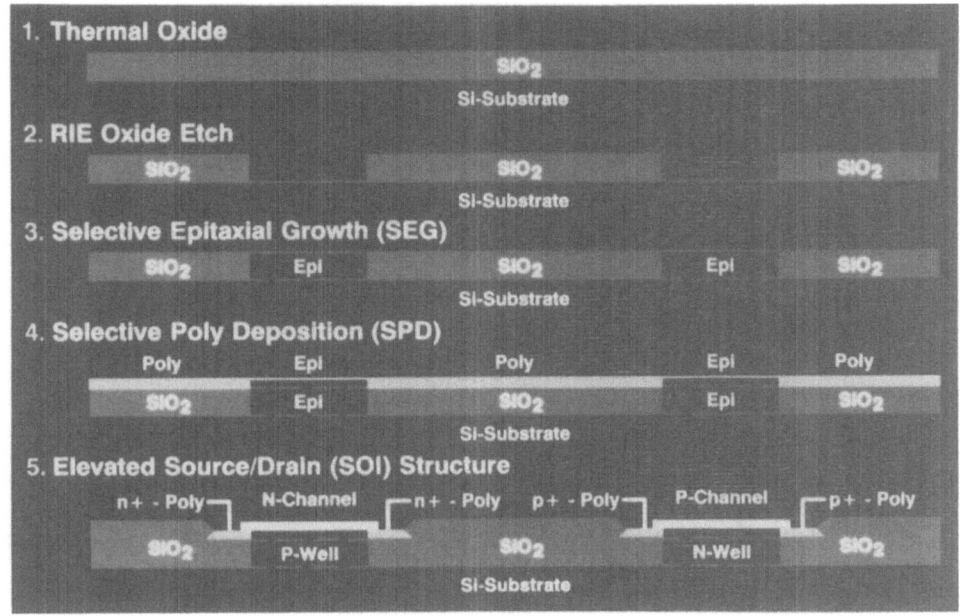

Figure 44. Elevated source/drain structure by SSPD.

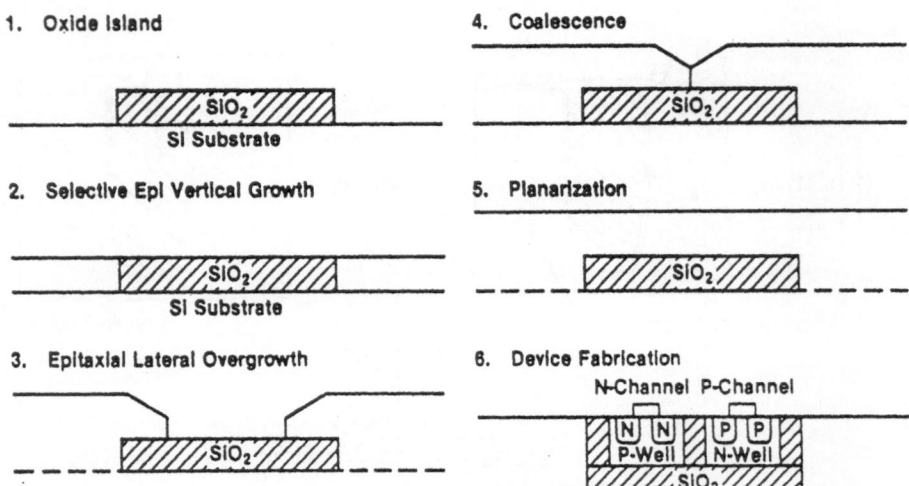

Figure 45. A totally isolated SOI CMOS structure using SEG
processing to achieve ELO (epi lateral
overgrowth)[40].

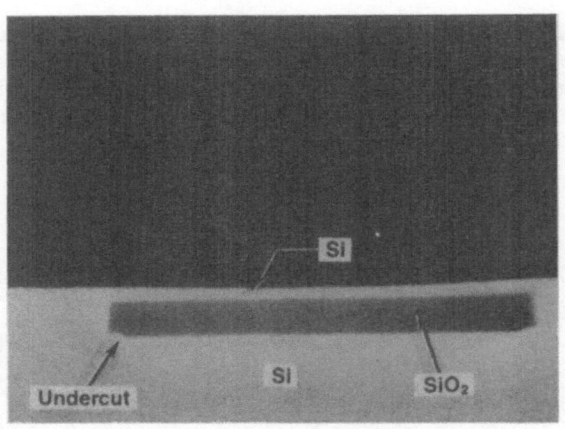

Figure 46. Cross-sectional SEM of an
ELO thinned sample[17].

SUMMARY/CONCLUSION

In summary, high quality low temperature/low pressure
epitaxial structures have been achieved at temperatures as
low as 775°C, 7.5 torr pressure in commerically available
cylindrical/barrel epi reactors. Advanced (novel) device
structures for CMOS, bipolar and BiCMOS technology are
possible through the use of SEG growth techniques which are
critical for next generation CMOS, bipolar and BiCMOS
device technology.

REFERENCES

1. P. Chatterjee and 4Mb dRAM Team, IEEE IEDM-86, section 6.1, p. 128, Dec. 1986.
2. Silicon Molecular Beam Epitaxy, the Electrochemical Society, PV 85-7, 1985.
3. R. Reif, private communications, Department of Electrical Engineering and Computer Science, MIT, Cambridge, Massachusetts.
4. B. Meyerson, E. Gain and D. Smith, Reduced Temperature Processing For VLSI, the Electrochemical Society, PV 86-5, p.285, 1986.
5. A. Ishitani, Y. Ohshita, K. Tanigaki and K. Takada, Journal of Applied Physics, vol. 61, p.2224, 15 March 1987.
6. K. Tanno, N. Endo, H. Kitajima, Y. Kurogi and H. Tsuya, Japanese Journal of Applied Physics, vol. 21, no. 9, p. L564, Sept. 1982.
7. E. Krullmann and W. Engl, IEEE Trans. Electron Devices, vol. ED-29, no. 4, p. 491, April 1983.
8. R. Pagliaro, J. Corboy, L. Jastrzebski and R. Soydan, Journal of the Electrochemical Society, vol. 134, no. 5, p. 1235, May 1987.
9. J. Borland and D. Schmidt, Technical Proceedings Semicon/East 1986, p.150, Sept. 1986.
10. R. Wise, Chemical Vapor Deposition 1987, the Electrochemical Society, PV 87-8, p.253,1987.
11. J. Borland, T. Thompson, V. Tagle and W. Benzing, Chemical Vapor Deposition 1987, the Electrochemical Society, PV 87-8, p.275, 1987.
12. J. Borland, presented at Semiconductor International Korea 1986, Applied Materials Technical Report HT-030, Feb. 1986.
13. M.J.P. Duchemin, M.M. Bonnet and M.F. Koelsch, Journal of the Electrochemical Society, vol. 125, no. 4, p. 637, April 1978.
14. S. Nagao, K. Higashitani, Y. Akasaka and H. Nakata, IEEE Transation on Electron Devices, vol. ED-33, no. 11, p. 1738, Nov. 1986.
15. J. Borland and C. Drowley, Solid State Technology, vol. 28, no. 8, p. 141, Aug. 1985.
16. K. Kugimiya and Y. Hirofuji, Japanese Journal of Applied Physics, vol. 24, no. 5, p. 518, May 1985.
17. S.T. Liu, L.S. Chan and J.O. Borland, Chemical Vapor Deposition 1987, the Electrochemical Society, PV 87-8, p.428, 1987.
18. M.R. Goulding and J.O. Borland, Semiconductor International, p.90, May 1988.
19. J.O. Borland, V. Murali and C.S. Wei, recent news paper #599 presented at the Electrochemical Society meeting, Atlanta, Georgia, May 1988.
20. J. Borland, R. Wise, Y. Oka, M. Gangani, S. Fong and Y. Matsumoto, Solid State Technology, vol.31, no.1, p.111, Jan.1988.
21. G.R. Srinivasan, Journal of Crystal Growth, vol. 70, nos. 1/2, p. 201, Dec. 1984.
22. H.R. Chang, Journal of the Electrochemical Society,

vol. 132, no. 1, p. 219, Jan. 1985.

23. G.W. Cullen and J.F. Corboy, Journal of Crystal Growth, vol.70, nos. 1/2, p. 230, Dec. 1984.

24. M.W.M. Graef, .J.H. Leunissen and H.H.C. de Moor, Journal of the Electrochemical Society, vol. 132, no. 8, p. 1942, Aug. 1985.

25. J.O. Borland, Chemical Vapor Deposition 1987, the Electrochemical Society, PV 87-8, p.389, 1987.

26. J.O. Borland, IEEE IEDM-87, section 2.1, p.12, Dec. 1987.

27. J. Chen, presented at Applied Materials Inc. sponsored "Innovations In VLSI/ULSI CMOS Technology" seminar, San Jose, CA, Dec. 3, 1986.

28. T. Yamaguchi, S. Morimoto, G. Kawamoto, H.Park and G. Eiden, IEEE IEDM-83, section 24.3, p. 522, Dec. 1983.

29. B.D. Joyce and J.A. Baldrey, Nature, vol. 195, no. 6, p. 485, Aug. 4, 1962.

30. A. Ishitani, N. Endo, and H. Tsuya, Japanese Journal of Applied Physics, vol. 23, no. 6, p. L391, June 1984.

31. A. Stivers, C. Ting and J. Borland, Chemical Vapor Deposition 1987, the ElectrochemicalSociety, PV 87-8, p.389, 1987.

32. J. Manoliu and J. Borland, IEEE IEDM-87, section 2.3, p. 20, Dec. 1987.

33. D.M. Jackson, Trans. Metall. Soc. AIME, vol. 233, p.596, March 1965.

34. Matsumoto, presented at Applied Materials Japan, "Advanced Epi Technology" seminar, Kyoto, Japan, May 29, 1987.

35. H. Voss and H. Kurten, IEEE IEDM-83, section 2.5, p. 35,Dec. 1983.

36. C. Ting, A. Stivers and J. Borland, Boston meeting of the Electrochemical Society, abstract no. 208, May 1986.

37. N. Endo, N. Kasai, A. Ishitani, H. Kitajima and Y. Kurogi, IEEE Transaction on Electron Devices, vol. ED-33, no. 11, p. 1659, Nov. 1986.

38. S. Nagao, K. Higashitani, Y. Akasaka and H. Nakata, IEEE Transation on Electron Devices, vol. ED-33, no. 11, p. 1738, Nov. 1986.

39. L. Jastrzebski, J. Corboy and R. Soydan, Chemical Vapor Deposition 1987, the Electrochemical Society, PV 87-8, p.334, 1987.

40. J. Borland, D. Schmidt and A. Stivers, Extended Abstracts of the 18th (1986 International) Conference on Solid State Devices and Materials, Tokyo, Japan, p. 53, Aug. 1986.

41. S. Fong and J. Borland, results from joint Monolithic Memories and Applied Materials,Inc. SEG trench refill development on the AMC-7810.

42. N. Kasai, N. Endo, A. Ishitani and H. Kitajima, IEDM-85, section 15.6, p. 419, Dec. 1985.

43. V. Silvestri, Chemical Vapor Deposition 1987, the Electrochemical Society, PV 87-8, p.366, 1987.

44. C. Chuang, D. Tang, G. Li, E. Hackbarth and R. Boedeker, Extended Abstracts of the 18th (1986 International) Conference on Solid State Devices and Materials, Tokyo, Japan, p. 267, Aug. 1986.

45. Private communicatios with Matsumoto of NEC 2nd LSI Division.

46. Private communications with D. Favreau of TI Huston Bipolar IV.

47. K. O, H. Lee and R. Reif, Extended Abstracts of the Electrochemical Society May 1987 Meeting, vol. 87-1, p. 407, abstract no. 279, 1987.

48. T. Sugii, T. Yamazaki, T. Fukano and T. Ito, 1987 Symposium on VLSI Technology, Karuizawa, Japan, section V-3, p. 35, May 1987.

49. F. Mieno, A. Shimizu, S. Nakamura, T. Deguchi, N. Haga, I. Matsumoto, Y. Furumura, T. Yamauchi, K. Inayoshi, M. Maeda and K. Yamagida, IEEE IEDM-87, section 2.2, p. 16, Dec. 1987.

50. D. Favreau and D. Hollingsworth, Extended Abstracts of the Electrochemical Society May 1987 Meeting, vol. 87-1, abstract no. 259, p. 376, 1987.

51. H. Binder, H. Geiger, R. Kakoschke, H. Muhlhuff and S. Rohl, Extended Abstracts of the 18th (1986 International) Conference on Solid State Devices and Materials, Tokyo, Japan, p. 299, Aug. 1986.

52. S. Samata, H. Shibata, T. Matsuno, H. Sasaki, Y. Matsushita and T. Ohta, Proceeding of the 31st Symposium on Semiconductor and Integrated Circuits Technology, the Electrochemical Society of Japan, Tokyo, Japan, Dec. 3&4, 1986.

53. H. Shibata, S. Samata, M. Saitoh, T. Matsuno, H. Sasaki, Y. Matsushita, K. Hashimoto and J. Matsunaga, 1987 Symposium on VLSI Technology, Karuizawa, Japan, p. 75, May 1987.

54. K. Nogami, T. Sakurai, K. Sawada, T. Wada, K. Sato, . Isobe, M. Kakumu, S. Morita, S. Yokogawa, M. Kinugawa, T. Asami, K. Hashimoto, J. Matsunaga, H. Nozawa and T. Iizuka, IEEE Jour. of Solid-State Circuits, vol. SC-21, no. 5, p. 662, Oct. 1986.

55. M.Matsui, T. Ohtani, J. Tsujimoto, H. Iwai, A. Suzuki, K. Sato, M. Isobe, K. Hashimoto, M. Saitoh, H. Shibata, H. Sasaki, T. Matsuno, J. Matsunaga and T. Iizuka, IEEE Journal of Solid-State Circuits, vol. SC-22, no. 5, p. 733, Oct. 1987.

56. T. Kubota, T. Ishijima, M. Sakao, K. Terada, T. Hamaguchi and H. Kitajima, IEEE IEDM-87, section 14.6, p. 344, Dec. 1987.

57. G. Bronner, N. Lu, T. Rajeevakumar, B. Ginsberg and B. Machesney, 1988 Symposium on VLSI Technology, San Diego, CA, section III-3, p. 21, May 1988.

PARTICIPANTS

H. AKBAS

Trakya Universitiesi
Fizik Bolumu
Fen-Edebiyat Fakultesi
Edirne, TURKEY

J. M. ALBELLA

Universidad Autonoma Cantoblanco
Cantoblanco, Madrid 28049
SPAIN

M. C. ARIKAN

Tubitak P. O. Box 74
Gabze-KocaeliTTURKEY

M. BAYHAN

Orto Dogu Teknik
Universitesi Fizik Bolumu
06531 Ankara
TURKEY

H. BENDER

IMEC Kapeldreef 75
B3030 Leuven BELGIUM

V. BOURLON

European Space Agency
Keplerlaan, 1
Noordwijk
THE NETHERLANDS

B. BOUYER

CENG, LETI
BP 85
38041 Grenoble
FRANCE

N. L. BRAGA

Departmento de Engenharia de
Electricidade da Escola
Politecnica da USP
Sao Paulo, Brazil

W. BUCHHOLTZ

Technical University of Aachen
Aachen West Germany

L. CALCAGNO

Universita di Catania
Dipartimento di Fisica
57 Corso Italia
I95129 Catania
ITALY

Y. CARATINI

CNET BP 98
38243 Meylan
FRANCE

R. PEREZ-CASERO

Universidad Autonoma
Instituto de Ciencia de Materiales
Cantoblanco, Madrid
SPAIN

Y. H. CHEN

Chung Hsing Road
195-4-K200
Sec. 4
Chu-Tung Hsin-Chu
Taiwan, CHINA

P. COECKELBERGHS

IMEC
Kapeldreef 75
B3030 Leuven
BELGIUM

H. S. DARENDELIOGLU

Selcuk University
Fizik Bolumu
Fen-Edebiyat Fakultesi
TURKEY

J. M. DILHAC

LAAS-CNRS
7 Avenue du Colonel Roche
31077 Toulouse
FRANCE

E. DOOMS

IMEC
Kapeldreef 75
B3030 Leuven
BELGIUM

H. EVERS

Institut Anorganische Chemie
Universitat Koln
5000 Koln 41
West Germany

H. S. S. GECIM

Hacettepe University
Department of Electrical Engineering
Ankara, TURKEY

K. HEYERS

Technical University of Aachen
Aachen West Germany

A. M. HODGE

Royal Signals & Radar Establishment
St. Andrews Road
Great Malvern
Worcster WR14 3PS ENGLAND

J. W. HONEYCUTT

North Carolina State University
Department of Material Science
Raleigh, NC 27965

H. M. KIZILYALLI

Ankara University
Department of Physics
06100 Tandogan
Ankara, TURKEY

T. KRUCK

Institut Anorganische Chemie
Universitat Koln
5000 Koln 41
West Germany

FRANCESCO LA VIA

Universita di Catania
Dipartimento di Fisica
57 Corso Italia
I95129 Catania
ITALY

A. LAFERLA

Universita di Catania
Dipartimento di Fisica
57 Corso Italia
I95129 Catania
ITALY

P. LANZA

SGS Microelectronica
Stratale Primsole
Catania 50-195100
ITALY

M. CLEMENT-LORENZO

Universidad Autonoma
Instituto de Ciencia de Materiales
Cantoblanco, Madrid
SPAIN

L. D. MADSEN

Northern Telecom Electronics Ltd.
P. O. Box 3511
Station C
Ottawa K1Y48H7, Canada

R. P. MOOS

AT&T Bell Laboratories
Crawfords Corner Road
Holmdel, NJ 07733

N. I. MORIMOTO

Departmento de Engenharia de
Electricidade da Escola
Politecnica da USP
Sao Paulo, Brazil

V. NAYAR

University College London
Department of Electrical Engineering
Torrington
London, U.K.

M. C. OZTURK

North Carolina State University
Department of Electrical Engineering
Raleigh, NC 27695

Y. PAULEAU

CNET
BP 98
38243 Meylan
FRANCE

B. RAICU

Monolithic Memories
2175 Mission College Blvd.
Santa Clara, CA 95054

F. ROOZEBOOM

Philips Research Laboratories
5600 JA Eindhoven
THE NETHERLANDS

J. C. ROSTAING

CNET
BP 98
38243 Meylan
FRANCE

G. A. ROZGONYI

North Carolina State University
Box 7916
Raleigh, NC 27695

O. L. RUSSO

NJIT
Physics Department
Martin Luther King Boulevard
Newark, NJ 07102

T. E. SEIDEL

Sematech
Austin, Texas 78759

U. SHARMA

Lehigh University
Sherman Fairchild Laboratory 6
Bethlehem, PA 18105

J. W. SWART .

Departmento de Engenharia de
Electricidade da Escola
Politecnica da USP
Sao Paulo, Brazil

D. TONNEAU

CNET
BP 98
38243 Meylan
FRANCE

P. VANDERNABEELE

IMEC
Kapeldreef 75
B3030 Leuven
BELGIUM

L. S. ZASNICOFF

University of Sao Paulo
Sao Paulo, Brazil

INDEX